TROPICAL FOREST CANOPIES

FORESTRY SCIENCES

Volume 69

Tropical Forest Canopies: Ecology and Management

Proceedings of ESF Conference, Oxford University, 12–16 December 1998

Edited by

K.E. LINSENMAIR
Department of Animal Ecology and Tropical Biology
Biozentrum, University of Wuerzburg, Germany

A.J. DAVIS
Green College, Oxford University, Oxford, UK

B. FIALA
Department of Animal Ecology and Tropical Biology
Biozentrum, Wuerzburg, Germany

M.R. SPEIGHT
St. Anne's College and Department of Zoology,
Oxford University, Oxford, UK

Reprinted from *Plant Ecology*, Volume 153, Nos. 1–2, pp. 1–370

KLUWER ACADEMIC PUBLISHERS
DORDRECHT / BOSTON / LONDON

A C.I.P. Catalogue record for this book is available from the Library of Congress.

ISBN 0-7923-7049-X

Published by Kluwer Academic Publishers,
P.O. Box 17, 3300 AA Dordrecht, The Netherlands.

Sold and distributed in North, Central and South America
by Kluwer Academic Publishers,
101 Philip Drive, Norwell, MA 02061, U.S.A.

In all other countries, sold and distributed
by Kluwer Academic Publishers,
P.O. Box 322, 3300 AH Dordrecht, The Netherlands

Printed on acid-free paper

Printed in the Netherlands

Table of Contents

Section 4 – Hydrology, Microclimate and Light

Section 5 – Architecture

Section 6 – Education, Management and Conservation

Lowland dipterocarp forest canopy in the Danum Valley Conservation Area, Sabah, E. Malaysia. Courtesy of A.J. Davis.

Scientists examining forest canopy from gondola in the Amazonas region, Venezuela (Surumoni site). Courtesy of A.J. Davis.

Plant Ecology **153:** 1–3, 2001.

Foreword

It was not until the end of the 80s that the scientific community slowly became aware of Terry Erwin's (1982) seminal paper on the immense diversity of insects in the crowns of a tropical forest tree, *Luehea seemannii*, and the conclusions he tentatively drew from his findings on potential global numbers of arthropod species. Gradually researchers began to realise that the canopy stratum – especially that of tropical forests – represents a vast blank patch on our map of biological knowledge. Before the publication of Erwin's article and some further timely studies by other pioneers in this field, not only were the dimensions of this stratum's richness grossly underrated by most biologists, but, in addition, this 'last biotic frontier' had barely been recognised as an especially diverse and functionally highly important region of the forest.

Initially, the lack of appropriate access systems halted any fast scientific progress and severely hampered the treatment of many open questions in ways that met the demands of quantitative experimental research work. Since then, however, a number of technical devices have been developed, currently further being improved, which allow an investigation of many important questions whilst also satisfying high scientific quality standards. Canopy science is now in a position to answer many of the outstanding questions and environmental problems that are a fundamental part of, and in many ways special to, canopy biology. Among these open problems are some of the most pressing environmental issues society is presently faced with (e.g., global climate change, biodiversity's functions in complex ecosystem processes and consequences of its loss, carbon and water cycling on the regional and the global scale, sustainable forest use, etc.). Thus, the emerging field of canopy research is now outgrowing its infancy, and is developing into a maturing field of scientific inquiry that is becoming more and more experimental and predictive.

In Europe, the European Science Foundation's (ESF) 'Tropical Canopy Programme' contributed to a positive development of this field. It added value to national programmes through intensifying the international exchange of ideas and by furthering the harmonisation of methods for collection, processing and access to data, by calling into life some additional new national programmes and by improving the status of already existing ones and by facilitating international co-operation.

This ESF programme originated from the German Research Council's (Deutsche Forschungsgemeinschaft, DFG) priority programme on the 'Mechanisms of the maintenance of tropical diversity'. It was Jan-Hendrik Kock, at that time representative in the ESF for the 'Life Sciences', who approached me and prompted an application to the ESF for an umbrella programme on tropical ecology. We owe much to him and his successors in charge, Niels Petersen and Annette Wiklund, for their special and untiring attention and support for our programme. We greatly appreciate their helpfulness and patience, which went far beyond bureaucratic routine.

In focusing this programme (financially supported via the ESF by Denmark, Germany, the Netherlands, Austria, Italy during the first two-thirds, and Great Britain during the last third of its duration) we very quickly reached the conclusion that tropical canopy research has – on the one hand – the highest innovative and developmental potential, while, on the other hand, it was clearly an especially underdeveloped subject within the realm of tropical ecology. Thus, we considered it a field that not only urgently deserved but also greatly earned special support on all levels: from the development of suitable methods of access and adequate sampling to defining the most relevant scientific questions and developing the guiding theory. Here, I extend my cordial thanks to Nigel Stork who acted temporarily as a co-chair while we were devising this programme and drafting the final application.

This volume summarises the currently achieved 'state of the art' of canopy biology, and gives a very authoritative overview over a majority of the important aspects of canopy research. Its content is based on lectures which programme participants and invited guests had presented and discussed during the final meeting of our ESF Tropical

Canopy Research Programme in 1998 at St. Anne's College in Oxford. We are extremely grateful to Andrew Davis and Martin Speight for the excellent organisation of this conference and for their acting as such superb hosts.

The organisers of the final conference and editors of this book thank all the speakers and authors for their contributions and also all those participants who actively took part in the lively discussions. We also gratefully acknowledge most authors for keeping all the deadlines we had to set. Very special thanks are also due to the many referees, who very carefully scrutinised all the papers and – by always very constructively criticising them – essentially contributed to their improvement. The following colleagues kindly acted as referees: J. Adis, M. Barker, D. Barthelemy, A. M. Bauer, B. Beehler, A. D. Bell, D. Benzing, F. Bouman, I. R. Calder, N. A. Chappell, D. B. Clark, A. M. Cleef, S. G. Compton, I. Cuthill, A. J. Davis, J. A. Endler, L. J. Fleishman, C. M. Francis, J. Ghazoul, F. Hallé, J. Hill, F. Houllier, E. Kalko, R. L. Kitching, W. E. Kunin, K. E. Linsenmair, M. D. Lowman, U. Lüttge, S. Marsden, A. Mitchell, N. Nadkarni, D. M. Newbery, R. A. A. Oldeman, C. Ozanne, F. E. Putz, D. Sabatier, M. R. Speight, N. E. Stork, R. M. Timm, H. Vester, R. P. D. Walsh, S. J. Willott, N. Winchester, J. H. D. Wolf.

We are especially indebted to Andrew Davis for his outstanding efforts and achievements during all steps of the editing process. Already during the first round of reviews, he had taken over an unproportionally heavy burden as referee and, in addition, he has scrutinised all revised contributions for remaining weaknesses in their contents, language usage and lack of conformance with the format of *Journal of Plant Ecology*.

This volume, with an introduction by Andrew Mitchell, who gives an overview of the history of canopy research and points out opportunities for the future, is subdivided into 6 sections. The first section *'Sampling and Access'* begins with an article by Stephen Sutton who examines the development of the technical side of accessing the crown stratum and the further needs in this respect. In a contribution by Martin Barker and Michelle Pinard problems of forest canopy research are discussed and some possible solutions suggested.

The second section is devoted to *'Plants in the Canopy'*. In accordance with their great importance in the crown layer of many tropical forests, epiphytes and hemiepiphytes feature prominently in this section and in other parts of the book. While Jürgen Nieder, Juliana Prosperi and Georges Michaloud discuss the contribution of epiphytes to canopy diversity and thus deal with a more general aspect, Gerold Schmidt, Sabine Stuntz and Gerhard Zotz treat a more particular issue of epiphyte biology, namely intraspecific size variability as an ecophysiological parameter. The two additional articles in this section are devoted to other important aspects, to pollination biology and flowering phenology (Arthur van Dulmen) and to general considerations into the actual plant related canopy research and its future development (Meg Lowman).

The third and largest section is concerned with *'Animals in the Canopy'*. Here articles devoted to invertebrates dominate which, in view of their ecological importance, their species diversity, and biomass, seems fully justified. Yves Basset took the trouble to screen most of the published literature to review the current state of knowledge on invertebrates in tropical rain forests and to pinpoint the most serious gaps and biases in regard to ecologically important higher taxa and biogeographical regions. The other contributions concerned with the fauna of arboricolous invertebrates focus on:

- spatial stratification (Christian Schulze, K. Eduard Linsenmair and Konrad Fiedler; Ulrich Simon and K. Eduard Linsenmair),
- seasonal changes in arboreal communities (Thomas Wagner), and
- disturbance gradients and their consequences for the structure and diversity of communities (Andreas Floren and K. Eduard Linsenmair).

Three articles deal with vertebrates in the canopy:

- Roland Kays and Allen Allison give a broad overview on studies published on arboreal vertebrates according to taxa and geographical regions during the last ten years;
- Hans Winkler and M. Preleuthner discuss how behaviour and morphology in canopy dwelling birds is adapted to the ecological demands of this stratum;
- Mike Shanahan and Stephen Compton present their results on fruit design characteristics, vertical stratification and corresponding guild structures in figure eating frugivores.

In section four, *'Hydrology, Microclimate and Light'*, issues of ecohydrology and micrometeorology of undisturbed and disturbed forests are discussed. Nick Chappell, Kawi Bidin and Wlodek Tych present their results on the biophysical controls of sub-canopy precipitation. Ian Calder treats some peculiarities of canopy hydrology, especially interception and transpiration and some other rain dependent processes in tropical forests. Jörg Szarzynski and Dieter Anhuf explored the day and night micrometereological conditions and the respective energy exchange between the canopy layer and the atmosphere and Martin Freiberg studied the influence of epiphytes on the small-scale micrometereology within tree crowns. Marc Théry gives an overview over the light climates in primary and disturbed tropical forests and their effects on plants and on habitat choice of animals.

Section five *'Architecture'* is devoted to the architecture of crowns and the overall canopy structure. Frank Sterck, Frans Bongers and David Newbery studied the intra- and intra-specific variability of the crowns in a number of rain forest tree species. Valérie Trichon tested whether species-specific characteristics of tree crowns can successfully be used for the identification of these species on large-scale aerial photographs. Frans Bongers gives an overview over methods for describing, measuring and analysing the three-dimensional structure of rain forests, and Philippe Birnbaum discusses the dynamical topography of the canopy surface viewed from different organisational and spatial levels of the forest ecosystem.

The final, sixth section *'Education, Management and Conservation'* contains contributions summarising research results of relevance for:

- education and for strengthening the sense for responsibility (Nalini Nadkarni),
- for general management aspects (Nigel Stork), and
- for more special management implications suited to mitigate anthropogenic disturbances, exemplified on one hand by mostly theoretical considerations of edge effects and their potential impact on forest canopies (by Andrew Foggo, Claire Ozanne, Martin Speight and Clive Hambler), and on the other hand by an empirical study that reports non-trivial direct and indirect effects of logging on the behaviour of important pollinators (Jaboury Ghazoul and Moray McLeish, and
- the development of applied conservation-oriented programmes (Elisabeth Kalko and Charles Handley).

I am very grateful to all the participants who actively and constructively participated in this programme, especially those who organised some of our meetings. Our thanks to the Royal Society of London, whose financial support via their SE Asia Rain forest Research Programme (RS SEARRP), made it possible to host this conference in the UK (in recognition of this support, this volume is research publication number A/322 of the RS SEARRP). Very special thanks, however, I owe Brigitte Fiala. Without her great help we (i.e., the steering committee and its chairperson) would not have been able to run the programme with the success we finally achieved.

I am confident that the added value which the national activities in our field gained through the ESF programme will not exhaust itself in what has been achieved in co-ordination, co-operation, and information exchange – and the first very promising steps towards a more comprehensive intercontinental programme have already been taken during the 'Forest Canopy Research Planning Workshop' on November 20–24, 1999 in Oxford, jointly funded by the ESF and the NSF.

Finally, I wish to express my appreciation to Kluwer Academic Publishers for supporting the publication of these seminal proceedings, and to Ursula Hertling, Noeline Gibson, and Arnold van der Valk, for their help and guidance in the realization of this project.

K. Eduard Linsenmair

Plant Ecology **153:** 5–11, 2001.

Introduction – Canopy science: time to shape up*

Andrew Mitchell
International Program Development, Earthwatch Institute, Oxford, UK (e-mail: a.mitchell@globalcanopy.on)

My first introduction to canopy science was as part of the large multinational Royal Geographical Society expedition to Mt. Mulu in Sarawak, Malaysia in 1977–1978. At that time I was inspired by the fact that the subjects I came to study, gibbons (*Hylobates mülleri*) and other species of primates remained invisible 30–40 m up, inhabiting a world I could only dream of. Thus challenged, I ventured a year later into the canopy world on an assortment of climbing ropes, platforms, and aerial walkways designed by myself and Stephen Sutton and built by teams of the Royal Engineers of the British Army with support from the Scientific Exploration Society. Our walkways were built in the rain forests of Panama, Papua, New Guinea, Sulawesi and Costa Rica and enabled some of the first comparative canopy studies across continents (Mitchell 1986).

Techniques of access have much improved since then. The late Alan Smith introduced me to the Smithsonian canopy crane in Panama and I provided modest input into Francis Hallé's 'Radeau des Cimes' prior to its spectacular launch beneath a hot air dirigible into the canopies of the world's forests through the 1990's. Sutton (2001) gives an excellent review of the rapid growth in canopy science since the 1950s and there have been a number of earlier valuable reviews (Moffett 1993; Nadkarni and Lowman 1995; Lowman & Wittman 1996). However, even as arboreal scientists have begun to solve the daunting task of reaching the rain forest roof, the environment we seek to study is almost vanishing before our eyes.

Two decades ago it took several days to reach our Gunung Mulu study area, mostly through thick dipterocarp forest. Indigenous Penan, most having had little or no contact with outsiders before, contributed immensely to the knowledge we gained. Today all the forest around our study site is gone, only an island, the Mulu National Park remains, soon to become a World Heritage Site. A handsome, international hotel graces the location near where our long house stood, airstrips provide easy access, and a huge golf course is planned. Of the Penan there is no sign, their nomadic existence outlawed, a culture compromised. Mulu is archetypal of the rapidly changing world in which we live and in which scientific research must now operate. At this third international conference on canopy science I was struck by three challenging drivers which I shall characterize as 'time', 'shape' and 'up', which I believe must propel canopy science in the next millennium.

Time

Time is not on our side. Despite all the efforts of conservation agencies around the world, forest conversion or clearance proceeds apace. I predict that about 5%–10% of the rain forest canopy that we had available when canopy science began, will remain intact by the end of this next century. The current extinction rate is now approaching 10 000 times the background rate over the last 65 million years and could rise to 100 000. This will mean that a mere $\frac{1}{3}$–$\frac{2}{3}$ of the species we have now will survive. The course of evolution is being massively altered. As Peter Raven said at the 1999 International Botanical Congress:

> we are living in an era when our great-grand children may live in a world in which more than half of the plant species that exist now will be known only as specimens.

Most scientists know this dilemma all too well, but agencies lack the hard data from forest science to convince policy makers to change. The questions we are asking need to be right and relevant. How valuable is an intact rain forest canopy versus other uses? Consider some ball park comparisons presented by Dr Nigel Stork at this conference:

*This paper is based on Andrew Mitchell's concluding keynote address to the Conference.

Income from timber extraction	\$100 ha^{-1}
Income from cattle farming	\$80 ha^{-1}
Calculation of goods and services from intact rain forest	\$2000 ha^{-1}
Cost of clearing 1 ton of CO_2 from the atmosphere	\$20 ha^{-1}
Carbon sequestration value of 1 h of Amazonian forest	\$1000–4000 ha^{-1}

The problem with figures such as those for environmental services is that decision makers often remain unconvinced that they are true and each day we fail to produce convincing data the world hurtles towards genetic bankruptcy. This must become a primary challenge of canopy science in the new millennium, lest Erwins 1980s and 1990s provocative estimates of canopy richness (Erwin 1991) become milestones of academic interest only and this new science of canopy biology rapidly finds itself without a study site.

Shape

Canopy biology is of course not a new science at all, it is merely a new environment in which to conduct science. The challenges of access that dominated its early years condemned many early studies to charges of being obscure or biased in the way they were carried out. New techniques using portable high strength materials for individuals to climb into crowns, or large equipment such as dirigibles or cranes to explore the region above the canopy, have largely overcome problems of access. The fact that the rain forest roof remains one of the most challenging environments to work in must no longer be a prime reason for going there. Canopy biology must increasingly shape itself to meet the demands of rigorous modern science which is replicable and tests hypotheses but is also relevant.

Papers presented at this conference in Oxford evidence the growing maturity of canopy science. Following on from the second international canopy conference held at The Marie Selby Botanical Gardens in 1998, this was the first in Europe on an international scale devoted to canopy science.

The following areas emerged as needing attention in shaping future study:

1. Rigor and replication: Challenges of access will not be accepted by the top journals as an excuse for small sample sizes. Without access to top journals canopy biology may continue to be regarded in some circles as obscure.

2. Standardization: Creating protocols acceptable worldwide enabling comparisons between study sites would be a major advance. At present lack of communication between sites, the pioneering and individualistic rather than collaborative nature of many studies, and the lack of a single journal in which to publish results is holding back the advances, though the advent of journals such as Biotropica has helped. Protocols in the fields of botany and animal science are discussed in volume 153 (Barker & Pinard; Basset; Bongers; Kays & Allison; Sutton; Thery; Trichon)

3. Communication: Pioneers are often not good communicators and conduct their art in isolation. Regular contact or integration between small or large scale canopy studies appeared lacking at this conference. This may cause unnecessary duplication of effort and certainly hinders progress fostered by shared experience. The European Science Foundation (ESF) Tropical Canopy Program has been a giant step forward. We need now to build on this success. Lowman and Whittman (1996) report 30 sites with regular canopy field studies and now nine sites exist with or will shortly have permanent cranes. What is needed is an integrated web site, perhaps managed by ICAN (the International Canopy Network) linking as many of these sites as possible and publishing new findings and techniques rapidly (see Nadkarni 2001). Modern technology should be embraced to promote this. An electronic journal feeding off this should also be considered which will post knowledge gained rather than scatter them across many outlets.

4. Digitize: Lack of government funding and collecting of knowledge into a relatively few Western dominated museums has almost reduced taxonomy to the role of hindrance rather than help. If canopy science is to understand biodiversity, new identification methods which are both faster and are more widely available are needed. Collections and keys need to be digitized and made available on line to developing nations as soon as possible thus saving costly journeys to distant collections. This would clearly be a major undertaking and specialist taxonomic expertise needs to lie at the heart of this process. A new methodology, such as that being developed by Robin Foster for plants at the Field Museum in Chicago, needs to be developed to assess biodiversity more rapidly. Teams of semi-skilled volunteers from Earthwatch Institute are assisting with this process at Kew and the Natural History Museum in London speeding up digitization. Other museums could follow suit. And finally, a new force of para-taxonomists needs to be trained with the capacity for using these systems accurately to speed the identification of areas with the highest biodiversity.

5. *Integration:* One of the most interesting areas for future study will be the extent to which forests are linked and produce global effects. It has been suggested that 50% of carbon dioxide flux between ecosystems and the atmosphere is produced by forests (Potter et al. 1993). Now that the techniques are largely in place, the opportunity for fully integrated studies from satellite, to crane, to climbing ropes in the forest biomes on all continents using standardized protocols and methodology exists. Clearly major funding and major effort is needed to achieve this. Such integrated studies should be linked to Long Term Ecological Research Sites (LTER) sites run by the National Science Foundation (NSF) and other long term studies by the Smithsonian, TROPENBOS, the Centre Nacional de Recherche Scientifique (CNRS) and other bodies. In addition, international organizations such as the IGBP (International Geosphere Biosphere Program), UNESCO's Man and Biosphere (MAB) program and ICSU's (International Council for Scientific Unions) Global Terrestrial Observation System (GTOS) program should be harnessed in support of such a global study.

Upping our game

It is time to up our game. Consider the significance of the environmental issues we are studying. The Thai nation has succeeded in reducing the number of children per couple from 7.5 to 2 in 25 years. Human population growth, though critically large, will level off. If properly tackled, pollution of the planet could be controlled and cleaned up in 100 years. Stabilization of global warming however, could take 1000 years. Depletion of biodiversity may take some 5 million years for evolution to restore – or more. Why is it then that funding for canopy studies is so hard to come by when governments, through the UNCED Conventions on Biological Diversity and Climate Change, are charged with stemming both the catastrophic loss of biodiversity the canopy contains and global warming? Also, climate change remains high on the political and economic agenda and this plus biodiversity are of great concern to the public on whose votes governments depend.

Consider some funding comparisons for major physical science projects under consideration within the ESF Large Facilities Program (ESF Communications 1998):

European Spallation Source (Project investigating the nature of neutrinos)	ECU 950 million
X-Ray Vision Project (A bio-molecular structure synchrotron)	ECU 150 million
100 Telsa High magnetic field magnet	ECU 50 million
ESF Tropical Canopy Research (Total budget estimate 1994–1998/9)	ECU 327,240

What kinds of questions and what international canopy research program might be developed which could address basic and applied science questions and which would assist governments to meet their UNCED Agenda 21 commitments? And further, how could such a program attract funding at similar levels to major physical science projects which traditionally out compete the natural sciences?

I propose the following five questions, put simplistically, which could form the basis of an integrated global canopy research program:

- What is the structure and function of forest canopies and how are they linked?
- How can we maintain biodiversity whilst also fostering its sustainable use?
- What is gained or lost through forest conversion in terms of economics, biodiversity, and biological services which maintain the planet and yes what about quantifying its value to the human spirit also?
- What is the global influence of pristine tropical and temperate forests in the sequestration of carbon and the production of oxygen?
- How can we use the canopy to train or influence the scientific, economic and political leadership of the future?

Not all of these questions are the traditional realm of the scientist but a fully integrated program needs to partner with other disciplines to achieve wide recog-

Figure 1. Proposed 'shell' design for a BIOTOPIA Research Centre. 25 researchers can be housed here, with laboratory facilities, seminar rooms, and full accommodation facilities. Inspired by a molluscan shell, it combines strength with a very small footprint in the forest. Hydroponic cladding encourages vegetation growth to lessen visual impact. Solar panels atop the structure provide power. Turbines in a nearby fast flowing river can do the same. Courtesy of Andrew Mitchell and Tazeer Amahd.

Figure 2. Canopy crane with 'torus' accommodation units. This imaginative design offers laboratory, accomodation and seminar area on three decks for up to 25 researchers. Each framed torus is wrapped in a retractable flexible membrane, enabling varying degrees of exposure. The habitable elements can be lowered to avoid exposure to high winds. Solar panels positioned along the crane's jib arm could provide sufficient power to drive it and raise water into four storage vessels at the top of the crane. A gravity fed turbine provides power at night. Courtesy of Andrew Mitchell & Tazeer Amahd.

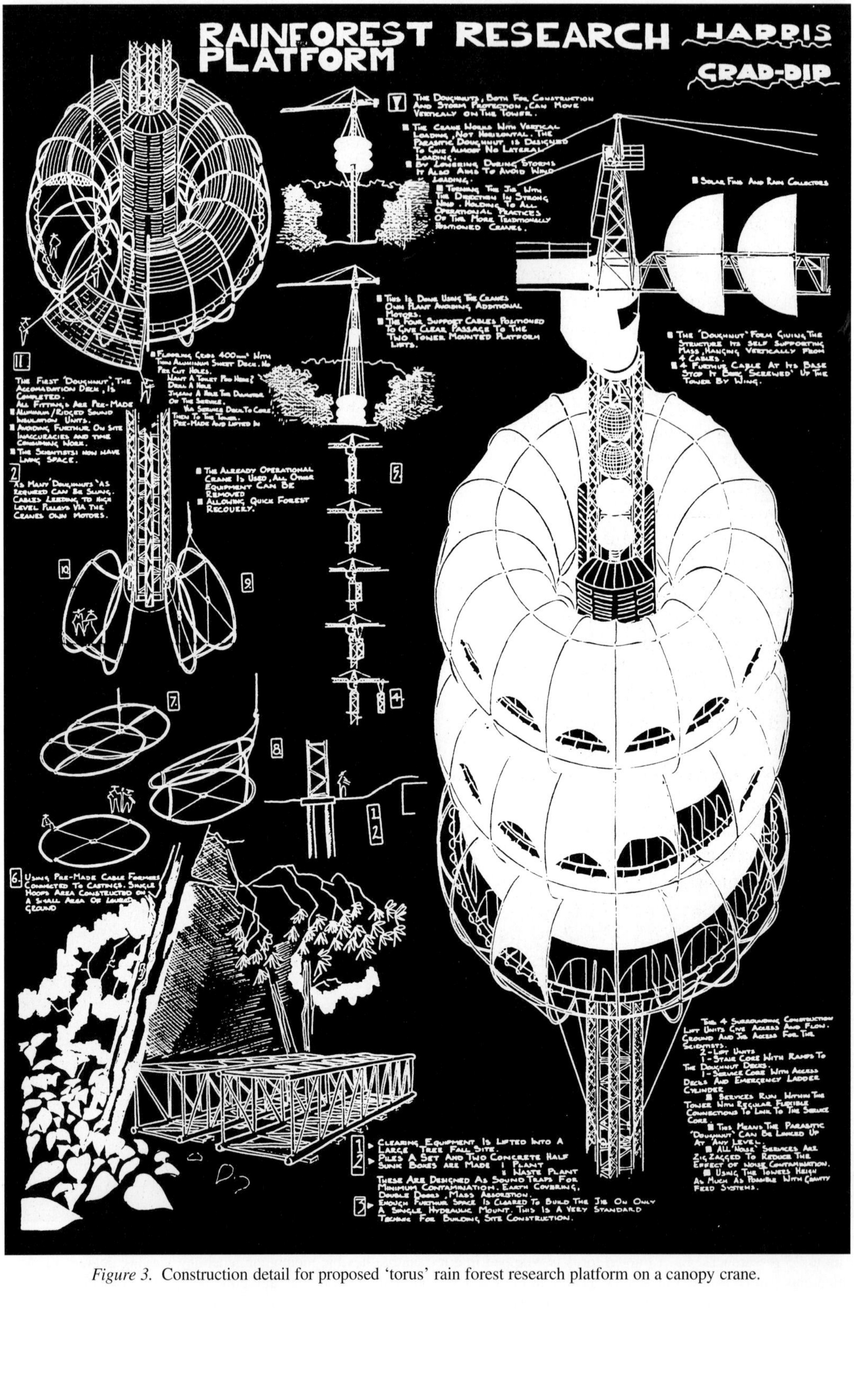

Figure 3. Construction detail for proposed 'torus' rain forest research platform on a canopy crane.

nition. Without such recognition key decision makers in government and bilateral aid agencies will fail to be influenced. Canopy science and the natural sciences need a big project to attract resources in the next millennium on the scale of the physical sciences.

An integrated research program requires an integrated facility which could address the issues that I have outlined above. At the first canopy conference in 1994 I proposed ideas for such a facility – which I termed 'BIOTOPIA'. This facility would use all the techniques currently available from climbing ropes to walkways, cranes, gondola cableways and lighter-than-air vehicles to maximize access to the forest. At its core, an inspirational field station would exist providing lab space, accommodations and seminar facilities for 25 researchers at canopy level in a field station with a minimal footprint (Figure 1). This should generate its own energy and recycle its own waste. I partnered with Tazeer Amahd and a team from the Oxford School of Architecture, Extreme Environments Laboratory to produce preliminary designs for such a facility. Professor Sir Ghillian Prance, director of the Royal Botanical Gardens at Kew, gave time and encouragement to the project along with other members of staff at Kew and other bodies.

The result has been a series of designs which challenge the imagination and provide a vision of the future (Figures 2 and 3). BIOTOPIA would use three cranes each at the apex of a triangle, with the field station at the centre, reachable by any of the cranes. There, up to 25 researchers could be accommodated year round. The cranes could be mounted on rails to extend their reach even farther. Calculations show that solar panels on the cranes could power them. A gondola system would further extend the study site and also provide supplementary income through tourism. Conference facilities at the BIOTOPIA Research Centre would attract politicians and business leaders who could not fail to be influenced by the normally unreachable environment in which they were conducting their deliberations. The Centre would attract funding for a valuable training and education program and would be on-line via the internet to centres of learning worldwide.

An advantage of the BIOTOPIA concept is that it takes existing tried technology and expands its scope significantly. At least three such centres should be contemplated, one in each of the major forest biomes. Funds need to be raised before a BIOTOPIA Field Station itself could be built. It is a concept designed to drive our thinking forward.

The next step

This conference was the first to bring together European researchers in the field of canopy science. Contributions from a number of US participants was welcome, but nothing was heard from Asian researchers, notably the Japanese who are at the forefront of canopy research in the dipterocarp forests of Borneo and co-workers from developing countries. A final discussion session chaired by Eduard Linsenmair produced a consensus for a forward thinking workshop to address some of the questions I have raised in this paper. Fortunately, the last remaining funds in the ESF Tropical Canopy Program were recently allocated to such a workshop held in Oxford on 20–24 November 1999. An NSF grant enabled 12 canopy researchers from the US to participate which made this a truly collaborative trans-Atlantic canopy science workshop dedicated to planning a global, integrated canopy research network. We have high hopes.

References

Erwin, T. L. 1991. How many species are there? Revisited. Cons. Biol. 5: 330–333.

ESF Communications (The journal of the European Science Foundation). Winter 1998/1999, No. 39.

Lowman, M. D. & Wittman, P. K. 1996 Forest canopies: methods, hypotheses, and future directions. Ann. Rev. Ecol. Syst. 27: 55–81.

Mitchell, A. W. 1986 The enchanted canopy. Collins, London.

Moffett, M. W. 1993. The high frontier. Harvard University Press, Cambridge, Massachusetts, 192 pp.

Nadkarni, N.M. 2001. Enhancement of forest canopy research, education, and conservation in the new millennium. Plant Ecol. 153: 361–367 (this volume).

Nadkarni, N. M. & Lowman, M. D. 1995. Forest canopies. Academic Press, San Diego.

Potter, C. S., Rounderson, J. T., Field, C. B., Matson, P. A., Vitousek, P. M. & Vitousek, P. M. 1993. Terrestrial ecosystem production: a process model based on global satellite and surface data. Glob. Biogeochem. Cycl. 7: 811–841.

Sutton, S.L. 2001. Alice grows up: canopy science in transition from Wonderland to Reality. Plant Ecology 153: 13–21 (this volume).

Plant Ecology **153:** 13–21, 2001.

Alice grows up: canopy science in transition from Wonderland to Reality

Stephen L. Sutton
Tropical Biology and Conservation Research Institute (IBTP), University of Malaysia Sabah, Locked Bag 2073, 88999 Kota Kinabalu, Sabah, Malaysia (E-mail sutton@pc.jaring.my)

Key words: Access, History, Rain forest canopy, Sampling

Abstract

This paper gives, for the first time, a complete history of rain forest canopy access, from the early years to the present day. This review is primarily from the European perspective, and explores the development of canopy access techniques, from 'low-tech' methods such as single rope technique, to 'hi-tech' approaches such as canopy cranes. In recent years, canopy science has moved away from pure exploration (the 'Wonderland' phase) to tackling the practicalities of rigorous canopy research ('Reality'), and the underlying emphasis is now shifting from access to the upper canopy *per se* to conducting replicative and manipulative science. The paper concludes by advocating the integration of many access techniques (both hi-tech and low-tech) at selected research sites, and certain neglected key areas of research are highlighted, including the comparison of adjacent primary and logged forests.

Introduction

This paper examines the way in which canopy access and sampling techniques have reached a new level of advancement from early pioneering work. This is analogous to the maturation of Alice Liddell (of Alice in Wonderland – Gardner 1970). At first, access to the upper canopy, and the act of discovery that went with it, were often ends in themselves (the 'Wonderland' phase), but this has given way to a phase where the early observations are replicated, quantified and rigorously tested. In this way, 'Wonderland' is being replaced by 'Reality'.

Pre-nineteen fifties

Native peoples have accessed the upper canopy of the tropical rain forest since their societies first developed. In many cases, sugar (in the form of honey) was the motivation, in others it was some particularly desirable fruit. An ability to climb into the canopy using vines, or by shinning up thin understorey trees is common in forest people, but the range of trees that can be tackled is limited; and for the researcher who wishes to target a particular population of trees, local climbers nearly always proves inadequate. This is even more critical today, when replicative and manipulative canopy science is more important than just the collection of specimens for study.

The European explorers of the nineteenth century used local climbers to make their canopy collections, but observation was almost entirely ground based. They were not slow to realise the potential for diversity of life and process in the upper canopy, but were too busy cataloguing and trying to understand rain forests at ground level to pursue studies above. Limited observations of canopy life were made, and whetted the imagination of the reading public. Somewhere out of the Victorian prose developed the anchor belief of canopy lore – that the upper canopy is where the action is, that plants and animals concentrate their vital functions of growth and reproduction in tree crowns, and there they often suffer the inevitable attentions of predators, parasites and pathogens. The idea of the forest floor as a side-show became deeply rooted. The importance of roots, of course, should have alerted everyone to the fallacious nature of this view. Only now, at the start of the new Millenium, is the truth and fiction embedded in this belief being systematically revealed.

Telescopes and binoculars were probably the first technological developments aiding canopy research. Ever since then, large, mobile animals such as primates and many birds, could be effectively observed from the ground using binoculars and this produces a great deal of data even now. Canopy science was not totally prevented from developing by the lack of access facilities, but was confined to very narrow tracks of enquiry. The first canopy measurements to be made were physical in nature, in the seasonal rain forests of Panama (Allee 1926).

It is perhaps surprising that it was not until 1929 that a concerted attempt was made to carry out biological work in the canopy. An observation platform was built by W R Hingston and colleagues at Moraballi Creek in British Guiana (now Guyana) and baited muslin traps for insects, birds and small mammals were hoisted (Hingston 1932). The procedure for accessing the canopy was graphically described by Max Nicholson (Mitchell 1986, and see Hingston 1932), but sadly no scientific account of the work was published.

Nothing much further happened in the 1930's and 1940's except for the enthusiastic writing on the potential for canopy research by William Beebe in the National Geographic magazine and in his own books (see Beebe 1925). The Second World War discouraged such things as canopy research while at the same time stimulating the development of high strength, lightweight metal and polymer materials which later became the keys to canopy access.

The nineteen fifties

The major development in the 1950's was the erection of a steel tower in the Mpanga Forest Reserve in Uganda. A tower provides a stable platform for studying gradients from forest floor to upper canopy. However these gradients are often influenced by the fact that the tower is built in a clearing that has very different properties, both physical and biological, from similar gradients within the forest. Nevertheless, towers offer an important form of canopy access. A team of entomologists based in East Africa realised this and found funds for a 40m tower for research on insect vectors of human diseases. They carried out a series of pioneering observations on such phenomena as insect swarming behaviour and the activity of insects at various height levels in relation to the daily microclimatic cycle. Moreover, this was (almost) all written up in a single volume (Haddow et al. 1961). It is ironic that funds from an applied science agency created this breakthrough in basic understanding of insect life in the canopy (funds for basic canopy science are perhaps not much easier to obtain now than they were then).

Another early tower-based study deserves mention because of its modern concept, manipulative aspect and elegance of design. Cachan (1964) placed standard samples of dead wood at different levels and analysed the height preferences of different mycetophagous Coleoptera. He produced clear evidence of height preferences between species of these fungus eating beetles (see Sutton 1989).

The nineteen sixties

The first use of the post war lightweight materials was in the early 1960s, when McClure (1966, 1977) attached aluminium ladders to a tree trunk at Gombak near Kuala Lumpur in Malaysia to track the daily activity of canopy vertebrates. The 50 m ladder stood for fifteen years until the tree came down to make way for a multi-lane trans-peninsular highway.

It was also in the 1960's that the first aerial walkway was built for the Institute of Medical Research at Bukit Lanjan in Malaysia to a design by US Army Engineers (Muul & Liat 1970). An aerial walkway is basically a suspension bridge from one tree crown to the next. To provide sufficient tension to keep it tolerably level, it has to be firmly anchored at the base of the crown. Walkways can be built to link a whole series of crowns, either linearly, or radially.

These were isolated initiatives, and there was no general commitment to canopy study. Access was perceived as too difficult to allow quantitative and experimental science and most scientists who found funding to work in tropical forest were pre-occupied with answering questions on the ground.

The nineteen seventies

All this changed in the 1970s, partly because of a revolution in means of access, and partly because the time was right to extend ground based, two dimensional studies into the third dimension. The key change in access technology was the development of single rope technique for climbing trees and the realisation that however frustrating firing lines over boughs 30–50 m above the ground could be, it was just about always successful if one persisted. Single rope technique (SRT) developed (at least outside the United States) from caving exploration (Montgomery 1977; Meredith 1979). Caves are wet, dark and often vertical, conditions which are similar to rain forests.

Perry (1978) not only developed SRT for accessing the centre of a tree crown, but also developed a system for access to the outer parts of canopies (Perry & Williams 1981). He generated enthusiasm and publicity for canopy research. SRT is relatively inexpensive, simple, safe, portable and operable by just one or two people. By the end of the 1970s it was in modest use in Central and South America and much more sparingly on other continents. It extended the geographical range of research activity as well as increasing the variety of studies at established sites. The value of SRT, on its own, can be over-emphasised. Gaining access to a substantial bough in the lower part of a tree crown does not in itself provide a passport to a successful piece of research. SRT is usually only the first step in the access process needed to confront canopy phenomena head on.

The introduction of SRT coincided with quite separate developments aimed at studying processes in the leafy parts of tree crowns. At Pasoh in Peninsular Malaysia Ashton and Appanah were perfecting the use of booms to study the pollination of flowers of dipterocarp trees (Appanah & Chan 1981). Walkways were built out on sloping ground to give height, and then booms were attached to anchor trees. Each boom had a bo'sun's chair (a suspended canvas seat) at the free end in which the researcher sat, and the boom could be manipulated by an assistant to allow observation and sampling of flowers. The work demonstrated that thrips (Thysanoptera) were the main pollinators for the species of dipterocarp under study. We now know, from the work of a Japanese/Malaysian team at the Lambir Hills site in Sarawak, that many other, less surprising insects, are involved in dipterocarp pollination (Momose et al. 1996).

In the 1970s aerial walkway construction began in earnest. Fritz Dieterlen built one in Zaire in 1971–1972 to study phenology and pollination biology (Dieterlen 1978) but funding constraints curtailed his study. This walkway (pictured in Mitchell 1982) was a primitive affair and not as easy to use as the earlier one put up at Bukit Lanjan in Malaysia; but it demonstrated considerable potential for canopy studies, and inspired others.

The period 1979–1980 saw the construction of multiple walkways in Panama, Papua New Guinea and Sulawesi (Indonesia) as part of the international expedition Operation Drake (Mitchell 1981, 1982; Sugden 1982). As much as 300 m of walkway was available at each site. Many preliminary studies were carried out, but the peripatetic nature of the expedition meant that the sites had to be abandoned before much of the potential of the access could be exploited. Much was learned about construction and maintenance, and how walkways could be used by researchers with very different interests. The major problem exists between those who need facilities for quiet observation (bird and mammal behaviour) and those who need to sample, although this is not exclusive to walkways.

Not all canopy access in the 1970s involved climbing. The first canopy fogging for insects using knockdown aerosols took place, using a fogger hauled into the canopy, with the affected insects falling onto sheets laid out at ground level. This later became an established part of the repertory of insect study in the canopy.

The nineteen eighties

The 1980s was a period of expansion and consolidation of the techniques introduced in the 1970s. SRT became widespread, although some researchers felt it was beyond them in terms of fitness or boldness. One should not have to be an athlete to conduct science: simple systems (e.g., fixed ladders with safety cages and platforms) will always have a place. SRT was made safer by adding additional gear and the alternative arborist method was introduced (Lowman & Wittman 1996, Barker 1997). Line placement underwent major improvement with the adoption of high performance fishing spools and more powerful crossbows capable of putting a line over a 50 m high bough. Barker describes how to rig the crown of a tree with SRT equipment to access individual leaves to measure photosynthetic activity (Barker 1997).

Canopy fogging flowered during the 1980's. It was improved by standardisation of the area of the collecting surfaces, the ease of actually collecting the insects and remote control of emission of the fogging agent. The technique received acclaim when Erwin published his estimate of 30 million insect species globally, based on canopy fogging (Erwin 1982). Fogging produced a wealth of information about the diversity of canopy insect faunas (Erwin 1995, reviewed in Stork et al. 1997a), while continuing to cause debate about precisely how to interpret the results. Problems of comparing the results have recently been addressed with the publication of a standard sampling protocol (Adis et al. 1998).

Perhaps the major feature of the 1980s was the publication of research begun in the mid- to late Seventies. The 1982 conference at Leeds University in the UK debated the ecology and management of rain forests including the importance of canopy studies. Excluding the management section, ten out of twenty five of the papers related to canopy research (Sutton et al. 1983).

Few of the efforts prior to 1990 to sample in the canopy resulted in publications in the top journals and in a number of cases no publications resulted at all. This is what I term the 'Wonderland' phase of canopy research. The effort to reach the canopy and the excitement of this achievement distracted people from the necessity of proper experimental design and sufficient replication. Often, too much time was consumed by the access process, leaving too little for repeated observations or manipulative experiments. Much of what was observed was new, and was written up as pilot studies. Pseudoreplication claimed many victims. Because of the effort required to access just one tree crown, the temptation was to do as much work as possible in it. For some studies one crown may be sufficient, but for most one tree crown represents inadequate replication (Barker & Pinard, 2001).

The nineteen nineties

This has been the decade of high-tech hardware, the further maturation of low-tech methods (e.g., Barker & Sutton 1997) and the building of networks. There have also been more publications in international journals. The support of the Marie Selby Botanic Gardens in Florida, USA, and the availability of the journal Selbyana for technical papers concerning canopy research, have been significant. Comprehensive reviews of forest canopy science (Lowman & Nadkarni 1995; Lowman & Wittman 1996) have also been beneficial. The International Canopy Network (ICAN) on the Internet has greatly facilitated the distribution of information. Two canopy conferences have been held in the US, and one in Europe (the source of volume 153). As part of a five year programme the European Science Foundation funded several workshops (e.g., Lucking et al. 1995, 2001)

Two non-technical but accurate books (Moffett 1993; Lowman 1999) have provided a very effective link between the scientists and the general public, and these have been augmented by a number of high quality television programmes. The National Geographic Society has supported canopy research for over 70 years, from the early observations of William Beebe (see above). Given the current high level of public interest in the rain forest canopy, the prospects for further funding for research may be good, although generally environmental interest needs to border on panic before funding becomes available.

Concern about global environment change has given a big boost to canopy research. It has led to the funding of the first rain forest canopy crane to study fluxes of gases between the canopy surface and the upper atmosphere, funded by the Smithsonian Tropical Research Institute and the United Nations Environment Programme. Situated near Panama City, this crane has been in operation for nine years and has proved its worth so well that it has now been joined by four others – another in Panama (in a much less seasonal region), Venezuela (as part of the European Science Foundation's programme), Oregon (USA) and Queensland (Australia) (Stork et al. 1997b). Another is planned for Sarawak, Malaysia (Lambir Hills) (see Stork 2001).

Canopy cranes carry one or more researchers in a gondola and have a 'footprint' of approximately one hectare or rather more. The 40 m high crane at Surumoni (Venezuela) runs on 110 m of rails, greatly increasing its footprint. With a boom reach of 40 m, the footprint of the crane is around 1.2 ha (Szarynski & Anhuf 2001). The Wind River crane in the Douglas fir forest of Oregon, USA, extends 85 m, resulting in a footprint of 2.3 ha and access to 320 individual tree crowns (Stork et al. 1997b).

Lighter-than-air (LTA) platforms, which may be un-powered balloons or airships, have already been used a number of times for canopy access. The airship providing the mobility for the French 'Radeau des Cimes' project has provided access in four expeditions to date. This major LTA platform development uses the airship to place an inflatable plastic raft over an area of several tree crowns, or to manoeuvre a suspended aerial sledge through the canopy. This 'ensemble' (e.g., Hallé & Pascal 1991) of dirigible, raft and sled has produced a number of interesting research findings (Hallé & Blanc 1990, Moffett 1993). Like most other access methods 'demonic intrusion' is a problem (Barker & Pinard 2001), where the very act of access affects some of the phenomena and organisms under study. 'Operation Canopèe' is also expensive to operate and at present the cost can only be met by commercial sponsorship. Such sponsorship relies on the 'Wonderland' aspect of canopy access to create the required publicity. This can conflict with the needs of scientists to carry out time consuming replicated research. High operating cost also translates into short field seasons, which limit the quality of the work generated. This is less of a problem with cranes, where operating costs are lower, but establishment costs are very high (although lower, hourly rates of crane use may still be prohibitive for individuals).

An ultralight, two seater, helium airship was tested by Dorrington at Danum Valley, Sabah, E. Malaysia in 1995 (Dorrington 1995). Some thirty test flights were made and demonstrated that this type of LTA platform has considerable potential. Use of the prototype was limited by the need for calm conditions, which only occurred above the canopy at dawn and dusk. With funding, it seems feasible to produce a small research airship powerful enough to plug a rather large gap in the spectrum of research facilities at present available for above canopy sampling and observation. Applications would include monitoring the dynamics of flowering of trees at the scale of tens or hundreds of hectares, or the spatial variation in gas fluxes.

In addition to these major hardware facilities, there is funding available in Germany for the construction of a balloon and cable system forming a triangular multi point access system. A similar principle was used by Perry in the 1970s to achieve multi-point access to a volume of canopy using an extension of SRT (Perry & Williams 1981). The balloon/cable system is derived from the Canopy Operation Permanent Access System or COPAS (Gottsberger & Doring 1995), which was too costly in its original form.

During the Nineties the steady development of remote sensing techniques has opened up a wide range of opportunities for canopy surface categorisation (see, for example, Hill & Foody (1994), Paradella et al. (1994) and Foody (1996) on the classification and separation of unlogged forest types, and Yanasse et al. (1997) and Luckman & Baker (1997) on radar remote sensing of the regenerative stages of forests following logging). It is now feasible to plot the grosser differences between forest stands (e.g., *Macaranga* dominated areas and tall forest remnants in selectively logged forest) and with ground referencing using GPS (Global Positioning Systems) detailed knowledge of the responses of large forest areas to extreme events like drought are probably now possible.

However, the potential has by no means yet been fully exploited. Satellite sensing has now reached a level of diversity, technical advance and high resolution that such problems as assessing the pattern of mast fruiting in a Bornean rain forest and other seasonal phenomena are just about feasible. As data on leaf physiology and gas exchange from the cranes and other canopy sites become available, there will be an increasing need to attempt inductive integration with remote sensing data, to provide area-wide values which will be useful for developing spatially explicit models. However, much caution is needed in interpreting satellite data. A detailed survey by Nepstad et al. (1999) suggests that more than half the damage by disturbance to rain forest in the Brazilian Amazon is being missed by satellite image analysis. The problem (Monastersky 1999) is due to a concentration on areas of clear felling rather than selective logging and surface fires.

Unfortunately, remote sensing analysis is an expensive, specialist task and obtaining the necessary images in many rain forest countries is not necessarily straightforward for reasons of security. In practice a remote sensing study is big science which must be undertaken by well funded international teams with good contacts. Given the high cost of such high-tech hardware, it is not surprising that low-tech approaches, available to most researchers, continue to be popular. There is not a clear dichotomy between high- and low-tech. In some types of forest, a tower may need only to be quite low and built cheaply of local materials, but in the tall dipterocarp forests of SE Asia an 80 m structure requires steel or some even more expensive, high strength, material.

The next five years: access

The watchword should be 'Integration': integration of different methods of access to allow more flexibility of access and study and to allow more time on data gathering. If a crane can deposit a person on a tree platform more suited to a particular study than riding in a gondola, then the person deposited will benefit and the gondola will be available for other studies. If the platform leads off to other platforms via a series of walkways, the effectiveness of the whole system is greatly improved.

An issue to be resolved is whether SRT techniques can be improved to give the necessary replication of trees. Hi-tech systems such as cranes and COPAS offer very good prospects for rapid replication, but are expensive. There is also a greater danger of pseudoreplication with these limited area techniques than with SRT, so it is unlikely that SRT will fall into disuse. What seems likely is that some field centres will continue the trend of developing a variety of canopy access facilities and that SRT will be used in conjunction with these to give the necessary replication.

Two systems of access seem to have fallen behind in exploitation. Walkways and booms are not exactly low tech in cost, but are relatively cheap compared to cranes and LTA platforms. As constructed in Malaysia, a high quality walkway providing a very stable platform will cost US\$250 per metre for a 100 m length at 1999 prices. Other walkway budgets are extremely modest (Lowman & Bouricius 1995). The comfortable sampling distance either side of a walkway (e.g., with a twig clipper) is 5 m, giving a 'footprint' per 100 m length of 1000 m^2.Thus 10 lengths of 100 m, totalling 1 km, will have a footprint of 1 ha for a cost of US\$250K. This compares with the cost of a crane, with the same footprint, of US\$750K or more. The advantage of the crane access is the ability to sample the canopy surface (or, within limits, the volume) on the spot at any point within the 1 ha footprint. The advantage of the walkway is to open up a much wider area of forest (with 1 km length) than the crane can cover. It also allows SRT and boom access to be installed. An integrated system of tree platforms with fixed ladder access, 1 km of walkway with booms and fixed ropes for SRT access could provide a new and versatile kind of canopy facility with relatively low operating costs, but with limited capability for sampling the actual canopy surface. The most promising development of integrated access at the moment seems to be the Japanese/Malaysian initiative at Lambir Hills in Sarawak (Inoue et al. 1995).

Funding costs for canopy walkways can, in some cases, be offset by tourist revenue. The walkway at Poring Hot Springs in Sabah, East Malaysia, designed by Illar Muul, operates as a tourist facility in the middle of the day, but as a research facility the rest of the time. The tourist presence precludes certain studies, but not others.

Booms can easily be installed on tree platforms and accessed from walkways, and can be raised well above the level of a platform and manoeuvred to access the canopy surface at a number of points. They could perhaps even be given a telescopic function and driven by solar powered batteries.

The next five years: experimental design and sampling

Effort should be prioritized to achieve replication levels and statistical rigour throughout canopy science (see Barker & Pinard 2001). Science in the canopy is merely an extension of science carried out elsewhere, and must be subjected to the same standards of analysis.

The 1990s were a welcome transition from purely descriptive canopy science to manipulative research. There is still much descriptive science to be done, including life history data and food plant range of practically all canopy insects. This prevents formulation of some key questions about how the canopy system functions. Where bionomics (life history) data are integrated with other approaches considerable progress can be made in answering key questions (e.g., Basset et al. 1996). Funding for a project needs to be partly for bionomics and partly for hypothesis testing by experimentation. Recent work in Puerto Rico points the way. The exclusion of *Anolis* lizards from tree crowns demonstrating an increase in insect prey species has shown the feasibility of the manipulative approach (Dial & Roughgarden 1995, Lawton 1995). However providing the necessary controls required by a fully experimental approach calls for considerable resources.

The next five years: standardisation

Major efforts to standardise sampling methods have been made over the years, but there is still a long way to go. Researchers are individuals and are often constrained by local conditions and resource availability. Studies of Lepidoptera, which are promising vehicles for research because of the relatively advanced state of their taxonomy, are bedevilled by the fact that there are practically as many survey methods as there are surveyors.

However, the establishment of protocols such as that for botanical studies (Lowman 2001) and canopy fogging (Adis et al. 1998) within the ESF canopy programme makes it far easier to gain the conformity necessary for direct comparisons between studies. Arguably, funding bodies have a right to insist on greater value from funded work by a guarantee of true comparability with other studies. If they pursued this policy explicitly, protocols would rapidly become effective.

The next five years: safety

Canopy access is potentially a dangerous business, and it is a tribute to the researchers that there have been very few serious accidents in the last thirty years. As more researchers spend more time in the canopy, hazardous situations will arise. While techniques such as SRT have built in safety to protect the operator, outside events such as a falling branch must be planned for. A number of climbers prefer the arborist method of tree climbing (e.g., Lowman & Wittman 1996; Barker 1997; Barker & Sutton 1997) because it allows more rapid descent. Safety procedures have to be thought through on the spot and provide *effective* solutions. It is one thing to insist that everyone using a walkway has a short safety line, but there needs also to be a realistic plan to enable safe descent in the event of a group of people marooned at canopy level.

The next five years: some key areas of investigation

Improved access and sampling in the next few years should fill some major gaps in our knowledge of the rain forest canopy. One of these involves the air space above the canopy. In 1927 R. Geiger published a book 'Das Klima der Bodennahen' (re-published as 'The Climate near the Ground', Geiger 1950) which provided a key stimulus to the infant discipline of microclimatology. Some similar work is now needed for what may some day come to be called the 'supra canopy zone', i.e., the surface of the canopy and the space 100 m above it. Establishing the physical parameters of this space is a key area of study in itself, and an essential basis for studies on biotic processes. Quantification of parameters such as the amount of gas and water vapour flux are necessary for the construction and validation of hydrological and climate models. For example, trace gas exchange between the atmosphere and forest canopies has a direct effect on the oxidising capacity of the atmosphere (Cao et al. 1997). The importance of this air above the canopy as a dispersal, circulating and feeding space for entities as diverse as pollen grains, bees and birds has yet to be assessed. The swifts (Apodidae) are birds of this space and as yet their aerial biology is unknown. An extreme gradient of temperature, humidity and light level exists at the canopy surface. Leaf physiology and its relation to this gradient can be assessed via cranes, but these are too expensive to explore the question fully. Furthermore stable, stationary or slow moving platforms to sample and observe the full 100 m column of air are lacking, because such platforms are extremely expensive. The development and proper funding of LTA platforms such as balloons and very small airships should be a matter of priority.

A second major 'lacuna' is the subject of bionomics (life history, including food and enemies). Victorian and subsequent naturalists conducted life history studies but only from ground level. Such studies are time consuming and therefore expensive, and are regarded by funding agencies as being almost as pedestrian as that other essential canopy science, taxonomy. For example, very few canopy insects have good life history data (but see Basset et al. 1996). Guild composition and the relative proportions of monophagy, oligophagy and polyphagy are crucial data for constructing food webs. Bionomics requires observation *in situ* and/or the collecting of live material for monitoring in captivity. This in turn requires comprehensive canopy access, ideally using a combination of techniques.

A third research gap relates to the impact of logging on diversity and forest processes.There are few papers in the primary literature directly comparing adjacent logged and unlogged areas specifically from the viewpoint of canopy loss (see Davis & Sutton 1998). Bearing in mind that there are (or rather were, before the fires and clearances of the 1990s) over 100 million ha of selectively logged forest in Indonesia alone, the hydrological and ecological impact of loss of canopy as a result of disturbance to primary forest is enormous. The hydrological consequences are perhaps better known than the ecological effects, but in both cases there is a dearth of papers in peer reviewed journals. Instead there exist many reports, workshop proceedings and other grey literature (e.g., SEARRC 1999) which have limited impact on decision makers.

Logging and clearance of forest can also result in fragmentation and isolation of remnant forest blocks. The main factors affecting biotic communities in forest fragments, namely the size and shape of the remnant, the degree of isolation, extent of edge effects and connectivity (e.g., Saunders et al. 1991; Murcia 1995) influence the canopy assemblages and processes as much as those below the canopy. There is wide speculation (although data are poor) that isolation and fragmentation-induced changes in forest structure may cause the disruption of biological processes such as pollination, seed dispersal, seed predation, and nutrient cycling (Didham et al. 1996). If data are lacking for tropical forest ecosystems in general however, the situation is particularly bad for the forest canopy. There has also been little development of the concept of canopy itself as 'forest-edge' habitat, using theory from forest fragmentation studies to examine changes in forest canopies following disturbance, and this may be a useful way forward for future research, particularly in the response of canopy systems to logging and fragmentation (see Davis & Sutton 1998). The very first study of the physical conditions within the canopy (Allee 1926) compared the parameters with those in open spaces. This work now needs to be taken up and extended, using adjacent areas of primary and disturbed forest. Specifically, there is a compelling need for comparative studies of primary forests with adjacent time series of logged areas.

With all these questions, funding is the key issue. The revenue derived from the harvesting of the timber from 100 million+ ha of forest is significantly greater than the paltry funding that has been put into investigating the ecological consequences of canopy loss. The next decade will see greater awareness of the consequences of neglecting environmental stewardship, but it probably needs fresh blood in the world's financial institutions and political hierarchies to take the necessary action. Meanwhile, canopy scientists should be ready for the call.

Acknowledgements

I am very grateful to Martin Barker for stimulating discussions and much help during the preparation of the talk from which this paper is derived, and to Andrew Davis, Nick Chappell and Terry Whitaker for much assistance in the preparation of the talk and the manuscript. Andrew Davis drafted the section on the need to study canopy fragmentation. I also thank Nick Chappell and Rory Walsh for searching out the latest work in the fields of atmospheric chemistry and remote sensing, respectively. Finally I would like to pay tribute to the European Science Foundation for funding the Tropical Forest Canopy Programme and to the Royal Society (UK) for facilitating UK involvement in this Programme.

References

Adis, J., Basset, Y., Floren, A., Hammond, P. & Linsemair, K. E. 1998. Canopy fogging of an overstory tree – recommendations for standardization. Ecotropica 4: 93–97.

Allee, W. C. 1926. Measurement of environmental factors in the tropical rain-forest of Panama. Ecology 7: 273–302.

Appanah, S. & Chan, H. T. 1981. Thrips: pollinators of some dipterocarps. Malay. For. 44: 234–252.

Barker, M. 1997. An update on low-tech methods for forest canopy access and on sampling a forest canopy. Selbyana 18: 61–71.

Barker, M & Pinard, M. A. 2001. Forest canopy research: sampling problems, and some solutions. Plant Ecol. 153: 23–38 (this volume).

Barker, M. & Sutton, S. L. 1997. Low-tech methods for forest canopy access. Biotropica 29: 243–247.

Basset, Y., Samuelson, G. A., Allison, A. & Miller, S. E. 1996. How many species of host-specific insects feed on a species of tropical tree? Biol. J. Linn. Soc. 59: 201–216.

Beebe, W. 1925. Studies of a tropical jungle; one quarter of a square mile of jungle at Kartabo, British Guiana. Zoologica 6: 4–193.

Cachan, P. 1964. Analyse statistique des pullulations de Scolytoidea mycetophages en forêt sempervirente de Cote d'Ivoire. Ann. Fac. Sci. Univ. Dakar 14: 5–70.

Cao, X. L., Boissard, C., Juan, A. J. & Hewitt, C. N. 1997. Biogenic emissions of volatile organic compounds from gorse (*Ulex europeus*): branch and canopy scale flux measurements. J. Geophys. Res. 102: 18 903–18 915.

Davis, A. J. & Sutton, S. L. 1998. The effects of rain forest canopy loss on arboreal dung beetles in Borneo: implications for the measurement of biodiversity in derived tropical ecosystems. Diversity Distrib. 4: 167–173.

Dial, R. & Roughgarden, J. 1995. Experimental removal of insectivores from rain forest canopy: direct and indirect effects. Ecology 76: 1821–34.

Didham, R. K., Ghazoul, J., Stork, J. E. & Davis, A. J. 1996. Insects in fragmented forests: a functional approach. Trends Ecol. Evol. 11: 255–260.

Dieterlen, F. 1978. Zur Phanologie des aequatorialen Regenwaldes im Ost-Zaire (Kivu), nebst Pflanzenliste und Klimadaten. Dissert. Bot. 47: 1–132.

Dorrington, G. 1995. Project Hornbill: evaluation of an ultralight airship for canopy science. Pp. 71–72. In: Marsh, C. (ed.), Danum Valley Rainforest Res. & Training Prog. Annual Rep. Rakyat Berjaya Sdn Bhd, Sabah Foundation, Sabah, E. Malaysia.

Erwin, T. L. 1982. Tropical forests: their richness in Coleoptera and other arthropod species. Coleopterists Bull. 36: 74–75.

Erwin, T. L. 1995. Measuring arthropod diversity in the tropical forest canopy. Pp. 109–127. In: Lowman, M. D. & Nadkarni, N. M. (eds), Forest canopies. Academic Press, San Diego.

Foody, G. M. 1996. Classification of tropical forest classes from Landsat TM data. Int. J. Remote Sensing 17: 2353–2367.

Gardner, M. 1970. The annotated Alice by Lewis Carroll. Penguin Books, London.

Geiger, R. 1950. The climate near the ground. Harvard University Press, Cambridge, Mass.

Gottsberger, G. & Doring, J. 1995. 'COPAS', an innovative technology for long-term studies of tropical rain forest canopies. Phyton (Horn, Austria) 35: 165–173.

Haddow, A. J., Corbet, P. S. & Gillet, J. D. 1961. Studies from a high tower in Mpanga Forest, Uganda. Trans. R. Ent. Soc. Lond. 113: 249–368.

Hallé, F & Blanc, P. (eds). 1990. Biologie d'une canopèe de forêt equatoriale. Institut Botanique, Montpellier.

Hallé, F. & Pascal, O. 1991. Biologie d'une canopèe de forêt equatoriale II. Rapport de mission: Radeau de Cimes a Reserve de Campo, Cameroun. Avec le parrainage du Ministere de la Rècherche et de la Technologie.

Hill, R. A. & Foody, G. M. 1994. Separability of tropical rain-forest types in the Tambopata-Candamo Reserved Zone. Int. J. Remote Sensing 15: 2687–2693.

Hingston, R. W. G. 1932. A naturalist in the Guiana forest. Edward Arnold & Co., London.

Hurlbert, S. H. 1984. Pseudoreplication and the design of ecological experiments. Ecol. Mon. 54: 187–211.

Inoue, T., Yumoto, A., Abang, A. H., Lee, H. S. & Ogino, K. 1995. Construction of a canopy observation system in a tropical rain forest of Sarawak. Selbyana 16: 24–35.

Lawton, J. H. 1995. Tabonuco trees, trophic manipulations & Tarzan: lizards in a tropical forest canopy. Trends Ecol. Evol. 10: 392–393.

Lowman, M.D. 1999. Life in the treetops. Yale University Press, New Haven, Connecticut.

Lowman, M.D. 2001. Plants in the forest canopy: some reflections on current research and future direction. Plant Ecol. 153: 39–50 (this volume).

Lowman, M. D. & Bouricius, B. 1995. The construction of platforms and bridges for forest canopy access. Selbyana 16: 179–184.

Lowman, M. D. & Nadkarni, N. (eds). 1995. Forest canopies. Academic Press, San Diego.

Lowman, M. D. & Wittman, P. K. 1996. Forest canopies: methods, hypotheses & future directions. Ann. Rev. Ecol. Syst. 27: 55–81.

Lucking, R., Freiburg, M., Lucking, A., Freiberg, E. & Gottsberger, G. 1995. Tropical forest canopies as an environment for arthropods and epiphytes with special reference to the phyllosphere. Proc. 2nd Int. ESF Workshop on tropical canopy research.

Luckman, A. J. & Baker, J. 1997. Interferometric coherence measurements of tropical, temperate and boreal forests. Pp. 275–278. In: Proc. 3rd Symp. on Space at the Service of our Environment, Florence, Italy.

McClure, H. E. 1966. Flowering, fruiting & animals in the canopy of a tropical rain forest. Malay For. 29: 182–203.

McClure, H. E. 1977. The secret life of a tree. Animal Kingdom 80 (4).

Meredith, M. 1979. Vertical caving. Westmorland Gazette, Kendal, UK.

Mitchell, A. W. 1981. Operation Drake: voyage of discovery. Severn House Publishers.

Mitchell, A. W. 1982. Reaching the rain forest roof. Leeds Symposium on Ecology & Management of Tropical Rain-Forest Special Publication.

Mitchell, A. W. 1986. The enchanted canopy. Collins, London.

Moffett, M. 1993. The high frontier: exploring the rain forest canopy. Harvard University Press, Cambridge, Mass.

Momose, K., Nagamitsu, T. & Inoue, T. 1996. The reproductive ecology of an emergent dipterocarp in a lowland rainforest in Sarawak. Am. J. Bot. 86: 732–743.

Monastersky, R. 1999. Hidden threats take toll in Amazon. Science News 155: 228.

Montgomery, N. 1977. Single rope techniques – a guide for vertical cavers. Sydney Speleological Society.

Murcia, C. 1995. Edge effects in fragmented forests: implications for conservation. Trends Ecol. Evol. 10: 58–62.

Muul, I. & Liat, L. B. 1970. Vertical zonation in a tropical forest in Malaysia: method of study. Science 196: 788–789.

Nepstad, D. C., Verrissimo, A & 11 others. 1999 Large scale impoverishment of Amazon forests by logging and fire. Nature 398: 505–508.

Paradella, W. R., Da Silva, M. F. F., Rosa, N. de A. & Kushigor, C. A. 1994. A geobotanical approach to the tropical rain forest environment of the Carajas Mineral Province (Amazon Region, Brazil) based on digital TM-Landsat and DEM data. Int. J. Remote Sensing 15: 1633–1648.

Perry, D. R. 1978. A method of access into the crowns of emergent and canopy trees. Biotropica 10: 155–157.

Perry, D. R. & Williams, J. 1981. The tropical rain forest canopy: a method providing total access. Biotropica 13: 283–285.

Saunders, D. A., Hobbs, R. J. & Margules, C. R. 1991. Biological consequences of ecosystem fragmentation: a review. Cons. Biol. 5: 18–32.

SEARRC (SE Asia Rain Forest Research Committee of the Royal Society, London). 1999. Research programme evaluation & review 1985–99 (CD_ROM). Whitaker, T. M. (compiling ed.), The Royal Society, London. http://www.royalsoc.ac.uk/rs.

Stork, N. E. 2001. The management implications of canopy research. Plant Ecol. 153: 313–317 (this volume).

Stork, N. E., Adis, J. & Didham, R. K. 1997a (eds). Canopy arthropods. Chapman & Hall, London.

Stork, N. E., Wright, S. J. & Mulkey, S. S. 1997b. Craning for a better view: the canopy crane network. Trends Ecol. Evol. 12: 418–420.

Sugden, A. 1982. Aerial walkways in rain-forest canopies. Construction and use in Panama, Papua New Guinea and Sulawesi. In: Gentry, A. (ed.), Symposium on the natural history of Panama. Ann. Missouri Bot. Gdns.

Sutton, S. L. 1989. The spatial distribution of flying insects. Pp. 427–436. In: Lieth, H. & Werger, M. J. A. (eds), Ecosytems of the World Vol. 14b; Tropical rainforest ecosystems. Elsevier.

Sutton, S. L., Whitmore, T. C. & Chadwick, A. C. 1983. Tropical rain forest: ecology and management. Blackwell Science, Oxford.

Szarynski, J & Anhuf, D. 2001. Micrometeorological conditions and canopy energy exchanges of a neotropical rain forest (Suromoni Crane Project, Venezuela). Plant Ecol. 153: 231–239 (this volume).

Yanasse, C. da C., Sant'Anna, J. S. S., Frery, A. C., Renno, C. D., Soares, J. V. & Luckman, A. J. 1997. Exploratory study of the relationship between tropical forest regeneration stages and SIR-C L and C data. Remote Sensing Env. 59: 180–190.

Plant Ecology **153:** 23–38, 2001.

Forest canopy research: sampling problems, and some solutions

Martin G. Barker[1] & Michelle A. Pinard[2]
[1]*Department of Plant and Soil Science, University of Aberdeen, Cruickshank Building, Aberdeen AB24 3UU, UK (E-mail: m.barker@abdn.ac.uk);* [2]*Department of Forestry, University of Aberdeen, MacRobert Building, Aberdeen AB24 5UA, UK*

Key words: Canopy access, Replication, Research methodology, Sample, Tree crowns

Abstract

Conducting research in forest canopies is accompanied by both opportunities and difficulties. Opportunities occur because the canopy is important in overall forest structure and function, for example in containing a high proportion of species richness, being the site of forest-atmosphere fluxes, and in strongly influencing understory microclimate. However, despite this importance, the canopy has been largely neglected by researchers, until recently. Many difficulties in forest canopy research apparently arise from problems related to access. In a survey of 112 canopy researchers, examples of problems cited include: bias in the selection of study species or sampling locations; difficulties in obtaining adequate replication; problems in collecting data in a three-dimensional, complex environment; logistical constraints in moving between canopy sampling points; and, disturbance, sometimes including interference of the object being studied (i.e., 'demonic intrusion') caused by access. Many of these problems have at least partial solutions. These include: using multiple access techniques to give greater sampling flexibility; identifying appropriate units of replication within the canopy; using ground-operated canopy sampling methods; and, using protocols for unbiased and systematic data collection in three-dimensional sampling space. Designing experimental protocols for work in the canopy requires at the outset the allocation of time and resources for anticipating and overcoming the particular sampling problems associated with the work.

Introduction

A tree's a tree. How many do you need to look at? (Ronald Reagan; speech, 12 September 1965).

Forest canopies have enormous importance in the overall functioning of forest ecosystems, and have been the focus of a rapid increase in research activity. Motivation for conducting canopy research includes the high species richness associated with the canopy (Allen 1996), the crucial role of the canopy in forest-atmosphere fluxes (Fitzjarrald & Moore 1995), and its dominating effect on understory microclimate (Baldocchi & Collineau 1994; Barker 1996). However, practical difficulties in reaching sampling points in tall forests or within deep crowns has been a major obstacle in some studies (Schowalter & Ganio 1998).

The growth of canopy research during the last two decades (Moffett & Lowman 1995) has been both a cause and a consequence of significant advances in canopy access techniques (Barker & Sutton 1997). As awareness and use of improved access methods have increased, research opportunities and output have grown (Nadkarni & Parker 1994; Nadkarni & Lowman 1995). Nevertheless, many researchers regard difficulty of access as a major obstacle to their research: 40% of researchers in a previous survey (Nadkarni & Parker 1994) regarded access as a 'very important' obstacle to the advancement of science, greater than any other obstacle. Even as access methodology is improved, it seems likely that all methods will continue to have inherent sampling limitations (e.g., Reagan 1995; Barker & Sutton 1997; Bongers 2001).

As problems of access are resolved, researchers need to determine what experimental work is possible or, more importantly, impossible, using available techniques for canopy access (Lowman & Wittman 1996). To establish the potentials and the limitations of experimental work in forest canopies, it is necessary to evaluate constraints to canopy sampling (Barker 1997). Relatively few publications refer to canopy sampling issues (but see Kapos et al. 1993; Koch et al. 1994; Baldocchi & Collineau 1994), however, aboveground forest sampling strategies have been receiving increased attention since the mid-1980s (Gregoire & Valentine 1996; Basson & Bazzaz 1997).

The principal constraints to research specifically imposed by limitations in canopy access are mainly problems associated with the choice of site or study species, achieving adequate replication, avoiding disturbance to the subject being studied, and working in a heterogeneous, three-dimensional environment. In this paper, we discuss these four possible problem areas in turn although, in practice, they tend to be interdependent.

Rather than attempting to review sampling limitations throughout the range possible of canopy studies, we focus instead on sampling issues emerging from comments made by 112 practising canopy scientists; many of these (including ourselves) work predominantly in plant sciences (68%) and in tropical countries (62%; see below); some of the sampling issues raised in the paper reflect this bias. However, many of the problems discussed in this paper are generic and extend beyond tropical plant-based studies.

Wherever possible, we cite examples from published sources to support the points made by the scientists in our survey. The literature was searched using major databases (Tree®, CAB® Agricola®, Forestry Abstracts®), resulting in citations of studies (cf., reviews) from 27 peer-reviewed journals. Where no such sources have been found, we have used personal communications. After all, our main contention in this paper is that sampling difficulties are rarely discussed in scientific papers, perhaps because the experimental design was modified to avoid such problems, or because the problems were not recognised. Although our main theme is in suggesting possible weaknesses in canopy research due to access-related problems, we do not identify examples of specific studies which we feel may have been undermined by such problems.

In evaluating possible consequences of canopy access on experimental practice, there is a risk of overemphasizing the limitations. Because of this, we have tried whenever possible to suggest possible solutions to apparent limitations, again, based on published material and personal communications with numerous canopy researchers. Inevitably, however, we will be presenting more problems than solutions.

The aim of this paper is to consider a simple but important question: does the choice or availability of canopy access method affect the science that is being done in the canopy? This question applies to both qualitative and, particularly, quantitative canopy science. The basic premise is that canopy scientists should apply the same rigor to their experimental design as do their terrestrial-based colleagues.

Sampling problems and solutions: a survey of canopy researchers

A survey was conducted among canopy researchers using e-mail addresses mainly from the current listing of the *International Canopy Network* (2103 Harrison NW, Suite 2612, Olympia, WA 98502-2607, USA) and the list of delegates at a European Science Foundation conference on Tropical Canopy Research (Oxford, UK; 12–16 December 1998). We cannot be sure how representative the respondents were of the wider community of canopy scientists, especially as those who responded were self-selected. However, we can at least conclude that those surveyed are interested or concerned in sampling issues, since they responded to requests for information. A limitation of the survey is that only networked scientists were contacted and, of these, 28% responded. Nonetheless, the numerous canopy sampling issues raised were frequently raised independently by more than one of the researchers who responded, and were often also supported by published material. For these reasons, we believe that responses in the survey are likely to be broadly representative of opinions held by active scientists who have considered possible canopy access influences on sampling. The number of respondents to the survey was intermediate in size between previous surveys of canopy scientists by Stork & Best (1994) and by Nadkarni & Parker (1994); both the previous studies focused on other issues in canopy research.

The survey (further details of which may be obtained from the corresponding author) was intended to elicit information without preempting responses. We asked researchers to choose which subject area(s) they worked in (from an initial list of 11 choices; Table 1), which canopy access method(s) they use (from an initial list of 11 alternatives; Table 2), and which country/countries they conduct canopy research in. The remainder of the survey consisted of the four questions (with three possible responses) shown in the captions for Figures 1–4. For each of these questions, we asked

Table 1. Distribution of canopy researchers surveyed across subject areas, with proportion of study sites in tropical or temperate countries. The number of canopy researchers working in each of the subject areas listed in the survey is presented, as is the total number of countries listed by researchers working in each subject area. Sixty-seven percent of researchers completing the survey listed more than one subject area and most listed more than one country. Work conducted in Brazil was assumed to be tropical and work conducted in the USA was assumed to be temperate. As alternatives to the 11 subject areas listed in the questionnaire, subjects marked * were suggested by some researchers.

Subject area	Number of researchers	% Study sites in tropical countries	% Study sites in temperate countries	Total number of countries listed
Plant ecophysiology	35	73	27	52
Plant ecology	32	48	52	58
Biodiversity	30	75	25	59
Animal ecology	24	69	31	48
Canopy-atmosphere fluxes	24	23	77	35
Nutrient cycling	19	42	58	26
Micrometeorology	17	50	50	34
Herbivory	16	60	40	25
Plant systematics	9	91	9	10
Forest structure*	9	40	60	23
Pollination/fruit development	8	92	8	14
Animal behaviour	8	86	14	12
Animal systematics*	1	–	–	1
Frugivory/seed dispersal*	1	–	–	1
Population genetics*	1	–	–	None listed
All areas combined	234	60	40	398

respondents reporting 'some limitations' or 'serious limitations' to provide further information including, if possible, means by which difficulties are overcome.

Contingency table analysis was used to compare the relative frequency of observations across categories for various groupings. Responses to the survey were received from 112 canopy researchers; 60% of the countries in which the respondents work are tropical, 40% temperate (Table 1). The number of study sites listed in temperate versus tropical countries varied by subject area (χ^2 tests, $p < 0.001$, using the eight subject areas listed most frequently in the survey). For most individual subject areas, the proportion of work being done in tropical versus temperate areas was not different from the overall division between tropical and temperate work. However, not surprisingly, for biodiversity studies proportionally more work was being done in tropical rather than temperate countries ($p < 0.03$). In contrast, proportionally more canopy atmosphere flux work was being done in temperate areas ($p < 0.001$). The relative number of plant ecophysiologists conducting work in the tropics was only marginally greater ($p = 0.07$) than expected. No tests were conducted for subject areas with fewer than 10 researchers.

A range of canopy access methods are used by the respondents to this survey (Table 2). A high proportion (66%) of canopy researchers use a combination of access methods (see also Nadkarni & Parker 1994). The most common combination of methods involved the use of rope (either in single rope technique (SRT) or arborist methods; see Barker 1997) and ground-based methods, consistent with previous observations (e.g., Lowman & Bouricius 1995). The relative frequency of canopy work in tropical/temperate sites varied with access method ($\chi^2 = 20.1$, df $= 8$, $p = 0.01$; using the eight access methods listed most frequently in the survey). Ropes were used in tropical sites more often than expected ($p = 0.03$), and scaffolding was used in temperate sites more often than expected ($p = 0.02$). Responses to the survey indicated that flexibility in the choice of canopy access methods is an important

Table 2. Distribution of canopy researchers surveyed across access methods, with proportion of study sites in tropical or temperate countries. Presented are the number of canopy researchers surveyed using different access methods, either in combination with another method, or as the main access method. As alternatives to the 11 methods listed in the questionnaire, access methods marked * were suggested by some researchers.

Access method	All users	Main users	% Study sites in tropical countries	% Study sites in temperate countries	Total number of countries listed
Ground access[1]	65	22	46	54	105
Rope	46	21	64	36	84
Tower/walkway	40	20	43	57	76
Ladder	24	5	60	40	40
Crane	20	8	56	44	39
Scaffolding	16	5	26	74	27
Remote sensing	14	0	33	66	21
Cherry picker / hydraulic lift	9	3	46	54	13
Spikes	5	1	77	23	13
Tree grippers[2]	4	0	86	14	7
Treefall / felling*	3	1	75	25	8
Free climbing [3,*]	3	2	83	17	6
Local climbers [3,*]	2	0	75	25	8
Dirigible / canopy raft	1	1	100	0	4
Helicopter*	1	1	Na	Na	1
Boat[4,*]	1	0	Na	Na	None listed
Totals	236	90	51	49	452

[1]Includes pole-pruners, binoculars, hemispherical photography, radio tracking, fogging.
[2]'Swiss tree bicycle'.
[3]With little or no specialised climbing equipment.
[4]In periodically flooded forest.

means of increasing sampling capability. The general advantages or disadvantages of different access techniques are described elsewhere (e.g., Lowman et al. 1993; Lowman & Moffett 1995; Barker 1997; Barker & Sutton 1997).

A comparison of most frequently used canopy access methods in relation to the most frequently cited subject areas indicated that the relative use of the different methods varies by subject area ($\chi^2 = 40.5$, df $= 20$, $p = 0.004$; animal ecology was not included in the analysis) (Table 3). The frequency of respondents using ropes was variable across the subject areas; ropes were used to a greater extent by researchers working in biodiversity and animal ecology, compared with other subject areas. Ropes presumably offer rapid and flexible access for sampling in these sorts of studies. In contrast, researchers studying canopy-atmosphere flux or plant ecophysiology appear to use ropes less frequently than other groups; we assume that this is because these workers mostly use, respectively, automated instruments installed on towers, or bulky or delicate equipment for which rope access is not suitable.

Choice of canopy access techniques can influence, and be influenced by, the objectives of the research. Science usually begins with a question or, more formally, a hypothesis. In canopy science, there are important, unresolved hypotheses awaiting attention from researchers (e.g. Smith 1973). However, in canopy research detailed experimental plans cannot be made until a decision has been taken on which canopy access methods are appropriate and available. Other factors affecting choice of access method will also be referred to in the following sections.

Table 3. Distribution of canopy researchers in each of five most popular subject areas using the five most popular access methods. Percentage of total researchers working in each subject area, using the listed access method (actual number of researchers by access method listed parenthetically).

Subject area	Tower or walkway	Ladder	Rope	Crane	Ground access	All other methods
Plant ecophysiology	21 (20)	8 (8)	10 (9)	15 (14)	20 (19)	26 (25)
Plant ecology	14 (10)	12 (9)	21 (15)	7 (5)	27 (20)	19 (14)
Biodiversity	13 (8)	8 (5)	32 (20)	8 (5)	26 (16)	13 (8)
Canopy-atmosphere fluxes	6 (3)	11 (6)	9 (5)	15 (8)	26 (14)	33 (18)
Animal ecology	12 (5)	5 (2)	33 (14)	2 (1)	29 (12)	19 (8)
All others	15 (35)	13 (29)	18 (41)	6 (14)	24 (55)	24 (54)

Effects of access method on choice of study site or study species

The location of access systems is often non-random and opportunistic. Sampling may also be arbitrary (referred to by Lowman (1997) as 'reach-and-grab'), possibly being restricted to accessible parts of a crown (Schowalter & Ganio 1999). Potential sample material may therefore be beyond reach (Elton 1973). Positions of sampling points may be unrepresentative, limited in number, or simply inconvenient. These constraints, which are often unavoidable, appear to apply to most canopy access methods. We will describe sampling issues associated with low- or high-tech access methods (*sensu* Barker & Sutton 1997) separately though the methods are often used in combination.

Low-tech canopy access methods

Low-tech methods such as bole-climbing, rope climbing and ground-based methods are generally highly mobile so, in theory at least, should not compel the researcher to operate in a particular forest (or part of forest), or to study a limited range of species. Survey results provide some support for this assertion for rope access (Figure 1), but less so for ladders. A high proportion of researchers (41% of those in our survey) use rope access methods, but climbing by ropes is limited to large, healthy trees with convenient branches, and access to the outer canopy is often limited (Nadkarni & Longino 1990; Longman et al. 1993; Moffett & Lowman 1995). However, there are techniques for extending the horizontal range of sampling from ropes (e.g., Perry & Williams 1981; Whiteacre 1981; Dial & Tobin 1994; Barker 1997). Vertical sampling range can be extended using pole-pruners (Mori 1995; Wieringa 1996). With low-tech access methods which require support from trees, size of tree is an issue; 'intermediate' sizes may be too tall to reach with ladders, but too small to support climbing by other methods (E. Braker, pers. comm.). The generally smaller size of mature trees in secondary forest can limit safe access using ropes (F. van Dunne, pers. comm.).

Ladders, though versatile and fairly mobile, often preclude studies within tree crowns. Sectional tree-climbing ladders are fairly bulky and heavy (e.g., each 3-m section weighing 5–8 kg) but have been used in Sarawak to access tops of trees 60–70 m, though portage of the sections from nearest road is a limiting factor (K. Reynolds, pers. comm.). Free-standing (guyed) ladders have been used in combination with pole-pruners to reach crowns up to 20 m (Morris 1955).

Ground-based techniques (see Table 2 for examples of these) provide complete two-dimensional mobility, and are often used as a supplement to methods giving direct access to the forest canopy. Many researchers (58% in our survey) also use remote sampling of the canopy from the ground as their main access method. Computer modelling allows estimations from ground-based measurements, validated by some direct sampling in the canopy (M. Hanus, pers. comm.). However, many (68%) respondents using ground-based methods reported constraints on choice of study species (Figure 1). The main problem seems to be in sampling or observing crowns of tall trees in multi-layered forests (e.g., Shaw & Bergstrom 1997). However, ground observations are easier if the site topography allows viewing crowns from nearby ridges (M. Shanahan, pers. comm.). In cases where actual sampling (cf., observation) is from the ground, there will be an influence on which trees (species or ages)

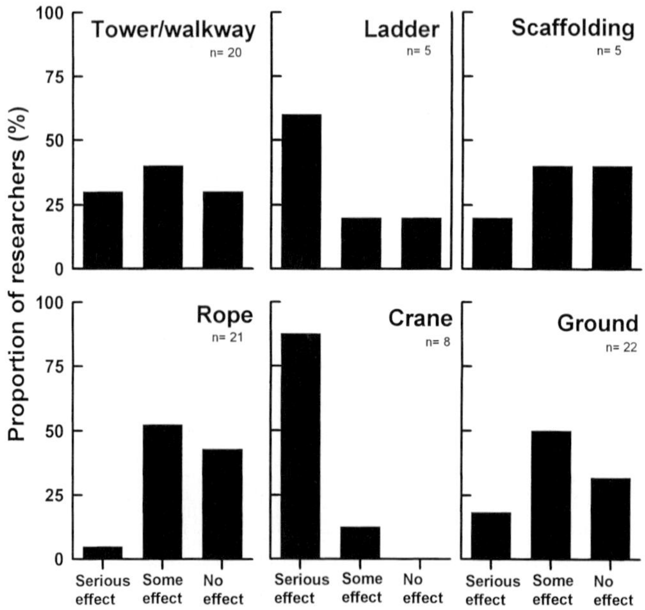

Figure 1. Effects of canopy access method on choice of study location or study species. The corresponding question in the survey was: do canopy access problems impose limitations on which species, forest type or forest location are included in your studies? Respondents were asked to choose one only of the three alternative responses: serious limitations; some limitations; no effect. Results are only for researchers using one of the techniques shown here as their *main* (possibly only) access method. Sample sizes (number of respondents) are shown.

or lifeforms are accessible, though the experimental design might not require higher access (e.g., Meinzer et al. 1995).

In extreme cases, access to the canopy using low-tech methods is limited to particular seasons. For example, ground-based access to the lower canopy in northern boreal forests may depend on, or at least be assisted by, increased snow depth at certain times of the year (Esseen & Renhorn 1996; Sorrensen-Cothern et al. 1993). In periodically flooded forest, access is provided by boat (S. Filoso, pers. comm.). Whether such seasonal phenomena constrain sampling regimes presumably is subject dependent.

High-tech canopy access methods

There are high-tech access systems (e.g., hydraulic lifts, scaffolding, booms) that can be moved to different locations within forests. In the case of scaffolding, at least, there are indications that there is little or no effect on choice of study site or study species (Figure 1). However, these systems are heavy and usually depend on road access. This presumably restricts access at some sites, or parts of sites, including those with soft substrate (Inoue et al. 1995; E. Middleton, pers. comm.), steep or unstable terrain, or in protected areas. In the case of smaller static systems, including some walkways, the use of light aluminium rather than traditional heavy construction materials allows access structures to be taken to more remote parts of forests (Inoue et al. 1995), which presumably allows a greater choice of potential study locations. Hydraulic lifts (cherry pickers) often cannot be moved away from roads, although they can be very effective in forest-edge studies.

As with some low-tech methods, there are upper height limits that may exclude some studies using mobile towers and lifts (e.g., ca. 12 m; Morris 1955; McCarthy 1988) and scaffolding towers (35 m; M. Ryan, pers. comm.). However, these systems have an advantage over static towers in allowing the height of sampling positions to be adjusted over time to keep up with height of fast-growing trees (e.g., in plantations) (J. Chambers, pers. comm.). Tall scaffolding can also support platforms at a range of heights, providing access throughout much of the forest profile (Bond et al. 1999).

There are differences in how sampling problems are dealt with by researchers using static versus dynamic high-tech methods. Static systems such as towers, walkways, and platforms occupy fixed locations within forests. Towers are inevitably located in canopy gaps (Parker et al. 1992). The location of walkways and platforms is often determined by the presence of large, healthy support trees (Reynolds & Crossley 1995) and is dependent on the structural integrity of the forest (Moffett & Lowman 1995). Other factors which influence the location of towers and walkways include topography and also the demands of tourism, since many of these large, expensive structures are built wholly or partly for use by forest visitors (e.g., Muul & Lim 1970; Sutton 2001). Once installed, static systems provide access to a limited number of trees (Nadkarni & Parker 1994) and (by definition) cannot be relocated to other parts of the forest, although they can be expanded laterally (Inoue et al. 1995; Lowman & Bouricius 1995). Further, the vertical sampling position of many access platforms is restricted to a particular vertical zone of the forest profile, neglecting some (e.g., lower) parts (e.g., Mori 1995). Given that some species are distributed unevenly throughout the vertical forest profile, sampling at a restricted range of heights may a problem for some types of studies. Since the location of sampling points from access facilities will not be random and (for tree-based studies) may restrict access to target or representative species, it is possible that sampling will be biased, or that some intended 'ecological' studies may in fact be 'autecological'.

Many researchers in the survey who use towers and walkways as their main access method do not report a strong effect on their selection of study site or study species (Figure 1). One explanation for this is that experiments are planned 'around' what species can be reached from the sampling point. Another possibility for minimising the limitations of fixed sampling points is by the use of additional access methods (e.g., rope, pole-pruners, temporary walkways) as well as the main tower or walkway. The precise dimensions of the footprint (sampling area) of micrometeorological towers may be unknown (Aber et al. 1996), but the tree species composition within the footprint are likely to influence biogenic fluxes and, during dry periods, water fluxes from the canopy (F. Mowry, pers. comm.); towers should presumably be positioned to include 'representative' trees within their footprints.

Although dynamic within their area of operation (typically ca. 1 ha), cranes are confined to fixed sites within forests resulting in limitations to choice of study site or species (see Figure 1). Researchers using cranes adjust their experimental designs to accommodate sites or species which are within reach of the gondola (e.g., F. Meinzer, pers. comm.), as well as planning their field operations around the needs of other researchers' priorities. Researchers who work from cranes in some cases also use rope access to reach the inner parts of tree crowns, where the gondola cannot penetrate (Gilbert 1995). Further, a new generation of cranes have a much increased sampling footprint: examples include the Surumoni (Venezuela) and Wind River (USA) cranes (see Sutton 2001; Szarzynski & Anhuf 2001). Other highly flexible high tech systems include COPAS, which operates using a fixed cable network (Sutton 2001).

There are a range of high-tech techniques (e.g., dirigibles, helicopters, balloons, remote sensing) which allow sampling the canopy from above. Of these, the dirigible technique allows both remote and direct canopy sampling (Lowman et al. 1993) and can also be used to deposit a 'canopy raft' onto the canopy. In addition, researchers using the raft can use rope techniques or pulleys to increase the vertical range of their sampling (e.g., Cosson 1995). Choice of study site and study species (of trees) will be strongly influenced by the requirements of this technique. For example, the canopy raft needs a structurally cohesive forest for support, and its placement is affected by the presence of emergents and gaps. Also, the location and duration of sampling from the raft may be affected by the different needs of other researchers. However, this versatile system provides invaluable access to the outer canopy, which many other techniques do not allow (Lowman et al. 1993).

Finally in this group, above-canopy methods do not impose any, or many constraints, on study site or study species, though they require validation from canopy- or ground-based work. Remote sensing may need simultaneous ground truthing at the same spatial scale, and it is difficult to get the required resolution (e.g., 10–30 m) by ground sampling (S. Bohlman, pers. comm). Aerial sketch-mapping (e.g., of defoliation) appears to be a reliable estimating technique when checked against ground-based observations (MacLean & MacKinnon 1996).

General problems relating to study site of species

Some additional general points spontaneously emerged from this part of the survey, in relation to possible canopy access problems on study species or forest type or forest location. Firstly, many canopy researchers minimise such constraints by using a combination of access techniques. However, choice of access method is itself limited. Cost is a major factor in determining what methods are actually used. Other factors (e.g., safety, convenience) were rarely mentioned.

Secondly, researchers are well aware of potential influences of access method on experimental designs. In some cases, while choice of study species/site may not be affected, access may strongly influence what questions can be asked (Bassow & Bazzaz 1997; J. Read, pers. comm.).

Effects of access method on replication and pseudoreplication

Canopy access methods, if chosen or used inappropriately, can have two effects on quantitative studies. Firstly, access methods may limit the number of replicate measurements which can be made (e.g., Zotz & Winter 1996). In the highly heterogeneous and complex canopy environment, especially in tropical forests, lack of replication may mean that spatial or temporal variability in the environment is not represented in the data (Lowman et al., 1993; Lowman 1997).

Secondly, access problems may result in statistical comparisons being made between groups of 'replicates' that are not truly independent, which is pseudoreplication (Hurlbert 1984). Sutton (2001) believes that pseudoreplication has 'claimed many victims' in canopy studies. However, the point at which replication becomes pseudoreplication is highly subject-dependent. Put another way, the unit of replication depends on the type of study being conducted (see Bongers 2001). For plant physiological ecology or invertebrate ecology studies, typical replicates are trees, while plant ecology or micrometeorological studies use replicates at the stand or community level.

Arthropod studies involve a range of possible sampling methods, which includes tree-level approaches such as fogging, where the orgins of sample material within the forest profile are not usually known, although yields per unit effort are comparatively high (Reynolds & Crossley 1995; Schowalter 1995). Branch-based sampling (unlike fogging and interception traps) is targetted and allows a standardised expression of arthropod abundance to be calculated (numbers of arthropods per unit of associated plant material), which permits comparison with other studies (Schowalter & Ganio 1998) at least for sessile species (Schowalter 1995). Ironically, sampling capability of canopy arthropods has now outpaced laboratory processing of material, perhaps by a factor of at least four times (Erwin 1995).

Achieving sufficient replication is difficult when sampling the outer canopy, mainly because of access constraints (Lowman et al. 1993). However, even when the canopy or sub-canopy can be easily reached, obtaining sufficient, representative samples can be a problem. This is because some access techniques effectively provide only one fixed sampling position (Schowalter 1995), either within a tree or within the forest as a whole. Aside from replication, working from a fixed position is a particular problem when using destructive sampling because the removal or study material can limit (at least temporarily) subsequent sampling (Reynolds & Crossley 1995) and can also alter local microclimate and microhabitats.

Several published canopy studies are based on a single tree. In such 'dendrocentric' cases (*sensu* Lieberman et al. 1989), within-tree samples such as leaves and branches have been used as replicates. Whether truly independent samples can be collected within a single tree is questionable, and depends on the type of study being conducted. For example, individual branches can sometimes be regarded as being autonomous for water relations (Sprugel et al. 1991) or carbon economy or herbivory (Watson & Casper 1984). In such cases, leaves taken from different branches can be treated as independent and not as pseudoreplicates (e.g., Barker & Booth 1996). However, the status of leaves as independent functioning units is debatable (Caldwell et al. 1986; Leverenz 1988).

In many ecological studies, replication occurs at the stand or community level, in which case individual trees are only sub-samples. In cases where samples taken from a single tree or forest stand are not truly independent in space or time, pseudoreplication has occurred (see Hurlbert 1984), and the sample size is effectively one.

Within-tree sampling is common in canopy studies. For example, researchers are often interested in understanding variability in tree crowns. Such sampling is repeated at a tree or stand level to give replication. Since canopies are known to be highly heterogeneous (e.g., Harper 1989), it is important to reduce or remove bias in within-tree sampling. Researchers have developed protocols for this, including randomized branch sampling to estimate total leaf area, biomass, or insect egg or larval populations (see Gregoire et al. 1995; Zandt 1994). Another example is the use of paired sampling, for example of 'sun' and 'shade' branches, because such sampling allows for

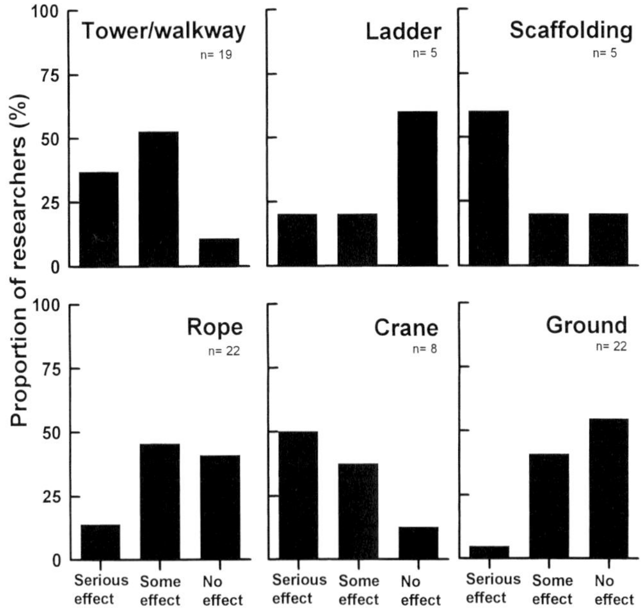

Figure 2. Effects of canopy access method on replication. The corresponding question in the survey was: do canopy access problems limit the number of independent replicates you include in your studies? For further details, refer to caption for Figure 1.

within crown variability and thus gives greater power in statistical tests (Brooks et al. 1996).

We believe that canopy researchers are generally well-aware of the problems of obtaining a sufficient number of independent replicates from fixed sampling points. In the survey reported here, effects of access method on sampling are reported particularly by researchers using methods involving large structures, whether static (tower, walkway) or even mobile (crane, scaffolding) within the forest (Figure 2). Furthermore, the canopy structure itself can obstruct potential sampling positions (Hollinger 1989). Sampling repeatedly from a tower platform or walkway cannot provide information on variation on a within forest scale. Variation also occurs among forest stands, for example for herbivory (Lowman 1997). However, stand-scale replication using multiple towers would need a huge budget (E. Miller, pers. comm.). For a walkway and platform system providing access to ca. 10 canopy trees and 25 undersory trees in Belize, Lowman & Bouricius (1995) report a cost (in 1994) of about US $32 000.

Cranes and scaffolding have a capability to access several trees, yet a high proportion (58, 80%) of researchers using these methods report 'serious effects' on replication. Cranes in particular have huge sampling capability (Zotz & Winter 1996); for example, access of one crane in Panama was 8000 m^2 on the upper canopy surface, 28 000 m^3 of forest volume, or 140 individual trees (Allen 1996). Thus, it could be argued that replication is only limited by time, which is itself constrained by budgets (W. Winner, pers. comm.).

However, the relative flexibility that cranes or movable scaffolding allow is often only within part of a single forest stand. Even within a tract of forest, there are logistical problems associated with cranes and scaffolding that can limit the number of trees which can be reached, especially if the study is focusing on a single species. Crane-based tree studies are generally restricted to the most abundant species (F. Meinzer, pers. comm.). However, in physiological measurements, species within functional groups may be used as replicates (R. Oren, pers. comm.).

Researchers in the survey mainly reported either some or no effect of rope access on obtaining adequate replication. Access by rope at any one time is usually limited to a single tree. However, access and therefore independent sampling can be increased by climbing a succession of trees over time. Alternatively, since rope access is a relatively inexpensive technique, multiple sets of rope equipment can be used to access several trees simultaneously by a group of researchers working together. Such temporal or spatial replication can of course include more than one forest stand or community. Achieving replication using rope access may be mainly limited by researchers' energy and time (E.

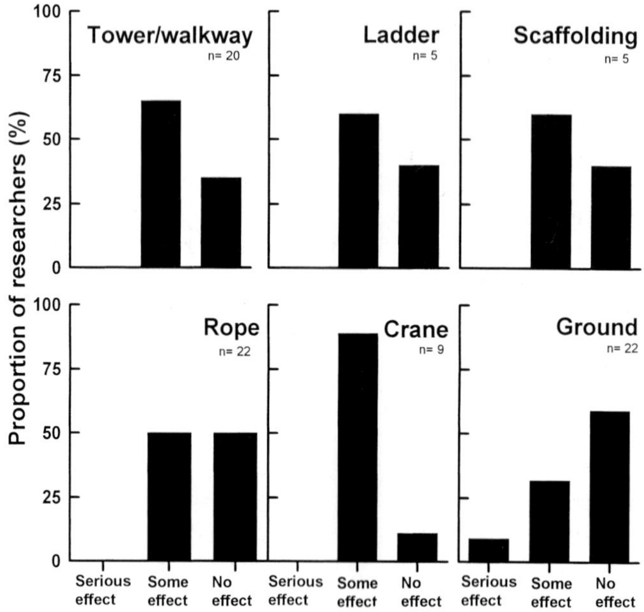

Figure 3. Effects of canopy access method on disturbance. The corresponding question in the survey was: do canopy access problems cause any disturbance to the subject which you are measuring/observing? For further details, refer to caption for Figure 1.

Olson, pers. comm.). Availability of sufficient field time is seen as a constraint by canopy researchers (Nadkarni & Parker 1994) and, for canopy arthropod studies, the time required for processing sampled material needs to be allowed for (Erwin 1995).

Even ground-based methods are regarded as having some effect on sampling by about 41% of researchers using this method (see Figure 2). For example, intervening vegetation (especially in tropical rain forests) can limit the number or quality of observations made from the ground in both plant and animal studies (Gillesberg & Carey 1991; Fintoura 1995; Fisher & Araujo 1995; Kaplin et al. 1998; Leite et al. 1996). Remote sampling of the upper canopy (e.g., by shotgun) from the ground may be restricted for similar reasons.

How can canopy researchers reduce the risk of bias and pseudoreplication in their studies? A common approach is to increase sampling flexibility by using more than one access technique. For example, several respondents to the survey who use crane or ground-based methods as their main means of canopy access also use ropes. Sampling resolution with any single access technique can also be increased spatially by an investment in a more extensive system (e.g., for walkways; Lowman & Bouricius 1995) or multiple access units (e.g., for rope methods; Lowman & Bouricius 1995). Another approach is to develop a common experimental design with other research groups, allowing shared datasets and increased replication (R. Fournier, pers. comm.).

An approach which appears to be used increasingly is modelling, making extrapolations from limited sampling in the canopy. Actual canopy measurements are necessary to calibrate or validate any model, that can often then be applied to a range of sites and further verified (E. Miller, pers. comm).

Several researchers in the survey comment on their desire to increase replication, though many indicate that it is time and costs rather than access *per se* which are limiting factors. Whatever the direct or indirect effects are of canopy access on sampling, it is important to develop appropriate an experimental design for each study. Also, sampling strategy should be clearly explained in the methods section of papers (Barker 1997; B. Kloeppel pers. comm.).

Disturbance and 'demonic intrusion'

Canopy access structures or activities can cause disturbance in the canopy. Although canopy access may impose constraints on the scope and quality of sampling, such limitations have generally only recently been explicitly discussed by researchers (Moffett & Lowman 1995, Barker 1997).

'Demonic intrusion' here refers to the specific effects of access activities on the variable being sampled or measured. In the survey, a majority (98%) of canopy researchers felt that their primary access method caused only some disturbance, or no disturbance, to the subject being measured or observed (Figure 3 shows this trend for six of the access methods). But disturbance, including damage, can also affect parts of the canopy not being studied. For example, branches broken by a crane's gondola or epiphytes dislodged by ropes may not be the subject of a particular study. However, the gradual degradation by access activities of a forest which is being intensively researched may have a significant impact on future research at the same site. This may be a particular problem in forests which are fragile (i.e., easily and/or irreversibly degraded) or in conservation areas, where the movement of heavy or bulky canopy access equipment through the forest is restricted.

A further problem of demonic intrusion is that the quality of data obtained may differ according to the access method being used. For example, the vertical distribution of irradiance within a forest may be differ considerably when measured from a tower or from a crane's gondola (see Chazdon et al. 1996). Such discrepancies introduce problems in comparing results among studies using different sampling protocols (e.g., Daw et al. 1998), or using different access techniques.

Examples of access-related disturbance

Disturbance and accidental damage to plants and animals by climbing structures or activities is a concern among many researchers. Examples include: spikes causing potentially fatal damage to trees (Donahue & Wood 1995; Mori 1995); carrying ladders, or moving a man-lift, through forests causing disturbance to understory vegetation (B. Lohr, pers. comm.; E. Middleton, pers. comm.); rope movements abrading lichens (W. Denison, pers. comm.); rope climbing causing nest abandonment (W. Ritchie, pers. comm.); canopy crane movements temporally scaring away larger birds (D. Shaw, pers. comm.); canopy walkways interfering with malaise trapping (Preisser et al. 1998); canopy walkways influencing the movements of arboreal animals (Perry & Williams 1981; but see Lowman & Bouricius 1995); climbing and branch sampling disturbing insects (Morris 1955); and, large access structures affecting microclimate (Parker et al. 1992; Koch et al. 1994; Moffett & Lowman 1995; Sutton 2001).

So, what are the solutions to such problems? Though not routinely referred to in scientific publications, it is clear from our survey that canopy researchers often have formulated protocols to minimise or avoid access-related damage and disturbance. For example, avoiding physiological measurements on branches that have been in contact with an access tower; periodically moving scaffolding to avoid sampling 'degraded' areas (B. Kloeppel, pers. comm.); placing towers carefully to avoid them being hit by branches moving in wind (P. Hanson, pers. comm.); avoiding interference during gas exchange measurements by turning off the engine of an aerial lift (Bassow & Bazzaz 1997); avoiding rope climbing at dawn or dusk, when birds most active (K. Nelson, pers. comm.); observing nesting behaviour with a remote-controlled video camera mounted on an adjacent tree (W. Ritchie, pers. comm.); reducing disturbance by repeated climbs by using pulley systems, allowing small mammal traps to be lowered for regular inspection (e.g., Taylor & Lowman 1996; Vieira 1998); reducing disturbance near the base of a crane to established paths or boardwalks (M. Ryan, pers. comm.); and, minimising damage to the undersory during walkway construction (Lowman & Bouricius 1995).

More generally, the risk of disturbance, including possible demonic intrusion, can be reduced by limiting the numbers of researchers entering study sites, ensuring good communication flow among project participants (R. Fournier, pers. comm.), by periodically relocating the sampling site (S. Thomas, pers. comm.) and by restricting intensive destructive sampling in sensitive habitats (C. Ozanne, pers . comm.). For all studies, data should be rejected if evidence of disturbance is obvious (J. Massheder, pers. comm.).

Logistics and hyperspace

Stork & Best (1994) suggested that lack of spatial replication is a major problem in nearly all canopy research. One explanation for this is that there are a number of problems in working in three-dimensional (3-D) space which affect the logistics of sampling (Nadkarni & Parker 1994), though such problems tend to be subject-specific. Planning sampling regimes within tree crowns needs to take into account the spatial variability of the parameter being measured. Such variability is, for example, large in studies of herbivory (Lowman 1997), arthropod distribution (Schowalter & Ganio 1998) and stomatal activity (Eliás 1979). If variables are statistically different (or at least distinct) between the upper and lower crown, sampling must allow for this; in practice, more sampling effort might be needed to ensure that upper crown data are collected (Schowalter & Ganio 1998), since this region is more difficult to reach.

Sampling in complex, 3-D space

The complexity of 3-D environments can make collecting data, and representing results, difficult (see Lowman 1997; Bongers 2001), including for ground-level observations (Leite et al. 1996). Actual sampling within the 3-D matrix causes particular problems. For example, the distribution of leaves within a crown is difficult to measure (Russell et al. 1989), though models are now being developed for describing leaf arrays (e.g., Pearcy. & Valladares 1999).

Another approach is to use manipulated or simulated canopy structures. Examples include the installation of branches (Renhorn et al. 1997), living tree leaves (Rowe & Potter 1996) or bryophytes (Clark et al. 1998) at particular positions within the forest structure, or simulated canopy branches (Schlesinger et al. 1993) or trees (Hilbert & Messier 1996), or actual branches (Parrish 1995) in laboratory conditions. An advantage of such approaches is presumably to simplify the 3-D canopy by isolating and defining the variable under investigation.

Various protocols have been developed for defining sampling points in 3-D canopy space (e.g., Ford & Newbould 1971; Hollinger 1989; Castellanos et al. 1992; Sumida 1995). Researchers also require techniques for analysing and representing data collected from 3-D sampling (Richards 1983; Popma et al. 1988; Nadkarni & Parker 1994; Barker 1997). The continuing trend in increased efficiency and lower costs of computing processing capability will make the analysis of such data more feasible (Weishampel et al. 1996; see Bongers 2001).

Another important consequence of working in 3-D space is that moving between sampling points often introduces a temporal dimension. Canopy sampling may thus occur in a four-dimensional hyperspace. This is likely to be a particular problem for studies involving transient changes in the object being measured or observed. For example, forest canopy microclimate variables typically show large temporal fluctuations (e.g., Baldocchi & Collineau 1994; Barker 1995) which can occur over small spatial scales (Chazdon et al. 1988). In real-time sampling, such fluctuations could be a problem given the added time required to move around in the canopy (Kapos et al. 1993). A result of this is that canopy measurements sometimes lack temporal resolution (Fitzjarrald & Moore 1995).

Temporal limitations of sampling the canopy may affect what type of research can be undertaken. For example, collecting samples (e.g., flowering parts) from trees takes much more time than collecting from shrubs or herbs, consequently trees may be under-represented in some systematic studies (J. Wieringa, pers. comm.).

Logistical problems and some solutions associated with particular access methods

The type of canopy access method has implications for the frequency and duration of sampling, and the type of equipment being used. For example, rope or bole-climbing techniques may be unsuitable for use with heavy or delicate equipment, and when the researcher needs to sample for long periods in the canopy. Bole climbing methods inevitably restrict sampling in the outer crown area (Hietz 1997). Even when access methods are used in combination, certain parts of the canopy may be difficult or impossible to reach, resulting in a possible under-representation of some sample types (e.g., Wright et al. 1997).

Responses in our survey indicated that problems associated with the logistics of sampling occur with all main canopy access methods, though relatively few researchers ($\leq 20\%$) in our survey felt that access imposed 'serious' effects (Figure 4); canopy cranes have large sampling capability (see above). Researchers can repeatedly revisit a sampling location, and can move around in the canopy with heavy or bulky equipment. However, temporal sampling may be constrained by the high costs involved (Nadkarni & Parker 1994; M. Ryan, pers. comm.) and because such facilities are shared among researchers.

Studies that involve climbing by the researcher are likely to impose time limitations on movements between sampling events. This may account for many of the logistical constraints reported by users of tower/walkway, ladder, scaffolding and rope systems (Figure 4). Logistical constraints on sampling were less evident among researchers using ladders and scaffolding. These methods are often used for sampling at relatively low heights. Furthermore, the equipment can be moved or duplicated to allow more of the forest area to be included (R. Teskey, pers. comm.).

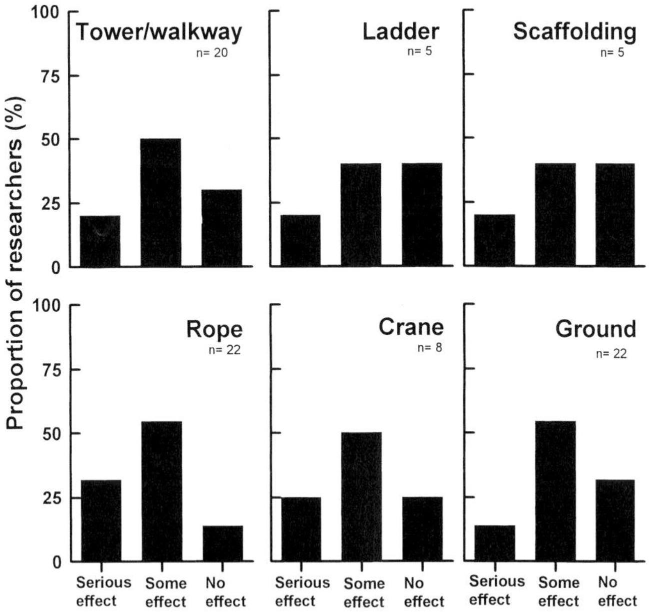

Figure 4. Effects of canopy access method on logistics. The corresponding question in the survey was: do canopy access problems limit the number of measurements or observations you can make in space or time? For further details, refer to caption for Figure 1.

Rope access is commonly used by canopy researchers (Nadkarni & Parker 1994; Lowman & Bouricius 1995; Table 2), but for some studies (and especially in tall forests) this method may impose substantial logistical constraints on sampling. For example, ropes do not allow sampling access throughout tree crowns (e.g., Hadwen et al. 1998). It is tiring and time-consuming to climb trees, and this limits the number of trees that can sampled per day (L. Risley, pers. comm.); in cases where extensive (e.g. diurnal) sampling is being conducted, rope access may only allow sampling from one tree per day (e.g., Barker & Pérez-Salicrup, 2000). However, sampling can be increased by the use of extra personnel during periods of intensive fieldwork (Lowman & Bouricius 1995; D. Coxson, pers. comm.). The effort needed for rigging trees with ropes can also have an indirect effect on sampling logistics. For example, sampling epiphytes using rope access involves time-consuming changes in the position of rope (O. Missa, pers. comm.). Distance between sample trees, using ladder access, can be a highly variable component of overall sampling effort; however, the number of trees climbed can be reduced by sub-sampling, without substantial loss of accuracy in the data (Caron & Fleming 1995).

Ground-based methods for canopy access can impose a temporal constraint on measurements. One reason for this is that moving through ground vegetation is often difficult, especially in luxuriant forests and when carrying sampling equipment. Canopy systems which are operated from the ground, for example pulley-operated mammal traps, can provide sampling resolution, but often demand a heavy time investment before sampling commences (McClearn et al. 1996) which needs to be factored in to experimental protocols. Some ground-based sampling is dependent on suitable weather conditions, such as lack of wind for fogging (H. Recher, pers. comm.). Similarly, the time of day in which hemispherical photos can be taken is limited (R. Hall, pers. comm.). However, 3-D canopy space can still be sampled effectively from the ground, for instance by using a balloon to measure microclimate (Parker et al. 1996).

Conclusions

There is an emerging consensus among canopy researchers that gaining access to the upper part of forests is not the only practical difficulty to be overcome in canopy research (Nadkarni & Parker 1994). To conduct rigorous science, it is necessary to be aware of access-related effects on the quantity and quality of sampling.

In some cases, it is possible to reduce or eliminate possible constraints of access on sampling. Many researchers use a combination of access techniques to extend sampling capability. Alternatively, where access-related limitations to experimental work cannot be avoided, a recognition and acceptance of this by canopy researchers is important. In such cases, experimental protocols need to be designed accordingly.

The introduction of unfamiliar access techniques may mean that researchers cannot easily anticipate problems associated with access and/or sampling. Ideally in such cases, pilot studies (see Lowman 1997) may be needed to determine what constraints are a problem and, if possible, how their effects can be reduced or eliminated.

Canopy research needs to be planned carefully to incorporate protocols for achieving effective sampling strategies. Research in forest canopies is a relatively young science; consequently, sampling protocols are still being developed (Bongers 2001). Where more than one access method is possible, decisions need to be taken about the best combination of methods to achieve good sampling. Budget limitations may require the use of cost/benefit analyses to determine choice of access method (sensu Zandt 1994; see also Elton 1973; Moffett & Lowman 1995; Barker 1997). In studies for which consistent sampling effort among trees is an issue, the effect of using different access methods on different trees needs to be evaluated (see Basset et al. 1996).

Increased communication among canopy researchers is essential in the development and implementation of standardized protocols for comparative studies (Nadkarni & Parker 1994; Lowman et al. 1995; Nadkarni & Lowman 1995; Schowalter 1995; Lowman & Wittman 1996). Increased communication among canopy researchers will hopefully allow further discussion and, if possible, resolution of some remaining problems associated with sampling and experimental design.

Finally, we feel that consideration should be given to the idea of establishing short in-forest training courses to promote safe and effective canopy access and sampling by both new and experienced researchers.

Acknowledgements

We are grateful for the feedback provided by the numerous canopy researchers who responded to the survey reported in this paper. In many cases, researchers contributed additional information, some of which is cited in the text as personal communications. We are also grateful for helpful comments by two anonymous reviewers. Funding for attendance (by MB) at the European Science Foundation conference on Tropical Canopy Research is acknowledged with thanks.

References

Aber, J. D., Reich, P. B. & Goulden, M. L. 1996. Extrapolating leaf CO_2 exchange to the canopy: a generalized model of forest photosynthesis compared with measurements by eddy correlation. Oecologia 106: 257–265.

Allen, W. H. 1996. Traveling across the treetops. BioScience 46: 796–799.

Baldocchi, D. & Collineau, S. 1994. The physical nature of solar radiation in heterogeneous canopies: spatial and temporal attributes. Pp. 21–71. In: Caldwell, M. M. & Pearcy, R. W. (eds), Exploitation of environmental heterogeneity by plants. Academic Press, San Diego.

Barker, M. G. 1996. Vertical profiles in a Brunei rain forest: I. Microclimate associated with a canopy tree. J. Trop. For. Sci. 8: 505–519.

Barker, M. G. 1997. An update on low-tech methods for forest canopy access and on sampling a forest canopy. Selbyana 18: 16–26.

Barker, M. G. & Booth, W. E. 1996. Vertical profiles in a Brunei rain forest: II. Leaf characteristics of *Dryobalanops lanceolata* Burck. J. Trop. For. Sci. 9: 52–66.

Barker, M. G. & Pérez-Salicrup, D. 2000 Comparative water relations of mature mahogany (*Swietenia macrophylla*) trees with and without lianas in a subhumid, seasonally dry forest in Bolivia. Tree Physiol. 20: 1167–1174.

Barker, M. & Sutton, S. 1997. Low-tech methods for forest canopy access. Biotropica 29: 243–247.

Basset, Y., Samuelson, G. A., Allison, A. & Miller, S. E. 1996. How many species of host-specific insects feed on a species of tropical tree? Biol. J. Linn. Soc. 59: 201–216.

Bassow, S. L. & Bazzaz, F. A. 1997. Intra- and inter-specific variation in canopy photosynthesis in a mixed deciduous forest. Oecologia 109: 507–515.

Bond, B. J., Farnsworth, B. T., Coulombe, R. A. & Winner, W. E. 1999. Foliage physiology and biochemistry in response to light gradients in conifers with varying shade tolerance. Oecologia 120: 183–192.

Bongers, F. 2001. Methods to assess tropical rain forest canopy structure: an overview. Plant Ecol. 153: 263–277 (this volume).

Brooks, J, R., Sprugel, F. G. & Hinckley, T. M. 1996. The effects of light acclimation during and after foliage expansion on photosynthesis of *Abies amabilis* foliage within the canopy. Oecologia 107: 21–32.

Caldwell, M. M., Meister, H.-P., Tenhunen, J. D. & Lange, O. L. 1986. Canopy structure, light microclimate and leaf gas exchange of *Quercus coccifera* L. in a Portuguese macchia: measurements in different canopy layers and simulations with a canopy model. Trees 1: 25–41.

Caron, G. E. & Fleming, R. A. 1995. A simple method for estimating the number of seed cones on individual black spruce. Can. J. For. Res. 25: 398–406.

Castellenos, V., Duran, R., Guzman, S., Briones, O. & Feria, M. 1992. Three-dimensional space utilization of lianas: a methodology. Biotropica 24: 396–401.

Chazdon, R. L., Williams, K. & Field, C. B. 1988. Interactions between crown structure and light environment in five rain forest *Piper* species. Amer. J. Bot. 75: 1459–1471.

Chazdon, R. L., Pearcy, R. W., Lee, D. L. & Fetcher, N. 1996. Photosynthetic responses of tropical plants to contrasting light environments. Pp. 5–55. In: Mulkey, S.S., Chazdon, R.L. & Smith, A.P. (eds), Tropical forest plant ecophysiology. Chapman & Hall, New York.

Clark, K. L., Nadkarni, N. M. & Gholz, H. L. 1998. Growth, net production. litter decomposition, and net nitrogen accumulation by epiphytic bryophytes in a tropical montane forest. Biotropica 30: 12–23.

Cosson, J.-F. 1995. Captures of *Myonycteris torquata* (Chiroptera: Pteropodidae) in forest canopy in South Cameroon. Biotropica 27: 395–396.

Daw, S. K., de Stefano, S. & Steidl, R. J. 1998. Does survey method bias the description of northern goshawk nest-site structure? J. Wildl. Manage. 62: 1379–1384.

Dial, R. & Tobin, S. C. 1994. Description of arborist methods for forest canopy access and movement. Selbyana 15: 24–37.

Donahue, P. K. & Wood, T. M. 1995. A safe, flexible and non-injurious technique for climbing tall trees. Selbyana 16: 196–200.

Eliás, P. 1979. Stomatal activity within crowns of tall deciduous trees under forest conditions. Biol. Plant. 21: 266–274.

Elton, C. S. 1973. The structure of invertebrate populations inside neotropical rain forest. J. Animal Ecol. 42: 55–104.

Erwin, T. L. 1995. Measuring arthropod biodiversity in the tropical forest canopy. Pp. 109–127. In: Lowman, M. D & Nadkarni, N. M. (eds), Forest canopies. Academic Press, San Diego.

Fisher, E. A. & Araujo, A. C. 1995. Spatial organization of a bromeliad community in the Atlantic rain forest, south-eastern Brazil. J. Trop. Ecol. 11: 559–567.

Fitzjarrald D. R. & Moore, K. E. 1995. Physical mechanism of heat and mass exchange between forests and the atmosphere. Pp. 45–72. In: Lowman, M. D & Nadkarni, N. M. (eds), Forest canopies. Academic Press, San Diego.

Fintoura, T. 1995. Distribution patterns of five Bromeliaceae genera in Atlantic Rain Forest, Rio de Janeiro State, Brazil. Selbyana 16: 79–93.

Ford, E. D. & Newbould, P. J. 1971. The leaf canopy of a coppiced deciduous woodland. J. Ecol. 59: 843–862.

Gillesberg, A.-M. & Carey, A. B. 1991. Arboreal nests of *Phenacomys longicaudus* in Oregon. J. Mamm. 72: 784–787.

Gregoire, T. G. & Valentine, H. T. 1996. Sampling methods to estimate stem length and surface area of tropical tree species. For. Ecol. Manage. 83: 229–235.

Gregoire, T. G., Valentine, H. T. & Furnival, G. M. 1995. Sampling methods to estimate foliage and other characteristics of individual trees. Ecology 76: 1181–1194.

Hadwen, W. L., Kitching, R. L. & Olsen, M. F. 1996. Folivory levels of seedlings and canopy trees in tropical and subtropical rain forests in Australia. Selbyana 19: 162–171.

Harper, J. L. 1989. Canopies as populations. Pp. 105–128. In: Russell, G. Ussell, G. Marshall, G.B. & Jarvis, P.G. (eds), Plant canopies: their growth, form and function. Cambridge University Press, Cambridge.

Heatwole, H. & Higgins, W. 1993. Canopy research methods: a review. Selbyana 14: 23.

Hietz, P. 1997. Population dynamics of epiphytes in a Mexican montane forest. J. Ecol. 85: 767–775.

Hilbert, D. W. & Messier, C. 1996. Physical simulation of trees to study the effects of forest light environment, branch type and branch spacing on light interception and transmission. Funct. Ecol. 10: 777–783.

Hollinger, D.Y. 1989. Canopy organization and foliage photosynthetic capacity in a broad-leaved evergreen montane forest. Funct. Ecol. 3: 53–62.

Hurlbert, S. H. 1984. Pseudoreplication and the design of ecological experiments. Ecol. Monogr. 54: 187–211.

Inoue, T., Yumoto, T., Abang, A. H., Lee H. S. & Ogino, K. 1995. Construction of a canopy observation system in a tropical rain forest in Sarawak. Selbyana 16: 24–35.

Kaplin, B. A., Munyaligoga, V. & Moermond, T. C. 1998. The influence of temporal changes in fruit availability on diet composition and seed handling in blue monkeys (*Cercopithecus mitis doggetti*). Biotropica 30: 56–71.

Kapos, V. G., Ganade, E., Matsui, & Victoria, R.L. 1993. $\delta^{13}C$ as an indicator of edge effects in tropical rain forest reserves. J. Ecol. 81: 425–432.

Koch, G. W., Amthor, J. S. & Goulden, M. L. 1994. Diurnal patterns of leaf photosynthesis, conductance and water potential at the top of a lowland rain forest canopy in Cameroon: measurements from the *Radeau des Cimes*. Tree Physiol. 14: 347–360.

Leite, Y. L. R., Costa, L. P. & Stallings, J. R. 1996. Diet and vertical space use of three sympatric opossums in a Brazilian Atlantic forest reserve. J. Trop. Ecol. 12: 435–440.

Leverenz, J.W. The effects of illumination sequence, CO_2 concentration, temperature and acclimation on the convexity of the photosynthetic light response curve. Physiol. Plant. 74: 332–341.

Lieberman, M., Lieberman, D. & Peralta, R. 1989. Forests are not just Swiss cheese: canopy stererogeometry of non-gaps in tropical forests. Ecology 70: 550–552.

Lowman, M. D. 1997. Herbivory in forests – from centimetres to megametres. Pp. 135–149. In: Watt, A.D., Stork, N.E. & Hunter, M.D. (eds), Forests and insects. Chapman & Hall, London.

Lowman, M. & Bouricius, B. 1995. The construction of platforms and bridges for forest canopy access. Selbyana 16: 179–184.

Lowman, M, Hallé, F., Bouricius, B., Coley, P., Nadkarni, N., Parker, G., Saterson, K. & Wright, S. 1995. What's up? – perspectives from the first international forest canopy conference at Sarasota, Florida, 1994. Selbyana 16: 1–11.

Lowman, M. & Moffett, M. 1995. The ecology of tropical rain forest canopies. Trends Ecol. Evol. 8: 104–107.

Lowman, M. & Moffett, M. & Rinker, H. B. 1993. A new technique for taxonomic and ecological sampling in rain forest canopies. Selbyana 14: 75–79.

Lowman, M. D. & Wittman, P. K. 1996. Forest canopies: methods, hypotheses, and future directions. Ann. Rev. Ecol. Syst. 27: 55–81.

MacLean, D. A. & MacKinnon, W. E. 1996. Accuracy of aerial sketch-mapping estimates of spruce budworm defoliation in New Brunswick. Can. J. For. Res. 26: 2099–2108.

McCarthy, B. C. 1988. A method of access into crowns of subcanopy and canopy trees. Can. J. For. Res. 18: 646–649.

McClearn, D., Roman, H. & Horan, S. 1996. Canopy mammal trapping on Barro Colorado Island, Panama: a six month survey. Selbyana 17: 27–35.

Meinzer, F. C., Goldstein, G., Jackson, P., Holbrook, N. M., Gutiérrez, M. V. & Cavelier, J. 1995. Environmental and physiological regulation of transpiration in tropical forest gap species: the influence of boundary layer and hydraulic properties. Oecologica 101: 514–522.

Moffett, M. & Lowman, M. D. 1995. Canopy access techniques. Pp. 3–26. In: Lowman, M.D. & Nadkarni, N.M. (eds), Forest canopies. Academic Press, San Diego.

Mori, S. A. 1995. Exploring for plant diversity in the canopy of a French Guianan forest. Selbyana 16: 94–98.

Morris, R. F. 1955. The development of sampling techniques for forest insect defoliators, with particular reference to the spruce budworm. Can. J. Zool. 33: 225–296.

Munn, C. A. & Loiselle, B. A. 1995. Canopy access techniques and their importance for the study of tropical forest birds. Pp. 165–177. In: Lowman, M. D. & Nadkarni, N. M. (eds), Forest canopies. Academic Press, San Diego.

Muul, I. & Lim, B. T. 1970. Vertical zonation in a tropical rain forest in Malaysia: methods of study. Science 169: 788–789.

Nadkarni, N. M. & Longino, J. T. 1990. Invertebrates in canopy and ground organic matter in a neotropical montane forest, Costa Rica. Biotropica 22: 286–289.

Nadkarni, N. M. & Parker, G. G. 1994. A profile of forest canopy science and scientists – who we are, what we want to know, and obstacles we face: results of an international survey. Selbyana 15: 38–50.

Nadkarni, N. M. & Lowman, M. D. 1995. Canopy science: a summary of its role in research and education. Pp. 609–613. In: Lowman, M.D. & Nadkarni, N.M. (eds), Forest canopies. Academic Press, San Diego.

Parker, G. G., Smith, A. P. & Hogan, K. P. 1992. Access to the upper forest canopy with a large tower crane. BioScience 42: 664–670.

Parker, G. G., Stone, P. J. & Bowers, D. 1996. A balloon for micro-climate observations within the forest canopy. J. Appl. Ecol. 33: 173–177.

Parrish, J. D. 1995. Effects of needle architecture on warbler habitat selection in a coastal spruce forest. Ecology 76: 1813–1820.

Pearcy, R. W. & Valladares, F. 1999. Resource acquisition by plants: the role of crown architecture. Pp. 45–66. In: Press, M. C., Scholes, J. D. & Barker, M.G. (eds), Plant physiological ecology. Blackwell Scientific, Oxford.

Perry, D. R. & Williams, J. 1981. The tropical rain forest canopy: a method providing total access. Biotropica 13: 283–285.

Popma, J., Bongers, F. & Meave del Castillo, J. 1988. Patterns in the vertical structure of the tropical lowland rain forest of Los Tuxtlas, Mexico. Vegetatio 74: 81–91.

Preisser, E. Smith, D. C. & Lowman, M. D. 1998. Canopy and ground level insect distribution in a temperate forest. Selbyana 19: 141–146.

Reagan, D. P. 1995. Lizard ecology in the canopy of an island rain forest. Pp. 149–164. In: Lowman, M. D & Nadkarni, N. M. (eds), Forest canopies. Academic Press, San Diego.

Renhorn, K.E., Esseen, P.-A., Pamqvist, K. & Sundberg, B. 1997. Growth and vitality of epiphytic lichens. I. Responses to microclimate along a forest edge-interior gradient. Oecologia 109: 1–9.

Reynolds, B. C. & Crossley, D. A. 1995. Use of a canopy walkway for collecting arthropods and assessing leaf area removed. Selbyana. 16: 21–23.

Richards, P. W. 1983. The three-dimensional structure of tropical rain forest. Pp. 3-10. In Sutton, S.L., Whitmore, T. C. & Chadwick, A.C. (eds), Tropical rain forest: ecology and management. Blackwell, London.

Rowe, W. J. & Potter, D. A. 1996. Vertical stratification of feeding by Japanese beetles within linden tree canopies: selective foraging or height per se? Oecologia 108: 459–466.

Russell, G., Jarvis, P. G. & Monteith, J. L. 1989. Absorption of radiation by canopies and stand growth. Pp. 21–39. In: Russell, G. Ussell, G. Marshall, G.B. & Jarvis, P.G. (eds) Plant canopies: their growth, form and function. Cambridge University Press, Cambridge, UK.

Schlesinger, W. H., Knops, J. M. H. & Nash, T. H. 1993. Arboreal sprint failure: Lizardfall in a California oak woodland. Ecology 74: 2465–2467.

Schowalter, T. D. 1995. Canopy invertebrate community response to disturbance and consequences of herbivory in temperate and tropical forests. Selbyana 16: 41–48.

Schowalter, T. D. & Ganio, L. M. 1998. Vertical and seasonal variation in canopy arthropod communities in an old-growth conifer forest in southwestern Washington, USA. Bull. Ent. Res. 88: 633–640.

Schowalter, T. D. & Ganio, L. M. 1999. Invertebrate communities in a tropical rain forest canopy in Puerto Rico following Hurricane Hugo. Ecol. Ent. 24: 191–201.

Shaw, J. D. & Bergstrom, D. M. 1997. A rapid assessment technique of vascular epiphyte diversity at forest and regional levels. Selbyana 18: 195–199.

Smith, A. P. 1973. Stratification of temperate and tropical forests. Am. Nat. 107: 671–683.

Sorrensen-Cothern, K. A., Ford, E. D. & Sprugel, D. G. 1993. A model of competition incorporating plasticity through modular foliage and crown development. Ecol. Monogr. 63: 277–304.

Sprugel, D. G., Hinckley, T. M. & Schaap, W. 1991. The theory and practice of branch autonomy. Ann. Rev. Ecol. Syst. 22: 309–334.

Stork, N. E. & Best, V. 1994. European Science Foundation – Results of a survey of European canopy research in the tropics. Selbyana 15: 51–62.

Sumida, A. 1995. Three-dimensional structure of a mixed broad-leaved forest in Japan. Vegetatio 119: 67–80.

Sutton, S. L., 2001. Alice grows up: canopy science in transition from wonderland to reality. Plant Ecol. 153: 13–21 (this volume).

Szarzynski, J. & Anhuf, D., this volume. Micrometerological conditions and canopy energy exchanges of a neotropical rain forest (Surumoni Crane Project, Venezuela). Plant Ecol. 153: 231–240.

Taylor, P. H. & Lowman, M. D. 1996. Vertical stratification of the small mammal community in a northern hardwood forest. Selbyana 17: 15–21.

Vieira, E. M. 1998. A technique for trapping small mammals in the forest canopy. Mammalia 62: 306–310.

Watson, M. A. & Casper, B. B. Morphogenetic constraints on patterns of carbon distribution in plants. Ann. Rev. Ecol. Syst. 15: 233–258.

Wieringa, J. J. 1996. Tree-climbing on a free rope. Pp. 819–821. In: van der Maesen, L.J.D., van der Burgt, X.M. & van Medenbach de Rooy, J.M. (eds). The biodiversity of African plants. Kluwer Academic Publishers, Dordrecht.

Weishampel, J. F., Ranson, K. J., Harding, D. J. 1996. Remote sensing of forest canopies. Selbyana 17: 6–14.

Whiteacre, D. F. 1981. Additional techniques and safety hints for climbing tall trees, and some equipment and information sources. Biotropica 13: 286–291.

Wright, D. D., Jessen, J. H., Burke, P. & dal Silva Garza, H. G. 1997, Tree and liana enumeration and diversity on a one-hectare plot in Papua New Guinea. Biotropica 29: 250–260.

Zandt, H. S. 1994. A comparison of three sampling techniques to estimate the population size of caterpillars in trees. Oecologia 97: 399–406.

Zotz, G. & Winter, K. 1996. Diel patterns of CO_2 exchange in rain forest canopy plants. Pp. 89–113. In: Mulkey, S.S., Chazdon, R.L. & Smith, A.P. (eds), Tropical forest plant ecophysiology. Chapman & Hall, New York.

Plant Ecology **153:** 39–50, 2001.

Plants in the forest canopy: some reflections on current research and future direction

Margaret D. Lowman
The Marie Selby Botanical Gardens, Sarasota, FL 34236, USA

Key words: Canopy access, Canopy biology, Epiphytes, Forests, Plants

Abstract

Plants are one of the sessile components of the forest canopy, and consequently quantitative studies of plant components are more widely available than for some of their mobile counterparts. From a global perspective, several exciting and innovative canopy access tools have been designed over the past few decades that have facilitated the expansion our understanding of canopy plants. These include a network of canopy cranes, the development of the French luge (or sled), and the construction of walkways and platforms for access into different levels of the canopy on a permanent basis. At the recent international forest canopy conference in Sarasota, Florida (*Forest Canopies 1998: Global Perspectives*, November 4–8, 1998), many canopy plant studies were presented that illustrated the achievements in this field. As co-chair at this event, I present here some reflections based on my observations of the development of canopy research between the first and second international conferences. A selection of case studies from the conference presentations are cited in this review, including: tree crown respiration studies from the crane in Panama; epiphyte ecology from walkways in Peru; insect-plant relationships in tree crowns of the USA using platforms; experimental studies of epiphytes in cloud forests using single rope techniques; and epiphyte diversity surveys in Africa. Ideas for the future are also mentioned such as the novel concept of canopy farming of orchids in Costa Rica, the linking of canopy processes to forest floor activities, and the construction of canopy walkways to provide a sustainable forest economy instead of the conventional practice of logging. The integration of research with sustainable use of forests provides a conservation theme for future canopy studies. Such new approaches to studies of canopy plants are important, as scientists increasingly play a role in global conservation policies.

Introduction

> *Botany needs help from the tropics. Its big plants will engender big thinking.*
> (E. J. H. Corner, 1939, Cambridge University)

We have come a long way in our studies of plants in the tropical rain forest since Professor Corner spoke these words over fifty years ago, but his philosophy holds true yet today. Our scientific interest in the tropics is manifested in part by the intrigue and mystery of the tall trees, and of their canopies. Ecological studies of tropical rain forests canopies are a fairly recent endeavour, and the progress over the past decade is impressive.

These are exciting times, and efforts to create successful access methods over the past decade are beginning to pay off. Over the past one hundred year history of forest exploration, plants were undoubtedly the major focus – for timber, for intrepid orchid collectors and sellers, for magic potions and cures, and probably for the sheer adventure of seeing enormous trees and complicated vines that soared above the heads of early explorers. Accounts such as these abound in the early botanical literature:

> Overhead, at a height, perhaps of a hundred feet, is an almost unbroken canopy of foliage formed by the meeting together of these great trees and their interlacing branches;.... It is a world in which man seems an intruder, and where he feels

overwhelmed..." (Alfred R. Wallace, 1878, in *Tropical Nature*)

Now, perhaps 125 years after this intrepid explorer first wrote about the wonders of plants in the canopy, we are just beginning to quantify the numbers of plants and aspects of their ecology in the upper reaches of the forests of the world, both temperate and tropical.

In the 1970s, several pioneers began canopy research in earnest (see Sutton 2001). Mitchell (1982) scaled tall trees in Indonesia and Perry (1986) in Costa Rica (reviewed in Moffett 1993). Soon after them, other pioneers such as Nalini Nadkarni and I began climbing trees using single rope techniques and publishing information on plants in the canopy (Lowman 1984; Nadkarni 1984). In the past 20 years, canopy research publications have increased tenfold (Nadkarni & Lowman 1995) as the numbers of canopy researchers expand and canopy access techniques become more widespread as part of conventional field equipment. Sites where ongoing canopy research takes place now range around the world (Table 1).

I would like to restrict this paper to the past 5 years, beginning with the review paper by Lowman & Moffett (1993) where we simplistically described the components of canopy research as follows: mobile, sedentary, and processes (Table 2). Using these three categories, we linked methodology to subject matter. We could do little else, since few long terms studies existed in the literature that reported extensive results. Shortly after this time, the first international canopy conference (titled as such) was held. Since that event, a virtual explosion of canopy research has occurred, including increased complexity of sampling and more ambitious international collaborations.

It is hard to believe that in 1993 we only had one canopy crane and two 'Radeau des Cimes' expeditions under our scientific belts, and there had never been a formal international gathering of canopy scientists. Yet now the network of canopy scientists, their collaborative research, and their findings are distributed worldwide. At the start of the new millennium, we can look forward to an age of intensive data collection, where canopy dynamics will become better understood. It is, alas, an urgent race against time to study many forest canopies before they disappear due to human activities.

Any review of plants in the canopy needs to begin (and probably end) with links to the forest floor. The canopy is not a distinct region, and it is not totally separate from the rest of the forest. When some of us first discussed the notion of canopy science, we voiced concern about making this a distinct discipline – none of us wish to separate the canopy from the forest, but rather to add the canopy as an integral portion of forest science. With that in mind, we need to acknowledge the importance of forest floor dynamics in shaping or structuring any forest of the world.

Seedlings and the dynamics of their survival form the future of the canopy (e.g., Connell et al. 1984). Studies of the maintenance of species diversity have been a major focus of many ecologists in recent years. The notion of species diversity perhaps also formed the basis for initial interest in tropical rain forests by many temperate-based scientists. Both Dan Janzen and Joe Connell championed theories of species equilibrium that sent many temperate botanists rushing to the tropics to expand their horizons (Connell 1978; Janzen 1970). Those two ecologists deserve thanks for opening up the eyes of many biologists to *see* tropical forests, not just temperate ones. And it was their enthusiasm for the tropical forests during the 1970s that led many biologists to 'look up'.

To become a tree in the canopy, a seedling must overcome many hurdles, and much of their success can be attributed to chance. We now know a lot more about seedlings and how they reach the canopy; factors that affect the recruitment of seedlings and the advanced regeneration of individuals can only be determined from long term studies such as those set up by Connell in Australia (Connell & Green 1999; Connell et al. 1997) and by Foster and Hubbell in Panama and elsewhere (Foster & Hubbell 1990; Hubbell & Foster 1990). These plots are producing important information about the ecology of the forest floor; similar long-term studies in the canopy (e.g., crane sites) will produce similar information about the dynamics of a forest beyond the forest floor. It is important to recognize that long-term studies are changing our impressions of regeneration and forest dynamics, and for canopy research, such long term research is only just underway.

Methods for studying plants in canopies

Seedlings and saplings have a distinct advantage for field research: they are accessible for study on the

Table 1. Canopy research sites throughout the world.

Region	Name of site	Location	Time frame	Habitat	Canopy research projects
North America					
	Wind River Canopy Cran Research Facility	Washington, USA 45°49′13.78″ N 121°57′06.88″ W	1930's – Exper forest 1994 – canopy studies canopy crane April 1995	Temperate coniferous forest Tree heights to 64 m	Forest ecology, structure & function, arboreal insects, dwarf mistletoe, avifauna, meteorology, remote sensing, forest health
	Hopkins Memorial	Forest Williamstown, USA	1990 – present	Temperature deciduous	Herbivory, phenology, mammals, insect diversity
	Milbrook Schoot	Milbrook, NY, USA 50° 75′ N 73° 37 26″ W	1995 – present	Temperature deciduous	Meteorology, phenology, insect diversity
	Coweeta Hydrologic Laboratory	North Carolina, USA 35° N 83°30′ W	1934 – forest hydrology 1968 – forest ecology	Temperate deciduous	Forest hydrology, forest ecology, nutrient fluxes, watershed experiments, vegetation dynamics, net primary production, decomposition, herbivory, streams, biotic regulation
	Marie Selby Botanical Gardens	Sarasota, FL, USA 27°19′ N 82°32′ W	1994 – present	Live oak canopy	Herbivory, throughfall, insect diversity, epiphytes
	My Akka Riverstate Park Sarasota, FL, USA	27° N 82° W	1992 – present	Palm canopy	Opipaytes, arboreal insects
	Penn State Research Forest	near University Park, PA, USA	1988 – present	Oak forest	Plant-insect relationships, leaf chemistry, phenology
	Hampshire College	Amherst, MA, USA	1993 – present	Temperate deciduous	Birds, phenology
	Findly lane	Kingcounty, Washington USA	1966 – present	Montane forest, west side of Cascade Mountains	Production ecology, limnology, physiological ecology, community ecology, canopy processes, soil genesis, remote sensing, biogenochemistry
	Smithsonian Environmental Research Center	Maryland, USA 38°58′ N 67°55′ W	1987 – present	Temperate deciduous	Canopy- atmosphere interactions, forest structure, phenology, hydrology, radiation, transpiration, canopy interception
	Harvard Forest	Petersham, MA, USA		Temperate deciduous	Canopy-atmosphere interactions, radiation, tree physiology
	Vermont Monitoring Cooperative	Mt. Mansfield, VT, USA 44°31′ N 72°52′ W Lye Brook Wild., VT, USA 43°06′ N 73°03′W	1990 – present	Temperate northern hardwood & montane confer forest	Forest health, atmospheric chemistry, within-canopy environmental gradients, insect, bird & amphibian biodiversity
	Andrews Experimental Forest	Oregon ~44-45° N	1970 – present	Old growth Douglas fir forest	Forest-atmosphere interactions, canopy-soil interactions, arthropods
	Wilamette National Forest	Central Oregon Cascade ~44-45° N	1992 – present	Old growth Douglas fir forest	Epiphyte diversity, long term growth of lichens
	Duke Forest	Durham, NC, USA	1995 – present	Temperate deciduous	Leaf area dynamics (FACE facility), canopy CO_2 & water flux
	Hakalau Forest National Wildlife Refuge	Hawaii, USA	1991–1992	Ohia forest	Insect diversity

Table 1. Continued.

Region	Name of site	Location	Time frame	Habitat	Canopy research projects
	Hawaik Volcanoes NP	Hawaii, USA	1971–1973	Ohia forest	Insect diversity
	Mesita del Buey	Los Alamos, NM, USA 34°30′ N 106°27′ W	1984 – present	pinyon-juniper woodland	Water balance, soil moisture, solar radiation, tree ecophysiology
	BOREAS	Prince Albert NP, Saskatchewan, Canada ~54° N ~99° W	1994 – present	Boreal forest	Carbon balance, gas exchange & energy flux, canopy architecture & remote sensing, solar radiation
	Vancouver Island Forests Canada	Vancouver Island, BC, 48°44′ N 124°37′ W	1990 – present	Old growth forest	Community composition, arthropod diversity
Central & South America					
	JASON V Project	Blue Creek, Belize 16°10′ N 89°5′ W	1994 – present	Tropical wet rain forest	Ant gardens, herbivory, arboreal insects
	Luquillo Experimental Forest	Puerto Rico 18°19′ N 65°45′ W	~1962	Tropical rain forest	Predator-prey relationship, Lizards, arthropods, herbivory
	Coffee Plantations	Central Valley, Costa Rica	~1992 – present	Coffee agroecosystem	Arthropods, agricultural transformation
	Finca, La Selva	La Selva, Costa Rica	1974–1983	Tropical rain forest	Natural history, canopy methods, polination
	Monteverde Cloud Forest Preserve	Monteverde, Puntarenas Province, Costa Rica 10°18′ N 84°48′ W	1980 – present	Leeward lower montane wet forest	Epiphyte ecology, nutrient cycling, bird use of epiphytes
	Smithsonian Tropical Research Institute	Barro Colorado Island, Panama 9°10′ N 79°51′ W	1972 – present	seasonally dry Tropical Moist forest	Insects, photosynthesis, epiphytes, phenology
	Parque Natural Metropolitano	near Panama City, Panama	1990 – present	seasonally dry rain forest	Photosynthesis, structure, herbivory, insects
	Operation Drake	near Punta Escoces, Darien Panama	1979–1980	Tropical rain forest	Insect diversity, walkway methods, forest structure, polination biology, bats
	Radeau de Cimes – Expedition I	French Guiana 5°4′36″ N 53°3′15″ W	1989	Tropical rain forest	Tree architecture & growth, plant & animal relationships, ecology, canopy-atmosphere interface
	Radeau de Cimes – Expedition II	Paracou, French Guiana 5°15′36″ N 52°56′15″ W	1996	Tropical rain forest	Tree architecture & growth, plant & animal relationships, ecology, canopy-atmosphere interface

Table 1. Continued.

Region	Name of site	Location	Time frame	Habitat	Canopy research projects
	French Guiana Research Station	Nouragues, French Guiana 4°5′ N 52°40′ W	1986 – present	Evergreen tropical forest	Gowth dynamics, gaps
	Smithsonian Fogging Studies	Pakitza, Peru	1988–1992	Lowland rain forest	Insect diversity
	ACEER	Napo River, Peru 3°15′ S 72°54′ W	1991 – present	Lowland tropical rain forest	Epiphytes, herbivory
	Explorama Reserves	Amazon & Napo Rivers, Iquitos, Peru	1994 – present	Tropical rain forest	Insect diversity
	Yacumama Preserve	Yarapa River, Peru 4°23′ S 73°30′ W	1998 – present	Lowland tropical rain forest	Insect diversity, epiphytes
	Tiputini Biodiversity Station	Tiputini River, Ecuador 0°40′ S 76°10′ W	1998 – present	Lowland tropical rain forest	Bats, birds, insect diversity
	Choco Canopy Walkway	Los Cedros, Ecuador 0°20′ N 78°50′ W	1997 – present	Cloud forest	Pollination, epiphytes
	Surumoni Research Project – European Science Foundation	Orinoco River, Venezuela	1995 – present	Lowland wet rain forest	Structure, epiphytes, ant gardens, birds, phenology
Asia & South Pacific					
	Lamington National Park	Queensland, Australia 28°13′ S 153°07′ E	1979 – present	Subtropical cool rain forest	Insect diversity, herbivory
	Dorrigo National Park	NSW, Australia 30°20′ S 153° E	1979–1990	Subtropical cool rain forest	Herbivory, phenology, insects
	New England National Park	NSW, Australia 30°30′ S 152° E	1979–1990	Cool temperate forest	Herbivory, phenology, insects
	Royal National Park	NSW, Australia 34°10′ S 151°30′ E	1979–1990	Warm temperate forest	Herbivory, phenology, insects
	CSIRO	Atherton, Queensland, Australia	~1980 – present	Wet tropical rain forest	Phenology, herbivory, reproductive biology
	Mt. Spec. National Park	Paluma, Queensland, Australia	~1980 – present	Lower montane tropical forest	Birds, insect diversity
	Cradle Mountain	Central Tasmania 41°35.4′ S 145°56.9′ E	1989 – present	Cool temperate rain forest	Invertebrate surveys (12 sites)
	Operation Drake (Bulolo Forestry College)	Buso, Morobe Province, Papua New Guinea	1979–1980	Tropical rain forest	Insect diversity, walkway methods, forest structure, pollination biology, borts

Table 1. Continued.

Region	Name of site	Location	Time frame	Habitat	Canopy research projects
	Wau Ecology Institute	Wau, Papua New Guinea 7°24′ S 146°44′ E	~1980 – present	Lower and mid-montane tropical rain forests	Herbivory, birds
	Operation Drake (Morowali Nature Reserve)	Sulawesi Tengah, Indonesia	1979–1980	Tropical rain forest	Insect diversity, walkway methods, forest structure, pollination biology, bats
	Gomback Watershed	35 km east of Kual Lampur, Malaysia	1960 – present	Hill, dipterocarp forest	Phenology, vertebrates
	Aerial Walkway	Bukit Lanjan, W. Malaysia	1968–1976	Tropical rain forest	Insect vectors, phenology
	Canopy Biology Program in Sarwak (CBPS)	Lambir Hills NP, Sarawak, Malaysia 4°20′ N 113°50′ E	1991 – present	Tropical rain forest	Phenelogy, insect abundance, plant/animal interactions in canopy layers
	Falealupo	Savai'i, Western Samoa	1996 –	Tropical island rain forest	Ethnobotany, plant taxonomy
	Padang	Padang, Indonesia 0°53′ S 100°21′ E	~1980's – present	Tropical rain forest	Foliage-canopy structure, height distribution of woody species
	Lake Rara National Park	Nepal 29°34′ N 82°5′ E	~1990	Alpine tree limit forest	Foliage-canopy structure, height distribution of woody species
	Yatsugatake	Yatsugatake, Japan 36°5′ N 136°21′ E	~1980's – present	Subalpine mixed forest	Foliage-canopy structure, height distribution of woody species
	Daisen	Daisen, Japan 35°21′ N 133°33′ E	~1980's – present	Cool temperate deciduous broad-leaved forest	Foliage-canopy structure, height distribution of woody species
	Yakushima Island	Yakushima Island, Japan 30°20′ N 130°24′ E	~1980's – present	Warm temperate evergreen broad-leaved forest	Follage-canopy structure, height distribution of woody species
	Hahajima Island	Hahajima Island, Japan 26° 39′ N 142°8′ E	~1980's – present	Subtropical evergreen broad-leaved forest	Foliage-canopy structure, height distribution of woody species
	Mengla Protected Nature Reserve	Southern Yunnan Province P.R. China	1989 –	Dipterocarp forest	Conservation
	Changbaishan Natural Reserve	Jilin, P.R. China 41°23′ N 126°55′ E	1996 –	Old-growth temperate spruce-fir forest	Canopy processes, structure & modeling
Europe					
	River Esk Woodlands	Midlothian, Scotland	1990 – present	Fagus woodlands	Phytophagous insects, woodland biodiveristy, land structure
	Gisbum Forest	Lancashire, England	1955–1992	Uplands mixed deciduous/conifer plantation	Tree growth, invertebrate populations, soil processes

Table 1. Continued.

Region	Name of site	Location	Time frame	Habitat	Canopy research projects
	Imperial College Field Station	Silwood Park, Ascot Berkshire, U.K.	~1977–1982	Temperate deciduous	Insect diversity and abundance
	Järvseija Towers	Järvseija, Estonia	1990s	Temperate deciduous	Plant physiology
Africa					
	Radeau de Cimes - Expediton II	Reserve de Fauna de Campo, Cameroon 2°30′ N 10°0′ E	1991	Lowland tropical rain forest	Tree architecture & growth, plant & animal relationships, ecology, canopy-atmosphere interface
	Radeau de Cimes - Expedition IV.	Gabon	1999	Lowland tropical rain forest	Tree architecture & growth, plant & animal relationships, ecology, insect vectors
	Irangi	Irangi, Congo 1°54′ S 28°27′ E	1993	Lowland rain forest	Distribution of canopy arthropods
	Cyamudongo	Cyamudongo, Rwanda 2°34′ S 25°59′ E	1993	Montane rain forest	Distribution of canopy arthropods
	Ibanda Makera	Ibanda Makera, Rwanda 2°9′ S 30°55′ E	1993	Gallery forest	Distribution of canopy arthropods
	Budongo Forest	Budongo Forest, Rwanda 1°45′ N 31°25′ E	1995 & 1997	Seasonal rain forest	Distribution of canopy arthropods
	Semilki Forest	Semliki Forest, Uganda 0°46′ N 26°57′ E	1997	Rain forest	Distribution of canopy arthropods
	East African Virus Research Institute	Mpanga Research Forest, west of Kampala, Uganda	1961 – present	Tropical forest	Insect vectors, meteorology

forest floor (Figure 1). Their canopy counterparts – tree foliage, vines, epiphytes – do not have such relative ease of access, but require careful design of hypotheses and safe methodology to collect data. Fortunately, such methods have been developed over the past decade that make all this possible. A range of safe techniques for canopy access (see also Barker & Pinard 2001; Sutton 2001) in order from simple to more complex, include:

- *Binoculars* – the conventional mode of examining the canopy that has been used worldwide. Although safe, this method has limitations for observations of small or cryptic organisms.
- *Single rope techniques (SRT)* – a relatively inexpensive method of scaling tall trees but requires a slingshot, bow, or cross-bow for setup. This method has probably been the most widely used, although it is limited to reaching one vertical transect along the rope at a time.
- *Ladders* – when affixed to a tree trunk, they offer access to branches in the mid (and sometimes upper) canopy.
- *Cherry pickers* – can be towed by a vehicle and transported to many trees in one day, although limited to dry forests (they can get bogged in moist forests) with road access.
- *Walkways* – the construction of modular units of bridges and platforms can provide permanent access to many elevated parts of a forest.
- *Hot-air balloons and inflatable rafts* – even the tallest tree or most remote forest stand can be reached using the dirigible and canopy raft in combination.
- *Construction cranes* – an excellent tool for reaching all the parts of a stand of trees within several hectares.
- *Satellites and aerial reconnaissance* – useful for mapping large regions of forest or surveying canopy disease vectors and/or forest distributions.
- *Towers* – used in climate studies and for mapping canopy topography.

Improved canopy access techniques, the network of cranes and walkways that facilitate comparative studies, and the increased communication among researchers have greatly expanded the potential for research on canopy plants.

Figure 1. Professor Joe Connell with 40-year old seedling of *Argyrodendron trifolialatum* in Australian rain forest (photograph by the author).

Advances in research on canopy plants – a few case studies

Improved canopy access has facilitated the initiation of experimental procedures in canopy field work, just as the advent of scuba promoted experimental manipulation of fish in and around patch reefs in marine biology. Once above the forest floor, the forest biologist is presented with a smorgasbord of opportunity. I have chosen several case studies presented at the recent Canopy Conference in Sarasota, Florida (Lowman 1999) that represent the diversity of current canopy research. Many other notable studies exist, but these examples are illustrative of the breath and scope of current projects. For a comprehensive report, refer to Lowman & Wittman (1996); this review included all studies of canopy plants that were submitted via the International Canopy Network (ICAN) Bulletin Board.

A major advancement in canopy plant research includes the improved accuracy of surveys to identify vascular plants in the canopy, including orchids, bromeliads, gesneriads, vines and canopy trees, many of whose flowers remained out of reach until improved canopy access. At Marie Selby Botanical Gardens in Florida, a major collection of epiphytes has been accessioned since 1975 including over 29 000 specimens of Orchidaceae and 8000 of Bromeliaceae, and these were available for the conference participants. Approximately 123 new species of bromeliads have been classified since 1990 in their Bromeliad Identification Center (H. Luther pers. com.). Each year, expeditions are carried out to collect plants in different equatorial regions of the world including Venezuela, Brazil, Cameroon, Peru and Ecuador. Sometimes, the more conventional techniques for collecting flowers in the canopy such as clippers (Figure 2) are not useful, but inventions such as the canopy sled enable botanists to explore and sample even the tallest emergent (Figure 3).

A second advancement in canopy plant studies is the initiation of ecological experimentation of epiphytes. Using single rope techniques (e.g., Nadkarni 1984, 2001), Nadkarni and colleagues have performed the first field experiments, moving epiphytes from one local climatic condition to another. To simulate global climate change, they shifted epiphytes in Costa Rica from moist montane tree crowns to lower valley situations where it is sunnier and drier. As predicted, the result was higher mortality of epiphytes and slower growth (Nadkarni 2001).

A third major advancement involves the use of construction cranes for canopy access that have greatly expanded the scope of intensive research within one forest plot. The Smithsonian Tropical Research Institute in Panama erected the first construction crane for canopy research in 1990 under the direction of the late Alan Smith (Moffett & Lowman 1995). This tropical dry forest was the site of pioneering work on forest canopy structure and whole tree physiology, especially phenology and photosynthesis (Kitajima et al. 1997a, b; Mulkey et al. 1995, 1996; Zotz 1993). In particular, the series of studies by Mulkey and Kitajima and associates have shown that some tropical tree leaves in the canopy live up to 700 days whereas leaves of other tree species live less than 100 days. This information is important for creating accurate models of productivity in forest canopies, as well as for understanding the natural history of tropical trees. In addition, they have carried out comprehensive photosynthetic measurements on populations of leaves with known age and micro-climatic conditions, using the canopy crane as their access tool. Because of the extensive information they can collect for each leaf, they can hypothesize about photosynthetic rates and water use efficiency for different leaf phenotypes (Kitajima & Mulkey 1998). With the excellent access provided the crane arm above the canopy, researchers can reach virtually all of the foliage of an entire tree crown for the purposes of whole tree measurements. It is also possible to haul fairly heavy and bulky equipment such as gas analyzers through the canopy in the gondola with relative ease. Since 1990, a series of cranes have been established in different forest types around the world: USA, Venezuela, Australia, Austria, a second one in Panama, and one pending in Malaysia. Similarly a network of canopy walkways is facilitating comparative studies of other canopy organisms, such as birds and their host trees (banding studies by Stokes

Figure 2. Conventional means of harvesting samples from canopy branches using clippers (photograph by Bruce Holst, reproduced with permission).

Figure 3. Non-conventional means of harvesting canopy branches using a hot-air dirigible and canopy raft (photograph by the author).

& Schultz 1995) and visitors to epiphytes (Lowman et al. 1996).

Herbivory is an important canopy process that directly affects the amount of leaf material available for photosynthesis. Not only is herbivory linked to the forest carbon balance, but herbivory affects other forest dynamics such as tree growth, soil processes, successional status, and nutrient cycling (Coley 1983; Coxson & Nadkarni 1995; Leigh & Smythe 1978; Lowman 1992, 1995). Forest herbivory has historically been measured by collecting samples of leaves growing in the understorey and estimating missing leaf area (Leigh & Smythe 1978; but see Lowman 1984). With the advent of improved methods of canopy access, estimates of herbivory in forests are more comprehensive, and include leaves that are stratified vertically throughout the canopy not just those close to the forest floor. Whereas original understorey measurements of herbivory usually estimated losses of 5–7% leaf surface area annually (Landsburg & Ohmart 1989; Wint 1983; Lowman 1984), measurements that included leaves from throughout different levels in the canopy ranged from 30% annual foliage loss in Australian temperate rain forests (Lowman 1992) to 300% in Australian dry sclerophyll woodlands where beetle outbreaks successively defoliated up to three new flushes of leaves (Lowman & Heatwole 1992). Improved canopy access has led to improved sampling designs for measuring herbivory, accounting for vertical stratification throughout the foliage of a crown, and spatial scaling from leaf to entire forest (Figure 4). With more comprehensive canopy access, ecologists can design sampling regimes that include different spatial and temporal scales, from leaf to branch to crown height to whole tree to entire stand to forest type.

Future directions

We need improved methods for quantifying the biodiversity observed in the canopy – not just streamlining the time and effort required to collect, classify and name species, but also advances in the technology for networking between herbaria, museums and data bases. Certain field studies have not yet been undertaken. For example, vine diversity, growth and abundance both within and among different tropical forests is still unknown, even though vines may be the largest component of biomass and stems in the forest (reviewed in Putz & Mooney 1991). Do vines create highways for herbivores or other canopy organisms? What is the extent of vine growth and does it limit the growth of host trees? Who pollinates vines?

More integrated studies of certain canopy processes are lacking, such as pollination of vines and tree crowns, nocturnal uses of epiphytes by different organisms, and dispersal of epiphyte seeds and establishment of juveniles. Increasingly, canopy researchers are seeking protocols to facilitate collaboration of results, which are integral to the future management of forests. The increased use of cranes, balloons, and walkways that enable groups of scientists to work in the canopy together, also promotes collaboration on both data bases and funding. The use of computer bulletin boards to communicate among botanists worldwide has increased our sharing of information, and effectively publicized different field sites. Table 1 lists all the canopy research sites (Nadkarni 2001) submitted through a survey conducted on the ICAN bulletin board. With the use of computer networking,

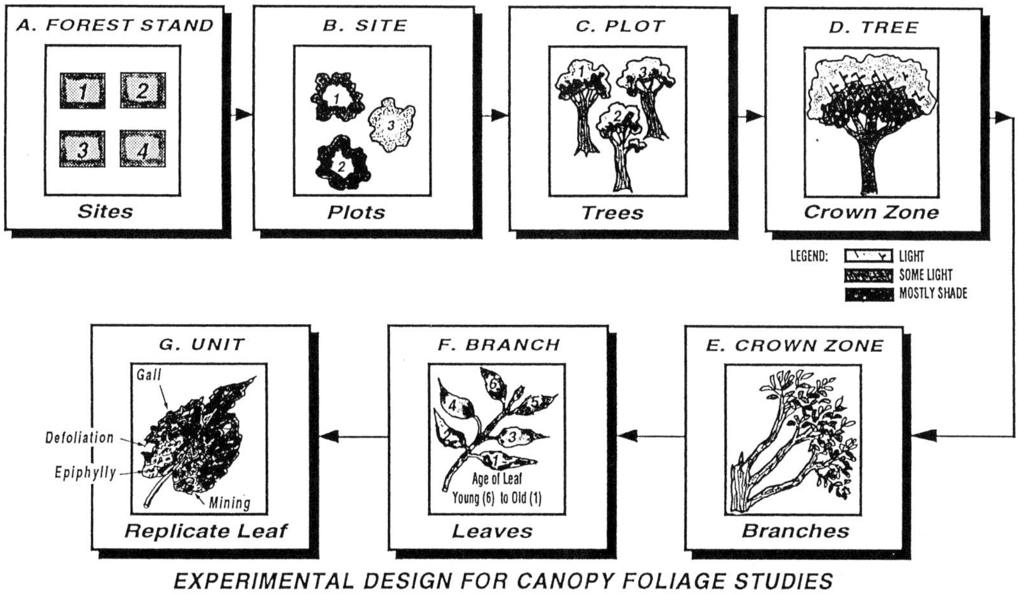

Figure 4. Illustration of scaling from small (leaf) to large (entire forest stand) in canopy sampling.

this list can continually be updated as scientists initiate studies at certain sites or establish new field projects at a global scale.

Canopy processes must be linked to other forest components instead of treating them as discrete units as we have done in the past. Studies to quantify frass fall and the herbivory pathway in conjunction with decomposers on the forest floor will enable us to map nutrient cycling comprehensively rather than keep them as separate components of the forest. Once more ecological information about canopies is quantified, then the sustainable management of forests may become more feasible. For example, canopy walkways are being used for both research and for eco-tourism, and serve to promote forest conservation. Walkways exist in Peru, Australia, Florida, Western Samoa, Costa Rica, Uganda, and Belize, with other structures in the planning phases.

Another innovative research arena involves canopy farming, whereby epiphytes could be grown and exported from the tropics as a sustainable harvest. The first canopy farming project in Costa Rica was described at the Second International Forest Canopy conference (Verhoeven & Beckers 1999) whereby orchids were propagated in natural canopy conditions for commercial purposes. Similar projects are underway in Cameroon (Nkongmeneck et al. 1998) and elsewhere.

Other ways to utilize canopy plants in forest conservation include the simple but obvious concept that climbing trees and exploring canopy life is exciting and fun. In particular, the enormous biodiversity and the wide array of interactions in the treetops are compelling subjects for lay persons to comprehend. Children's books, semi-popular texts, videos and films on forest canopies are promoting the cause of conservation in a very effective fashion. At the Sarasota conference, keynote speaker Tom Lovejoy strongly advocated that all scientists pay serious attention to translating their scientific results into text that can be disseminated to policy makers and the voting public, not just a handful of experts (see also Stork 2001). The use of canopies as a tool for effective conservation education cannot be underestimated, if we are to retain intact forests for the next generation.

Acknowledgements

I am grateful to Andrew Davis and Martin Speight for their invitation to participate in the excellent conference *Tropical Forest Canopies* at Oxford University, December 10–17, 1999. I am also indebted to my research assistant, Tom Roberts, for editing; to Steve Mulkey, Kaoru Kitajima, Bruce Holst, Harry Luther, John Atwood, Nalini Nadkarni, Koen Voorhoeven, R.A.Oldeman, Gabriel Beckers, Neville Winchester, Phil Wittman, and Bernard Nkongmeneck for the use of their important canopy projects in my review. And to all the 263 participants at the Second International Forest Canopy Conference in Sarasota, Florida (November 1999), I am grateful for your participation and dialogue, that formulated the basis for this personal assessment. This review is not meant to be comprehensive, but represents a collation of several main points highlighted by delegates at the 1998 conference.

References

Barker, M. & Pinard, M. A. 2001. Forest canopy research: sampling problems, and some solutions. Plant Ecol. 153: 23–38 (this volume).

Coley, P. D. 1983. Herbivory and defensive characteristics of tree species in a lowland tropical forest. Ecol. Monogr. 53: 209–233.

Connell, J. H. 1978. Diversity in tropical rain forests and coral reefs. Science 199: 1302–1310.

Connell, J. H. & Green, P. 1999 The dynamics of germination and recruitment of Planchonella australis. Ecology.

Connell, J. H, Lowman, M. D. & Noble, I. R. 1997. Subcanopy gaps in temperate and torpical forests. Austr. J. Ecol. 22: 163–168.

Connell, J. H., Tracey, J. G. & Webb, L. J. 1984. Compensatory recruitment, growth, and mortality as factors maintaining rain forest tree diversity. Ecol. Monogr. 154 (2): 141–164.

Coxson, D. S. & Nadkarni, N. M.1995. Ecological roles of epiphytes in nutrient cycles of forest ecosystems. Pp. 495–543. In: Lowman, M. D. & Nadkarni, N. (eds), Forest canopies. Academic Press, San Diego, California.

Foster, R. B. & Hubbell, S. P. 1990. The floristic composition of the Barro Colorado Island forest. Pp. 85–99. In: Gentry, A. (ed.), Four neotropical rain forests. Yale University Press, New Haven, Connecticut.

Gehring, C. A. & Whitham, T. G. 1994. Interactions between above-ground herbivores and the mycorrhizal mutualists of plants. Trends Ecol. Evol. 9: 251–255.

Hubbell, S. P. & Foster, R. B. 1990. Structure, dynamics, and equilibrium status of old-growth forest on Barro Colorado Island. Pp. 522–542. In: A. Gentry (ed.), Four neotropical rain forests. Yale University Press, New Haven, Connecticut.

Janzen, D. H. 1970. Herbivores and the numbers of tree species in tropical forests. Am. Nat. 104: 501–528.

Kitajima, K., Mulkey, S. S. & Wright, S. J. 1997a. Decline of photosynthetic capacity with leaf age in relation to leaf longevities for five tropical canopy tree species. Am. J. Bot. 87: 702–708.

Kitajima, K., Mulkey, S. S. & Wright, S. J. 1997b. Seasonal leaf phenotypes in the canopy of a tropical dry forest: photosynthetic characteristics and associated traits. Oecologia 109: 490–498.

Kitajima, K. & Mulkey, S. S. 1998. Ecophysiology of canopy leaves; constraints on resource allocations in time and space. Selbyana 19 (2): 283–284.

Landsburg, J. & Ohmart, C.P. 1989. Levels of defoliation in forests: patterns and concepts. Trends Ecol. Evol. 4: 96–100.

Leigh, E. & Smythe, N. 1978. Leaf production, leaf consumption, and the regulation of folivory on Barro Colorado Island. Pp. 33–50. In: Montgomery, G.G. (ed.), The ecology of arboreal folivores. Smithsonian Press, Washington DC.

Lowman, M. D. 1984. An assessment of techniques for measuring herbivory: is rain forest defoliation more intense than we thought? Biotropica 16: 264–268.

Lowman, M. D. 1992. Leaf growth dynamics and herbivory in five species of Australian rain forest canopy trees. J. Ecol. 80: 433–447.

Lowman, M. D. 1995. Herbivory as a process in rain forest canopy trees. Pp. 431–455. In: Lowman, M. D. & Nadkarni, N. M. (eds.) Forest canopies. Academic Press, San Diego, California.

Lowman, M. D. 1999 Life in the treetops. Yale University Press, New Haven, Connecticut.

Lowman, M. D. & Heatwole. H. 1992. Spatial and temporal variability in defoliation of Australian eucalypts and its consequence for the dieback syndrome. Ecology 73: 129–142.

Lowman, M. D. & Moffett, M. 1993. The ecology of tropical rain forest canopies. Trends Ecol. Evol. 8: 104–108.

Lowman, M. D. & Wittman, P. K. 1996. Forest canopies: methods, hypotheses, and future directions. Ann. Rev. Ecol. Syst. 27: 55–81.

Lowman, M. D., Wittman, P. K. & Murray, D. 1996. Herbivory in a bromeliad of the Peruvian rain forest canopy. J. Brom. Soc. 46: 52–55.

Mitchell, A. W. 1982. Reaching the Rain Forest Roof. UNEP publication, Leeds Philosophical and Literary Society, Leeds, United Kingdom.

Moffett, M. 1993. The high frontier. Harvard University Press, Boston Massachusetts.

Moffett, M. W. & Lowman, M. D. 1995. Canopy access techniques. Pp. 3–26. In: Lowman, M. D. & Nadkarni, N. M. (eds), Forest canopies. Academic Press, San Diego, California.

Mulkey, S. S., Kitajima, K. & Wright, S. J. 1995. Photosynthetic capacity and leaf longevity in the canopy of a dry tropical forest. Selbyana 16: 169–173.

Mulkey, S. S., Kitajima, K., & Wright, S. J. 1996. Plant physiological ecology of tropical forest canopies. Trends Ecol. Evol. 11: 408–412.

Nadkarni, N. 1984. Epiphyte biomass and nutrient capital of a neotropical elfin forest. Biotropica 16: 249–257.

Nadkarni, N. M. 2001. Enhancement of forest canopy research, education, and conservation in the new millennium. Plant Ecol. 153: 361–367 (this volume).

Nadkarni, N. & Lowman, M. D. 1995. Canopy science: a summary of its role in research and education. Pp. 609–615. In: Lowman, M. D. & Nadkarni, N. M. (eds), Forest canopies. Academic Press, San Diego, California.

Nkongmeneck, B., Lowman, M. D. & Atwood, J. 1998. Epiphytes in Cameroon: a survey of host trees and human activity. Selbyana 19 (2): 290.

Perry, D. 1986. Life above the jungle floor. Simon & Schuster, New York, NY.

Putz, F. E. & Mooney, H. A. (eds). 1991. The biology of vines. Cambridge University Press, Cambridge, UK.

Stokes, A. E. & Schultz, B. B. 1995. Mist netting birds from canopy platforms. Selbyana 16: 144–146.

Stork, N.E. 2001. The management of canopy research. Plant Ecol. 153: 313–317 (this volume).

Sutton, S. L. 2001. Alice grows up: canopy science in transition from Wonderland to Reality. Plant Ecol. 153: 13–21 (this volume).

Verhoeven, K, & Beckers, G. J. L. 1999. Canopy farming - a conservation alternative in Costa Rica. Selbyana 20: 191–193.

Wint, G. R. W. 1983. Leaf damage in tropical rain forest canopies. Pp. 229–241. In: Sutton, S.L., Whitmore, T.C. & Chadwick, A.C. (eds). Tropical rain forest: ecology and management. Blackwell Scientific Publication, Cambridge, UK.

Zotz, G. & Winter, K. 1993. Short-term photosynthesis measurements predict leaf carbon balance in tropical rain forest canopy plants. Planta 191: 40.

Plant Ecology **153:** 51–63, 2001.

Epiphytes and their contribution to canopy diversity

Jürgen Nieder[1], Juliana Prosperí[2] & Georges Michaloud[2]
[1]*Botanisches Institut der Universität Bonn, Meckenheimer Allee 170, 53115 Bonn, Germany;* [2]*Université de Montpellier II, CNRS–ISEM–UMR 5554, Laboratoire de Botanique, 34090 Montpellier, France*

Key words: Biodiversity, *Ficus*, Neotropical and Paleotropical epiphytes, Rain forest, Vascular plants

Abstract

About ten percent of all vascular plant species world-wide are epiphytes and they are almost exclusively found in tropical forests. Therefore, they constitute a large part of the global plant biodiversity (10% of all species), and in tropical countries represent up to 25% of all vascular plant species. Focusing on the differences between epiphytes in the strict sense or 'holo-epiphytes' (non-parasitic plants that use other plants – usually trees – as growing sites all through their life-cycle) and 'hemi-epiphytes' ('half' epiphytes which only spend part of their life as epiphytes until their aerial roots become connected to the ground), horizontal and vertical distribution patterns of both in relation to some of their ecological requirements are compared. Vertical ecological gradients (i.e., insolation and humidity differences from the forest floor to the canopy surface) are relevant for the composition of the holo-epiphytic vegetation. For hemi-epiphytes, however, ecological differences between distinct forest habitats (i.e., horizontal gradients) are relevant, but not primarily the canopy structure, as the individual host tree structure is more important. The scale-dependence of epiphyte diversity assessment (relatively small study areas for holo-epiphytes, large study areas for hemi-epiphytes) is mainly due to the striking differences in plant sizes and related mechanical and physiological requirements.

Introduction

Although much has been published on epiphytes (Watson et al. 1988), the question of the global importance of epiphytes for the biodiversity of tropical forests in general, and the canopy in particular, has not been explicitly addressed. This paper is based on our own studies in Africa and South America, and uses our results and a review of literature data in order to point out the relative importance of holo-epiphytes (epiphytes in the strict sense, i.e., living their whole life cycle as epiphytes) and hemi-epiphytes (which at some stage of their life cycle grow on the ground) for plant species richness on a local, regional and global scale. The differences between the two life-habits (holo-epiphytism and hemi-epiphytism) in terms of ecological requirements, life history and reproduction, will be addressed and used for a comparison of vertical and horizontal distribution patterns in the tropical forest habitat.

Our own results on holo-epiphyte diversity come from studies in Venezuela (Carbonera Forest, 71°21′ W, 8°37′ N; 2200–2700 m a.s.l.), Ecuador (Otonga Reserve, 79°01′ W, 0°25′ S; 1680–2251 a.s.l.; Río Guajalito, 78°49′ W, 0°14′ S; 1780–2200 m a.s.l.) and Bolivia (Sehuencas, 65°16′ W, 17°30′ S; 2100–2300 m a.s.l.). The results are based on various doctoral and diploma theses (see Rauer 1995; Rudolph 1995; Ibisch 1996; Nowicki 1998; Engwald 1999). Additionally, the floras and checklists of Florida, Panama, Peru, Java, the Malaysian Peninsula, Sri Lanka and Japan were evaluated and computer data- sets were compiled in order to perform statistical analyses (Boegner 1999).

Hemi-epiphytic *Ficus* were studied in lowland rain forests in north-eastern Gabon (M'passa Reserve, near Makokou, by the Ivindo river, 0°04′ N, 12°52′ E; 500 m a.s.l.), in south-west Ivory Coast (The Taï Reserve, 5°13′ N, 6°55′ W), and this genus as well as *Clusia* and *Coussapoa* were also studied in a lowland rain forest of French Guyana (Piste de Saint Elie, 5°30′ N, 53°40′ W).

Three study sites were selected within the Gabonese rain forest area, each representing a different habitat within the same forest block: 100 ha of a forest plateau, 19 ha on a neighbouring island and 12.8 km of the Ivindo river bank (12.8 km length, 3.2 m width, corresponding to 4 ha ground area) located partly along the two previous habitats.

The term 'diversity' in the following is synonymous with 'species diversity', since assessment of other aspects of biodiversity (i.e., taxonomic diversity and genetic diversity, see Heywood 1995) is non-existent in the case of vascular epiphytes.

Epiphyte diversity in the context of global plant biodiversity

Global biological diversity is concentrated in the tropics, where nine out of 12 so-called 'megadiversity countries' lie (Groombridge 1992). Although the species richness of plants is probably 'only' about 270 000 (animal species numbers are currently estimated to be a 100 times higher), neither their exact number nor their global diversity pattern is known. A map of the global diversity of vascular plants (Barthlott et al. 1996, 1999) on the basis of flora data gives an overview. The centres of plant diversity in the Neotropics are the Andes, north-west Amazonia and the coastal rain forest of Brazil, where epiphytes are known to be concentrated in abundance and diversity.

It has been estimated that 24 000 or more vascular plant species are epiphytes (Kress 1986), which constitute a major part (perhaps as much as 10%) of the global plant diversity. Holo-epiphytism degrees (i.e., the percentages of epiphyte species in the total flora of a country or region) in tropical countries confirm this tentative figure. Ecuador has 25.8% vascular holo-epiphytes (Møller Jørgensen 1999), Java approximately 12%, Panama 13%, Peru 11%, West-Malaysia 11% (data from Ibisch et al. 1996; and Boegner 1999). Africa on the whole has less epiphyte species than America or Asia, probably due to aridity and the lack of refuge areas in the late Pleistocene (Johansson 1989), but there are no figures of epiphytism degrees available.

Hemi-epiphytes represent about 0.8% of all vascular plant species with almost 2000 species (Putz & Holbrook 1986; Gentry & Dodson 1987). In Ecuador (the only country for which we have the data), 2% of all vascular plants are hemi-epiphytes (Møller Jørgensen 1999).

Systematically, hemi-epiphytism is less widespread than holo-epiphytism. According to Prosperi (1998), 31 plant families have hemi-epiphytes, whereas epiphytism (including hemi-epiphytism) occurs in 83 families of vascular plants (Kress 1986).

Vascular epiphyte species richness on a local scale

Holo-epiphytes

On a small spatial scale, the diversity of tropical forests is largely based on the diversity of epiphytic plants. Sixty-six epiphyte species were found on one tree of *Decussocarpus rospigliosii* (Pilg.) de Laub. in the Carbonera Forest, Venezuela (Engwald 1999). The amazing concentration of epiphytes on single trees has attracted the naturalist's attention many times (e.g., Longman 1917; Dunsterville 1961). Epiphytes need little space in order to develop a high diversity. In the Ecuadorian mountain forest Otonga Reserve 109 epiphyte species were found on just 20 m^2 of branch surface (single branches from several neighbouring trees), compared to only 67 terrestrial plant species on a 100 m^2 ground plot of elongated shape in the immediate vicinity of the phorophyte (Nowicki 1998).

In order to put epiphyte diversity into the context of plant diversity in a given study area, it is necessary to compare inventories of epiphytes and non-epiphytes. The 'epiphyte quotient' (a term introduced by Hosokawa 1950), sometimes called 'epiphytic index' (Schnell 1952), is by definition the percentage of epiphyte species in the total number of vascular plants. Table 1 shows the size, elevations and epiphyte numbers in a variety of sites, and in three cases the epiphyte quotients (of our study sites only). The quotients of Sehuencas, Río Guajalito and Carbonera show that, on a local scale, epiphytes are major components of the plant diversity. The quotients approximately match those of other neotropical forests, e.g., a premontane forest of Río Palenque, Ecuador (22%; Gentry & Dodson 1987), a lowland forest at La Selva, Costa Rica (26%; Hartshorn & Hammel 1994) and a montane rain forest near Mérida, Venezuela (50%; Kelly et al. 1994).

African forests, in contrast, seem to be much poorer in relative epiphyte species richness. For Ghana, a typical epiphyte quotient in forest plots is 8% (Hall 1976, 1978 cited after Martin 1989). For several forest types in central Africa (Rwanda and Zaire) the combined degree of epiphytism was only 2.5% (Biedinger & Fischer 1996).

Table 1. Study sites of the epiphyte project of the Botanical Institute, Bonn (Germany): site characteristics and number of epiphyte species (epiphyte quotient = % epiphyte species in the total number of species of a study site).

Study site	Elevation (m a.s.l.)	Precipitation (mm y^{-1})	Number of epiphyte species	Epiphyte quotient (area of reference)
Sehuencas, Bolivia	2100–2300	5000	230	37% (0.1 ha)
Otonga, Ecuador	1700–2200	2500	196	– (400 ha)
Río Guajalito, Ecuador	1800–2200	2700	166	28% (400 ha)
Carbonera, Venezuela	2100–2300	1500	192	45% (360 ha)
Surumoni, Venezuela	100	2800	53	– (1.5 ha)

Hemi-epiphytes

The relative diversity and abundance of hemi-epiphytes may vary between forests. According to Williams-Linera & Lawton (1995), they are not present in dry forests. At La Selva (Costa Rica; 1500 ha area size), about 2% of all vascular plant species are hemi-epiphytes (Hartshorn & Hammel 1994). In a 6 ha plot in Panama, 9.8% of the trees were colonised by hemi-epiphytic figs (Todzia 1986); in Venezuela 13% of all trees carried both hemi-epiphytic *Clusia* and *Ficus* (Putz 1983); 17% in French Guyana, on a 1 ha plot (*Clusia*, *Coussapoa* and *Ficus* combined together; Prosperi 1998). In Africa, Guy (1977) found in Zimbabwe that 12.6% of trees were bearing figs, and Michaloud & Michaloud-Pelletier (1987) found 20.6% in the Ivory Coast along 7 km long, 30 m wide transects.

In the old forest part of Barro Colorado, 1.5 hemi-epiphytic fig plants per ha were found (Todzia 1986), and Prosperi (1998) censused 2 fig plants per ha in a Guyana forest. In Africa, in Ivory Coast Michaloud & Michaloud-Pelletier (1987) censused 4.6 figs per ha and 1.5 in Gabon (Michaloud 1988). The similarity between the above densities shows that we may indeed compare figures from both continents.

Environmental factors and epiphyte diversity

The canopy offers epiphytes more light compared to the dark understory (see Lüttge et al. 1986), but a number of physiological constraints are relevant: for example the relative scarcity of nutrients due to the smaller amount of suspended soil in the canopy compared to the forest floor (not necessarily a lower relative nutrient contents, see Coxson & Nadkarni 1995); the instability of the substrate; and above all the results of water stress (see Lüttge 1989). This explains why vascular epiphytes are largely only found in the subtropics and tropics. In temperate regions, frost dryness kills vascular epiphytes.

The amount of rainfall is a good indicator of epiphyte abundance, but with some exceptions which are illustrated by the Venezuelan study sites La Carbonera (Engwald 1999) and Surumoni (Morawetz 1998; Nieder et al. 2000). Whereas at La Carbonera rainfall is low (1500 mm yr^{-1}) and epiphyte abundance and species richness were very high (over 5000 individuals on 800 m^2 of study plot area, 192 species), the Surumoni site had less individuals (775) with only 53 species on 1.5 ha, with 2800 mm yr^{-1} rainfall (Table 1).

At La Carbonera a daily cloud cover around noontime allows the condensation of water on the epiphyte surfaces and decreases evapo-transpiration. At Surumoni, on the other hand, the increased insolation and consequent evaporation and a short dry period up to 3–4 months probably are the cause of a relatively scarce epiphyte vegetation.

In hemi-epiphytes, different genera exhibit different ecological requirements within the canopy. For example *Clusia* and *Ficus* often colonise different niches (Prosperi 1998). This can also be the case within the same genus and African fig species may be habitat specific or generalists (Table 2). Ecological differences among hemi-epiphyte genus can partly be explained by physiological and morphological adaptations, such as xeromorphic leaves, thick cuticula, depressed stomates, and water accumulation in the hypoderm (Putz & Holbrook 1986). Such adaptations may even vary through the developmental stages, as observed for several fig species in Venezuela and Costa Rica where

Table 2. Hemi-epiphytic *Ficus* species encountered within and between forest habitats in Gabon (after Berg et al. 1984). In each column, the first number corresponds to the number of trees, and the second to the relative density.

Species	Plateau (100 ha)	Island (19 ha)	River bank (4 ha)
F. craterostoma	32;0.32	6;0.32	12;3.00
F. lingua	19;0.19	2;0.11	3;0.75
F. wildemaniana	17;0.17	1;0.05	1;0.25
F. barteri	12;0.12	10;0.53	2;0.50
F. elasticoïdes	5;0.05	4;0.21	1;0.25
F. ovata	1;0.01	5;0.26	5;1.25
F. sp.	3;0.03	2;0.10	8;2.00
F. macrosperma	7;0.07	1;0.05	–
F. cyathistipuloïdes	6;0.06	4;0.21	–
F. kamerunensis	4;0.04	2;0.10	–
F. conraui	3;0.03	5;0.26	–
F. ottoniifolia	2;0.02	2;0.10	–
F. artocarpoïdes	1;0.01	1;0.05	–
F. bubu	1;0.01	1;0.05	–
F. louisii	2;0.02	–	5;1.25
F. mangiferoïdes	–	1;0.05	1;0.25
F. thonningii	15;0.15	–	–
F. densistipulata	11;0.11	–	–
F. burretiana	4;0.04	–	–
F. pringsheimiana	3;0.03	–	–
F. kimuenzensis	1;0.01	–	–
F. adolfi-friderici	1;0.01	–	–
F. preussii	–	3;0.15	–
F. dryepondtiana	–	1;0.05	–
F. natalensis leprieuri	–	–	128;32.00
F. subcostata	–	–	8;2.00
F. lutea	–	–	2;0.50
F. cyathistipula	–	–	1;0.25

water and nutritional elements contained in leaves at the epiphytic stage decreased once the aerial roots had reached the ground (Putz & Holbrook 1989; Holbrook & Putz 1996). Different growth forms are also responsible for the differences in ecology between *Clusia* and *Ficus* in Amazonia (Prosperi 1998), where *Clusia* can better adapt to shadowy canopy sites than *Ficus*. These differences also have a consequence on population size. Indeed, Prosperi (1998) showed that in French Guyana, *Clusia* was the most frequent genus among hemi-epiphytes, followed by *Coussapoa* and then *Ficus*.

Host tree structure is also responsible for the distribution of hemi-epiphytes and constitutes a limiting factor for their settlement (Michaloud & Michaloud-Pelletier 1987). Eight host tree parameters (see legend of Figure 1) were measured and estimated respectively, by two observers. From Figures 1 and 2 we see that there is a positive correlation between the potential host tree structure and colonisation by hemi-epiphytic figs in a rain forest in Ivory Coast. As a consequence Michaloud & Michaloud-Pelletier (1987) observed that there was no positive correlation between the different abundance of potential host tree species and the number of hemi-epiphytic figs hosted. For instance they showed that *Parinari excelsa* (Rosaceae) with a mean density of 1.26 tree per ha was hosting 18 fig individuals, and *Coryanthe pachyceras* (Rubiaceae) with a mean density of 19 trees per ha was hosting only one fig individual (on 20.9 ha censused). In Guyana, Prosperi (1998) also found a highly significant positive correlation be-

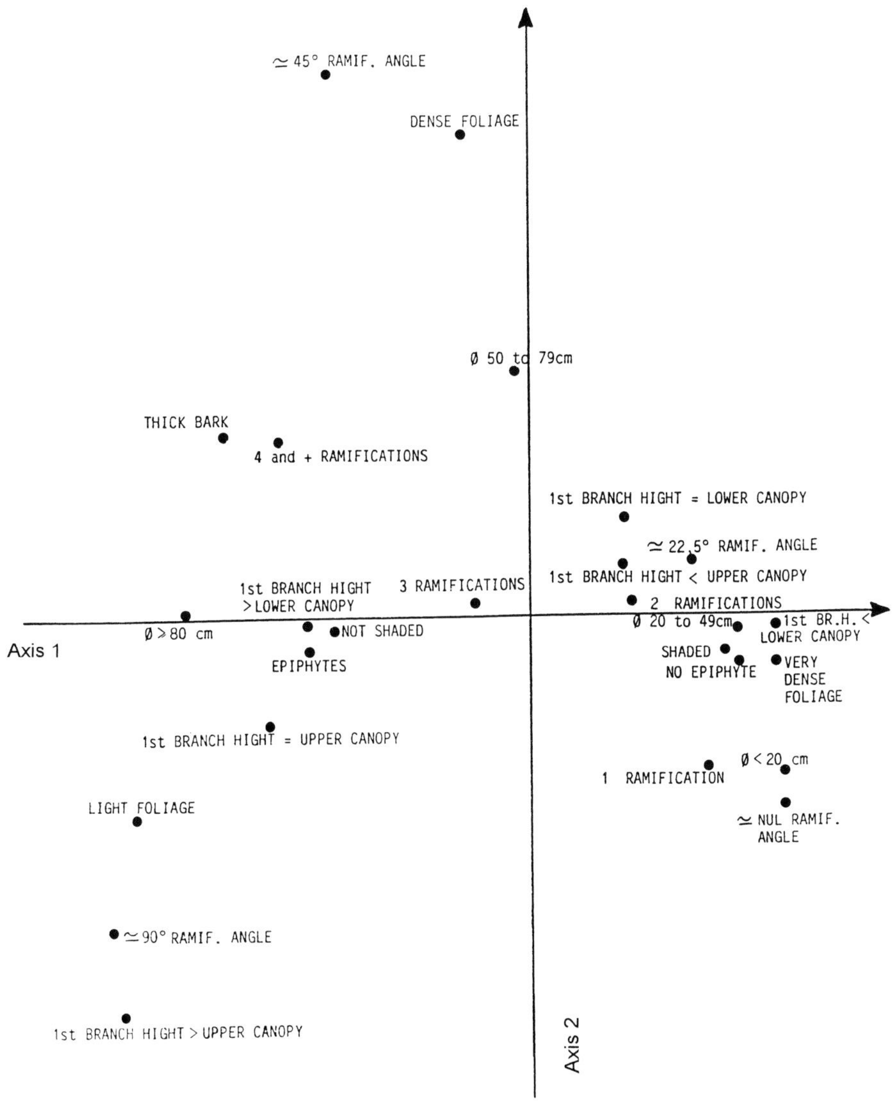

Figure 1. Multivariate analysis of the structural parameters of 39 colonised and 113 non colonised trees belonging to the hemi-epiphytic fig host species (see Figure 2 for list of species analysed). Axis 1 explains 26% of the interactions, and axis 2, 10%. The tree structure parameters which are analysed are the following: four classes of trunk diameter (DBH); two classes of bark thickness; according to lower and upper canopy strata, three classes for the height of the first main branch respectively; number of main branch ramifications (four classes from one to four and above); the ramification angle formed with the trunk (four classes from zero, when branches are almost parallel to the trunk at 90°); foliage density (very dense, dense and light); tree position (dominating or dominated) and last, presence or absence of ferns and (or) orchids (from Michaloud & Michaloud-Pelletier 1987, with permission).

tween the rate of colonisation and host diameter above 50 cm (DBH), but no positive correlation with hemi-epiphytic potential host species abundance in a 1 ha plot.

In the Ivory Coast, Michaloud and Michaloud-Pelletier (1987) showed that quality of aerial substratum was also important, and that the correlation between the settlement of hemi-epiphytic figs and the presence of epiphytic ferns and orchids was highly significant.

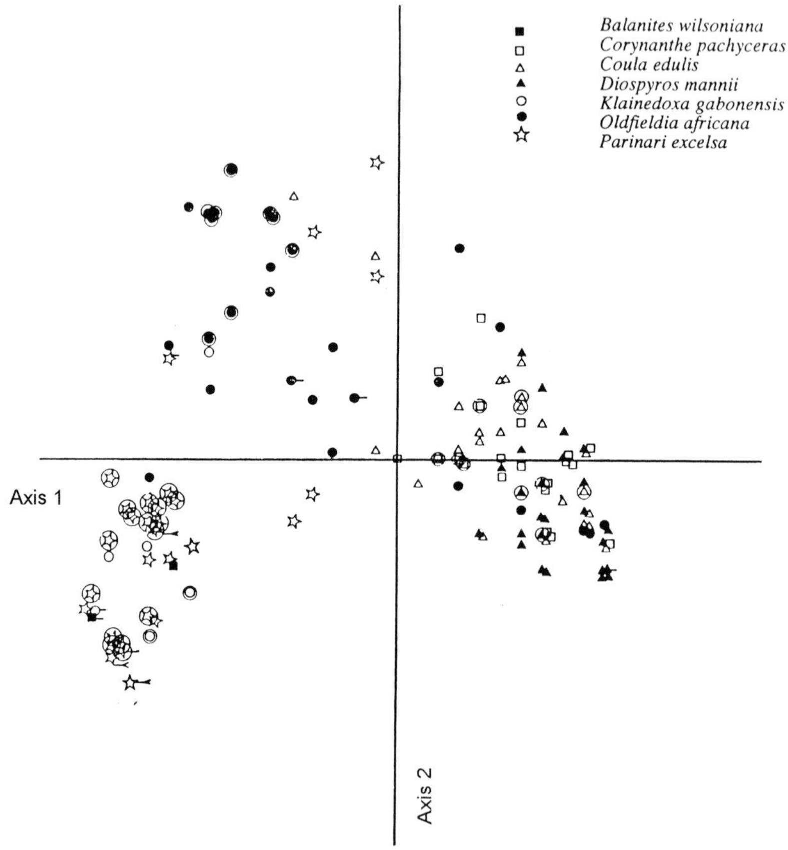

Figure 2. Distribution of trees analysed in Figure 1 according to their structural parameters between Axis 1 and Axis 2. The presence of a circle around each species symbol indicates that the tree in question bears an adult hemi-epiphytic fig. A single line added to a circle indicates that one young epiphytic fig is also present on the tree in question, and two lines added to a circle (forming a horizontal 'Y') indicates that there are also two young epiphytic figs present (from Michaloud & Michaloud-Pelletier 1987, with permission).

Scale dependence of relative epiphyte diversity

Holo-epiphytes

At Sehuencas (Bolivia), the epiphyte quotient decreased with the increase of plot size (Ibisch 1996), due to the increase of the number of epiphytic plant species with area being lower than that of terrestrial plants (see Figure 3).

On a larger scale the quotient is considerably smaller. At Río Palenque this represents a reduction from 35% to 22%, moving from plot size (1000 m^2) to study site size (1.7 km^2) (Gentry & Dodson 1987). At Barro Colorado Island in Panama (15 km^2) the epiphyte quotient is approximately 16% (Foster & Hubbel 1990), that of the whole country Panama (75 650 km^2) is 12% (Boegner 1999).

One factor contributing to this phenomenon is that epiphytes in many cases have a lower rate of endemism than terrestrial plants. This is the case for bromeliads and aroids (among others) in Peru, with the exception of some genera of epiphytic orchids (Ibisch et al. 1996).

This phenomenon also seems to hold true for African epiphytes: Sanford (1969) found that his study area contained a third of all the epiphytic orchid species of Nigeria, but less than a fifth of the tree species.

There is, however, a considerable species turnover between mountain forest sites: the study sites Río Guajalito and Otonga in western Ecuador had only 35% species in common, although these two forests are only 30 km away from each other and altitude, precipitation, soil characteristics and census methods were similar.

The species number of the Río Surumoni site was much lower (53 spp.), but the sampling of additional plots in the surroundings of the crane site yielded an-

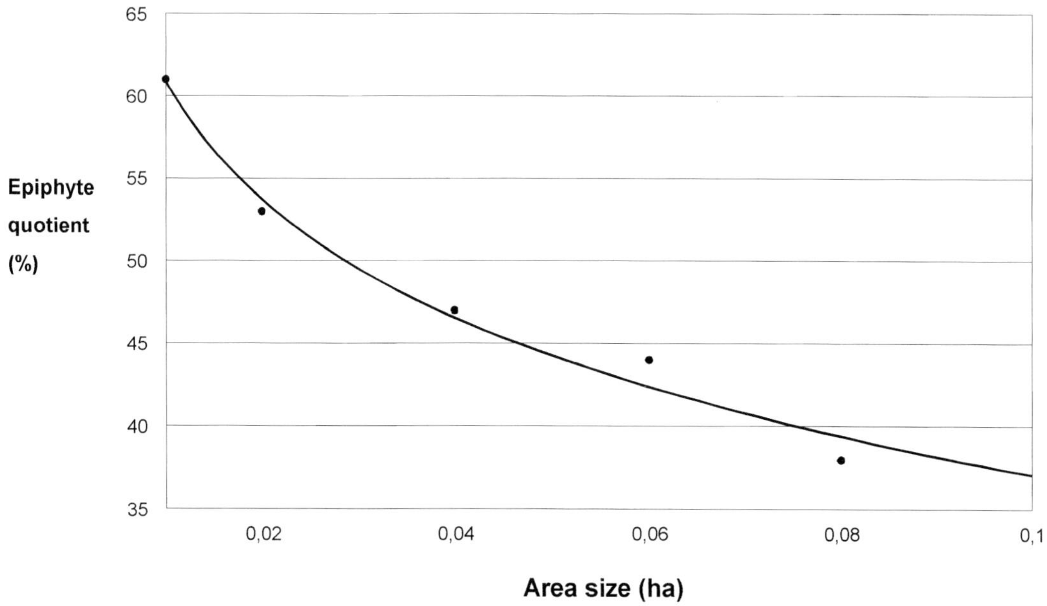

Figure 3. Decrease of the epiphyte quotient with sample plot size (0.01–0.1 ha) at Sehuencas, Bolivia (after Ibisch 1996).

other 59 species. We conclude that species area curves of epiphytes are strikingly different for mountain rain forests and Amazonian lowland forests. Whereas the first tend to rise in 'steps', the increase in lowland epiphyte species numbers is gradual.

The horizontal distribution of epiphyte species richness has consequences for sampling protocols: where epiphytes are abundant, even small study plots can be sufficient to cover a considerable proportion of the epiphyte vegetation of a study area. On Barro Colorado Island, sampling only five trees of *Annona glabra* yielded 10% of all epiphyte individuals and 82% of all species present on 1000 trees (Nieder & Zotz 1998).

Hemi-epiphytes

In north-eastern Gabon, when comparing species richness (Figure 4) between the three studied habitats at the level of the minimum common area (i.e., the river bank rain forest at 4 ha), we see a different situation to that obtained from using absolute species richness. A scale increase from 4 to 100 ha raises the fig species richness from 4 to 20 species on the forest plateau. Note that the different trend seen between river bank and plateau is due to the fact that densities per ha are always below one individual for each of these fig species on the plateau and island, and often below one individual for 10 ha, whereas the species density is higher by the riverside, thus giving a more aggregated pattern at smaller scale. In such conditions, it is obvious that one needs a large census scale to include most of the available species. In order to register 70% of the species of hemi-epiphytic figs in a given area of lowland rain forest, a 100-ha plot is needed. As seen above, a comparable amount of the species richness of epiphytes of this area can be covered by sampling a small number of host trees.

Vertical and horizontal ecological gradients affect epiphyte diversity

Holo-epiphytes

Holo-epiphyte occurrence in the forest shows vertical distribution patterns. The key environmental factors that determine their distribution are microclimatic gradients (Freiberg 1997) and substrate differences in the form of different branch diameters. Canopy soil thickness varies on phorophyte stems and branches, and the phorophytes can be divided into 'zones', following Johansson (1974) (see also Nieder & Zotz 1998).

Vascular epiphytes are usually most abundant in the 'Johansson zone 4', where relatively large branches offer suitable attachment sites (Ter Steege & Cornelissen 1989). Branch inclination is always a determining factor in epiphyte abundance and clearly inversely correlated to the number of epiphyte individuals (Rudolph et al. 1998).

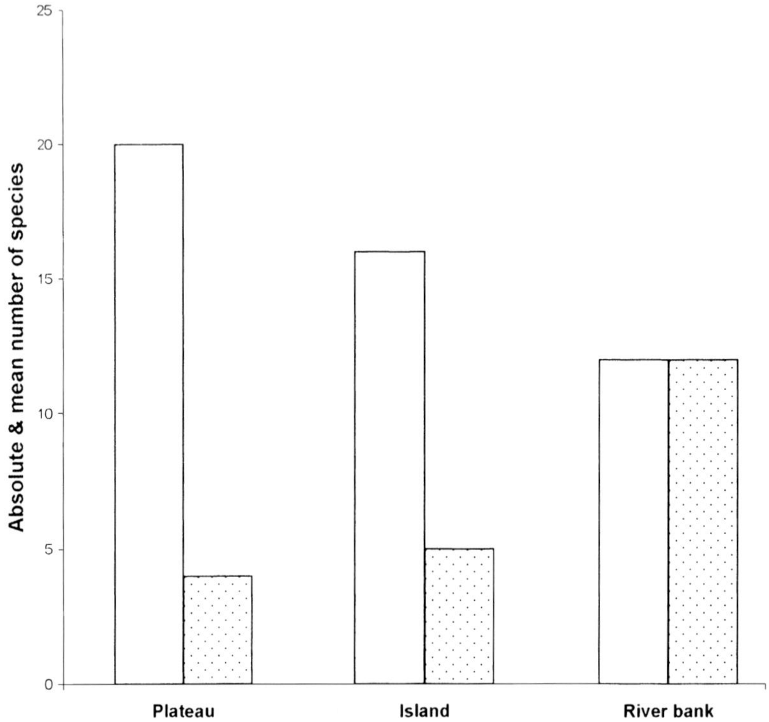

Figure 4. Comparison between the absolute number of species (open bars) of hemi-epiphytic figs encountered on the plateau (100 ha), island (19 ha) and river bank (4 ha), and the mean number of hemi-epiphytic fig species (shaded bars) per 4 ha unit (the minimum common scale) in each of these habitats.

Epiphyte species often occur in determined strata of the canopy (corresponding, but not identical to, dominant structural features of the phorophytes such as branch diameter). There is usually a relatively species poor stratum near the canopy surface (roughly equivalent to Johansson zone 5), a species rich middle stratum (Johansson zones 2–4) (e.g., Papulin et al. 1995; Ibisch 1996), and a fern-rich zone in the lower canopy (Johansson zone 1) (Kelly 1985). At Río Surumoni, typical height occurrences of different epiphyte species were identified on the bases of significantly different mean heights of particular species above ground level (Nieder et al. 2000).

Hemi-epiphytes

In contrast, in north-eastern Gabon, the patterns of distribution of hemi-epiphytic figs were mostly arranged horizontally (as exemplified by Figures 5 and 6), although some species colonize host trees at different vertical levels (Figure 7).

This pattern can be also explained by dispersal factors. The mature fruit size of *Ficus thoningii* (Figure 5) is about 1 cm in diameter, and that of *F. densistipulata* (Figure 6) about 3 cm in diameter. They are not eaten by the same mammals, whose different home range and feeding behaviours could partly explain the different distribution patterns of these hemi-epiphytes.

Although it is obviously more difficult for hemi-epiphytic figs to settle abundantly in the forest canopy than it is for the much smaller non woody vascular epiphytes, the scarcity of woody hemi-epiphytic figs is not only dependent on the relative abundance of host trees presenting the typcial host tree structure mentioned above (Figures 1 and 2), but also due to seed dispersal mode. Indeed, if *F. l. natalensis* is so much more abundant than the other species (Table 2), this is mainly due to the hydrochorous dispersal from which about 90% of these individuals originate and which is favoured by the increase of the river water level which can be up to 5 m during some rainy seasons (Michaloud 1988). Also, as a consequence, its density was very much lower along the narrow Ivindo river streams located between some of the islands and the forest plateau where floating items (including figs) were much less transported than along the main river bed (Michaloud 1988).

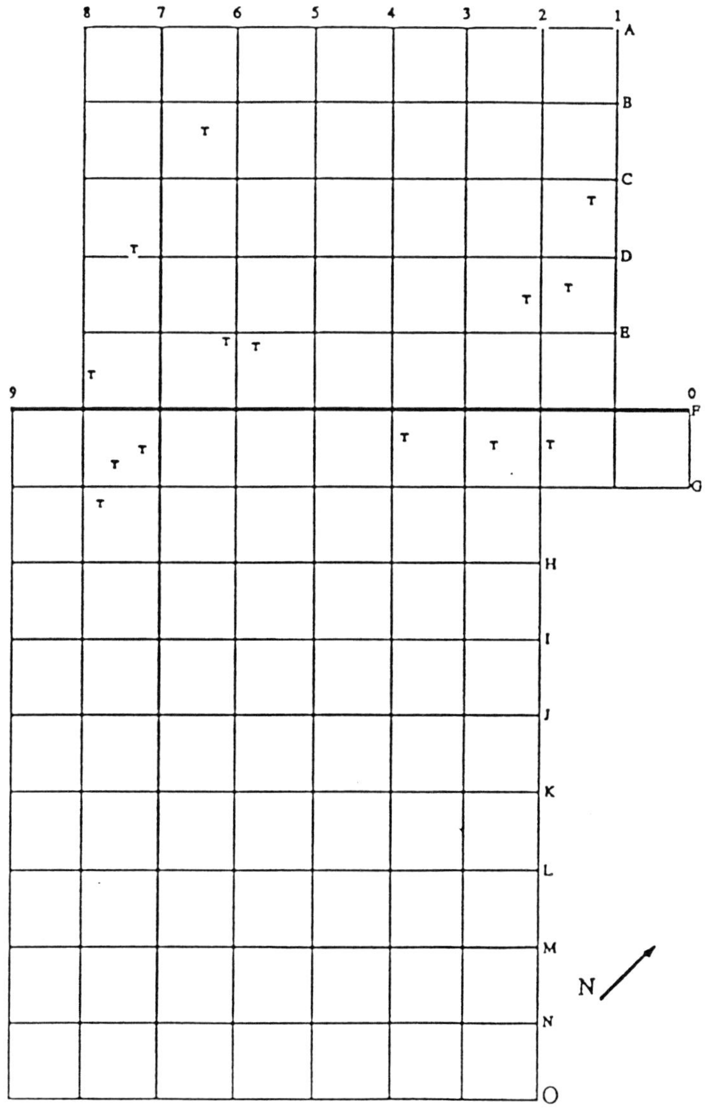

Figure 5. Distribution of individuals of *Ficus thonningii* occurring in our squares of study (100 × 100 m) on the rain forest plateau.

The dispersal mode argument can be illustrated by the following fact. Although figs do not constitute keystone resources for primates in that area (Gautier-Hion & Michaloud 1989), they do for the largest African bat *Hypsignathus monstrosus* (Bradbury 1977, 1981). However, this species was often observed flying by to drink the river water along the fruiting trees of *F. n. leprieuri*, but it was never observed eating figs of this species which mature fruit size does not exceed 1 cm, nor were these fig species seeds found in the *H. monstrosus* excrements. Indeed, germination of defecated fig seeds and cultivation of the plantules proved that this species of bat was feeding on figs whose fruit diameter was about a minimum diameter of 3 cm. However, many other smaller frugivores could be dispersers, and some were indeed; but although crops were small (around 500 on average) it was very common to find the fruit dry on the trees, unconsumed, and zoochory seemed a problem (if we exclude dispersal by fish which we did not control).

From an evolutionary point of view, it is obvious that conditions of settlement of hemi-epiphytic figs in the canopy, often leading to very sparse distribution, is rendered possible because of the reproductive species specific mutualism between figs and associated fig wasps (Berg & Wiebes 1992), that allows pollen to

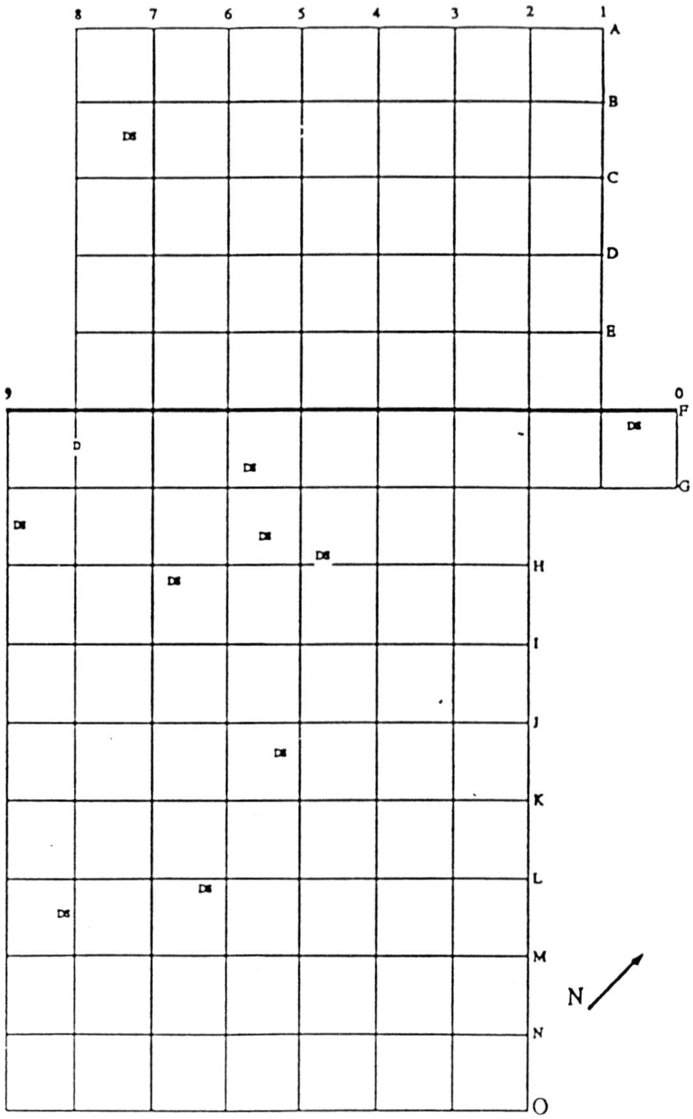

Figure 6. Distribution of individuals of *Ficus densistipulata* occurring in our squares of study (100 × 100 m) on the rain forest plateau.

be transferred between isolated individuals. The distribution of hemi-epiphytic figs through a variety of habitats is also a result of the home range of some seed dispersers (such as *Hypsignathus monstrosus* flying 10 to 15 km per night (Bradbury 1981), and may lead to habitat shift of figs and associated fig wasps and set ecological barriers, providing the necessary conditions for speciation and co-speciation (Michaloud et al. 1996).

Epiphytes as indicators of forest dynamics

Epiphyte life-histories are very different between species: for example twig-orchids have a short life-span, whereas many larger orchids grow slowly and reach old individual ages, e.g., 50 years (Zotz 1995). Because epiphytes tend to accumulate on a host tree with tree age, older (and larger) trees bear a more abundant epiphytic vegetation than younger phorophytes (Kikuchi 1992; Hietz-Seifert et al. 1996). The development of the epiphytic vegetation on a phorophyte represents a succession with different stages (Dudgeon 1923; Tewari et al. 1985; Catling & Lev-

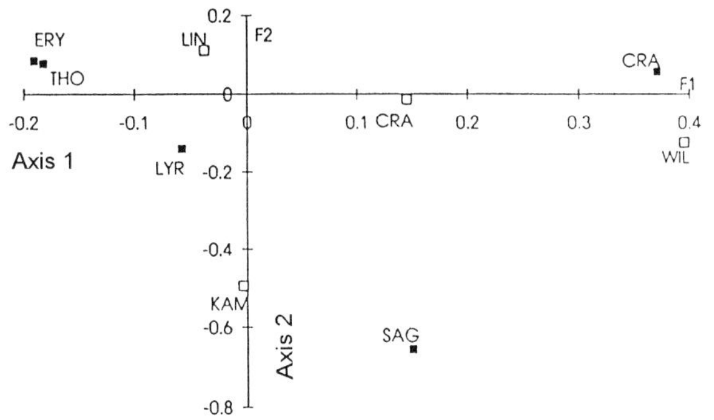

Figure 7. Multivariate analysis of the vertical distribution of *Ficus craterostoma* (CRA), *F. wildemaniana* (WIL), *F. lingua* (LIN), *F. kamerunensis* (KAM), *F. sagittifolia* (SAG), *F. lyrata* (LYR), *F. thonningii* (THO) and *F. eriobotryoides* (ERY) synonyma of *F. saussureana* (Berg & Wiebes 1992). Axis 1 (horizontal) explains 54% of the interactions, and Axis 2, 23%. The parameters which are represented by Axis 1 are height of the first host branch, height of the host, height of the *Ficus* foliage and height of germination. Positive values on Axis 1 correspond to a fig position above host canopy, and the negative ones below it. The Axis 2 represents the values of the host tree diameter and the height of the surrounding forest. The dark squares correspond to individuals in the Taï forest of Ivory Coast, and the white squares to individuals in Gabon, both observed by Michaloud and Michaloud-Pelletier and analysed in collaboration with David (in David 1993).

kovitch 1989). In Bolivia, the number of epiphyte species on *Alnus acuminata* increased with branch diameter (Ibisch 1996). This shows that epiphyte communities develop slowly, in this case reaching a saturation only after some 25 years of branch age. In general, epiphyte diversity and abundance are therefore positively correlated to the successional stage of a forest. However, this does not hold true for the special case of ant gardens. On the crane plot of Río Surumoni, 51% of all epiphyte individuals found in 1997 were obligate ant garden epiphytes. All these epiphytes (of only 4 species, whereas the other 49% of individuals represented 49 species!) were especially frequent in the drier and lighter gap sections of the plot. These observations confirm that ant-gardens are typically common in disturbed forest patches, such as frequently flooded habitats or treefall zones (Davidson 1988), or the early stages of gap vegetation development. They greatly contribute to epiphyte abundance, but not necessarily to epiphyte diversity.

Hemi-epiphytic figs may colonise trees only after a succession of non vascular and vascular organisms (Van Oye 1921; Dudgeon 1923). These ideas are supported by our observations correlating significantly the presence and absence of hemi-epiphytic figs with the presence and absence of ferns and (or) orchids (Michaloud & Michaloud-Pelletier 1987), and Sanford's suggestion that nutrients freed by orchids, and other epiphytes could be used by others. Richards (1952) mentioned that some epiphytes could prepare the way for others. This suggests that trees colonized by hemi-epiphytic figs are old enough to exhibit these successional stages. In fact, Michaloud and Michaloud-Pelletier (1987) showed that in the Taï forest of Ivory Coast 81% of the 98 hemi-epiphytic figs studied had germinated on damaged and rotten parts of the host trees, and that all of them were hosted by partly decayed trees (trees of the past; *sensu* Oldeman 1974). Therefore we suggest that in rain forests, hemi-epiphytic figs require high trees with large diameters and wide opened branchings (Figures 1 and 2; height and diameter often being correlated), so they can access light through the canopy, and probably also require a good feeding substrate until their roots can reach the ground. If a forest meets all these criteria, it would suggest that the ecosystem is in its late turnover stages, and may almost be in a climax condition.

Conclusions

There are characteristic differences between holo- and hemi-epiphytes in species diversity on a global, regional (e.g., Ecuador) and local (e.g., La Selva, Costa Rica) scale. The relative proportion of hemi-epiphyte species in the total number of epiphyte species is in all three cases 8%. Both habits are especially diverse in the Neotropics, where a varied topography, especially the altitudinally complex Andes, and the presence of refuge areas in the Pleistocene allowed for the evolution and conservation of a rich epiphytic flora.

The differences in distribution between holo- and hemi-epiphytism (mainly vertical vs. mainly horizontal patterns) have their causes in very different plant sizes and ecological requirements. Whereas holo-epiphytes have to be adapted to the restrictions of the epiphytic habitat (limited nutrient supply, water stress) through their complete life-cycle, hemi-epiphytes overcome these physiological limitations once they root in the ground. But because settlement is crucial for the population size of hemi-epiphytes, they depend on the presence of suitable hosts (sufficiently large and offering substantial amounts of nutrient-rich substrate) in their habitat. Because of larger plant sizes, their demands in respect to canopy structure and canopy soil quantity are greater than those of holo-epiphytes. Together with limitations due to obligate fruit dispersal by vertebrates in the case of some species, all these factors account for the relative scarcity of hemi-epiphytes in comparison to holo-epiphytes.

Acknowledgements

We would like to thank the generous funding of our projects by Deutsche Forschungsgemeinschaft (Ba 605/6-1 and 6-2) and Volkswagen Stiftung (I/74 390 and I/70 734). Without the untiring support of our partners at the Pontificia Universidad Católica in Quito (Ecuador) the Ecuadorian studies would not have been possible. The Surumoni and La Carbonera projects were only possible with the help of the Österreichische Akademie der Wissenschaften (in particular that of the project coordinators W. Morawetz and H. Winkler), the Centro Jardín Botánico of the Universidad de los Andes in Mérida (Venezuela), the Ministerio del Ambiente y de los Recursos Naturales Renovables (MARNR) and SADA Amazonas in Venezuela. The Venezuelan ambassador in Germany, His Excellency E. Becker-Becker and his embassy staff were extremely helpful in all diplomatic and consular matters.

The Spanish airline Iberia helped by carrying abundant extra luggage without charging for it.

We are also most thankful to ESF and the organising committee of the final 'Tropical Canopy Research' Conference in Oxford which allowed us to develop the present topic.

The studies on the hemi-epiphytic part were supported by ORSTOM (French Guyana), and by the Université de Montpellier II and CNRS UMR 5554 (ISEM 2000-086).

References

Barthlott, W., Lauer, W. & Placke, A. 1996. Global distribution of species diversity in vascular plants: towards a world map of phytodiversity. Erdkunde 50: 317–327.

Barthlott, W., Biedinger, N., Braun, G., Feig, F., Kier, G. & Mutke, J. 1999. Terminological and methodological aspects of the mapping and analysis of global biodiversity. Acta Bot. Fenn. 162: 103–110.

Berg, C. C. & Wiebes, J. T. 1992. African fig trees and fig wasps. Kon. Nederl. Akad. Wet., Verhand. Afd. Natuurk., Tweede Reeks.

Biedinger, N. & Fischer, E. 1996. Epiphytic vegetation and ecology in central African forests (Rwanda, Zaire). Ecotropica 2: 121–142.

Boegner, A. 1999. Diversität und Phytogeographie der epiphytischen Angiospermen ausgewählter paläo- und neotropischer Gebiete. Unpublished diploma thesis. Botanisches Institut, Universität Bonn, Bonn.

Bradbury, J. W. 1977. Lek mating behavior in the hammer-headed bat. Z. f. Tierpsychol. 45: 225–255.

Bradbury, J. W. 1981. The evolution of lek. Pp. 138–169. In: Alexander, R. D. & Tinckle, D. W. (eds), Natural selection and social behavior. Chiron, New York.

Catling, P. M. & Levkovitch, L. P. 1989. Associations of vascular epiphytes in a Guatemalan cloud forest. Biotropica 21: 35–40.

Coxson, D. S. & Nadkarni, N. M. 1995. Ecological roles of epiphytes in nutrient cycles of forest ecosystems. Pp. 495–543. In: Lowman, M. D. & Nadkarni, N. M. (eds), Forest canopies. Academic Press, San Diego.

David, P. 1993. Etude comparative des préférences écologiques de quelques espèces de Ficus hémi-épiphytes des forêts tropicales. DEA Evolution et Ecologie, Université de Montpellier II, Pp. I–VI.

Davidson, D. W. 1988. Ecological studies of neotropical ant gardens. Ecology 69: 1138–1152.

Dudgeon, W. 1923. Succession of epiphytes in the *Quercus incana* forest at Landour, western Himalayas, preliminary note. J. Ind. Bot. Soc. 3: 270–272.

Dunsterville, G. C. K. 1961. How many orchids on a tree? Am. Orchid Soc. Bull. May 1961: 362–363.

Engwald, S. 1999. Ökologie und Diversität der Epiphyten eines Tiefland- und eines Bergregenwalds in Venezuela. Libri-Books on Demand. Hamburg.

Foster, R. B. & Hubbel, S. P. 1990. The floristic composition of the Barro Colorado Island forest. Pp. 99–11. In: Gentry, A. H. (ed.), Four neotropical rain forests. Yale University Press, New Haven and London.

Freiberg, M. 1997. Spatial and temporal pattern of temperature and humidity of a tropical premontane rain forest tree in Costa Rica. Selbyana 18: 77–84.

Gautier-Hion, A. & Michaloud, G. 1989. Are figs always keystone resources for tropical frugivorous vertebrates? A test in Gabon. Ecology 70: 1826–1833.

Gentry, A. H. & Dodson, C. H. 1987. Diversity and biogeography of neotropical vascular epiphytes. Ann. Miss. Bot. Gar. 74: 205–233.

Groombridge, B. (ed.) 1992. Global biodiversity: status of the earth's living resources. Chapman & Hall, London.

Guy, P. R. 1977. Notes on the host species of epiphytic figs (*Ficus* spp.) on the flood plain of the Mana pools game reserve, Rhodesia. Kirkia 10: 559–562.

Hall, J. B. 1976. Classification and ecology of closed-canopy forest in Ghana. J. Ecology 64: 913–951.

Hall, J. B. 1978. Check-list of the vascular plants of Bia National Park and Bia Game Production Reserve. In: Martin, C. (ed.), Management plan for the Bia Wildlife Conservation Areas, Part I. Wildlife and National Parks Division, Ghana Forestry Commission. Final Report IUCN/WWF Project 1251.

Hartshorn, G. S. & Hammel, B. E. 1994. Vegetation types and floristic patterns. Pp. 73–89. In: McDade, L. A., Bawa, K. S., Hespenheide, H. A. & Hartshorn, G. S. (eds), La Selva. The University of Chicago Press, Chicago.

Heywood, V. H. (ed.) 1995. Global biodiversity assessment. Cambridge University Press, Cambridge.

Hosokawa, T. 1950. Epiphyte-quotient. Bot. Mag. Tokyo 63: 18–19.

Hietz-Seifert, U., Hietz, P. & Guevara, S. 1996. Epiphyte vegetation and diversity on remnant trees after forest clearance in southern Veracruz. Biol. Cons. 75: 103–111.

Ibisch, P. 1996. Neotropische Epiphytendiversität - das Beispiel Bolivien. Martina Galunder-Verlag, Wiehl.

Ibisch, P., Boegner, A., Nieder, J. & Barthlott, W. 1996. How diverse are neotropical epiphytes? An analysis based on the 'Catalogue of the flowering plants and gymnosperms of Peru'. Ecotropica 2: 13–28.

Johansson, D. 1974. Ecology of vascular epiphytes in West African rain forest. Acta Phytogeographica Suecica 59: 1–136.

Johansson, D. R. 1989. Vascular epiphytism in Africa. Pp. 183–194. In: Lieth, H. & Werger, M. J. A. (eds), Tropical rain forest ecosystems. Elsevier, Amsterdam.

Kelly, D. L. 1985. Epiphytes and climbers of a Jamaican rain forest: vertical distribution, life forms and life histories. J. Biogeog. 12: 223–241.

Kelly, D. L., Tanner, E. V. J., Nic Lughada, E. & Kapos, V. 1994. Floristics and biogeography of a rain forest in the Venezuelan Andes. J. Biogeog. 21: 421–440.

Kikuchi, T. 1992. Communities of epiphytic vascular plants on a Himalayan mountainside in far eastern Nepal. Ecol. Rev. 22: 121–128.

Kress, W. J. 1986. The systematic distribution of vascular epiphytes: an update. Selbyana 9: 2–22.

Longman, H. A. 1917. The flora of a single tree. Proc. Royal Soc. Queensland 29: 64–69.

Lüttge, U. (ed.) 1989. Vascular plants as epiphytes. Springer-Verlag, Berlin.

Lüttge, U., Ball, E., Kluge, M. & Ong, B. L. 1986. Photosynthetic light requirements of various tropical vascular epiphytes. Physiol. Vég. 24: 315–331.

Martin, C. 1989. Die Regenwälder Westafrikas: Ökologie - Bedrohung - Schutz. Birkhäuser, Basel.

Michaloud, G. 1988. Aspects de la reproduction des figuiers monoïques en forêt équatoriale africaine. Unpublished PhD thesis, University of Montpellier II, Montpellier.

Michaloud, G. & Michaloud-Pelletier, S. 1987. Ficus hémi-épiphytes (Moraceae) et arbres supports. Biotropica 19: 125–136.

Michaloud, G., Carrière, S. & Kobbi, M. 1996. Exceptions to the one : one relationship between African fig trees and their fig wasp pollinators: possible evolutionary scenarios. J. Biogeog. 23 : 513–520.

Møller Jørgensen, P. & León-Yánez, S. (eds). 1999. Catalogue of the vascular plants of Ecuador. Missouri Botanical Garden Press, St.Louis, Missouri.

Nieder, J. & Zotz, G. 1998. Methods of analyzing the structure and dynamics of vascular epiphyte communities. Ecotropica 4: 33–39.

Nieder, J., Engwald, S., Klawun, M. & Barthlott, W. 2000. Spatial distribution of vascular epiphytes in a lowland Amazonian rain forest in southern Venezuela. Biotropica 32: 385–396.

Nowicki, C. 1998. Diversität epiphytischer und terrestrischer Pflanzen eines ecuadorianischen Bergnebelwaldes im Vergleich. Unpublished diploma thesis. Botanisches Institut, Universität Bonn. Bonn.

Papulin, F., Bianchi, E., Germani, M., Pedruzzi, D. & Wagner, A. 1995. Orchid diversity and distribution on a tree at Reserva Forestal de San Ramón, Costa Rica. Brenesia 43–44: 47–54.

Prosperi, J. 1998. Biologie du développement des hémi-épiphytes ligneux. Unpublished PhD thesis, University of Montpellier II, Montpellier.

Putz, F. E. 1983. Liana biomass and leaf area of a "Tierra Firme" forest in the Rio Negro basin, Venezuela. Biotropica 15: 185–189.

Putz, F. E. & Holbrook, N. M. 1986. Notes on the natural history of hemiepiphytes. Selbyana 9: 61–69.

Putz, F. E. & Holbrook, N. M. 1989. Strangler fig rooting habits and nutrient relations in the Llanos of Venezuela. Am. J. Botany 76: 781–788.

Rauer, G. 1995. Epiphytische Orchidaceae eines westandinen Bergregenwaldes in Ecuador. Unpublished diploma thesis. Botanisches Institut, Universität Bonn, Bonn.

Richards, P. W. 1952. The tropical rain forest. Cambridge University Press, Cambridge.

Rudolph, D. 1995. Vaskuläre Epiphyten eines westandinen Bergregenwaldes in Ecuador. Unpublished diploma thesis. Botanisches Institut, Universität Bonn, Bonn.

Rudolph, D., Rauer, G., Nieder, J. & Barthlott, W. 1998. Distributional patterns of epiphytes in the canopy and phorophyte characteristics in a western Andean rain forest in Ecuador. Selbyana 19: 27–33.

Sanford, W. W. 1969. The distribution of epiphytic orchids in Nigeria in relation to each other and to geographic location and climate, type of vegetation and tree species. Biol. J. Linnean Soc. 1: 247–285.

Schnell, R. 1952. Végétation et flore de la région montagneuse des Nimba. Mem. I.F.A.N. 22: 1–356.

Ter Steege, H. & Cornelissen, J. H. C. 1989. Distribution and ecology of vascular epiphytes in lowland rain forest of Guyana. Biotropica 21: 331–339.

Tewari, M., Upreti, N., Pandey, P. & Sing, S. P. 1985. Epiphytic succession on tree trunks in a mixed oak-cedar forest (Kumaun, Himalaya). Vegetatio 63: 105–112.

Todzia, C. 1986. Growth habits, host tree species, and density of hemiepiphytes on Barro Colorado Island, Panama. Biotropica 18: 22–27.

Van Oye, P. 1921. Influence des facteurs climatiques sur la répartition des épiphytes à la surface des troncs d'arbres à Java. Rev. Gén. Bot., 33: 161–176.

Watson, J. B., Kress, W. J. & Roesel, C. S. 1988. A bibliography of biological literature on vascular epiphytes. Selbyana 10: 1–23.

Williams-Linera & Lawton, R. O. 1995. The ecology of hemiepiphytes in forest canopies. Pp. 255–283. In: Lowman, M. D. & Nadkarni, N. M. (eds), Forest canopies. Academic Press, San Diego.

Zotz, G. 1995. How fast does an epiphyte grow? Selbyana 16: 150–154.

Plant Ecology **153:** 65–72, 2001.

Plant size: an ignored parameter in epiphyte ecophysiology?

Gerold Schmidt[1], Sabine Stuntz[1] & Gerhard Zotz[1,2]
[1]*Lehrstuhl für Botanik II der Universität Würzburg, D-97082 Würzburg, Germany*
(E-mail: gerold.schmidt@gurx.de); [2]*Smithsonian Tropical Research Institute, Apdo 2072 Balboa, Panama*

Key words: Allometry, Inter-/Intraspecific variability, Nitrogen, Photosynthetic capacity

Abstract

A central objective of many ecophysiological investigations is the establishment of mechanistic explanations for plant distributions in time and space. The important, albeit mostly ignored, question arises as to the nature of the organisms that should be used as representative in pertinent experiments. We suggest that it is essential to use a 'demographic approach' in physiological ecology, because physiological parameters such as photosynthetic capacity (PC, determined under non-limiting conditions with the oxygen electrode) may change considerably with plant size. A survey of ten vascular epiphyte species from Central Panama revealed an up to fivefold continuous increase in PC from small to large conspecifics. Moreover, the *intraspecific* variation in PC was almost always higher than the *interspecific* variation when comparing only large individuals. Theoretical considerations and additional evidence from other ecophysiological experiments corroborate the conclusion that size must be taken into account in the design of ecophysiological experiments with vascular epiphytes. In the past, most studies have ignored plant size, which may now lead to complications in the hitherto unambiguous interpretation of their results.

Introduction

Most vascular epiphytes show clear habitat preferences, both between ecosystems or within a given forest (Benzing 1990). Numerous authors have tried to correlate physiological properties of epiphytic species with their habitat preferences (e.g., Benzing & Dahle 1971; Benzing & Renfrow 1971; Griffiths et al. 1984; Earnshaw et al. 1987; Zotz & Ziegler 1997; Hietz & Briones 1998). However, reviewing all available publications on the ecophysiology of vascular epiphytes of the past 80* years reveals that 75% of the 153 articles did not specify the size of the study organisms at all. Sixteen percent gave vague descriptions like 'adult' and 'mature' or noted the age of the plants they used. Only 13 studies, i.e., about eight percent, provided a clear description of the actual size of the study organisms, e.g., plant height, length of the longest leaf or plant dry mass. In other words, most authors unwittingly treated individuals of unspecified size as representative for a given species.

Recent studies by Zotz (1997) and Zotz & Ziegler (1999) with the epiphytic orchid, *Dimerandra emarginata*, question this tacit assumption by showing that important physiological parameters such as photosynthetic capacity (PC, as determined in an oxygen electrode under non-limiting conditions) or carbon isotope discrimination change with plant size. Both studies, however, were conducted on only one species. To allow a generalisation of their conclusions, we repeated one of these studies with a range of species: PC was determined on individuals of nine additional species. Within a given species various specimens were considered, covering the entire size range of the species. Although the studied organisms comprised species from a variety of habitats, e.g., species typically found in the outer canopy, at mid-level growing sites, or in the understory, we expected to find the same pattern of size-related changes in PC. This expectation arises from the following argument. Assuming a decrease of surface/volume-ratios with increasing plant size and similar transpiration rates irrespective of size, internal water reservoirs of smaller plants will be depleted faster than those of larger conspecifics, resulting in

*A detailed list of the reviewed literature can be obtained from the corresponding author.

a less favourable water status in smaller plants under otherwise identical conditions. The optimal partitioning theory (Bloom et al. 1985) predicts that plants respond to a limiting environmental resource in such a way as to make all resources equally limiting. If smaller plants experience water as a limiting factor earlier and more often than larger conspecifics, they should invest less of other important resources such as nitrogen (Chapin 1980) in comparatively short-lived photosynthetic tissues than larger conspecifics. Given the well documented correlation between leaf nitrogen content and PC (e.g., Field & Mooney 1986; Evans 1989), a size-related change in PC should be the consequence. Key parameters of these theoretical considerations are plant surface/volume ratio, leaf nitrogen content and PC. Determining these parameters for individuals that represent the entire size range of a species will either support or contradict the proposed theory.

Methods

Field site and study organism

The study was conducted on Barro Colorado Island (BCI, 9°10′ N, 79°51′ W), a 1500-ha reserve within the Barro Colorado Nature Monument, Republic of Panama. Mean annual rainfall at this tropical moist forest site (Holdridge et al. 1971) is approximately 2600 Kg m^{-2} with a pronounced dry season from late December to late April (Leigh & Wright 1990).

Epiphytes and hemiepiphytes comprise about 16% of the BCI flora, totalling 207 species (Croat 1978; Zotz & Ziegler 1997). The ten species used for this study (nine taxa plus *Dimerandra emarginata*, Zotz 1997) were chosen from all strata of the forest and represent all major epiphytic taxa, e.g., orchids, bromeliads, aroids and ferns. Information on photosynthetic pathway and preferred habitat type is given in Table 1. Species names of flowering plants and ferns follow the *Flora of Panama Checklist and Index* (D'Arcy 1987) and the *Flora of Barro Colorado Island* (Croat 1978), respectively.

Within each species we studied 9–18 individuals of different size, trying to include the entire size range of the taxon from the smallest to the largest specimen encountered in the natural habitat. We used maximal leaf length (non-orchids) or shoot length (orchids) as a measure for size. In the four species we tested, these parameters were highly correlated (i.e., $r^2 > 0.95$) with a more rigorous measure for size, i.e., plant dry mass (data not shown). Adopting this non-destructive measure for size allowed us collecting a large number of conspecifics from the same host tree without destroying the population. Within a given species we avoided collecting samples of conspecifics growing under obviously differing microclimatic conditions. Furthermore, whenever it proved impossible to find all individuals, e.g., in the outer canopy on a single tree, we collected the specimens from exposed sites on several host trees (phorophytes).

Gas exchange and leaf nitrogen content

Oxygen exchange was measured in the laboratory using a Hansatech LD2 Leaf Disc Electrode (Hansatech Ltd., King's Lynn, UK). From each selected study plant, we collected the youngest fully expanded leaf and brought it to the laboratory in a moist zip-lock bag. Within three hours after collection, we started the measurements by enclosing the freshly punched, weighed leaf disc in the cuvette of the leaf disc electrode unit (kept at 30 °C using a circulating water bath). Subsequently, the cuvette was charged with ca. 5% CO_2. After calibration, the leaf discs were induced in six light-dark cycles of 5 min (ca. 300 μmol m^{-2} s^{-1}) – 5 min (0 μmol m^{-2} s^{-1}). To determine saturating but not inhibiting light conditions for each species we first studied the response of O_2 evolution to PFDs of up to 1800 μmol m^{-2} s^{-1} with specimens of intermediate size. Using a fixed light intensity for each species (range: 950–1500 μmol m^{-2} s^{-1}) the photosynthetic capacity of leaf discs was determined. Samples were subsequently dried at 60 °C for 4–7 days, weighed and nitrogen content determined at the University of Würzburg using a CHN-O Element Analyser (Heraeus, Hanau, Germany).

Shoot water content and plant leaf area

Shoot water content was calculated as the difference of fresh mass and dry mass of individuals, which were similar to the plants used for leaf gas exchange measurements. Roots, however, were discarded because of their strongly varying water contents, which depended on the preceding weather conditions. Transpiring plant leaf surface was determined using a LI-3100, Li-Cor leaf area meter (Lincoln, Nebraska, USA). Since none of the study organisms is amphistomatal, only one side of the leaf was considered.

Table 1. Characterisation of study species. Given are photosynthetic pathway (compare Griffiths & Smith 1983; Zotz & Ziegler 1997) and habitat preferences (*sensu* Zotz & Ziegler 1997): canopy sites in the outer crown of trees (c), mid-level growing sites (m), and the understory (u).

Species	Family	Pathway	Habitat
Caularthron bilamellatum (Reichb.f.) Schult.	Orchidaceae	CAM	c
Dimerandra emarginata (G. Meyer) Hoehne	Orchidaceae	C_3	c
Polypodium crassifolium L.	Polypodiaceae	C_3	c
Vriesea sanguinolenta Cogn. & Marchal	Bromeliaceae	C_3	c
Anthurium brownii Mast.	Araceae	C_3	m
Epidendrum nocturnum Jacq.	Orchidaceae	C_3	m
Guzmania monostachya (L.) Rusby ex Mez	Bromeliaceae	C_3-CAM	m
Polystachya foliosa (Lindl.) Reichb.f.	Orchidaceae	C_3	m
Anthurium salviniae Hemsl.	Araceae	C_3	u
Aspasia principissa Reichb.f.	Orchidaceae	C_3	u

Table 2. Ratio of shoot water content and plant leaf area (SWC/PLA, in mg H_2O m^{-2}) as a function of plant size (x = maximal leaf length [non-orchids] or maximal shoot length [orchids], in cm) for different species. Regression equations (y = SWC/PLA), as determined by Pearson product-moment correlation, coefficients of determination (r^2), sample size (n), and significance levels (p) are given.

Species	Linear regression	r^2	n	p
Dimerandra emarginata	$y = 60.7 + 2.97x$	0.70	34	< 0.001
Epidendrum nocturnum	$y = 53.2 + 2.67x$	0.55	14	< 0.01
Guzmania monostachya	$y = 22.3 + 1.36x$	0.84	15	< 0.001
Polypodium crassifolium	$y = 42.1 + 0.33x$	0.38	14	< 0.05
Polystachya foliosa	$y = 30.3 + 3.02x$	0.56	11	< 0.01
Vriesea sanguinolenta	$y = 60.3 + 0.93x$	0.65	13	< 0.001

Results

A basic assumption of our argument is a size-related change in the ratio of shoot water content (SWC) to transpiring leaf surface of the plant (PLA). Table 2 shows that in all six species for which data were available, this ratio scaled with plant size. The linear increase of SWC/PLA from smallest to largest individual was as low as ca. 60% in *P. crassifolium* to almost 300% in *D. emarginata*. Thus, changes in plant size correspond to substantial changes in surface/volume-ratios.

Leaf nitrogen content (leaf N) changed with plant size in nine of ten taxa (Table 3). Likewise the well-established positive relationship between leaf N and photosynthetic capacity (PC; e.g., Field & Mooney 1986) also applied to these taxa (data not shown). Not surprisingly, we also found size-related changes in the area-based PC in nine species (Figure 1). The increase in *E. nocturnum* was non-significant though only marginally ($p = 0.055$). Changes in PC from smallest to largest individuals ranged from ca. 110% (*P. crassifolium*) up to almost 500% (*G. monostachya*). Our finding of a size-related increase of PC is not merely a consequence of changes in leaf anatomy, e.g., an increase in leaf thickness. Dry mass based PC showed clear increases in seven of the investigated species (data not shown), although in two cases (*G. monostachya, A. salviniae*) trends were marginally non-significant, ($p = 0.056$ and $p = 0.058$, respectively).

The magnitude of the interspecific and intraspecific variation in PC is illustrated in Figure 2. Comparing the variability in PC of individuals within a species (expressed as the deviation of the extreme values from the mean in percent), with the variability between the

Table 3. Leaf nitrogen content (N, in g N m^{-2}) as a function of plant size (x = maximal leaf length [non-orchids] or maximal shoot length [orchids], in cm) for different species. Regression equations ($y = N$), as determined by Pearson product-moment correlation, coefficients of determination (r^2), sample size (n), and significance levels (p) are given.

Species	Linear regression	r^2	n	p
Anthurium brownii	$y = 0.69 + 0.012x$	0.39	12	< 0.05
Anthurium salviniae	$y = 0.61 + 0.009x$	0.43	9	0.056
Aspasia principissa	$y = 0.43 + 0.015x$	0.58	12	< 0.01
Caularthron bilamellatum	$y = 0.94 + 0.052x$	0.52	17	< 0.01
Dimerandra emarginata	$y = 0.59 + 0.014x$	0.61	15	< 0.001
Epidendrum nocturnum			14	n.s.
Guzmania monostachya	$y = 0.13 + 0.008x$	0.57	18	< 0.001
Polypodium crassifolium	$y = 0.41 + 0.005x$	0.49	15	< 0.01
Polystachya foliosa	$y = 0.35 + 0.028x$	0.44	12	< 0.05
Vriesea sanguinolenta	$y = 0.28 + 0.003x$	0.66	14	< 0.001

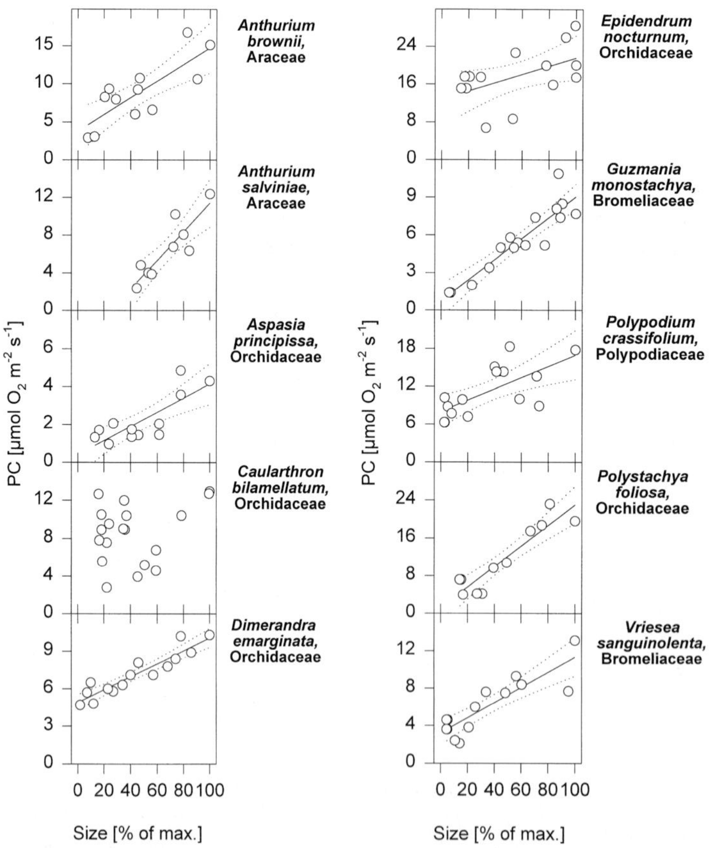

Figure 1. Relationship of photosynthetic capacity (PC, in μmol O_2 m^{-2} s^{-1}) and plant size (in % of maximal of the species) in ten epiphyte species. Each circle symbolises an individual plant. Coefficients of determination range from 0.47 to 0.86 (Pearson product-moment correlation), regression lines (solid) and 95% confidence intervals (dotted lines) are given for species with $p \leq 0.05$. Data for *D. emarginata* are from Zotz (1997).

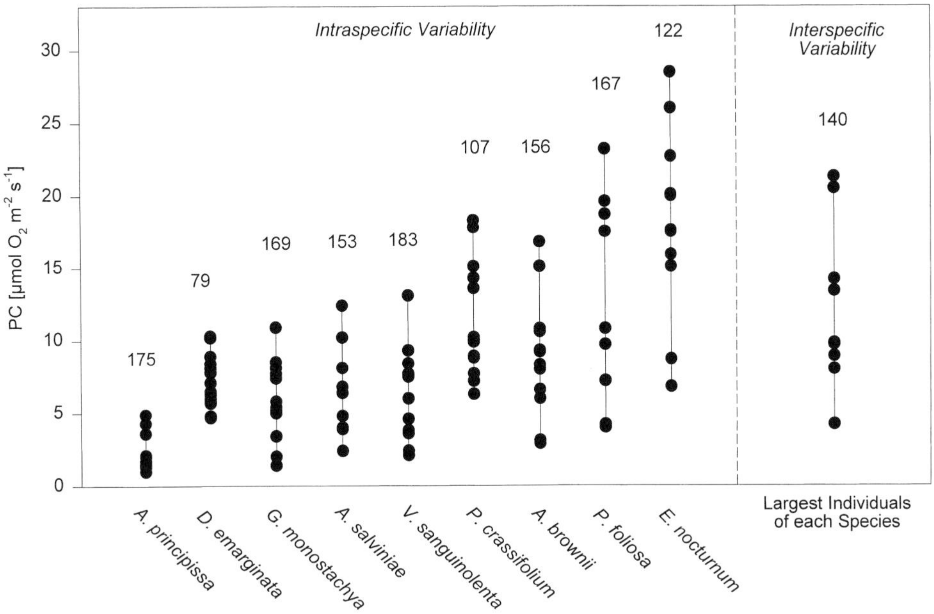

Figure 2. Comparison of inter- and intraspecific variability of photosynthetic capacity (PC). Numbers above bars are relative deviations (in %) of the extremes from the mean. Data for *D. emarginata* are from Zotz (1997). Left panel: *intraspecific* variability of PC (in μmol O_2 m^{-2} s^{-1}). Only those species were considered in which PC scales with plant size (compare Figure 1). Given are ranges of PC (solid line) with each circle symbolising one plant. Right panel: *interspecific* variability of PC with each point representing one taxon. Species values were calculated as the mean PC of the largest individuals (80–100% of maximal size) of each species.

larger individuals of different species yields a surprising result: *intraspecific* variability equals or even exceeds *interspecific* variability in most cases.

Discussion

Our study strongly suggests that the finding of a size-related change in PC is not limited to an exceptional case, i.e., *Dimerandra emarginata* (Zotz 1997), but is observed in almost all taxa investigated to date (Figure 1). Physiological results, i.e., PC values, from larger individuals are clearly not representative for smaller conspecifics, and vice versa. Hence, the concept of a species-specific PC for the members of a given species growing under similar climatic and nutritional conditions does not necessarily apply to epiphytes. The tightness of the correlation between plant size and PC is surprising considering that samples were collected in the field: in almost all cases, plant size alone explained more variation in PC than all other sources such as possible small scale differences in microclimate or differences in genotype (i.e., $r^2 > 0.5$; Figure 1). Moreover, the intraspecific variability may even exceed interspecific differences (Figure 2). As a consequence, we suggest to use a 'demographic' approach in ecophysiological studies with vascular epiphytes. When seeking physiological explanations for plant distributions, it is not sufficient to exclusively study larger individuals or individuals of unspecified size, and interpret correlations of ecophysiological comportment and habitat preference as meaningful for "*the*" species. Although numerous studies in the past have yielded apparently reasonable relationships (e.g., Benzing & Dahle 1971; Medina 1974; Griffiths et al. 1984; Earnshaw et al. 1987; Zotz & Ziegler 1997; Hietz & Briones 1998), the fact that smaller individuals may function quite differently from larger conspecifics now calls for caution in the interpretation of these studies in an ecological context. The key to finding an ecophysiological basis of epiphyte distribution patterns is certainly the study of physiological properties of different individuals representing the entire size range of a species.

As yet, this statement is based mostly on oxygen electrode measurements (Zotz 1997; this study). To put measurements of O_2 exchange in an ecologically more meaningful framework, we should relate them to *in situ* CO_2 gas exchange. Zotz et al. (1997) – in a study on tropical hemiepiphytes – reported a close correlation of PC (as determined under non-

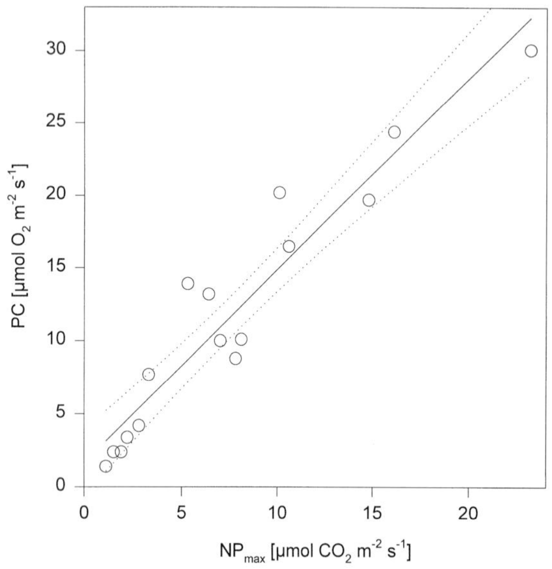

Figure 3. Relationship of maximal photosynthesis rates (NP_{max}, IRGA, in μmol CO_2 m^{-2} s^{-1}) and photosynthetic capacity (PC, oxygen electrode, in μmol O_2 m^{-2} s^{-1}). Each circle symbolises a taxon (or a size category of a taxon) for which both parameters were available. Given are regression line (solid) and 95% confidence intervals (dotted lines). The linear regression is: $PC = 1.32\ NP_{max} + 1.69$, $r^2 = 0.91$, $n = 16$, $p < 0.001$ (Pearson product-moment correlation). Data for individual species are listed in Appendix 1.

limiting conditions in the oxygen electrode) and the maximum rate of net photosynthesis under favourable field conditions (NP_{max}, determined with an infrared gas analyser, IRGA, at ambient CO_2): NP_{max} was consistently lower by about 20–40%. We extended this analysis to C_3 epiphytes by using unpublished data or by including values available in the literature (Appendix 1). Although parameters in the field while determining NP_{max} can not be assumed to be uniform and although the exact definitions of NP_{max} differed in the cited studies (Appendix 1), the correlation of PC and NP_{max} is surprisingly robust (Figure 3). This suggests that size-dependent changes in PC do correspond to similar changes in the maximal rates of *in situ* CO_2 exchange. Moreover, given the correlation of NP_{max} and diel carbon gain (Zotz & Winter 1993), our results can even be used as an indicator of potential 24-h leaf carbon budgets.

However, the maximum rate of CO_2 uptake is dependent on plant water status (Larcher 1994). Here, smaller plants should strongly differ from larger conspecifics (Table 2; Zotz & Andrade 1998). Plant size may not only affect tissue water content, but also the efficiency of external water storage as in water impounding tanks of many bromeliads (Benzing 1980). Using a model approach, Zotz & Thomas (1999) showed for two epiphytic bromeliads, *Guzmania monostachya* and *Tillandsia fasciculata*, that smaller individuals experienced empty tanks more often than larger conspecifics. Thus, a smaller plant is not only expected to show a lower PC (and therefore also a lower NP_{max}; Figure 3) but the frequency with which maximal assimilation rates can be performed should also be lower. The integrated long-term carbon gain under field conditions should therefore be even more reduced in smaller plants than indicated by the differences in PC. Strictly speaking, this argument is only valid for the leaf type measured, i.e., the youngest fully expanded leaf. Considering that this developmental stage of a leaf is usually the most productive (Tichá 1985) and that water availability should affect the investment strategies of the entire organism, we suggest that similar size-related changes could be seen in older leaves as well. Currently, we are collecting data to develop a whole-plant model for predicting long-term carbon gain for the C_3 tank bromeliad, *V. sanguinolenta*. This model will incorporate all the different aspects mentioned above, and also address the problem of size-related architectural differences, which might change the relative importance of self-shading.

We do not yet have a conclusive, experimental proof of the actual mechanism behind our results. Thus, other explanations cannot be ruled out. For example, given the nutrient-poor habitat, smaller and hence normally younger plants may simply have not had the time to take up sufficient amounts of, e.g., nitrogen. Until experiments prove or disprove our proposed scheme, we have to rely on circumstantial evidence and the consistency of our data with theoretical considerations, i.e., the optimal partitioning theory (Bloom et al. 1985). Irrespective of the actual mechanism, the basic conclusion retains its validity: size-related changes in ecophysiological parameters do exist and have to be adequately addressed in future studies.

Acknowledgements

This research was supported by a grant of the Deutsche Forschungsgemeinschaft ('Sonderforschungsbereich 251'). The study could not have been conducted without the logistical support of the Smithsonian Tropical Research Institute, Panama. S. Stuntz wants to thank Elisabeth Schaller for financial support.

Appendix 1. NP_{max} (IRGA, in μmol CO_2 m^{-2} s^{-1},) and PC (oxygen electrode, in μmol O_2 m^{-2} s^{-1}) for epiphytic and hemiepiphytic C_3-species. 'Dev' is the deviation of NP_{max} from PC (in %). If available, specifications on plant size are provided. Grouped in size categories (SC) maximal leaf length (LL_{max}) served as a measure of plant size.

Species	Family	Specification on plant size	NP_{max}	PC	Dev	References
Aspasia principissa	Orchidaceae	SC A (LL_{max}=5-10cm)	1.1[1]	1.4	21	Schmidt unpubl. data *
Aspasia principissa	Orchidaceae	SC B (LL_{max}=15-20cm)	1.5[1]	2.4	37	Schmidt unpubl. data *
Aspasia principissa	Orchidaceae	SC C (LL_{max}>28cm)	2.2[1]	3.4	35	Schmidt unpubl. data *
Catasetum viridiflavum	Orchidaceae	none; max. size	6.4[1]	13.2	52	Zotz & Winter 1994; Stuntz & Zotz 2001
Cosmibuena macrocarpa	Rubiaceae	none	16.1[1]	24.4	34	Zotz et al. 1997
Coussapoa magnifolia	Moraceae	none	23.2[1]	30.1	23	Zotz et al. 1997
Coussapoa panamensis	Moraceae	none	14.8[1]	19.7	25	Zotz et al. 1997
Dimerandra emarginata	Orchidaceae	max. size	7.0[1]	10.0	30	Zotz & Tyree 1996; Zotz 1997
Ficus citrifolia	Moraceae	none	10.6[1]	16.5	36	Zotz et al. 1997
Guzmania lingulata	Bromeliaceae	adult; max. size	1.9[2]	2.4	21	Benzing & Friedman 1981; Stuntz & Zotz 2001
Havetiopsis flexilis	Clusiaceae	none	8.1[1]	10.1	20	Zotz et al. 1997
Polypodium crassifolium	Polypodiaceae	none; max. size	7.8[1]	8.8	11	Zotz & Winter 1994; Stuntz & Zotz 2001
Souroubea sympetala	Marcgraviaceae	none	10.1[1]	20.2	50	Zotz et al. 1997
Vriesea sanguinolenta	Bromeliaceae	SC B (LL_{max}=5-10cm)	2.8[1]	4.2	33	Schmidt unpubl. data *
Vriesea sanguinolenta	Bromeliaceae	SC C (LL_{max}=20-25cm)	3.3	7.7	57	Schmidt unpubl. data *
Vriesea sanguinolenta	Bromeliaceae	SC D (LL_{max}>70cm)	5.3[1]	13.9	62	Schmidt unpubl. data *

[1]Maximal values from diel courses in the field.
[2]Values from light response curves at saturating light intensities.

References

Adams, W. W. & Martin, C. E. 1986. Physiological consequences of changes in life form of the Mexican epiphyte *Tillandsia deppeana* (Bromeliaceae). Oecologia 70: 298–304.

Benzing, D. H. 1980. The biology of bromeliads. Mad River Press, Eureka, California.

Benzing, D. H. 1990. Vascular epiphytes. General biology and related biota. Cambridge University Press, Cambridge.

Benzing, D. H. & Dahle, C. E. 1971. The vegetative morphology, habitat preference and water balance mechanisms of the Bromeliad *Tillandsia ionantha* Planch. Am. Midl. Nat. 85: 11–21.

Benzing, D. H. & Friedman, W. E. 1981. Patterns of foliar pigmentation in Bromeliaceae and their adaptive significance. Selbyana 5: 224–240.

Benzing, D. H. & Renfrow, A. 1971. The significance of photosynthetic efficiency to habitat preference and phylogeny among Tillandsioid Bromeliads. Bot. Gaz. 132: 19–30.

Bloom, A. J., Chapin, F. S. & Mooney, H. A. 1985. Resource limitation in plants - an economic analogy. Annu. Rev. Ecol. Syst. 16: 363–392.

Chapin, F. S. 1980. The mineral nutrition of wild plants. Annu. Rev. Ecol. Syst. 11: 233–260.

Croat, T. 1978. Flora of Barro Colorado Island. Stanford University Press, Stanford.

D'Arcy, W. G. 1987. Flora of Panama: Checklist and Index. Missouri Botanical Gardens, St. Louis.

Earnshaw, M. J., Winter, K., Ziegler, H., Stichler, W., Cruttwell, N. E. G., Kerenga, K., Cribb, P. J., Wood, J., Croft, J. R., Carver, K. A. & Gunn, T. C. 1987. Altitudinal changes in the incidence of crassulacean acid metabolism in vascular epiphytes and related life forms in Papua New Guinea. Oecologia 73: 566–572.

Evans, J. R. 1989. Photosynthesis and nitrogen relationships in leaves of C_3 plants. Oecologia 78: 9–19.

Field, C. & Mooney, H. A. 1986. The photosynthesis-nitrogen relationship in wild plants. Pp. 25–55. In: Givnish, T. J. (ed.), On the economy of plant form and function. Cambridge University Press, Cambridge.

Griffiths, H. & Smith, J. A. C. 1983. Photosynthetic pathways in the Bromeliaceae of Trinidad: relations between life-forms, habitat preference and the occurrence of CAM. Oecologia 60: 176–184.

Griffiths, H., Smith, J. A. C., Bassett, M. E. & Griffiths, N. M. 1984. Ecophysiology and habitat preference of Trinidad epiphytic bromeliads. Pp. 117–122. In: Chadwick, A. C. & Sutton, S. L. (eds), Tropical Rain-Forest: The Leeds Symposium. Leeds Philosophical and Literary Society, Leeds.

Hietz, P. & Briones, O. 1998. Correlation between water relations and within-canopy distribution of epiphytic ferns in a Mexican cloud forest. Oecologia 114: 305–316.

Holdridge, L. R., Grenke, W. C., Hatheway, W. H., Liang, T. & Tosi, J. A., Jr. 1971. Forest environments in tropical life zones: a pilot study. Pergamon Press, Oxford.

Larcher, W. 1994. Ökologie der Pflanzen. Ulmer, Stuttgart.

Leigh, E. G., Jr. & Wright, S. J. 1990. Barro Colorado Island and tropical biology. Pp. 28–47. In: Gentry, A. H. (ed.), Four Neotropical forests. Yale University Press, New Haven and New York.

Medina, E. 1974. Dark CO_2 fixation, habitat preference and evolution within the Bromeliaceae. Evolution 28: 677–686.

Stuntz, S. & Zotz, G. 2001. Photo-synthesis in vascular epiphytes – a survey of 27 specfies of diverse taxonomic origin. Flora, in press.

Tichá, I., Catsky, F., Hodánová, D., Pospísilová, F. & Sesták, Z. 1985. Gas exchange and dry matter accumulation during leaf development. Pp. 157–216. In: Z. Sesták (ed.), Photosynthesis during leaf development. Tasks for vegetation science. Junk Publisher, Dordrecht.

Zotz, G. 1997. Photosynthetic capacity increases with plant size. Bot. Acta 110: 306–308.

Zotz, G. & Andrade, J. L. 1998. Water relations of two co-occurring epiphytic bromeliads. J. Plant Physiol. 152: 545–554.

Zotz, G., Patiño, S. & Tyree, M.T. 1997. CO_2 gas exchange and the occurrence of CAM in tropical woody hemiepiphytes. Flora 192: 143–150.

Zotz, G. & Thomas, V. 1999. How much water is in the tank? Model calculations for two epiphytic bromeliads. Ann. Bot. 83: 183–192.

Zotz, G. & Tyree, M.T. 1996. Water stress in the epiphytic orchid, *Dimerandra emarginata* (G. Meyer) Hoehne. Oecologia 107: 151–159.

Zotz, G. & Winter, K. 1993. Short-term photosynthesis measurements predict leaf carbon balance in tropical rain-forest canopy plants. Planta 191: 409–412.

Zotz, G. & Winter, K. 1994. Annual carbon balance and nitrogen use efficiency in tropical C_3 and CAM epiphytes. New Phytol. 126: 481–492.

Zotz, G. & Ziegler, H. 1997. The occurrence of crassulacean acid metabolism among vascular epiphytes from Central Panama. New Phytol. 137: 223–229.

Zotz, G. & Ziegler, H. 1999. Size-related differences in carbon isotope discrimination in the epiphytic orchid, *Dimerandra emarginata*. Naturwissenschaften 86: 39–40.

Plant Ecology **153:** 73–85, 2001.

Pollination and phenology of flowers in the canopy of two contrasting rain forest types in Amazonia, Colombia

Arthur van Dulmen
Hugo de Vries Laboratory, University of Amsterdam, Kruislaan 318, 1098 SM Amsterdam, The Netherlands (E-mail: dulmen@bio.uva.nl)

Key words: Neotropics, Plant community, Pollination spectra, Sexual systems, Seasonally inundated forest, *Tierra firme* forest

Abstract

The main objective of this investigation was to study the pollination characteristics of two types of Amazonian rain forest at plant community level. Seasonally inundated forest was compared with upland (*tierra firme*) forest. The study focused on plant species in the canopy. The pollination spectra show that in both forests most canopy trees and lianas are pollinated by small bees, large bees, butterflies or by small, relatively unspecialized insects. In the upland forest small bees are the most important pollinators (32% of all species of trees and lianas are pollinated by them), whereas large bees are predominant in the floodplain (22%). Other pollinators, like hummingbirds, bats, moths, and beetles are less common (< 10%), but always somewhat more important in the flood plain than in the upland forest. Bees are the most common pollinators of epiphytes. In the flood plain forest, flies are also important as epiphyte pollinators (19%), whereas in the upland forest hummingbirds pollinate more epiphytes. The phenological patterns are quite similar in both the upland and the flood plain. We found a peak in flowering in the transition period between the wet and the dry season. Flowering activity was lowest during the wet season. Differentiation in sexual systems was correlated with life form. Dioecy and monoecy were found mostly among tree species. Most species of all life forms though were hermaphroditic. No difference with respect to the relative importance of sexual systems was found between the two forest types.

Introduction

In the last few decades there has been an increasing number of investigations concerning pollination at the plant community-level (e.g., Frankie 1975; Stiles 1975; Heithaus 1979; Opler et al. 1980; Ramirez 1989; Ramirez & Britto 1992; Ramirez 1993; Morellato & Leitão-Filho 1996). However, relatively few studies have focused on the relative importance of pollinator groups in different vegetations types and any such information is very scarce for wet tropical forest vegetation types (Bawa et al. 1985a; Kato 1996; Momose et al. 1998). There is hardly any information available concerning pollination of rain forest plant communities in the Amazon basin.

The phenological patterns of plant species may be an important aspect of community organization (Hartshorn 1983; Newstrom et al. 1994). The flowering phenology of a plant may be the outcome of competition for pollinators or of facilitation processes in which various plant species together maintain a population of pollinators (Proctor et al. 1996), although these processes are not necessarily the only ones responsible for the observed patterns. The phenological cycles of plants and the life cycles and the behavioural patterns of animals are often closely correlated (Frankie 1975; Stiles 1975). Frankie (1975) found a relationship between the relative importance of different pollinators groups and different seasons. Large bees, for example, were especially common in the mid dry season, whereas small bees or opportunistic insects were more common in the wet season.

Community level patterns of breeding systems are another important factor of community structure and organization. Different sexual systems will have different effects on the rates of outcrossing, as well as on pollination mechanisms (Faegri & Van der Pijl 1966; Bawa 1980; Bawa et al. 1985a, b; Dafni 1992). Dioecy ensures outbreeding. Other systems like monoecy or andromonoecy promote outcrossing, although they cannot completely avoid self pollination. There seems to be a relation between dioecy and pollination by small bees or small opportunistic insects (Bawa 1980; Bawa et al. 1985a,b). Studies on the distribution of sexual systems have been conducted as a first assessment of the rates of gene flow in plant communities (Bawa 1974; Bawa et al. 1985b; Bullock 1985).

The present study aims to describe the differences in the pollination characteristics, sexual systems and phenology of two contrasting north west Amazon rain forest communities: a seasonally inundated forest in the floodplain of a white-water river is compared with an upland *tierra firme* community. Gentry (1983) hypothesised that in neotropical plant communities one can distinguish patterns in pollination and dispersal strategies. These might be predicted by abiotic factors such as precipitation and substrate, and are largely independent of the species represented in a particular plant community. The distinct environmental differences between the flooded and the upland forest were expected to be reflected not only in a different species composition, but also in different reproduction strategies of both vegetation types. We concentrated on the canopy of these forests, since it is here that most tree species, lianas and epiphytes form their leaves, flowers and fruits.

Methods

Study area

The study was conducted near Peña Roja, a small indigenous settlement, in the center of Colombian Amazonia along the Caquetá River. The region is situated at 0°37′ S and 72°20′ W at an altitude of about 150–200 m. Mean daily temperature is 25.7 °C, which does not fluctuate much throughout the year (Figure 1). Annual rainfall averages 3059 mm. The wettest months are April, May and June. In this period the water level of the river rises to reach its highest point in July – which is 6 to 8 m above the lowest level – inundating large parts of the adjacent forest. Two relatively dry periods can be distinguished, a larger one from December to March and another, shorter and

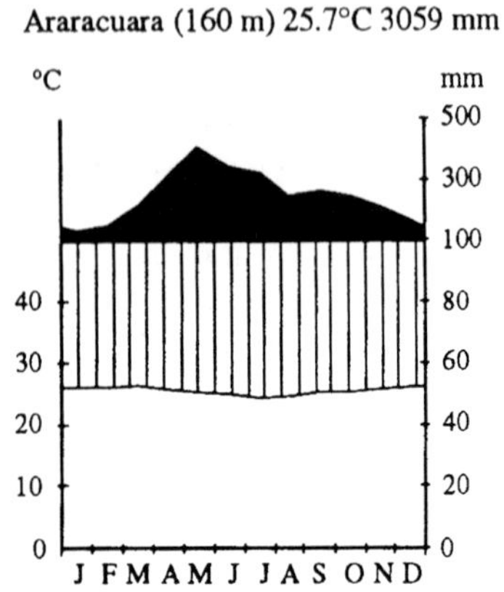

Figure 1. Climate diagram of the study area. Temperature is on the left axis; the right axis gives mm rainfall per month. The line seperating the white and striped area is the temperature curve, the topmost line shows the rainfall throughout the year. The scales are chosen in such a way that the temperature curve also gives a rough estimate of potential evaporation (10 °C = 20 mm). Whenever rainfall exceeds this estimate, this is indicated by striping. Mean monthly precipitation > 100 mm is indicated in black (and then also the scale is reduced to 1/10). Rainfall data are from the period 1979–1990, temperature data are from the period 1980–1989.

weaker one, from the end of August to the end of September. Rainfall is always higher than 100 mm a month though.

For more detailed information regarding the climate, vegetation and soils of the study area see the work of Tobón Marin (1999) and Duivenvoorden & Lips (1993).

Data collection and analysis

Fieldwork was carried out in different periods: from August to December 1993, from May 1994 to May 1995, and from October 1995 to May 1996. Research was done in two different forest types. The upland *tierra firme* forest belonged to the *Goupia glabra – Clathrotropis macrocarpa* floristic community according to Duivenvoorden & Lips (1993). This forest community is very abundant in the area and is characteristic for well-drained *Tierra firme* sites. Common tree species in this community are *Sloanea* aff. *maroana, Nealchornea yapurensis, Pouteria.spp.*, *Scleronema micranthum, Heterostemon conjugatus* and *Virola calophylloidea.* The seasonally inundated forest belonged to the *Brownea grandiceps – Ireartea deltoidea* community. Common tree species of this community are *Brownea grandiceps, Virola duckei* and *Inga thibaudiana*. This community is characteristic for well drained floodplain sites. Both forest were classified as high forests, with a high to intermediate biomass (Duivenvoorden & Lips 1993). To enable the observations in the canopy two aerial walkways were constructed, one in the *tierra firme* forest and the other in the seasonally inundated forest. Both walkways were about 50 m long and had a height of 25 m. Observations on visitors of flowers of trees and lianas could be made with reasonable reliability within a distance of about 30 m from the walkway. Around each walkway a rectangular research plot was laid out of 60 m wide and 110 m long. Occasionally single trees in or near the plots were climbed for more detailed information. Herbaceous epiphytes and hemi-epiphytes were sampled in a smaller subplot. All herbaceous epiphytes growing within a distance of 5 m from the canopy walkway were included in this study. Woody hemi-epiphytes, such as the various species of *Ficus* (Moraceae) or *Souroubea* (Marcgraviaceae), were included either as 'trees' or 'lianas', depending on their growth form. They were sampled in the complete (0.66 ha) research plots. For the data analysis palms were included in the 'tree' category.

The canopy of both forests is about 30 m high, with occasional emergents of 35–40 m. In the canopy of the upland forest plot 52 tree species, 25 liana species and 23 epiphyte species were found flowering during the study period (Appendix 1). Common trees were *Buchenavia* aff. *viridiflora*, *Scleronema micranthum*, *Goupia glabra*, *Clathrotropis macrocarpa* and *Parkia multijuga*, which all had 2 or more individuals in the plot. In the flooded forest plot 49 tree species, 31 liana species and 31 epiphyte species were found in flower during the research period. The plot was characterised by *Campsiandra angustifolia*, *Euterpe precatoria*, *Aspidosperma* cf. *megalocarpon*, *Eschweilera juruensis* and species of *Inga*. The two plots had very few species in common. *Parkia multijuga* was the only tree found in both plots. As far as lianas are concerned, only *Derris* aff. *negrensis* and *Securidaca hostmanii* occurred both in the upland forest and the flooded forest. *Aechmea angustifolia* and *Codonanthe uleana* were the only epiphyte species shared by both plots.

There has been much discussion whether or not it is possible to distinguish layers of vegetation in tropical humid forests (Richards 1952; Hartshorn 1983; Terborgh 1992). For the purpose of distinguishing between the canopy and the understorey we made a vegetation profile in both plots. Using this profile and making observations climbing trees in the forest around the walkways we decided to place a border at about eighteen m high in both forests, which seemed to coincide with the tops of the highest understorey trees and the lowest branches of the bigger tree members of the canopy. All trees, climbers and epiphytes with flowers above that line were included in the analysis.

The most probable pollinators were determined for every plant species, that was found flowering in the canopy of one of the research plots. Information on pollinators was collected by means of direct observations in the field with the aid of binoculars and a 20–60× telescope. Criteria for establishing a species as pollinator were: (i) regularity and duration of visit is, and (ii) the behaviour of the animal correlated with the morphology of the flower (stigma and anther touch). The time spent observing the visits of possible pollinators to each plant species ranged from a couple of hours to several days (or nights). For example *Dicorynia paraensis* is swarmed by bees once it is flowering and its pollinators are easy to establish, whereas we never saw any visits to flowers of species of Hippocrateaceae, despite several days and nights observing them. Night-time observations were made using a night vision device. These occasional observations were done whenever we expected night-time visits to a certain species, based on its flowering time and/or flower characteristics (flower syndrome), or when we did not observe any visitors during the day. Because of the difficulty of night-time observations, nocturnal pollinators may be subestimated in the final analysis. In a few cases where it was not possible to identify the pollinator of a particular species by direct observations, we obtained data from the literature (Faegri & Van der Pijl 1966; Croat 1978; Janzen 1983; Endress 1994). We also used so-called flower syndromes: combinations of characteristics of particular flowers which give an indication of the probable pollinator (Faegri & Van der Pijl 1966; Real 1983). The data were used to construct pollination spectra for the different life forms in the different plots. For the analysis we combined the results of trees and lianas. They were studied in the same sample area and their pollination spectra were rather similar. Epiphytes were treated separately, since they were studied in a smaller subplot of the total study area.

Approximately every 25 days we counted the number of flowering species and individuals in both plots. Although phenological data were collected during the whole study period, in this paper we only present the data from May 1994 to April 1995, since these represent a period of one whole year. The weather during this period was fairly typical for the climate. Total rain fall was 3388 mm. The wettest months were May and June (411 and 395 mm of rainfall, respectively) and the driest months were January and February (171 and 106 mm rainfall respectively) (Tobón Marin 1999).

In addition, observations were made on the sexual systems of the plant species present in the plots. We used the sexual systems as defined by Bawa and Beach (1981) and revised by Dafni (1992), although we restricted ourselves to the systems based on the spatial distribution of male and female organs over the flowers and individual plants of a population, since these could easily be observed within the framework of this study. We did not investigate the occurrence of incompatibility systems or the occurrence of dichogamy (separation in time of the ripening of the anthers and the stigma within a single flower, thus preventing self pollination), although we observed dichogamy in several plant species.

Specimens of collected plant species were deposited at the 'Herbario Amazónico' in Santafé de Bogotá and in the herbarium of the University of Utrecht. Nomenclature of plants followed Mabberley (1989) and Sánchez (1997)

Results

In this study we recognised eleven different pollination classes. If a plant species was pollinated by several different vectors it was scored more than once for the calculation of the pollination spectrum. So, a species like *Forsteronia affinis* (Apocynaceae), the flowers of which are visited by both butterflies and small bees, contributed to both pollination categories. The small diverse insect category (SDI) encompasses the species visited by an assemblage of small insects such as flies, beetles, wasps, small bees and small butterflies. Although both wasps and flies were observed as flower visitors, they were not common enough on any tree or liana species to be included as a separate pollination class. *Ficus* spp., which are pollinated by the highly specialized wasps of the Agaonidae, have their own category. These wasps are ecologically so different from the other pollinators that it seems proper to treat them separately. Species which could not be assigned with reasonable certainty to any of the categories, were classified as 'unknown'.

Figure 2 shows the pollination spectra for the canopy trees and lianas. In both forest types pollination by small and large bees, small diverse insects and butterflies are the most important. Almost all plants depend on animal vectors for the pollination of their flowers. Pollination by bats, hummingbirds, moths and beetles is less important in both forest types, but always somewhat more important in the flood plain than in the upland forest. We found a marked difference between the two forests with respect to the frequency of pollination by different groups of bees. While in the floodplain large bees are the most predominant pollinators (with a frequency of 22% as compared to only 15% for small bees), small bees form the dominant group of pollinators in the *Tierra firme* (frequency 32%, in contrast to only 15% for large bees). Only two species were found to be anemophilous (wind pollinated).

The pollination spectrum of the epiphyte species in both forest types is given in Figure 3. Bees are the most important pollinators of the epiphytes in both the floodplain and the *tierra firme*. We decided to combine the large and small bee category in a single category of bee pollinated plants, since we could not gather enough information on the pollination of *Anthurium* spp. and various Orchidaceae species, neither in the field nor in the literature. Dressler (1968) and Croat (1978) mentioned that some species of *Anthurium* were pollinated by scent gathering Euglossine bees, but they did not mention the precise species. During this study no Euglossine bees were observed visiting *Anthurium* spp. Very small bees of the stingless bee tribe Trigonini (*Nannotrigona* spp.) were observed collecting pollen from *Anthurium clavigerum* and *Anthurium gracile*, but we cannot be certain they are proper pollinators, as although we observed them filling their pollen baskets completely before flying off, they aparently had no need to visit another *Anthurium* before emptying their pollen baskets at the nest. Many of the orchid species observed are thought to be bee pollinated (Van der Pijl & Dodson 1966; Croat, 1978), but based only on the flower morphology no clear conclusion could be drawn whether this should be a large or a small bee. Orchids were the only species which were found to be pollinated by flies, and these orchids were only found in the floodplain plot. More epiphytes were pollinated by birds in the upland forest plot than in the seasonally inundated forest plot. Birds also showed a relation with one taxonomical group. All bird pollinated epiphytes belonged to the Bromeliaceae.

We expected a large difference in flowering phenology between the two plots, because of the influence of the annual flooding of the floodplain forest. However, flowering phenology showed a more or less similar pattern in both forest types (Figure 4). The number of flowering tree and liana species was relatively low

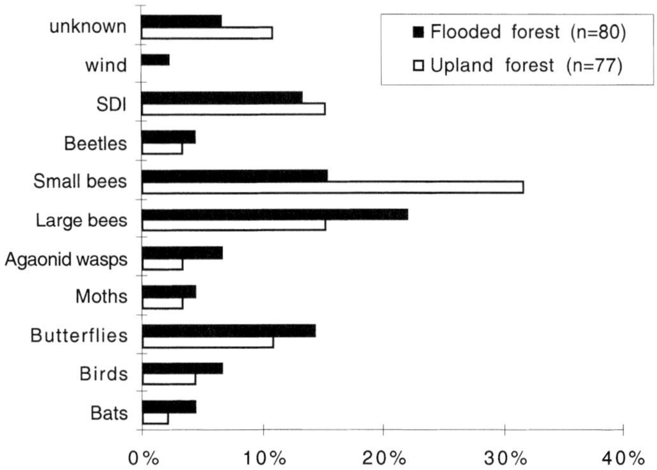

Figure 2. Relative importance of different pollination systems of the trees and lianas in each plot. SDI = small diverse insects; n = number of species.

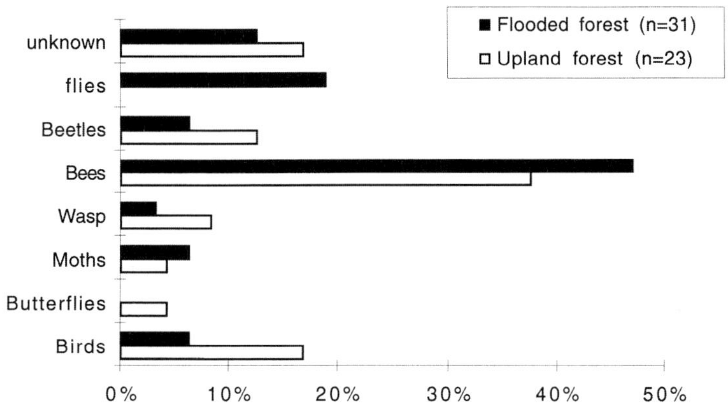

Figure 3. Relative importance of different pollination systems of the epiphytes in each plot. n = number of species.

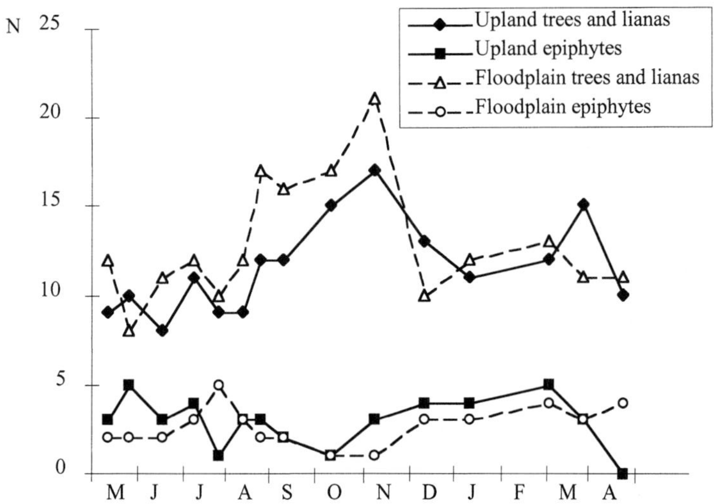

Figure 4. Flowering phenology of the plant species in each plot. n = number of species. Figure shows the number of species flowering at specific moments through a 12-month period.

during the rainy season (which almost coincides with the period of inundation in the floodplain forest). From the end of August until the beginning of December a peak of flowering occured in both the seasonally inundated and the upland forest. This peak corresponds to a transition period from the wet to the dry period. The number of flowering species was relatively low during the dry season from mid-way through December until the end of March. The number of flowering epiphyte species was too low to show marked patterns. However, it seems that flowering of epiphytes was especially low during the period of peak flowering of trees and lianas.

Table 1 shows the relative proportions of sexual systems in the different vegetation types. Twenty-nine percent of the tree species in the upland forest and 21% of the tree species in the seasonally inundated forest were dioecious. Epiphytes were mostly hermaphroditic (with 87% and 94% in the upland and floodplain forest respectively). Some monoecious species were found (all belonging to the aroid genus *Philodendron*), but not a single dioecious epiphyte. Lianas were more or less intermediate between trees and epiphytes. A small set of andromonoecious species were found among trees and lianas of both forest types. In both plots there was a significant association between life form and sexual system. The χ^2 for the upland forest is 14.079 ($0.05 < p < 0.01$) and the χ^2 for the floodplain forest is 15.612 ($p < 0.001$). Since in both plots we found a significant association, we also executed a heterogeneity test to check whether both plots are homogeneous in the distribution of sexual systems over the different life forms, which would allow pooling of the results of both types of forest. The heterogeneity χ^2 is 1.845, indicating that there is no difference between the two plots. The χ^2 of the pooled data of the two plots is 31.536 ($p < 0.001$), confirming the strong association between life form and sexual system.

Discussion

Hymenoptera, and in particular bees, are generally found to be the most important pollinators of lowland tropical ecosystems (Endress 1994; Proctor et al. 1996). Structurally complex vegetation types show more diverse spectra of pollinators than less complex vegetation types. In open and relatively species poor vegetations in Venezuela wasps were found to be much more important pollinators than they are in species rich rain forests (Ramirez 1989; Ramirez & Britto 1992; Bawa 1985a). The same holds true for flies as pollinators. Bees were also important in these open vegatation types, but overall the diversity of different pollination systems seemed much lower than in tropical rain forest. According to Ramirez and Brito (1992) the pollinator fauna in an area may be strongly correlated with vegetation structure and plant species diversity. The results of our study can to some degree be compared with the results of the study by Bawa et al. (1985a) on the tree species of a wet forest site at La Selva, Costa Rica. At La Selva, large bees, small bees and small diverse insects were among the most important pollinator groups, just as was found in this study. The percentages of plants pollinated by bats, hummingbirds, beetles, wasps and wind were also quite similar. There is a big difference, however, as far as butterflies and moths are concerned. In La Selva butterflies were responsible for the pollination of only 5% of the plant species studied (11% and 14% in the upland and floodplain forest respectively in this study), whereas moths were much more important (16% at La Selva vs. 4% in our seasonally inundated forest and 3% in the *tierra firme* forest). The high percentage of moths in the La Selva study may be explained by the fact that subcanopy and understorey trees were also included in the investigation, and moth pollination seemed to be more common in lower vegetation strata. Butterflies also were more common in the subcanopy at La Selva, making it more difficult to find a good explanation for their low frequency compared with our study. The results of Momose et al. (1998), where pollination systems in a dipterocarp forest in Sarawak, Malaysia, were investigated, show that in in Malaysia a much higher proportion of plants is pollinated by social bees or beetles than in our study. This is mostly because beetles and *Apis dorsata*, a social bee, were found to be able to exploit the mast flowering period (in which over 80% of the canopy species flower in a short, irregularly occurring ‘boom’) of which dipterocarp forests are famous. This type of flowering phenology does not occur in the neotropical rain forests. (Momose et al. 1998).

Epiphytes showed a lower diversity in pollination spectra than trees and lianas. No epiphytes were found to be pollinated by an array of small diverse insects. Instead the pollination systems seem to be more specific. This might be correlated by the fact that epiphytes occur often highly dispersed throughout the forest in low densities (Benzing 1990), which might require more precise pollen transfer between individ-

Table 1. Frequencies of the sexual systems in the different vegetation types.

	Upland forest			Flooded forest		
	Trees (n = 52)	Climbers (n = 25)	Epiphytes (n = 23)	Trees (n = 48)	Climbers (n = 31)	Epiphytes (n = 31)
Hermaphroditic	28 (54%)	21 (84%)	20 (87%)	27 (56%)	25 (81%)	29 (94%)
Monoecious	7 (13%)	1 (4%)	3 (13%)	9 (19%)	2 (6%)	2 (6%)
Dioecious	15 (29%)	2 (8%)	0 (0%)	10 (21%)	2 (6%)	0 (0%)
Andromonoecious	2 (4%)	1 (4%)	0 (0%)	2 (4%)	1 (3%)	0 (0%)
Polygamomonoec.	0 (0%)	0 (0%)	0 (0%)	0 (0%)	1 (3%)	0 (0%)

uals. The specialized pollination interactions we see among many rain forest epiphytes are thought to have evolved in environments which have remained stable over long periods of time (Proctor et al. 1996).

The difference in importance of small and large bees as pollinators between the seasonally inundated forest site and the upland forest is notable. De Dijn (in press) found that the stingless bee fauna of a swamp forest was impoverished in comparison with an adjacent *tierra firme* forest in Surinam. He hypothesized that this might be correlated with a smaller availability of nest sites because of the lack of suitable nesting substrates at ground level. The swamp forest he studied in Surinam was structurally less diverse than the *tierra firme* forest, which might also limit nesting possibilities. This is certainly not the case in the floodplain forest in our study. It might be the case that during the annual inundation relatively few species in the understorey are flowering. Many of the smaller herbs and shrubs would be submerged, and many of the smaller trees synchronize their fruiting with the yearly inundations, so as to profit from fish as seed dispersers (Gottsberger 1978; Goulding 1983, 1985). Although they do not exclusively rely on nectar and pollen, this seasonally low availability of resources in the understorey may also contribute to constraining the diversity of stingless bee species in the floodplain forest, as such species need to maintain a large nest all year round (Janzen 1983; Roubik 1989). Large bees may be less affected by these seasonal phenomena, as they are strong flyers, often traplining for food and able to come flying in from the adjacent *tierra firme* forest.

Flowering in tropical ecosystems has often been found to show a clear seasonality. In the wet forest of La Selva, Costa Rica, with a relatively small dry period in February and March peak in flowering begins in the dry period and extends until halfway through the wet period (Frankie 1975). Croat (1978) on Barro Colorado Island, Panama, found a peak in flowering of trees and lianas in the dry season. In the area of this study in both plots the peak in flowering occurs during the transitional wet-dry season. That might be because the pattern of decrease in rainfall between the wet and the dry season is not as pronounced as in La Selva and Barro Colorado.

Ashton (1969) in evergreen forest in Malaysia, Bawa et al. (1985b) in wet tropical forest in Costa Rica, and Bullock (1985) in tropical deciduous forest in Mexico, all found very similar distributions of sexual systems among trees. The frequency of hermaphroditic species is always around 60%, the frequency of monoecious species around 12%, and the frequency of dioecy around 24%. Bawa et al. (1985b) hypothesize that the upper limit of the number of dioecious species in a community may be limited by the pollinator fauna, because of the association of dioecy with pollination by small generalist insects. Indeed, of the 29 dioecious plants found in this study 69% are pollinated by small bees or by small diverse insects. Of all the small bee pollinated plants 67% are hermaphroditic and 16% dioecious. Fifty percent of the plants pollinated by small diverse insects are hermaphroditic and 46% are dioecious. By contrast, of all the plants pollinated by large bees only 4% (2 species) were found to be dioecious, against 94% hermaphroditic plants. However, pollinators such as large bees and butterflies are known to function as pollinators of dioecious plants. The relationship between opportunistic insects and dioecy is thus not an exclusive one. Bawa (1980) and Bawa et al. (1985b) mention a relation between dioecy and trees. Also the results of Bullock (1992) clearly show an association between sexual systems and life forms. Dioecy was found to be associated with trees, whereas epiphytes were found to be mostly hermaphroditic, just as was found in this study.

The combined results of this study indicate that there is some basic underlying pattern of phenology, pollination systems and breeding systems, that applies to both forest types. Superimposed on this basic pattern are some small differences between the forests that may be related to the specific characteristics of the vegetation types, such as the seasonal inundation of the floodplain forest. The differences between the species composition of the two plots seem to have no effect on the patterns observed. However the relationship we observe between certain pollination systems and particular plant taxa does suggest that species composition is not entirely without its influence.

Acknowledgements

This study was made possible by the financial support of the Tropenbos Foundation and WOTRO (Netherlands Foundation for the Advancement of Tropical Research). I would like to thank all the specialists responsible for the identification of plant material, especially those of the herbaria in Bergen, Kew, Missouri Botanical Gardens, New York and Utrecht.

References

Ashton, P. S. 1969. Speciation among tropical forest trees: some deductions in the light of recent evidence. Biol. J. Linn. Soc. 1: 155–196.

Bawa, K. S. 1974. Breeding systems of tree species of a lowland tropical community. Evolution 28: 83–92.

Bawa, K. S. 1980. Evolution of dioecy in flowering plants. Annu. Rev. Ecol. Syst. 11: 15–39.

Bawa, K. S. & Beach, J. H. 1981. Evolution of sexual systems in flowering plants. Ann. Missouri Bot. Gard. 68: 254–274.

Bawa, K. S., Bullock, S. H., Perry, D. R., Coville, R. E. & Grayum, M. H. 1985a. Reproductive biology of tropical lowland rain forest trees II. Pollination systems. Amer. J. Bot. 72: 346–356.

Bawa, K. S., Perry, D. R. & Beach, J. H. 1985b. Reproductive biology of tropical lowland rain forest trees I. Sexual systems and incompatibility mechanisms. Amer. J. Bot. 72: 331–345.

Benzing, D. H. 1990. Vascular epiphytes. Cambridge University Press, Cambridge.

Bullock, S. H. 1985. Breeding systems in the flora of a tropical deciduous forest in Mexico. Biotropica 17: 287–301.

Croat, T. B. 1978. Flora of Barro Colorado Island. Stanford University Press, Stanford.

Dafni, A. 1992. Pollination ecology. A practical approach. Oxford University Press, Oxford.

De Dijn, B. P. E. (in press). Stingless bee communities: are they poorer in more constrained ecosystems? J. Apic. Res.

Dressler, R. L. 1968. Pollination by Euglossine Bees. Evolution 22: 202–210.

Duivenvoorden, J. F. & Lips, J. M. 1993. Ecología del Paisaje del Medio Caquetá. Tropenbos series: Estudios en la Amazonia Colombiana 3A/B. Tropenbos Colombia, Santafé de Bogotá.

Endress, P. K. 1994. Diversity and evolutionary biology of tropical flowers. Cambridge University Press, Cambridge.

Faegri, K. & van der Pijl, L. 1966. The principles of pollination ecology. Pergamon Press, Oxford.

Frankie, G. W. 1975. Tropical forest phenology and pollinator plant coevolution. Pp. 192–209. In: Gilbert, L. E. & Raven, P. H. (eds). Coevolution of animals and plants. University of Texas Press, Austin.

Gentry, A. H. 1983. Dispersal ecology and diversity in neotropical forest communities. Sonderbd. naturwiss. Ver. Hamburg 7: 303–314.

Gottsberger, G. 1978. Seed dispersal by fish in the inundated regions of Humaita, Amazonia. Biotropica 10: 170–183.

Goulding, M. 1983. The role of fishes in seed dispersal and plant distribution in Amazonian floodplain ecosystems. Sonderbd. naturwiss. Ver. Hamburg 7: 271–283.

Goulding, M. 1985. Forest fishes of the Amazon. Pp. 267–276. In: Prance, G. T. & Lovejoy, T. E. (eds). Key environments. Amazonica. Pergamon Press, Oxford.

Hartshorn, G. S. 1983. Plants. Pp. 118–350. In: Janzen, D. H. (ed.), Costa Rica natural history. The University of Chicago Press, Chicago.

Heithaus, E. R. 1979. Community structure of neotropical flower visiting bees and wasps: Diversity and phenology. Ecology 60: 190–202.

Janzen, D. H. 1983. Costa Rican Natural History. University of Chicago Press, Chicago.

Kato, M. 1996. Plant-pollinator interactions in the understorey of a lowland mixed Dipterocarp forest in Sarawak. Amer. J. Bot. 83: 732–743.

Mabberley, D. J. 1989. The plant book. A portable dictionary of the higher plants. Cambridge University Press, Cambridge.

Momose K., Yumoto, T., Nagamitsu, T., Kato, M., Nagamasu, H., Sakai, S., Harrison, R. D., Itioka, T., Hamid, A. A., & Inoue, T. 1998. Pollination biology in a lowland dipterocarp forest in Sarawak, Malaysia. I. Characteristics of the plant pollinator community in a lowland dipterocarp forest. Am. J. Bot. 85: 1477–1501.

Morellato, P. C. & Leitão-Filho, H. F. 1996. Reproductive phenology of climbers in a southeastern Brazilian forest. Biotropica 28: 180–191.

Newstrom, L. E., Frankie, G. W. & Baker, H. G. 1994. A new classification for plant phenology based on flowering patterns in lowland tropical rain forest trees at La Selva, Costa Rica. Biotropica 26: 141–159.

Opler, P. A., Baker, H. G. & Frankie, G. W. 1980. Plant reproductive characteristics during secondary succession in neotropical lowland forest ecosystems. Biotropica 12, suppl. Tropical succession: 40–46.

Proctor, M., Yeo, P. & Lack, A. 1996. The natural history of pollination. Harper Collins Publishers, London.

Ramirez, N. 1989. Biología de polinización en una comunidad arbustiva tropical de la Guyana Venezolana. Biotropica 21: 319–330.

Ramirez, N. 1993. Reproductive biology in a tropical shrubland of Venezuelan Guyana. J. Veg. Sci. 4: 5–12.

Ramirez, N. and Brito, Y. 1992. Pollination biology in a palm swamp community in the Venezuelan Central Plains. Bot. J. Linn. Soc. 110: 277–302.

Real, L. (ed.). 1983. Pollination biology. Academic Press, New York.

Richards, P. W. 1952. The tropical rain forest: An ecological study. Cambridge University Press, Cambridge.

Roubik, D. W. 1989. Ecology and natural history of tropical bees. Cambridge University Press, Cambridge.

Sánchez, M. 1997. Catálogo preliminar commentado de la Flora del Medio Caquetá. Tropoenbos series: Estudios en la Amazonia Colombiana 12. Tropenbos Colombia, Santafé de Bogotá.

Stiles, F. G. 1975. Ecology, flowering, and hummingbird pollination of some Costa Rican *Heliconia* species. Ecology 56: 285–301.

Terborgh, J. 1992. Diversity and the tropical rain forest. Scientific American Library, New York.

Tobón Marin, C. 1999. Monitoring and modeling hydrological fluxes in support of nutrient cycling studies in Amazonian rain forest ecosystems. Tropenbos series 17. Stichting Tropenbos, Wageningen.

Van der Pijl, L. & Dodson, C. H. 1966. Orchid flowers. Their pollination and evolution. University of Miami Press, Coral Gables.

Appendix I

Life form, pollinators and sexual system of the plants found in the canopy of an upland rain forest and a seasonally inundated rain forest in Amazonia, Colombia. Life form: T = tree, P = palm, L = liana, E = epiphyte, HT = hemi-epiphytic tree or shrub, HL = hemi-epiphytic liana. Forest type: F = flooded forest, U = upland forest. Pollinator: SDI = small diverse insects.

Family and species	Life form	Forest type	Pollinator	Sexual system
indet.	L	U	unknown	bisexual
ANNONACEAE				
Annona hypoglauca Mart.	L	F	beetles	bisexual
Ephedranthus amazonicus R.E.Fr.	T	U	beetles	bisexual
Guatteria modesta Diels	T	F	beetles	bisexual
Oxandra xylopioides Diels	T	F	beetles	bisexual
Pseudoxandra leiophylla (Diels) R.E.Fr	T	F	beetles	bisexual
APOCYNACEAE				
Aspidosperma cf. megalocarpon Muell. Arg.	T	F	moths	bisexual
Forsteronia affinis Muell. Arg.	L	U	small bees, butterflies	bisexual
Forsteronia sp.	L	U	small bees, butterflies	bisexual
Odontadenia macrantha (Roem. & Schult.) Markgr.	L	F	large bees	bisexual
ARACEAE				
Anthurium clavigerum Poeppig.	E	F	bees	bisexual
Anthurium gracile (Rudge) Lindl.	E	F	bees	bisexual
Anthurium sp. 1	E	U	bees	bisexual
Anthurium sp. 2	E	F	bees	bisexual
Anthurium sp. 3	E	F	bees	bisexual
Anthurium sp. 4	E	F	bees	bisexual
Philodendron cf. linnaei Kunth	E	U	beetles	monoecious
Philodendron panduriforme (H.B.K.) Kunth	E	F	beetles	monoecious
Philodendron sp. 1	E	U	beetles	monoecious
Philodendron sp. 2	E	F	beetles	monoecious
ARECACEAE				
Astrocaryum aculeatum G. Meyer	P	U	beetles	monoecious
Desmoncus polyacanthos Mart.	L	U	beetles	monoecious
Euterpe precatoria Mart.	P	F	beetles, small bees	monoecious
Ireartea deltoidea Ruiz & Pavón	P	F	small bees	monoecious
BIGNONIACEAE				
Adenocalymna cf. purpurascens Rusby	L	U	large bees	bisexual
Arrabidea inaequalis (DC. ex Splitg.) K. schum.	L	U	large bees	bisexual
Distictella sp.	L	U	large bees	bisexual
Memora sp.	L	F	large bees	bisexual
Paragonia pyramidata (L.C.Richard) Bureau	L	F	large bees	bisexual
BOMBACACEAE				
Quararibea cf. guianensis Aubl.	T	F	bats	bisexual
Scleronema micranthum (Ducke) Ducke	T	U	moths, bats	bisexual
BORAGINACEAE				
Cordia sp.	T	U	butterflies	bisexual
BROMALIACEAE				
Aechmea angustifolia Poepp. & Endl.	E	F, U	birds	bisexual
Aechmea cf. servitensis André	E	U	birds	bisexual

Appendix 1. Continued

Family and species	Life form	Forest type	Pollinator	Sexual system
Aechmea mertensii (Meyer) Schultes	E	U	birds	bisexual
Aechmea woronowii Harms	E	F	birds	bisexual
Vriesia pleiosticha (Grisebach) Gouda	E	U	birds	bisexual
BURSERACEAE				
Protium divaricatum Engl.	T	U	SDI	dioecious
Trattinnickia peruviana Loes.	T	U	small bees	dioecious
CECROPIACEAE				
Cecropia ficifolia Snethlage	T	F	Wind	dioecious
CELASTRACEAE				
Goupia glabra Aubl.	T	U	unknown	bisexual
CHRYSOBALANACEAE				
Couepia racemosa Benth. ex Hook	T	U	large bees, small bees	bisexual
Hirtella elongata Mart. & Zucc.	T	F	butterflies	bisexual
Licania licaniiflora (Sagot) Blake	T	U	small bees, butterflies	bisexual
Licania sp.	T	F	small bees	bisexual
Licania unguiculata Prance	T	U	small bees, butterflies	bisexual
COMBRETACEAE				
Buchenavia aff. Viridiflora Ducke	T	U	small bees	bisexual
CONNARACEAE				
Connarus ruber (Poepp.) Planch.	L	F	SDI	bisexual
Pseudoconnarus macrophyllus (Poepp.) Radlk.	L	U	SDI	bisexual
Rourea sp.	L	U	small bees	bisexual
CONVOLVULACEAE				
Maripa cf. glabra Choisy	L	U	large bees, moths	bisexual
Maripa paniculata Barbosa Rodrígues	L	F	large bees	bisexual
CUCURBITACEAE				
Psiguria sp.	L	F	butterflies	monoecious
CYCLANTHACEAE				
Ludovia lancifolia Brogniart	E	U	beetles	monoecious
DICHAPETALACEAE				
Tapura acreanea (Ule) Rizzini	T	F	small bees	bisexual
DILLENIACEAE				
Doliocarpus sp.	L	F	small bees	bisexual
Pinzonia coriacea Mart. & Zucc.	L	U	small bees	bisexual
EUPHORBIACEAE				
Conceveiba guianensis Aubl.	T	U	unknown	dioecious
Drypetes variabilis Uittien	T	F	SDI	dioecious
Hevea guianensis Aubl.	T	U	unknown	monoecious
Hyeronima alchorneoides Allemao	T	U	SDI	dioecious
Nealchornia yapurensis Huber	T	U	unknown	dioecious
Omphalea diandra L.	L	F	unknown	monoecious
FLACOURTIACEAE				
Homalium guianense (Aubl.) Oken	T	F	SDI	bisexual
GESNERIACEAE				
Codonanthe uleana Fritsch	E	F, U	large bees	bisexual
Drymonia cf. quitobeo Wiehler	E	F	large bees	bisexual
GUTTIFERAE				
Clusia aff. insignis Mart.	HT	F	large bees, small bees	dioecious
Clusia araracuarae Pipoly	HL	U	small bees	dioecious
Clusia penduliflora Engl.	HL	U	large bees, small bees	dioecious
Clusia sp.	HT	U	small bees	dioecious
HERNANDIACEAE				
Sparattanthelium sp.	L	U	SDI	bisexual
HIPPOCRATEACEAE				
Cheiloclinium hippocrateoides (Peyr.) A.C.Sm.	L	U	unknown	bisexual
Cheiloclinum anomalum Miers	L	F	unknown	bisexual
Hylenaea comosa (Sw.) Miers	L	F	unknown	bisexual
Salacia multiflora (Lam.) DC.	L	F	unknown	bisexual
ICACINACEAE				
Icac. indet.	T	U	small bees	bisexual
Pleurisanthes flava Sandw.	L	U	unknown	bisexual

Appendix 1. Continued

Family ans species	Life form	Forest type	Pollinator	Sexual system
LAURACEAE				
Ocotea cf. rubrinervis Mez.	T	U	SDI	bisexual
Ocotea longifolia HBK.	T	U	SDI	bisexual
Ocotea sp.	T	U	SDI	bisexual
LECYTHIDACEAE				
Eschweilera coriacea (A.P. DC.) Mori	T	F	large bees	bisexual
Eschweilera juruensis R. Knuth	T	F	large bees	bisexual
Eschweilera rufifolia Mori	T	U	large bees	bisexual
Gustavia hexapetala (Aubl.) J.E.Smith	T	U	small bees	bisexual
LEGUMINOSAE, CAESALPINIOIDAE				
Caesap. indet.	T	F	large bees	bisexual
Cynometra spruceana Benth.	T	U	small bees, large bees	bisexual
Dialium guianense (Aubl.) Sandw.	T	U	SDI	bisexual
Hymenea oblongifolia Huber	T	U	large bee, birds	bisexual
Macrolobium angustifolium (Benth.) Cowan	T	F	butterflies	bisexual
Dicorynia paraensis Bentham (C)	T	F	small bees, large bees	bisexual
LEGUMINOSAE, MIMOSOIDAE				
Campsiandra angustifolia Spruce ex Benth.	T	F	butterflies, bird	bisexual
Enterolobium schomburgkii (Benth.) Benth.	T	U	small bees	bisexual
Inga sp. 1	T	U	large bees, butterflies birds	bisexual
Inga sp. 2	T	F	butterflies	bisexual
Inga sp. 3	T	F	large bees, butterflies moths, birds, bats	bisexual
Mimosa annularis Benth.	L	U	small bees	bisexual
Parkia multijuga Benth.	T	F, U	bats	andromon.
Zygia inaequalis (Willd.) Pittier	T	F	butterflies	bisexual
LEGUMINOSAE, PAPILIONIDAE				
Clathrotropis macrocarpa Ducke	T	U	small bees	bisexual
Clitoria javitensis (Kunth) Benth.	L	U	large bees	bisexual
Dalbergia riedelii (Benth.) Sandw.	L	U	small bees	bisexual
Derrisaff. negrensis Benth.	L	U, F	small bees	bisexual
Swartzia racemosa Benth.	T	F	large bees	bisexual
Taralea oppositifolia Aubl.	T	F	large bees	bisexual
Dalbergia sp.	L	F	small bees	bisexual
Swartzia cf. loevicarpa Amsh.	T	F	large bees	bisexual
Vatairea guianensis Aubl.	T	F	large bees	bisexual
LINACEAE				
Hebepetalum humirifolium (Planch.) Benth.	T	U	small bees	bisexual
Roucheria sp.	T	U	unknown	bisexual
LOGANIACEAE				
Strychnos cf. mitscherlichii Ri. Schomburgk.	L	F	butterflies, moth	bisexual
LORANTHACEAE				
Phthirusa retroflexa (R.&P.) Kuyt	E	U	small bees, wasp, butterflies	bisexual
Phthirusa stelis (L.) Kuijt	E	F	small bees, wasps	bisexual
MALPIGHIACEAE				
Banisteriopsis wurdackii B. Gates	L	F	large bees	bisexual
Tetrapteris sp.	L	F	large bees	bisexual
MARCGRAVIACEAE				
Marcgravia sp.	HL	F	bats	bisexual
Norantea guianensis Aubl.	HL	F	birds	bisexual
Souroubea corallina (Mart.) De Roon	HL	F	birds	bisexual
Souroubea crassipetala De Roon	HL	F	birds	bisexual
Souroubea guianensis Aubl.	HL	U	birds	bisexual
MELASTOMATACEAE				
Blakea brasiliensis Cogn.	HT	U	large bees	bisexual
MELIACEAE				
Cedrela odorata L.	T	F	moths	monoecious
Guarea cf. purusana C. DC.	T	U	butterflies	dioecious
Guarea sp.	T	U	moths	dioecious
Trichilia sp.	T	F	butterflies	dioecious

Appendix 1. Continued

Family ans species	Life form	Forest type	Pollinator	Sexual system
MENISPERMACEAE				
Abuta imene (Mart.) Eichl.	L	F	SDI	dioecious
Sciadotenia toxifera Krukoff & A.C.Sm.	L	F	SDI	dioecious
MORACEAE				
Coussapoa cf. tessmannii Mildbr.	HT	F	Wind	dioecious
Ficus paraensis (miq.) Miq.	HT	F	agaonid wasps	monoecious
Ficus pertusa L.F.	HT	F	agaonid wasps	monoecious
Ficus cf. aripuenensis C.C.Berg & F. Kooy	HT	U	agaonid wasps	monoecious
Ficus cf. guianensis Desv.	HT	U	agaonid wasps	monoecious
Ficus cf. pallida Vahl	HT	F	agaonid wasps	monoecious
Ficus cf. trigona L.F.	HT	F	agaonid wasps	monoecious
Ficus sp. 1	HT	U	agaonid wasps	monoecious
Ficus sp. 2	T	F	agaonid wasps	monoecious
Ficus sp. 3	T	F	agaonid wasps	monoecious
Naucleopsis cf. glabra Spruce ex Pittier	T	U	unknown	dioecious
Pourouma sp.	T	F	unknown	dioecious
Moracea indet.	T	U	small bees	dioecious
MYRISTICACEAE				
Iryanthera sp.	T	U	SDI	dioecious
Iryanthera tricornis Ducke	T	U	SDI	dioecious
Virola calophylla Warb.	T	U	SDI	dioecious
Virola sp. 1	T	U	SDI	dioecious
Virola sp. 2	T	F	SDI	dioecious
Virola sp. 3	T	F	SDI	dioecious
Virola sp. 4	T	F	SDI	dioecious
OPILIACEAE				
Agonandra silvatica Ducke	T	U	SDI	bisexual
ORCHIDACEAE				
Orchid indet. 1	E	U	Wasps	bisexual
Orchid indet. 2	E	U	unknown	bisexual
Orchid indet. 3	E	U	bees	bisexual
Orchid indet. 4	E	F	unknown	bisexual
Orchid indet. 5	E	U	bees	bisexual
Bifrenaria sp.	E	U	unknown	bisexual
cf. Epidendrum	E	U	moths	bisexual
Epidendrum anceps Jacq.	E	F	moths	bisexual
Epidendrum rigidum Jacq.	E	F	moths	bisexual
Epidendrum cf. immatophyllum Lindl.	E	F	large bees	bisexual
Lepanthes sp.	E	F	unknown	bisexual
Masdevallia cf. Wendlandiana Rchb. F.	E	F	flies	bisexual
Maxillaria camaridii Rchb.f.	E	F	bees	bisexual
maxillaria cf. Purpurea (Spreng) Ames & Correll	E	F	bees	bisexual
maxillaria sp. 1	E	U	bees	bisexual
maxillaria sp. 2	E	F	bees	bisexual
maxillaria sp. 3	E	U	bees	bisexual
maxillaria sp. 4	E	F	bees	bisexual
maxillaria sp. 5	E	U	bees	bisexual
maxillaria sp. 6	E	U	bees	bisexual
Octomeria sp.	E	U	unknown	bisexual
Pleurothallis cf. picta Lindl.	E	F	flies	bisexual
Pleurothallis sp. 1	E	F	flies	bisexual
Pleurothallis sp. 2	E	F	flies	bisexual
Pleurothallis sp. 3	E	F	flies	bisexual
Pleurothallis sp. 4	E	F	flies	bisexual
Polystachya foliosa (Hook.) Reichb.f.	E	F	small bees	bisexual
Scaphyglottis sp.	E	U	small bees	bisexual
Schomburgkia aff. gloriosa Rchb.f.	E	F	large bees	bisexual
PASSIFLORACEAE				
Passiflora sp. 1	L	U	birds	bisexual
Passiflora sp. 2	L	F	birds	bisexual

Appendix 1. Continued

Family ans species	Life form	Forest type	Pollinator	Sexual system
PIPERACEAE				
Peperomia sp. 1	E	U	unknown	bisexual
Peperomia sp. 2	E	F	unknown	bisexual
Peperomia sp. 3	E	F	unknown	bisexual
POLYGALACEAE				
Securidaca hostmanii Miq.	L	F, U	large bees	bisexual
POLYGONACEAE				
Cocoloba sp.	T	U	small bees	monoecious
RHAMNACEAE				
Gouania aptera DC.	L	F	small bees, butterflies	bisexual
RUBIACEAE				
Parachimarrhis breviloba Ducke	T	F	unknown	bisexual
Uncaria guianensis (Aubl.) Gmel.	L	F	butterflies	bisexual
Warscewiczia coccinea (M.Vahl) Klotzsch	T	U	butterflies	bisexual
SABIACEAE				
Meliosma herbertii Rolfe	T	U	unknown	bisexual
SAPINDACEAE				
Sapind. indet. 1	T	U	small bees	andromon.
Sapind. indet. 2	T	F	small bees	unknown
Paullinia sp. 1	L	U	small bees	andromon.
Paullinia sp. 2	L	F	small bees, butterflies	andromon.
SAPOTACEAE				
Micropholis guyanensis (A.DC.) Pierre	T	U	small bees	dioecious
Micropholis venulosa (Mart. & Eichler) Pierre	T	U	small bees	monoecious
Pouteria cladantha Sandwith	T	F	SDI	bisexual
Pouteria putamen-ovi Pennington	T	F	SDI	bisexual
Pouteria sp. *(sectie Pouteria)*	T	U	SDI	bisexual
SIMAROUBACEAE				
Simarouba amara Aubl.	T	F	small bees	dioecious
SOLANACEAE				
Solanum sp.	L	F	large bees	bisexual
TILIACEAE				
Apeiba membranacea Spruce ex. Benth.	T	F	large bees	andromon.
ULMACEAE				
Celtis iguanaeus (Jacq.) S. g.	L	F	SDI	polygamomon.
VERBANACEAE				
Aegiphila smithii Mold.	L	U	small bees, butterflies	bisexual
VITACEAE				
Cissus sp.	L	F	SDI	bisexual
VOCHYSIACEAE				
Erisma uncinatum Warm.	T	U	large bees, small bees	bisexual
Qualea paraensis Ducke	T	F	large bees	bisexual
Vochysia sp.	T	F	small bees, butterflies	bisexual

Plant Ecology **153:** 87–107, 2001.

Invertebrates in the canopy of tropical rain forests How much do we really know?

Yves Basset
Smithsonian Tropical Research Institute, Apartado 2072, Balboa, Ancon, Panama (E-mail: bassety@tivoli.si.edu)

Key words: Biodiversity, Biogeography, Collecting methods, Conservation, Predator-prey relationships, Species richness

Abstract

The current state of knowledge of canopy invertebrates in tropical rain forests is reviewed using data drawn, without bias toward taxon, collecting method or biogeographical region, from 89 studies concerned with mass-collecting (>1000 individuals). The review is intended to identify the most serious gaps and biases in the distribution of higher taxa among forest types and biogeographical regions. With respect to knowledge, biogeographical regions can be ranked as Neotropical > Australian > Oriental > Afrotropical. The canopy of lowland wet and subtropical forests has been studied in greater detail, whereas the canopy of lowland dry and montane forests is much less well known. Collecting techniques influence greatly the present knowledge of canopy invertebrates. Invertebrates other than arthropods, often abundant in epiphytic habitats, phytotelmata and perched litter, are virtually unknown. The abundance of several groups, such as Acari, Collembola and Isoptera, is almost certainly seriously underestimated. Densities of invertebrate individuals in the canopy of tropical rain forests appear to be lower than in temperate forests, although invertebrate abundance is dissipated by the high standing-biomass of rain forests. Coleoptera, particularly Staphylinidae, Curculionidae and Chrysomelidae, along with Hymenoptera, Lepidoptera and Araneae appear to be the most speciose taxa in the canopy, and it is probable that this reflects their range of feeding habits and exploitation of rain forests habitats. The distribution of individuals among the major arthropod orders and across the studies examined is complex and depends on many factors. The amount of variance that can be directly explained by biogeography, forest types (subtropical, wet, dry or montane), or collecting methods appears to be about 11%. The explained variance increases when considering major families of Coleoptera (28%) or subfamilies of Chrysomelidae (40%). In all cases, the variance explained by the type of forest is much higher than by that explained by biogeography. These conclusions are similar when considering various prey-predator relationships in the canopy. This suggests that, at the higher taxa level, the composition of the invertebrate fauna in the canopy may vary comparatively more across forest types than across biogeographical regions and this is discussed briefly from a conservation viewpoint.

Introduction

Several reviews focusing on the canopy of tropical rain forests have considered, more or less directly, canopy invertebrates with, for example, particular reference to canopy biology (Lowman & Wittman 1996), forest ecology (Lowman & Moffett 1993), ecosystem dynamics (Stork 1996), herbivory (e.g., Coley & Aide 1991; Lowman 1995; Coley & Barone 1996), global species richness (e.g., Stork 1988; Hammond 1992; May 1994; Erwin 1995), sampling methods (e.g., Erwin 1989; Basset et al. 1997) or canopy access (Moffett & Lowman 1995). Other detailed sources of information on canopy invertebrates can be found in thorough studies of particular tropical locations (e.g., Knight & Holloway 1990; Leigh et al. 1996; Junk 1997) or in proceedings of various meetings (e.g., Sutton et al. 1983b; Stork et al. 1997).

However, to date, there has been no attempt to bring together most of the studies on canopy invertebrates without bias as to collecting methods, taxa or biogeographical regions. The aim of this article is to identify the main biases and gaps in the knowledge of canopy invertebrates in tropical and subtropical rain forests, to review the information available in terms of composition, abundance and diversity, to identify broad patterns in the distribution of higher canopy taxa and prey-predator ratios, and to discuss future prospects in the study of canopy invertebrates.

Methods and their limitations

The literature published up to September 1998 was collated, using various sources and with additional information from colleagues. Studies were retained only if they:

- concentrated on the canopy of tropical or subtropical rain forests;
- included mass collecting of at least 1000 invertebrates (with some rare exceptions below this threshold), preferably spanning over several orders;
- included quantitative data; and
- were not concerned primarily with only one or several pest species.

This search yielded 89 studies, which are summarized in Appendix 1. For sake of brevity, only a key reference is given per study, giving access to other relevant references.

There are several obvious problems with such a compilation. First, the same data set may have been used by several workers to study different taxa and topics. These 'pseudoreplicates' (from a statistical viewpoint) can be easily identified and grouped with the original data set. Second, there is a bias against agricultural, medical and forestry entomology studies. Third, the data collated were sorted to varying levels of resolution, or concentrated only on specific taxa. The analyses consider the distribution of individuals, rather than species richness, for which data are scarce. Estimates of species richness for particular taxa were derived considering species accumulation within the number of individuals collected, or an index of 'species × individuals^{-2}' averaged over the relevant studies (the higher the index the more speciose). This crude index permitted the comparison of studies with different sampling techniques, focal taxa and sampling effort.

Lastly, the data collated were collected with different methods, over different time-span and number of habitats, etc. Even for a technique such as fogging, the application of different insecticides may result in different results (Erwin 1995). To reduce the effect of sample size (collecting for longer time or in different habitats, etc.), individuals collected of a particular taxon were expressed in % of the total catch during the study. This permitted the consideration of studies not reporting data as numbers of individuals (e.g., biomass or average no. individual per m^2 of fogging tray, etc.). Percentages were then averaged across all studies amenable to analysis (e.g., ordinal signatures, $n = 42$), to derive relative estimates of invertebrate abundance. These estimates were further multiplied with modal classes of insect body weight, as reported in Zug & Zug (1979), to derive corresponding percentages of biomass. In a separate analysis (see below), collecting methods were coded into a variable for each study, and the specific contribution of this variable to overall variance was estimated by canonical correspondence analysis.

Analyses to identify broad patterns in the distribution of higher taxa were performed at three taxonomic levels:

- Major invertebrate orders/groups present in the canopy: Araneae, Acari, other Arthropoda, Collembola, Blattodea, Isoptera, Dermaptera, Orthoptera, Psocoptera, Homoptera, Heteroptera, Thysanoptera, Coleoptera, Diptera, Lepidoptera, Neuroptera, Hymenoptera and Formicidae.
- Most common families of Coleoptera present in canopy samples: Anthribidae, Carabidae, Cerambycidae, Chrysomelidae, Cleridae, Coccinellidae, Corylophidae, Curculionidae, Lathridiidae, Mordellidae, Phalacridae, Scolytidae, Staphylinidae, Tenebrionidae (see recent discussion of beetle taxa in the canopy in Hammond et al. 1996).
- Major subfamilies/groups of Chrysomelidae: Eumolpinae, Galerucinae, Alticinae and 'Others'.

Formicidae were also included in the first level, as most workers readily sorted ants in their samples. The choice of beetles and chrysomelids for the other levels of analyses was dictated by the nature of the data, as most workers who sorted their samples at lower taxonomic level concentrated on beetles.

The total of ecological publications that resulted from the original data was used to estimate the amount of accessible knowledge related to each study (total number of publications considered = 266; taxonomical publications were too difficult to track with an equal

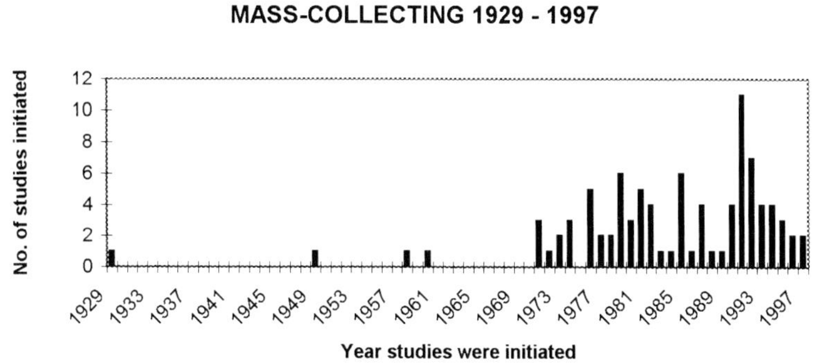

Figure 1. Mass collecting of invertebrates in the canopy of tropical rain forests: distribution of years in which the studies examined in this contribution were initiated.

accuracy for each study). To partition the respective effects of sampling methods, biogeographical effects and forest types on the data, a correspondence analysis (CA) and a canonical correspondence analysis (CCA) were performed with ADE 3.4 (Chessel & Dolédec 1992) for each level of anaysis, above, with relevant data sets. The variables for the CCA included:

– sampling methods: these were ordered along a gradient from more systematic methods to those targeting particular taxa or habitats, using Basset et al. (1997) as a guide; methods were ordered as follows: several methods used; fogging; light traps; other traps; hand collecting and related; and litter extraction;
– biogeographical regions, ordered from east to west: Australian, Oriental, Afrotropical and Neotropical;
– forest types, ordered according to a climatic gradient: subtropical forest; lowland (seasonal) dry forest; lowland wet forest; montane wet forest;
– knowledge, number of publications related to canopy invertebrates at each site; this tested for independence between the data and the effort applied to study the material collected at each site.

In order to investigate prey-predator ratios in the canopy and the relationships between ants, spiders and several taxa, similar CA and CCA were also performed. However, in this case, the taxa matrix was filled with 'prey'-predator ratios (expressed as number of 'prey' available to a particular individual of predator), instead of the number of individuals. Two separate analyses were performed, one with spiders and one with ants. Both included the following taxa as 'prey': Collembola, Thysanoptera, Psocoptera, Homoptera and all arthropod prey (minus spiders and ants).

Results and discussion

A recent field of investigation

Mass-collecting of invertebrates in the canopy of tropical rain forests is a recent field of investigation (Figure 1). The oldest attempt to collect quantitatively invertebrates in the canopy of tropical rain forests appears to be the pioneering efforts lead by O. W. Richards, who hoisted light traps up in the canopy in 1929, during the Oxford University expedition in Guyana (Hingston 1932; Sutton 2001). With the exception of the erection of towers and sporadic insect collection from these late in the 50's, insect mass-collecting from the canopy did not progress notably until the development of fogging (e.g., studies by H. R. Roberts, W. C. Gagné, T. L. Erwin) and light-trapping techniques (e.g., N. Smythe, H. Wolda), in the early 1970s. The field of research is becoming increasingly popular (Figure 1) as new methods of access in the canopy (cranes, rafts, single rope techniques, etc.) permit the experimentation with a variety of collecting methods (Basset et al. 1997; Sutton 2001). It should be noted that, since a lag of 4.6 years (on average) is observed between the initiation of the collecting and the date of the first publication studies of canopy arthropods (Erwin 1995), the trends in number of studies being initiated cannot be interpreted safely past 1993.

Bigeographical biases and biases by forest types

The sites of mass-collecting of invertebrates in the canopy are mapped in Figure 2, weighted by an estimate of knowledge (number of ecological publications relevant to each study site). The areas best studied appear to be Panama, Costa Rica, Manaus and Sulawesi (included in the 'Australian' region for the present analysis). If we compare with the actual distribution of

Table 1. Taxonomic distribution of the major phyla, classes and orders of invertebrates in the canopy of tropical rain forests, with estimates of the number of species described worldwide and of relative abundance and biomass in the canopy (average % of individuals/biomass collected, see Methods).

Taxa	No. species[a]	Abund.	Biomass	Notes	Examples
Platyhelminthes	20000	?	?	Parasites of canopy animals	Bakarr et al. 1991
Nematoda	25000	?	?	Various habitats	Chabaud & Bain 1990, Stoner 1996
Arthropoda	1085000				
Arachnida	100000				
Araneae	37000	2.85	3.47		
Pseudoscorpiones	3200	<0.30	<0.45	TRU, EPI	Morais et al. 1997; Guilbert 1997; Cotgreave et al. 1993
Opiliones	5000	<0.30	<0.45	TRU, EPI	Paoletti et al. 1991; Cotgreave et al. 1993; Nicolai 1989
Acari	45000	2.71	0.21	Various habitats, UND	Walter et al. 1994; Walter & Behan-Pelletier 1999; Franklin et al. 1997
Crustacea	40000				
Isopoda	4000	<0.30	<0.45	LIT, EPI, UND	Nadkarni & Longino 1990; Nicolai 1989; Kitching et al. 1997
Amphipoda	6000	<0.30	<0.45	LIT, EPI, UND	Nadkarni & Longino 1990; Guilbert 1997
Chilopoda	3000	<0.30	<0.45	LIT, TRU, EPI	Paoletti et al. 1991; Guilbert 1997; Kitching et al. 1997
Symphyla	120	<0.30	<0.45	LIT, TRU, EPI	Adis et al. 1997b; Nadkarni & Longino 1990
Diplopoda	10000	<0.30	<0.45	LIT, TRU, EPI	Paoletti et al. 1991; Hopkin & Read 1992; Kitching et al. 1997
Hexapoda					
Collembola	8000	6.35	0.97	Various habitats, UND	Kitching et al. 1993, 1997; Guilbert 1997; Palacios-Vargas et al. 1998
Insecta	950000	88.89	94.90		
Blattodea	4000	5.32	24.29		
Isoptera	2300	6.20	0.47	UND	Martius 1994; Eggleton et al. 1996
Dermaptera	1800	0.36	0.14		
Orthoptera	20000	1.14	2.17		
Psocoptera	3000	2.80	1.28		
Hemiptera	50000	10.44[2]	8.23	[2]Homoptera = 8.61, Heteroptera = 1.83	
Thysanoptera	5000	2.83	0.86		
Neuroptera	5000	0.50	0.69		
Coleoptera	300000	12.36	18.82		
Diptera	250000	15.42	9.39		
Lepidoptera	140000	4.44	8.78		
Hymenoptera	100000	25.99[3]	19.79	[3]Formicidae = 16.36	
Mollusca					
Gastropoda	70000	Low	Low	UND	Schowalter 1994; Guilbert 1997
Annelida	12000				
Oligochaeta	3000	Low	?	LIT, EPI, UND	Paoletti et al. 1991; Nadkarni & Longino 1990; Adis & Righi 1989
Hirudinea	500	Low	Low	LIT, EPI, UND	Paoletti et al. 1991

[a]Wells et al. 1983; Hammond 1995; J. Adis, pers. comm. UND = abundance/biomass underestimated; TRU = on trunks and limbs; LIT = in perched litter; EPI = in epiphytes.

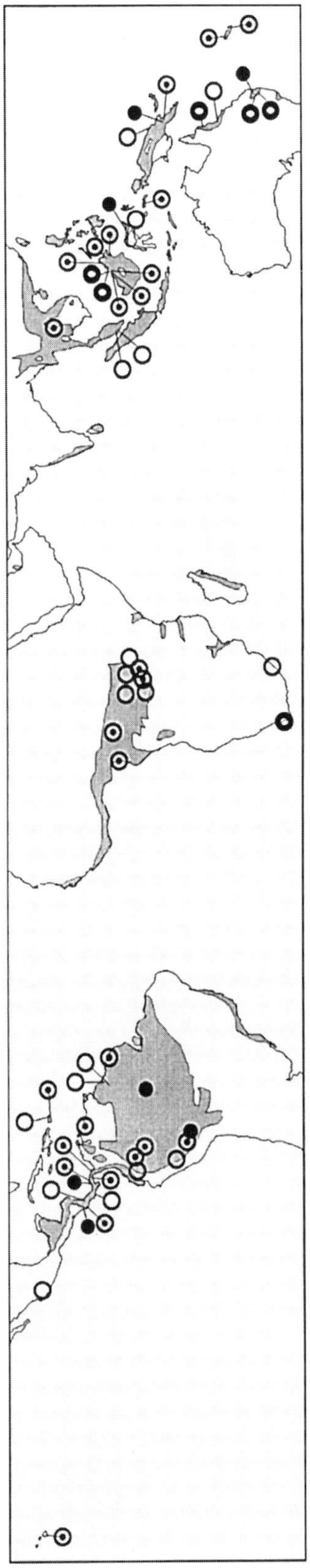

Figure 2. Location of the major sites examined in this contribution, weighted by an estimate of knowledge (see text). ○ = 1 publication, ⊙ = 2–5 publications, **O** = 6–10 publications, ● = > 10 publications. A schematic distribution of rain forests is indicated in grey (after Lieth & Werger 1989).

tropical and subtropical rain forests, where the area in km^2 of rain forests can be classified as Neotropical > Oriental > Afrotropical > Australian, there are some discrepancies with regard to the location of the major studies of canopy invertebrates. If the distribution of studies is considered either by sites or by knowledge (Figures 3a and 3b), biogeographical regions can be ranked as Neotropical > Australian > Oriental > Afrotropical. Although the Neotropical area overall appears to be the best studied, when we take account of the area of rain forests available there, it become apparent that about 40% of the information originates from Central America – from Panama and Costa Rica. The forests of the Amazon basin are comparatively less known. The contribution of the Australian region appears disproportionately high, with regard to the area covered by rain forests there. This results from the expedition of the Royal Entomological Society in 1985 in Sulawesi. In contrast, knowledge of the canopy fauna of the Afrotropical region (André et al. 1992) and of mainland SE Asia is fragmentary.

Unsurprisingly, the canopy of lowland wet forests has been studied in greater detail, whereas the canopy of lowland dry and montane forests is much less well known (Figure 4). Ten years ago, Janzen (1988a) stressed that tropical lowland dry forests represented the most endangered tropical ecosystem and that the conservation and study of this ecosystem should be given priority. To date, this plea seems to have been only partly followed by canopy entomologists. Knowledge of subtropical forests is disproportionately high, perhaps since logistics and resources, such as insect collections, in countries such as Australia are very good and permit efficient studies of canopy habitats and of the material collected.

Methodological biases

Fogging is the most popular technique among the studies surveyed, followed by combinations involving several collecting methods, light traps and hand collecting (Figure 5). Since each of these methods have biases with regard to particular taxa (e.g., Adis et al. 1984; Basset et al. 1997), it is difficult to compare the data without accounting specifically for these effects in the analyses (see below). Interestingly, if the number of publications related to each study is compared by collecting method, ratios of number of publications per study are 4.9, 3.7, 3.0 and 2.0 for several methods, hand collecting, fogging and light traps, respectively. The high score of studies relying on several methods to

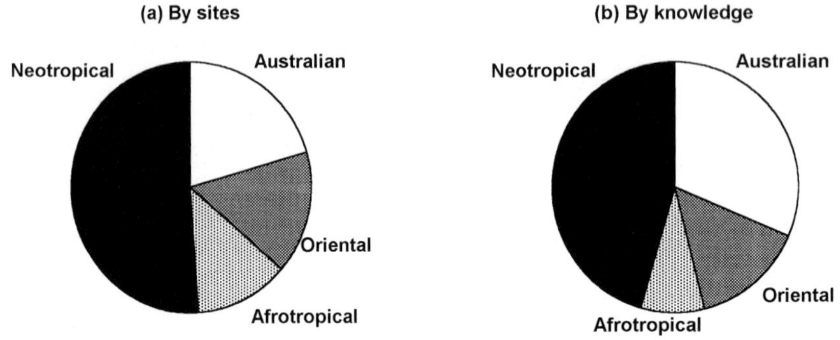

Figure 3. Distribution of studies, ordered by biogeographical regions, either by (a) sites or (b) knowledge (number of publications relevant to each site).

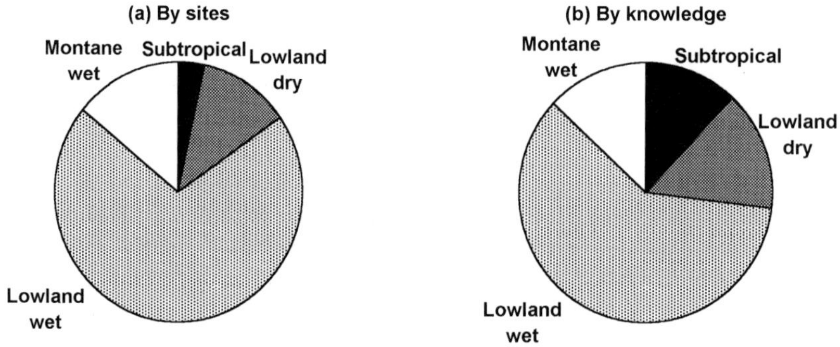

Figure 4. Distribution of studies, ordered by forest type, either by (a) sites or (b) knowledge (number of publications relevant to each site).

collects invertebrates may suggest that this strategy is also more likely to result in data directly interpretable and/or that teamwork resulted in a better knowledge of the material collected. The high score for hand collecting may result from the targeting of a particular habitat in the canopy (foliage) and particular taxa (usually insect herbivores), a strategy more amenable to analysis and publication of results than general arthropod mass-collecting.

The current knowledge of the taxonomic distribution of higher taxa in the canopy greatly depends on data collected by fogging, and of biases inherent to this and other methods commonly used by canopy workers (Figure 5). For example, inhabitants of arboreal phytotelmata (e.g., water-filled tree holes, bromeliads, ferns, etc.) represent diverse taxa which are often dominated by Diptera (e.g., Lounibos 1981; Louton et al. 1996) but cannot be sampled efficiently with fogging. Common inhabitants of bromeliads may include Oligochaeta, Opiliones and Pseudoscorpiones (e.g., Paoletti et al. 1991; Cotgreave et al. 1993; Fragoso & Rojas-Fernandes 1996). Other phytotelmata may also include some Rotifera, Polychaeta, Ostracoda or Copepoda (Kitching 1987; in press). Although crabs may be common in mangrove canopies, an endemic Jamaican crab, *Metopaulias depressus* Rathbun, is only found in arboreal bromeliads above 300 m asl (Hartnoll 1964). Similarly, the suspended litter in the canopy is often rich in Acari, Collembola, Isopoda, Myriapoda, etc. (e.g., Kitching et al. 1997; Behan-Pelletier et al. 1993) that can be best surveyed with litter extraction or related techniques. Epiphytes in the canopy often support more insects than those closer to the ground (e.g., Paoletti et al. 1991; Kitching et al. 1997; but see counter-example in Nadkarni & Longino 1990). Another important habitat in the canopy that has rarely been studied and is difficult to sample by fogging is that provided by bryophytes (e.g., Young 1986). A rich fauna of Rotifera, Tardigrada, Copepoda, Acari, Collembola, Diptera, to cite a few taxa, may be associated with arboreal mosses.

Further, some invertebrate taxa may only be present temporarily in the canopy and a fogging snapshot is unlikely to detect their presence and reveal their true abundance and interactions with other organisms in the canopy. For example, many soil taxa migrate up in the canopy of Central Amazonian inundation forests, as a response to flooding (e.g.,

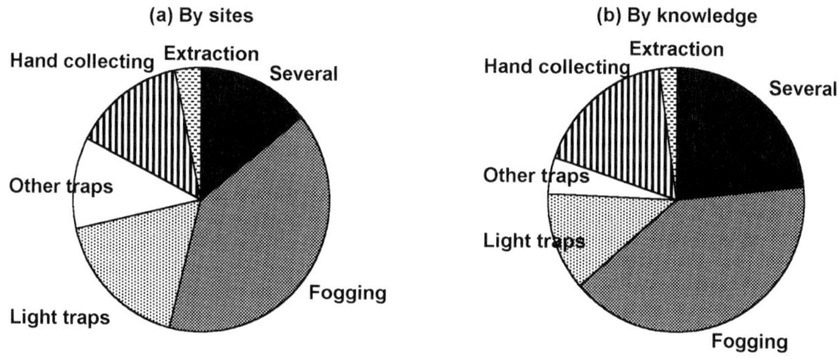

Figure 5. Distribution of studies, ordered by collecting method, either by (a) sites or (b) knowledge (number of publications relevant to each site).

Adis 1997). This may include taxa not often seen in the canopy, such as Annelida (e.g., Adis & Righi 1989), Pseudoscorpiones (e.g., Morais et al. 1997), Symphyla, Chilopoda (Adis 1997) or Scorpiones (Lourenço 1988).

Taxonomic distribution and abundance of canopy invertebrates

Current knowledge of canopy invertebrates suggests that the most abundant taxa may be Hymenoptera (mostly represented by Formicidae), Diptera, Coleoptera, Hemiptera (mostly Homoptera) and Lepidoptera (Table 1). The abundance of ants in the canopy has been reported by many studies and appears to be a characteristic of tropical rain forest canopies (review in Davidson 1997). Although often neglected in taxonomic studies, Diptera play a much more important role in arboreal community interactions than implied by their traditional designation as 'tourists' in the canopy (Didham 1997). However, a critical reading of the data presented in Table 1 is needed.

There are virtually no data about the abundance and diversity of Platyhelminthes and Nematoda in the canopy, with the exception of scattered records. These invertebrates cannot be sampled with the methods commonly used in the canopy, with the exception of litter extraction for free-living nematodes. Plants and animals in the canopy have their loads of parasitic worms (see, for example, references in Table 1). Judging by the worldwide diversity of these groups, and by the variety of life-histories of nematodes, it would be reasonable to assume that these groups must be relatively abundant in the canopy, particularly nematodes. Further, canopy animals, either vertebrates or invertebrates (e.g., Clastrier & Delecolle 1997), may also support a rich fauna of ectoparasites (Siphonaptera, Mallophaga, Anoplura, etc.).

The abundance in the canopy of other non-arthropod groups, such as Gastropoda and Annelida, may also be underestimated. Both groups are more or less confined to epiphytes, suspended soils and accumulations of litter in the canopy. The proper estimation of their abundance may require sampling their preferred habitats selectively. Further, land snails are particularly diverse on islands, partly due to the absence or low abundance of vertebrate predators there (Wells et al. 1983), and many species are truly arboreal (see, for example, Cowie 1992, on Partulidae and Achatinellidae). Studies of forest canopies on tropical islands, such as in Puerto Rico or New Caledonia, even when performed with fogging, suggest that arboreal snails may be more abundant there (e.g., Garrison & Willig 1996; Guilbert 1997).

As for non-insect arthropods, there is no *a priori* reason why the relative abundance of spiders should be greatly understimated, since they are often of similar or larger body size than many foliage insects, and can be readily collected with a variety of methods. Although Pseudoscorpiones, Opiliones, Chilopoda, Symphyla and Diplopoda may be more or less restricted to cryptic habitats in the canopy, they are, with the exception of the first order, rather large. If their abundance in the canopy is underestimated, it should not be by a high factor. The abundance of terrestrial crustaceans (mainly Isopoda and Amphipoda), often found in epigean and humid habitats, may well prove to be more seriously underestimated.

However, the most serious biases are likely to be related to microarthropods which exploit a variety of canopy habitats, such as, particularly, Acari and Collembola (see references in Table 1). Most techniques used by canopy workers are inadequate for sampling these organisms, which often represent a distinct fauna from that at ground level (e.g., Behan-Pelletier et al. 1993; Walter et al. 1994 & 1998; Wunderle 1992; Rodgers & Kitching 1998; Palacios-Vargas et al. 1998). For example, Walter and O'Dowd (1995) estimate that an individual rain forest tree in Australia may easily contain about 380 000 mites.

Numerous other insect orders other than those listed in Table 1 are present in the canopy but they are not likely to be abundant (this is also true of many other invertebrate taxa). The relative importance of the major insect orders as listed in Table 1 should be reasonably close to current knowledge, with some exceptions. Isoptera may be the second most important insect group to ants in tropical rain forests (e.g., Beebe 1925). Many species have arboreal nests and their sampling in the canopy is difficult, particularly with conventional methods (Martius 1994; Eggleton et al. 1996). Termites collected in the canopy often include alates caught by traps, not workers representing the bulk of populations and which rarely circulate beyond their galleries. The abundance of photophobe Blattodea and minute Thysanoptera may also be questioned. Thrips may be very seasonal, often associated with flowers, and the precise estimation of their abundance will require long-term surveys.

Incidentally, one common feature of many of the compiled studies is the low incidence of Apoidea in the canopy, which may appear to be surprising. Although there has been claims that good fliers such as bees may escape the fog of insecticide in the canopy, recent studies on bee stratification (e.g., Roubik 1993) suggest that relatively few species forage preferentially in the canopy.

The remainder of this contribution discusses mainly arthropods since non-arthropod data are very limited.

Biomass and density of canopy invertebrates

Blattodea, Hymenoptera (particularly ants), Coleoptera and Diptera appear to represent most of the invertebrate biomass in the canopy (Table 1). Still, estimates of percentage biomass reported in Table 1 might be imprecise and as good as those of invertebrate body weights. Although these figures may be far from reality, they draw attention to the biomass of certain taxa, such as Blattodea, which may be non-negligible in the canopy. The contribution of other groups in terms of biomass (particularly Annelida) in perched litter/epiphytes, and Isoptera in arboreal nests, may well be underestimated and needs further investigation.

Invertebrate densities in the canopy have been quantified either as number of individuals or biomass per ha of forest (e.g., Stork 1988, 1996), or as number of individuals per leaf area, particularly for insect herbivores (e.g., Basset et al. 1992). In both cases data are rare and await refinement. For example, Stork (1988) estimated that more than 42 million arthropods, equivalent to 30 kg of dry weight, were present in one ha of forest in Seram, but admitted subsequently that these figures may be too low (Stork 1996). The canopy habitat contributed more than a quarter of the individuals present and more than one third of the biomass present. Diplopoda represented most of the biomass in the soil. These calculations were derived from the numbers of arthropods collected in fogging trays of 1 m^2 and were summed over 1 ha of forest. It is not clear whether this, and other attempts to compare densities of invertebrates per m^2 of fogging trays among different sites, represent valid procedures, as the number of arthropods in the trays may be influenced greatly by the amount of leaf area in the column of foliage above the trays. Reporting data as volume of foliage suffers from the same deficiency, as it says little about leaf area, which often varies from one tree species to another. Fogging data reported as number of specimens per trays may be difficult to compare unless, for the purpose of extrapolation to a whole site, a certain number of tree species representative of the plot have been fogged (see Adis et al. 1998, for recent guidelines as to how to standardize the technique).

The data collected in Seram (Stork 1988) sparked a lively debate as to the relative contributions, either in terms of individuals, biomass or species richness of the canopy versus the soil/litter fauna in tropical rain forests (e.g., Stork 1988; Hammond 1990; André et al. 1992). Intensive fogging by T. L. Erwin in the late 1970s (e.g., Erwin 1982) revealed a rich canopy fauna and led to speculations that this fauna may be more diverse than in the soil. To date, the evidence seems to be contrary to these views (e.g., Hammond 1990, 1995; André et al. 1992; Walter et al. 1998). However, caution is required since most methods used in the canopy are inadequate to sample Acari and Collembola, which represent the dominant groups in the soil of rain forests (e.g., Stork 1988).

Invertebrate densities obtained from foliage samples in the canopy appear low, but higher than in the understorey. For example, a conservative 24 arthropod individuals in the canopy against 7 individuals in the understorey for a sample size of 0.85 m^2 of foliage were reported from a rain forest in Cameroon (Basset et al. 1992). In a nutrient-poor forest with monodominant stands in Guyana, 2.1 and 3.2 insect herbivores per sample of 1.5 m^2 of leaf area were obtained in the understorey and canopy, respectively (Basset et al. 1999). Preliminary analyses of material collected in French Guiana also suggest that insects are more abundant in the canopy than in the understorey (Lowman et al. 1998). About 1025 insect individuals per g of dry foliage were reported from a canopy in Puerto Rico (Schowalter 1994). These densities, with the exception of the data from Puerto Rico, appear to be lower than comparable data from temperate forests (Basset et al. 1992) and this confirms the suggestion that invertebrates are scarce in tropical rain forests (e.g., Elton 1975). Note that these low densities do not imply that sampling the tropical canopy for arthropods is necessarily going to be a frustrating exercise, but rather, that the abundance of arthropods is diluted into the vast standing biomass of the rain forest. Many invertebrate herbivores have specific food requirements that render most rain forest foliage unsuitable for them, being not the proper host, tissue or proper developmental stage. Hence, herbivores tend to aggregate on certain food resources and this complicates the proper estimation of their abundance, unless sampling schemes are comprehensive enough to account for such variance, using high numbers of spatial and temporal replicates.

Species richness in the canopy

Current knowledge of the species richness of canopy invertebrates would tend to indicate that Coleoptera, particularly Staphylinidae, Curculionidae and Chrysomelidae, along with Hymenoptera, Lepidoptera and Araneae, are the most speciose taxa in the canopy (Figure 6). Since in none of the studies compiled were all invertebrates sorted to species or morphospecies (Stork 1991, represented probably the highest taxonomic effort overall), one must be cautious with these figures. Further, they refer to species accumulation in samples of individuals, not to the actual number of species. Taxa that are relatively rare, such as Heteroptera, may have a steep accumulation of species in collection, but overall may not be substantially more diverse than other taxa. Species accumulation rates for Araneae and Orthoptera may also be inflated because typically many juveniles are also collected from the canopy and these are difficult to incorporate in taxonomic studies (and, therefore, in species accumulation).

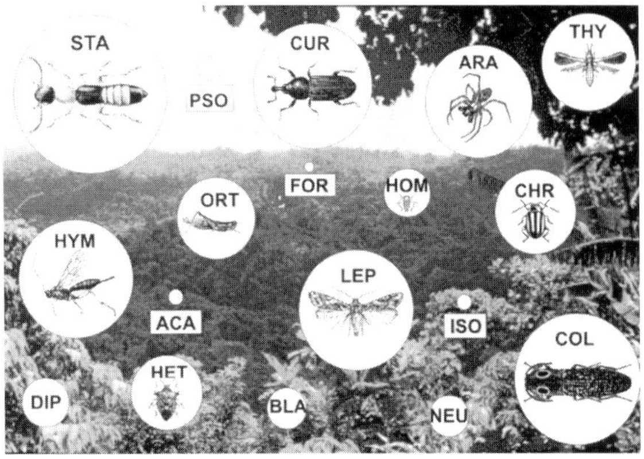

Figure 6. Species-scape in the canopy of a fictitious rain forest. The size of the circles are proportional to the putative species richness (measured as average species × individuals^{-2}) of the taxa, which are abbreviated as follows: STA: Staphylinidae (19.70); COL: other Coleoptera (15.96); CUR: Curculionidae (15.15); LEP: Lepidoptera (14.94); HYM: Hymenoptera (14.22); ARA: Araneae (13.45); THY: Thysanoptera (11.88); CHR: Chrysomelidae (10.60); ORT: Orthoptera (10.18); HET: Heteroptera (9.31); DIP: Diptera (5.66); HOM: Homoptera (5.60); NEU: Neuroptera (5.12); BLA: Blattodea (4.78); PSO: Psocoptera (3.40); ISO: Isoptera (2.27); ACA: Acari (1.80); FOR: Formicidae (1.56).

Since knowledge of Hymenoptera and Diptera in the canopy certainly do not compare with that of Coleoptera or Lepidoptera, estimates of species richness for the former two groups are suspect at best. Close to 20% of the studies compiled targeted Coleoptera so that strict comparisons with other taxa should assume a level of knowledge similar to that of the Coleoptera, if this is possible at all. The Acari, for example, are likely to contribute significantly to animal diversity in the canopy since they have various life-histories and functional roles in ecosystems and, globally, represent a diverse taxon, despite being often neglected (e.g., André et al. 1992). Walter and O'Dowd (1995) suggested that mites may be the most diverse and abundant canopy arthropods.

Massive sampling in several strata of a rain forest in Sulawesi with 1.1 million of specimens collected showed that, overall, Staphylinidae, Curculionidae and Tenebrionidae were the most species-rich families of beetles (Hammond 1990). In the canopy, Chrysomelidae seems also to be diverse, as attested by many of the compiled studies. The fact that, overall, Staphylinidae and Curculionidae may be more diverse in the canopy than Chrysomelidae is interesting. This may reflect their range of feeding habits and exploitation of rain forest habitats, since in this regard these groups can be ranked as follows: Staphylinidae (predators, fungal feeders, scavengers) > Curculionidae (wood-eaters, herbivores on roots, foliage and seeds) > Chrysomelidae (mainly herbivores on roots and foliage). Further, the species richness of these groups may also be a result from their intrinsic evolutionary dynamics, as these groups are often diverse in a variety of habitats not limited to the canopy.

The relative steep accumulation of species of Thysanoptera warrants further research, as many of them may be associated with flowers and potential pollinators. Conversely, ants seem not to be particularly diverse as their species accumulation rates are decreased by the enormous populations sustained by colonies. Wilson (1987) commented on the species-richness of ants using Peruvian data, although the species richness of other taxa within the same samples could not be compared. Indeed, the most species-rich insect families sorted from samples obtained in Borneo by Stork (1991) and collaborators included, in order of decreasing importance, Eulophidae, Encyrtidae, Aphelinidae, Curculionidae, Staphylinidae, Chrysomelidae, Formicidae, Phlaeothripidae, Cicadellidae, Chloropidae, Ceratopogonidae and Theridiidae.

Patterns of distribution of higher taxa in the canopy

The distribution of individuals among the major arthropod orders and across the studies examined was complex and depended on many factors (graph of eigenvalues of the CA, Figure 7a). Taxa of small body size, such as Collembola, Acari and Thysanoptera, scored highly on the first axis of the CA (Figure 7a). This suggests that the formation of this axis was induced by collecting methods, as well as by the specific habitat that was targeted in the canopy (e.g., epiphytes, perched litter). Interpretation of the other axes is difficult. Collectively, the variables constraining the CCA explained only 11% of the total variance (ratio of inertia of the CCA to the CA = 0.2436/2.2121). In the CCA (Figure 7b) the first axis, which explained 53% of the (explained) variance, was best correlated with collecting methods ($r = 0.862, p < 0.001$). The second canonical axis which explained 26% of the variance was best correlated with forest type ($r = 0.594$, $p < 0.05$), whereas biogeography loaded best on the third axis ($r = 0.936$, $p < 0.001$), which explained 18% of the variance. These analyses confirm that current knowledge of canopy invertebrates is strongly influenced by collecting methods. However, the proportion of variance explained by the type of forest is higher than that explained by biogeography. Collembola seem to be particularly sensitive to forest type (Figure 7b) and this may be related to the high accumulation of organic matter in the canopy or slower decomposition rates, as discussed by Kitching et al. (1997) or Palacios-Vargas et al. (1998). For example, Kitching et al. (1997) reported higher abundances of Collembola and Acari in subtropical than in tropical forests in Australia.

The situation was similar when analyses considered only particular beetle families (Figure 8). The constraining variables included in the CCA explained 26% of the overall variance (ratio of inertia = 0.2536/0.9919). The formation of the first axis of the CA was difficult to interpret. The formation of the second axis of the CA appeared to be induced by either corticolous and wood-related families (upper part of Figure 8a) vs rather folivorous or floricolous families (lower part of Figure 8a). This may result from different amounts of wood or corticolous habitats being available in different forest types, as suggested by the results of the CCA. Of the explained variance in the CCA, the first axis represented 68% of the variance and was best related to methods ($r = 0.702, p < 0.01$), whereas the second axis, which explained 17% of the variance, was best correlated with forest type ($r = 0.795$, $p < 0.001$). Beetle families particularly sensitive to the effect of forest type included Lathridiidae, Scolytidae, Tenebrionidae and Carabidae (Figure 8). Several authors, such as Hammond (1990), Stork and Brendell (1990) or Wagner (1997), reported differences in the occurrence of beetle families in different forest types.

More variance (40%) was explained by the constraining variables in the analyses considering subfamilies of Chrysomelidae (ratio of intertia of CCA to CA = 0.1552/0.3840; Figure 9). The higher amount of variance explained with analyses at the lower taxonomic rank probably results from related taxa responding in a similar way to the variables considered in the analyses than is the case with higher taxa. The significance of the first axis of the CA for chrysomelid subfamilies was obscure, with no obvious influence of either biogeography, forest type, methods or knowledge. Most (52%) of the variance explained in the CCA was best related to forest type ($r = 0.779$, $p < 0.001$), which loaded highly on the first axis. The second canonical axis contributed to 43% of the variance and was best correlated with collecting methods ($r = -0.660$, $p < 0.05$). Galerucinae appeared to

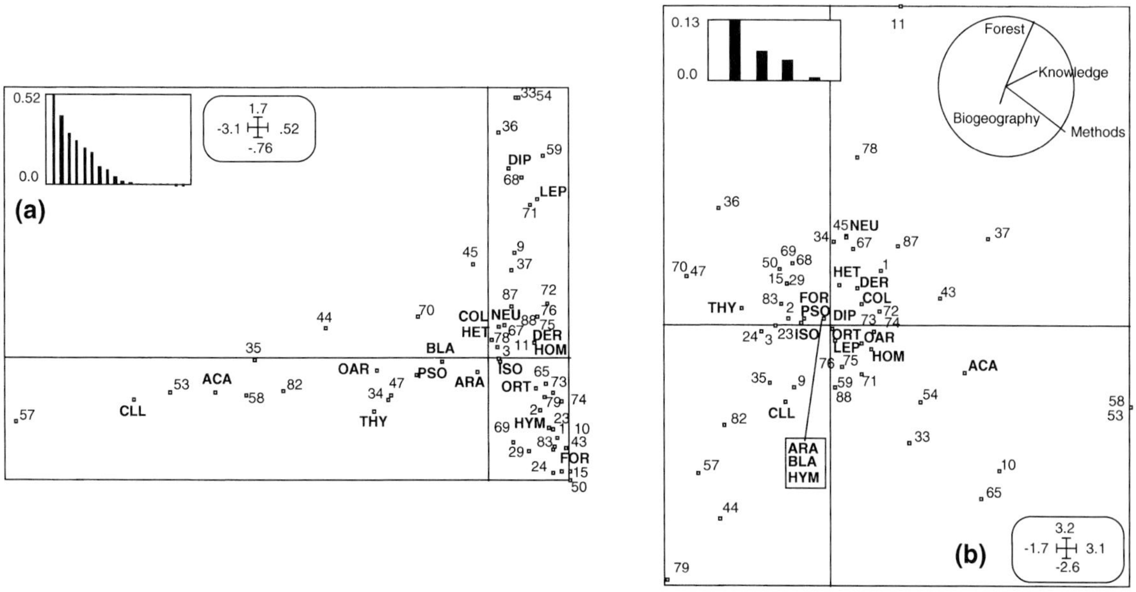

Figure 7. Ordinations of 18 higher insect taxa across 42 canopy studies amenable to analysis. Plots of the species into axes 1 and 2 of the (a) CA and (b) CCA. Inset boxes detail the graphs of eigenvalues and the circle the vectors representing the constraining variables. Studies are coded with numbers as in Appendix 1. Taxa codes: ACA = Acari, ARA = Araneae, BLA = Blattodea, CLL = Collembola, COL = Coleoptera, DER = Dermaptera, DIP = Diptera, FOR = Formicidae, HET = Heteroptera, HOM = Homoptera, HYM = Hymenoptera, ISO = Isoptera, LEP = Lepidoptera, NEU = Neuroptera, OAR = Other arthropods, ORT = Orthoptera, PSO = Psocoptera, THY = Thysanoptera.

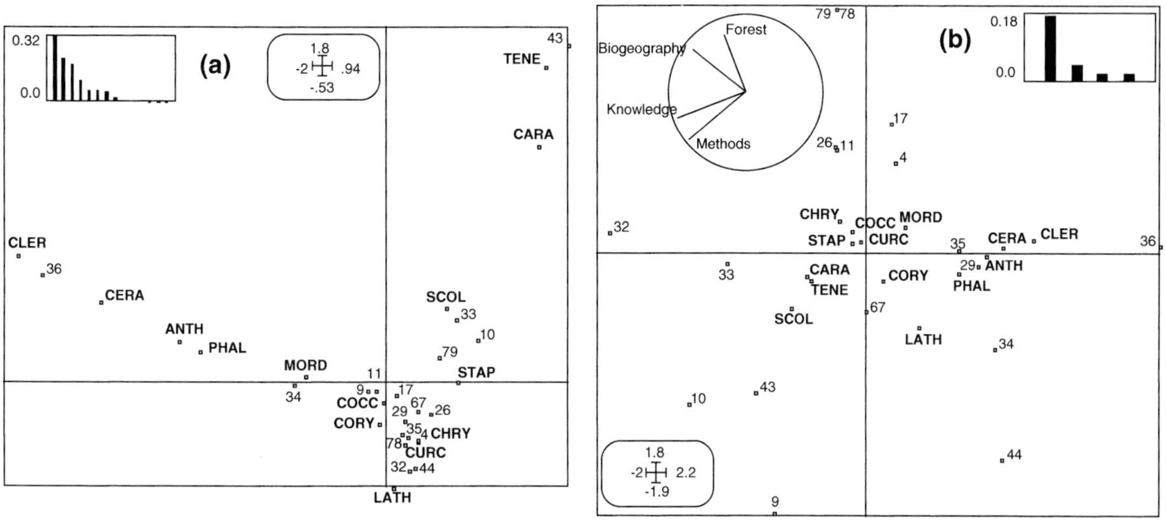

Figure 8. Ordinations of 14 beetle families across 17 canopy studies amenable to analysis. Plots of the species into axes 1 and 2 of the (a) CA and (b) CCA. Inset boxes detail the graphs of eigenvalues and the circle the vectors representing the constraining variables. Studies are coded with numbers as in Appendix 1. Taxa codes: ANTH = Anthribidae, CARA = Carabidae, CERA = Cerambycidae, CHRY = Chrysomelidae, CLER = Cleridae, COCC = Coccinellidae, CORY = Corylophidae, CURC = Curculionidae, LATH = Lathridiidae, MORD = Mordellidae, PHAL = Phalacridae, SCOL = Scolytidae, STAP = Staphylinidae, TENE = Tenebrionidae.

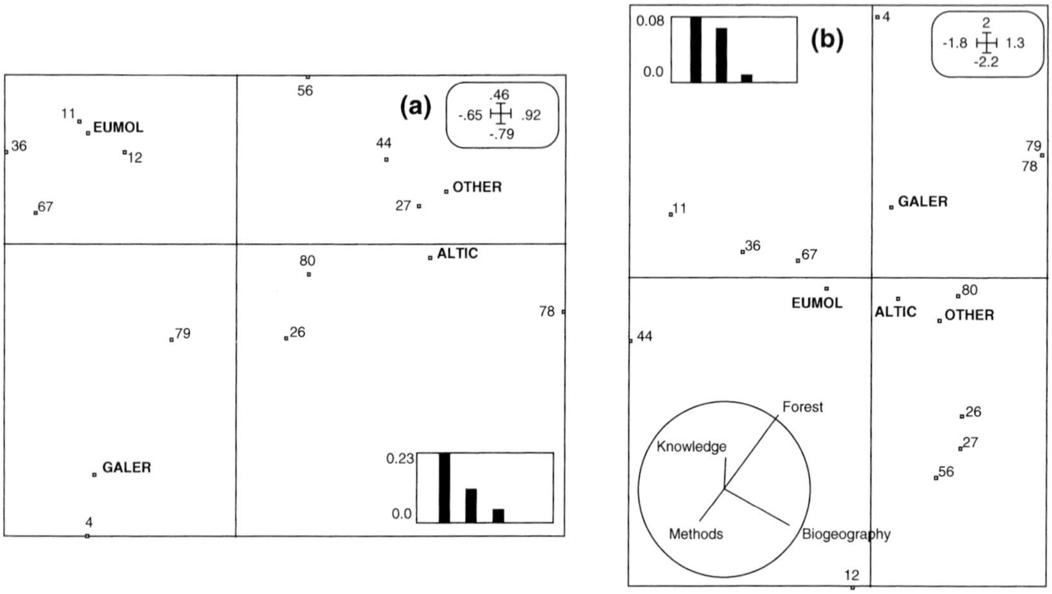

Figure 9. Ordinations of 4 chrysomelid taxa across 12 canopy studies amenable to analysis. Plots of the species into axes 1 and 2 of the (a) CA and (b) CCA. Inset boxes detail the graphs of eigenvalues and the circle the vectors representing the constraining variables. Studies are coded with numbers as in Appendix 1. Taxa codes: ALTIC = Alticinae, EUMOL = Eumolpinae, GALER = Galerucinae, OTHER = other Chrysomelidae.

be more sensitive to the effect of forest type than were other chrysomelid subfamilies, but the biological interpretation of this observation is difficult.

Prey-predator ratios and relationships between ants and invertebrates in the canopy

There were significant regressions between spiders and several taxa most likely to be their potential prey (Collembola, Thysanoptera, Psocoptera, arthropod prey, Homoptera, etc., all with $p < 0.05$). The relationships between spiders and their potential prey in the canopy appeared to be most influenced by forest type (Figure 10a, CCA plot not presented for sake of brevity). The constraining variables explained 35% of the total variance (ratio of intertia of CCA to CA = 0.3648/1.0509) and most of the explained variance (74%) was related to forest type (correlation with canonical axis 1, $r = -0.951, p < 0.001$). Subtropical and, particularly, dry forest canopies appeared to support high ratios of spiders to Collembola (expressed in numbers of Collembola available to each spider individual), and of spiders to arthropod prey. In contrast, in lowland wet forests, the proportion of ants to spiders tended to be higher (Figure 10a).

In contrast, most of regressions calculated between ants and potential prey were not significant. This confirms that ants have a variety of feeding habits and ecological roles in the canopy, which are not restricted to predatory activities (e.g., Tobin 1991). The relationships between ants and several taxa appear to be influenced mostly by forest type (Fig 10b, CCA plot not presented). The constraining variables explained 46% of the total variance (ratio of inertia of CCA to CA 0.2471/0.5404), and 61% of the explained variance was related to forest type (correlation with canonical axis 1, $r = -0.748, p < 0.001$). Ratio of ants to other taxa seemed to decrease from the montane forests to lowland wet forests (cf., Stork & Brendell 1990). Montane rain forest may be too cold to allow ants to forage efficiently or for their larvae to develop fast enough (Brown 1973), or the resource base may be too low to allow large populations of their homopteran associates to exist. In lowland forests, high ratios of ants to Homoptera are more common and this may be partly related to mutualism and the tending of homopterans in the canopy in favourable situations. This is consistent with the recent views (e.g., Tobin 1991; Davidson 1997) that most arboreal ants in tropical forests, particularly the dominant forms, are not predators, but obtain their carbohydrates mainly from plant and homopteran exudates.

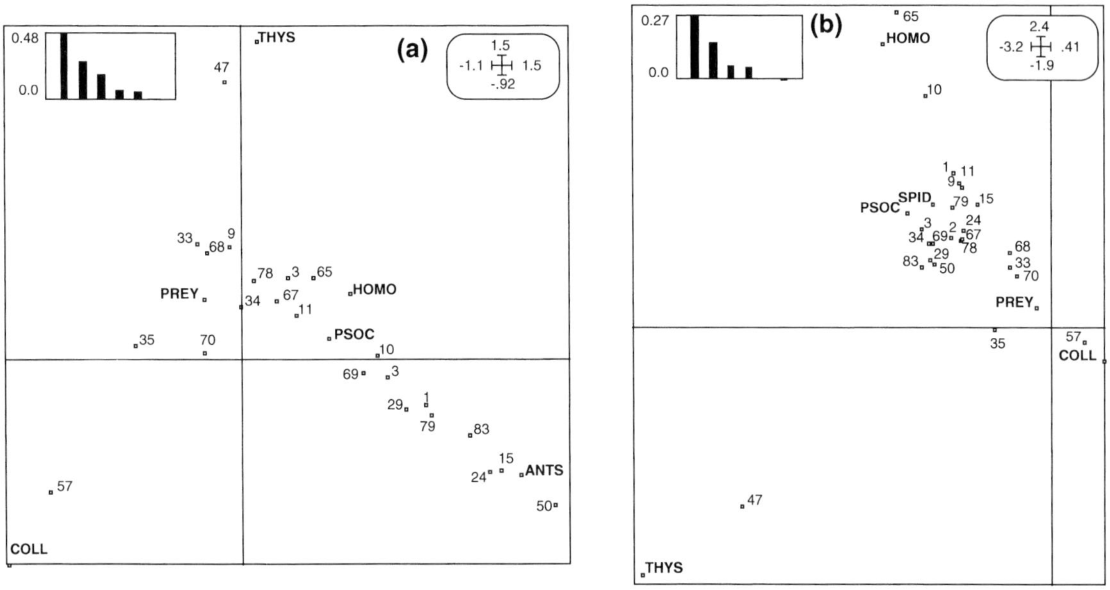

Figure 10. Correspondence analyses of prey-predator ratios across 23 canopy studies amenable to analysis. Plots of the species into axes 1 and 2 of the (a) prey-spider ratios and (b) arthropod-ant ratios. Studies are coded with numbers as in Appendix 1. Ratio codes: ANTS = Formicidae; COLL = Collembola; HOMO = Homoptera; PREY = all arthropod prey; PSOC = Psocoptera; SPID = Araneae; THYS = Thysanoptera.

Conclusions: prospects in the study of canopy invertebrates

The overall picture of the canopy as habitat for invertebrates is strongly influenced by collecting methods, a common situation in entomology. For general surveys of the canopy, it would be preferable to use a panoply of techniques rather than a particular method (Basset et al. 1997), and to standardize samples obtained with these methods (e.g., 'sampling package': Gadagkar et al. 1990; Stork 1994). Although this is common knowledge in entomology, time and financial constraints often result in only one method being applied, often fogging. The studies of Hammond (1990) and Noyes (1989) in Sulawesi, who collected Coleoptera and Hymenoptera respectively, with several complimentary methods, are exemplary and show how much can be accomplished with this strategy.

Whether the canopy should be studied on its own or conjointly with other forest habitats, such as soil and litter, is debatable. Many insect herbivores, such as some chrysomelids and curculionids, feed on roots as larvae and later migrate in the canopy to feed as adults on leaves. Although it is relatively easy to report differences in the occurrence of particular species of beetles in the adult stage either in the soil or in the canopy, our understanding of the relationships between the canopy and soil should also proceed by assessing how many insect species depend on the soil/litter habitat during their juvenile stages and on the canopy during their adult phase. Understanding the distribution of adult insects in the canopy may require solid data on their distribution as larvae in the soil (Basset & Samuelson 1996). Further, comparison between the litter and canopy faunas may emphasize specific adaptations of arboreal invertebrates which may be important from a conservation viewpoint.

For studies of specific taxa and/or topics, many specific methods have been devised and as the study of canopy invertebrates progress, new methods are likely to surface, as entomologists represent a rather ingenious breed. Recently, arboreal bait traps have been used to study dung beetles in the canopy (Davis & Sutton 1998), selective fogging with a cotton roof to study recolonization dynamics and disturbance (Floren & Linsenmair 1997; Floren & Linsenmair 2001), and bioacoustic monitoring to study sound-producing arthropods (Riede 1997), to cite a few examples.

Studies such as these, at a finer scale in the canopy, are imperative to understand better insect distribution patterns within and between tree crowns, or movements of invertebrates between different canopy habitats. Ultimately, distribution data are important to estimate the diversity, host specificity and endemicity of canopy invertebrates and how these taxa are affected by forest fragmentation and destruction. In the 1980s there have been formidable claims that the tropical canopy may harbour most of the arthropod species on Earth and that these species are highly host-specific (Erwin 1982). To date, these claims have not been substantiated as either the soil fauna appears as rich or richer than the canopy fauna (e.g., Stork 1988; Hammond 1990), or canopy insects appear much less specialised than originally thought (Basset et al. 1996a; Kitching et al. 1997). If anything, more data are needed to substantiate this. Killing insects during a series of fogging snapshots may represent a first step towards this direction, but may not, in the end, resolve the debate. Mass-rearing of live insects, obtained from various rain forest habitats, appears much more promising toward this goal.

The study of insect host specificity in tropical rain forests and in their canopy appears to be related to at least three critical issues: (1) sample size, (2) number of singleton and rare species and (3) aggregation patterns of arthropods. Since the vegetation is highly diverse in rain forests, the sample size needed to estimate the true range of a species of herbivorous insect must be high, although no guidelines exists at the moment. Sufficient spatial and temporal replicates need to be combined with natural history data, rather than analysing the distribution of dead arthropods within a few snapshot samples, even with robust mathematics. Insufficient sampling and the mass effect described by Shmida and Wilson (1985) partly explain why so many species are represented by singletons in canopy samples (e.g., Morse et al. 1988; Allison et al. 1997; Basset 1997). In tropical rain forests, the distribution of many insect herbivores is aggregated on the foliage, even for generalist species (Novotny & Basset 2000; Basset 2000). This is reflected in their apparent host specificity and rarity at low sample size (Novotny & Basset 2000). These spatial and temporal considerations are not trivial in canopy science, since canopy access is often difficult, partial and constrained. With regard to invertebrate samples obtained *in situ*, seasonal aggregation may be better studied with construction cranes (e.g., Wright 1995; Ødegaard 1999; H. Barrios pers. comm.), whereas spatial aggregation may be better studied with mobile devices such as the canopy raft or the canopy sledge (e.g., Ebersolt 1990; see Basset et al. 1992; 1997).

That the higher taxonomic composition of the invertebrate fauna and various prey-predator relationships in the canopy vary more in particular forest types than across different biogeographical regions is interesting both from the ecological and conservationist viewpoints. From the former, this suggests that similar forces act on canopy invertebrates in different biogeographical regions to result in predictable communities of canopy invertebrates, but that these forces are different across forest types. Both the nutrient status and weather conditions are likely to contribute much to higher taxonomic differences in the fauna across forest types.

From the conservation viewpoint, this suggests that canopy invertebrates inhabiting certain types of rain forests may be particularly at risk, depending on the selection pressures at work in these forests. The relationships between soil and canopy may be important in shaping these selection pressures. In dry (seasonal) rain forests, many tree species are deciduous or partly deciduous and it is probable that nutrient cycling there is quicker than in wet rain forests, where the foliage turnover of evergreen trees is slower. Canopy arthropods may present a variety of adaptations to cope with gradual or sudden leaf exchange and increase in irradiance and water stress, or they may migrate to other locations, as it is known for a variety of moths (e.g., Janzen 1988b). As such, resilience to natural or man-made disturbance may be greater in wet rain forests than in dry forests (see Janzen 1988a) and the inhabitants of the former may be particularly at risk from disturbance.

In montane rain forests, the trunks of trees often bear a thick cover of mosses and epiphytes and the canopy is often lower than in lowland rain forests. Thus, the continuity between the soil/litter and canopy in montane forests may be greater than in lowland forests, and this could explain the higher occurrence in the fomer of taxa well-represented in the soil/litter habitat, such as Collembola, Acari and Arachnida (e.g., Stork & Brendell 1990). It is possible that the resilience of a rain forest after disturbance may depend in part on the continuity between the soil/litter and canopy habitats and the ecological processes that occur there. If this is the case, the resilience after disturbance of lowland rain forests may be greater than in montane rain forests. Testing this (and the previous proposition) would be challenging, but it may provide powerful indications of which taxa in the canopy may be particularly at risk from man-induced disturbance and how best to slow down the extinction of species in these habitats.

A long-term comparative study of invertebrates in the canopy of two different forest types would represent an excellent unifying study theme, as well as a fascinating challenge, for entomologists interested in the canopy of tropical forests. Ideally, the protocol should collect invertebrates with a wide array of collecting methods in most canopy habitats (targeting soil/litter habitats less extensively but appropriately), at locations offering good canopy access and infrastructure. The taxonomic study of the material collected in such conditions is likely to be a formidable challenge, but at least the initial burden of collecting, preparing and pre-sorting this material could be greatly facilitated by working with local parataxonomists specifically trained for this purpose (e.g., Janzen 1993; Novotny et al. 1997; Basset et al. 2000). Although Lawton et al. (1998) noted the huge scale of the biological effort required to provide inventories of tropical diversity, their view was over-pessimistic as they did not conceive recourse to local parataxonomists. Further, recent advances in digital photography and computing power are likely to help greatly in this task (Novotny et al. 1997; Basset et al. 2000). As far as possible, this collaborative effort should be planned with other zoologists and botanists, perhaps taking advantage of permanent botanical plots (Godfray et al. 1999).

The present recommendation rallies earlier pleas for local, long-term and extensive inventories of biodiversity (e.g., André et al. 1992; Janzen 1993; Stork 1994), which have been only partly implemented by the scientific community due to lack of funding and truly international enthusiasm and cooperation among scientists. To date, only three arthropod inventories in tropical forests are on-going and extensive (in terms of scope, time and methods), but they are all situated in the Neotropics and do not target specifically the canopy (Guanacaste area: Janzen 1998; La Selva: Longino & Colwell 1997; Manaus: Fonseca et al. 1998). Two of these inventories result from routine work of local parataxonomists. This type of study is not currently fashionable and may often be neglected by funding agencies, in comparison with reductionist studies of a handful of organisms (e.g., Lawton 1991). However, it appears to be one good strategy of research – if not the only one – towards a better understanding and conservation of the myriad animal and plant interactions in tropical rain forests.

Acknowledgements

I would like to express my thanks to the organisers of the Tropical Canopy Programme at the European Science Foundation to give me the impetus to write this review. Many colleagues and friends helped to collate the information presented here, and often supplied information unpublished or in press: Joachim Adis, Hector Barrios, Aimé and Patrick Basset, Robert Cowie, Giulio Cuccodoro, Andrew Davis, Andreas Floren, Milton Garcia, Eric Guilbert, Roger Kitching, Scott Miller, Olivier Missa, Frode Ødegaard, Maurizio Paoletti, Neil Springate, Thomas Wagner and Henk Wolda. Scott Miller, Vojtech Novotny, Neil Springate and Henk Wolda commented on and improved the MS, which was written while in receipt of a Tupper fellowship at the Smithsonian Tropical Research Institute in Panama.

References

Adis, J. 1997. Terrestrial invertebrates: survival strategies, group spectrum, dominance and activity patterns. Pp. 318–330. In: Junk, W. J. (ed.), The Central Amazon Floodplain. Ecological studies, Vol. 126. Springer-Verlag, Berlin.

Adis, J., Basset, Y., Floren, A., Hammond, P. M. & Linsenmair, K. E. 1998. Canopy fogging of an overstorey tree – Recommendations for standardization. Ecotropica 4: 93–97.

Adis, J., Lubin, Y. D. & Montgomery, G. G. 1984. Arthropods from the canopy of inundated and Terra firme forest near Manaus, Brazil, with critical consideration on the Pyrethrum-fogging technique. Stud. Neotrop. Fauna Environ. 19: 223–236.

Adis, J., Paarmann, W., da Fonseca, C. R. V. & Rafael, J. A. 1997a. Knockdown efficiency of natural pyrethrum and survival rate of living arthropods obtained by canopy fogging in Central Amazonia. Pp. 67–81. In: Stork, N. E., Adis, J. & Didham, R. K. (eds), Canopy arthropods. Chapman & Hall, London.

Adis, J. & Righi, G. 1989. Mass migration and life cycle adaptation – a survival strategy of terrestrial earthworms in Central Amazonian inundation forests. Amazonia 11: 23–30.

Adis, J., Scheller, U. Wellington, J. de M., Rochus, C., Gomes Rodrigues, J. M. 1997b. Symphyla from Amazonian non-flooded upland forests and their adaptations to inundation forests. Ent. Scand. Suppl. 51: 307–317.

Allison, A., Samuelson, G. A. & Miller, S. E. 1997. Patterns of beetles species diversity in *Castanopsis acuminatissima* (Fagaceae) trees studied with canopy fogging techniques in mid-montane New Guinea rain forest. Pp. 224–236. In: Stork, N. E., Adis, J. & Didham, R. K. (eds), Canopy arthropods. Chapman & Hall, London.

Amedegnato, C. 1997. Diversity of an Amazonian canopy grasshopper community in relation to resource partitioning and phylogeny. Pp. 281–319. In: Stork, N. E., Adis, J. & Didham, R. K. (eds), Canopy Arthropods. Chapman & Hall, London.

André, H. M., Lebrun, P. & Noti, M.-I. 1992. Biodiversity in Africa: a plea for more data. J. Afr. Zool. 106: 3–15.

Bakarr, M. I., Gbakima, A. A., Bah, Z. 1991. Intestinal helminth parasites in free-living monkeys from a west African rain forest. Afr. J. Ecol. 29: 170–172.

Barrios, H. 1997. Fluctuacion poblacional de curculionidos (Coleoptera: Curculionidae) capturados en trampa de luz en la isla Barro Colorado. Scientia 12: 55–68.

Basset, Y. 1991. The taxonomic composition of the arthropod fauna associated with an Australian rain forest tree. Aust. J. Zool. 39: 171–190.

Basset, Y. 1997. Species – abundance and body size relationships in insect herbivores associated with New Guinea forest trees, with particular reference to insect host-specificity. Pp. 237–264. In: Stork, N. E., Adis, J. & Didham, R. K. (eds), Canopy arthropods. Chapman & Hall, London.

Basset, Y. 2000. Insect herbivores foraging on seedlings in an unlogged rain forest in Guyana: spatial and temporal considerations. Stud. Neotrop. Fauna Environ., 35: 115–129.

Basset, Y., Aberlenc, H.-P. & Delvare, G. 1992. Abundance and stratification of foliage arthropods in a lowland rain forest of Cameroon. Ecol. Entomol. 17: 310–318.

Basset, Y., Charles, E. C. & Novotny, V. 1999. Insect herbivores on parent trees and conspecific seedlings in a rain forest in Guyana. Selbyana 20: 146–158.

Basset, Y., Novotny, V., Miller, S. E. & Pyle, R. 2000. Quantifying biodiversity: Experience with parataxonomists and digital photography in New Guinea and Guyana. BioScience 50: 899–908.

Basset, Y. & Samuelson, G. A. 1996. Ecological characteristics of an arboreal community of Chrysomelidae in Papua New Guinea. Pp. 243–262. In: Jolivet, P. H. A. & Cox, M. L. (eds), Chrysomelidae Biology. Volume 2: Ecological studies. SPB Academic Publishing, Amsterdam.

Basset, Y., Samuelson, G. A., Allison, A. & Miller, S. E. 1996a. How many host-specific insect species feed on a species of tropical tree? Biol. J. Linn. Soc. 59: 201–216.

Basset, Y., Samuelson, G. A. & Miller, S. E. 1996b. Similarities and contrasts in the local insect faunas associated with ten forest tree species of New Guinea. Pac. Sci. 50: 157–183.

Basset, Y., Springate, N. D., Aberlenc, H.-P. & Delvare, G. 1997. A review of methods for sampling arthropods in tree canopies. Pp. 27–52. In: Stork, N. E., Adis, J. & Didham, R. K. (eds), Canopy arthropods. Chapman & Hall, London.

Beebe, W. 1925. Studies of a tropical jungle; one quarter of a square mile of jungle at Kartabo, British Guiana. Zoologica 6: 4–193.

Behan-Pelletier, V. M., Paoletti, M. G., Bissett, B. & Stinner, B. R. 1993. Oribatid mites of forest habitats in northern Venezuela. Trop. Zool., Special Issue 1: 39–54.

Broadhead, E. & Wolda, H. 1985. The diversity of Psocoptera in two tropical forests in Panama. J. Anim. Ecol. 54: 739–754.

Brown, W. L. Jr. 1973. A comparison of the Hylean and Congo-West African rain forest ant faunas. Pp. 161–185. In: Meggers, B. J., Ayensu, E. S. & Duckworth, W. D. (eds), Tropical forest ecosystems in Africa and South America: a comparative review. Smithsonian Institution Press, Washington.

Casson, D. S. & Hodkinson, I. D. 1991. The Hemiptera (Insecta) communities of tropical rain forest in Sulawesi. Zool. J. Linnol. Soc. 102: 253–275.

Chabaud, A. G. & Bain, O. 1990. Three new filariae from African rain forest birds. Bull. Mus. Nat. Hist. Nat. A 12: 9–18.

Chessel D. & Dolédec, S. 1992. ADE Version 3.4. HyperCard stacks and quick basic Microsoft programme library for the analysis of environmental data, user's manual, modules graphiques, fiches pratiques, URA CNRS 1451, Université Lyon 1, Villeurbane, France.

Chey, V. K., Holloway, J. D., Hambler, C. & Speight, M. R. 1998. Canopy knockdown of arthropods in exotic plantations and natural forest in Sabah, north-east Borneo, using insectidal mist-blowing. Bull. Ent. Res. 88: 15–24.

Clastrier, J. & Delecolle, J. C. 1997. Description de *Forcipomya* (*Trichohelea*) *roubaudi* n. sp., ectoparasite d'un Hétéroptère Reduviidae capturé dans la canopée de la forêt guyanaise (Diptera, Ceratapogonidae). Bull. Soc. Ent. France 102: 379–383.

Coley, P. D. & Aide, T. M. 1991. Comparison of herbivory and plant defense in temperate and tropical broad-leaved forests. Pp. 25–49. In: Price P. W., Lewinsohn, T. M., Fernandes, G. W. & Benson, W. W. (eds), Plant-animal interactions: evolutionary ecology in tropical and temperate regions. Wiley, New York.

Coley, P. D. & Barone, J. A. 1996. Herbivory and plant defenses in tropical forests. Annu. Rev. Ecol. 27: 305–335.

Corbet, P. S. 1961. Entomological studies from a high tower in Mpanga Forest, Uganda. VI. Nocturnal flight activity of Culicidae and Tabanidae as indicated by light traps. Trans. Roy. Ent. Soc. Lond. 113: 301–314.

Cotgreave, P., Hill, M. J. & Middleton, D. A. J. 1993. The relationship between body size and population size in bromeliad tank faunas. Biol. J. Linn. Soc. 49: 367–380.

Cowie, R. H. 1992. Evolution and extinction of Partulidae, endemic Pacific island land snails. Phil. Trans. Roy. Soc. Lond. B 335: 167–191.

Davidson, D. W. 1997. The role of resource imbalances in the evolutionary ecology of tropical arboreal ants. Biol. J. Linn. Soc. 61: 153–181.

Davies, J. G., Stork, N. E., Brendell, M. J. D. & Hine, S. J. 1997. Beetle species diversity and faunal similarity in Venezuelan rain forest tree canopies. Pp. 85–103. In: Stork, N. E., Adis, J. & Didham, R. K. (eds), Canopy arthropods. Chapman & Hall, London.

Davis, A. J. & Sutton, S. L. 1998. The effects of rain forest canopy loss on arboreal dung beetles in Borneo: implications for the measurement of biodiversity in derived tropical ecosystems. Diversity Distrib. 4: 167–173.

Dejean, A., Belin, M. & McKey, D. 1992. Les relations plantes-fourmis dans la canopée. Pp. 76–80. In: Hallé, F. & Pascal, O. (eds) Biologie d'une canopée de forêt équatoriale – II. Rapport de Mission: radeau des cimes octobre novembre 1991, Réserve de Campo, Cameroun. Fondation Elf, Paris.

Dejean, A., Orivel, J., Corbara, B., Delabie, J. & Teillier, L. 1998. La mosaïque des fourmis arboricoles. Pp. 140–153. In: Hallé, F. (ed) Biologie d'une Canopée de Forêt Equatoriale – III. Rapport de la Mission d'Exploration Scientifique de la Canopée de Guyane, Octobre – Décembre 1996. Pro-Natura International & Opération Canopée, Paris.

DeVries, P. J., Murray, D. & Lande, R. 1997. Species diversity in vertical, horizontal and temporal dimensions of a fruit-feeding butterfly community in an Ecuadorian rain forest. Biol. J. Linn. Soc. 62: 343–364.

Dial, R. & Roughgarden, J. 1995. Experimental removal of insectivores from rain forest canopy: direct and indirect effects. Ecology 76: 1821–1834.

Didham, R. K. 1997. Dipteran tree-crown assemblages in a diverse southern temperate rain forest. Pp. 320–343. In: Stork, N. E., Adis, J. & Didham, R. K. (eds), Canopy arthropods. Chapman & Hall, London.

Ebersolt, G. 1990. Bilan technique partie radeau. Pp. 15–26. In: Hallé, F. & Blanc, P. (eds), Biologie d'une canopée de forêt équatoriale. Rapport de Mission. Radeau des Cimes Octobre-Novembre 1989, Guyane Française. Montpellier II and CNRS-Paris VI, Montpellier/Paris.

Eggleton, P., Bignell, D. E., Sands, W. A., Mawdsley, N. A., Lawton, J. H., Wood, T. G. & Bignell, N. C. 1996. The diversity, abundance and biomass of termites (Isoptera) under differing levels of forest disturbance in the Mbalmayo Forest Reserve, southern Cameroon. Phil. Trans. Roy. Soc. Lond. B 351: 51–68.

Elton, C. S. 1975. Conservation and the low population density of invertebrates inside neotropical rain forests. Biol. Cons. 7: 3–15.

Erwin, T. L. 1982. Tropical forests: their richness in Coleoptera and other arthropod species. Col. Bull. 36: 74–75.

Erwin, T. L. 1983. Beetles and other insects of tropical forest canopies at Manaus, Brazil, sampled by insecticidal fogging. Pp. 59–76. In: Sutton, S. L., Whitmore, T. C. & Chadwick, A. C. (eds), Tropical rain forest: ecology and management. Oxford, Blackwell.

Erwin, T. L. 1989. Canopy arthropod biodiversity: a chronology of sampling techniques and results. Rev. Per. Ent. 32: 71–77.

Erwin, T. L. 1995. Measuring arthropod biodiversity in the tropical forest canopy. Pp. 109–127. In: Lowman, M. D. & Nadkarni, N. M. (eds), Forest canopies. Academic Press, San Diego.

Erwin, T. L. & Scott, J. C. 1980. Seasonal and size patterns, trophic structure and richness of Coleoptera in the tropical arboreal ecosystem: the fauna of the tree *Luehea seemannii* Triana and Planch in the Canal Zone of Panama. Col. Bull. 34: 305–322.

Farrell, B. D. & Erwin, T. L. 1988. Leaf-beetle community structure in an amazonian rain forest canopy. Pp. 73–90. In: Jolivet, P., Petitpierre, E. & Hsiao, T. H. (eds), Biology of chrysomelidae. Kluwer Academic Publishers, Dordrecht.

Fisk, F. W. 1983. Abundance and diversity of arboreal Blattaria in moist tropical forests in the Panama Canal area and Costa Rica. Trans. Am. Ent. Soc. 108: 479–490.

Floren, A. & Linsenmair, K. E. 1997. Diversity and recolonization dynamics of selected arthropod groups on different tree species in a lowland rain forest in Sabah, Malaysia with special reference to Formicidae. Pp. 344–381. In: Stork, N. E., Adis, J. & Didham, R. K. (eds), Canopy arthropods. Chapman & Hall, London.

Floren, A. & Linsenmair, K. E. 2001. Changes in arboreal arthropod communities along a disturbance gradient. Plant Ecol. 153: 153–167 (this volume).

Fonseca, C. R. V. da, Adis, J. & Martius, C. 1998. Mechanisms that maintain tropical diversity – a project of Teuton-Brazilian cooperation 1991-1996. Acta Amazon. 28: 205–215.

Fragoso, C. & Rojas-Fernandez, P. 1996. Earthworms inhabiting bromeliads in Mexican tropical rain forests: ecological and historical determinants. J. Trop. Ecol. 12: 729–734.

Franklin, E., Adis, J., Woas, S. 1997. The oribatid mites. Pp. 331–349. In: Junk, W. J. (ed.), The Central Amazon floodplain. Ecological Studies, Vol. 126. Springer-Verlag, Berlin.

Gadagkar, R., Chandrashekara, K. & Nair, P. 1990. Insect species diversity in the tropics: sampling methods and a case study. J. Bombay Nat. Hist. Soc. 87: 337–353.

Gagné, W. C. 1979. Canopy-associated arthropods in Acacia Koa and Metrosideros trees communities along an altitudinal transect on Hawaï Island. Pac. Ins. 21: 56–82.

Galindo, P., Trapido, H., Carpenter, S. J., Blanton, F. S. 1956. The abundance cycles of arboreal mosquitoes during six years at a sylvan yellow fever locality in Panama. Ann. Ent. Soc. Am. 49: 543–547.

Garcia, M. N. J. 1999. Estructura y dinamica de los insectos del orden Coleoptera en la copa del arbol *Luehea seemannii* Triana y Planch 1862 (Tiliaceae), en el dosel del bosque del Parque Metropolitano, Panama. Tesis de Maestria en Ecologia y Conservación, Universidad Santa Maria La Antigua, Panama.

Garrison, R. W. & Willig, M. R. 1996. Arboreal invertebrates. Pp. 183–245. In: Reagan, D. P. & Waide, R. B. (eds), The food web of a tropical rain forest. The University of Chicago Press, Chicago.

Godfray, H. C., Lewis, O. T. & Memmott, J. 1999. Studying insect diversity in the tropics. Phil. Trans. Roy. Soc. Biol. Sci. 354: 1811–1824.

Guilbert, E. 1997 Arthropod biodiversity in the canopy of New Caledonian forests. Pp. 265–277. In: Stork, N. E., Adis, J. & Didham, R. K. (eds), Canopy arthropods. Chapman & Hall, London.

Hammond, P. M. 1990. Insect abundance and diversity in the Dumoga-Bone National Park, N. Sulawesi, with special reference to the beetle fauna of lowland rain forest in the Toraut region. Pp. 197–254. In: Knight, W. J. & Holloway, J. D. (eds), Insects and the rain forests of South East Asia (Wallacea). The Royal Entomological Society of London, London.

Hammond, P. M. 1992. Species inventory. Pp. 17–39. In: Groombridge, B. (ed.), Global biodiversity, status of the earth's living resources. Chapman & Hall, London.

Hammond, P. M. 1995. Magnitude and distribution of biodiversity. Pp. 113–138. In: Heywood, V. T. & Watson, R. T. (eds), Global biodiversity assessment. Cambridge University Press, Cambridge.

Hammond, P. M., R. L. Kitching & N. E. Stork 1996. The composition and richness of the tree-crown Coleoptera assemblage in an Australian subtropical forest. Ecotropica 2: 99–108.

Hartnoll, R. G. 1964. The freshwater graspid crabs of Jamaica. Proceedings of the Linnean Society of London 175: 145–169.

Hill, C. J. & Cermak, M. 1997. A new design and some preliminary results for a flight intercept trap to sample forest canopy arthropods. Aust. J. Ent. 36: 51–55.

Hingston, R. W. G. 1932. A naturalist in the Guiana forest. Edward Arnold & Co., London.

Höfer, H., Brescovit, A. D. Adis, J. & Paarman, W. 1994. The spider fauna of neotropical tree canopies in Central Amazonia: first results. Stud. Neotrop. Fauna Environm. 29: 23–32.

Holloway, J. D. 1987. Macrolepidoptera diversity in the Indo-Australian tropics: geographic, biotopic and taxonomic variations. Biol. J. Linnol. Soc. 30: 325–341.

Hopkin, S. P. & Read, H. J. 1992. The biology of millipedes. Oxford Science Publications, Oxford.

Janzen, D. H. 1988a. Tropical dry forests. The most endangered major tropical ecosystem. Pp. 130–137. In: Wilson, E. O. (ed), Biodiversity. National Academy Press, Washington.

Janzen, D. H. 1988b. Ecological characterization of a Costa Rican dry forest caterpillar fauna. Biotropica 20: 120–135.

Janzen D. H. 1993. Taxonomy: Universal and essential infrastructure for development and management of tropical wildland biodiversity. Pp. 100–113. In: Sandlund, O. T. & Schei, P. J. (eds), Proceedings of the Norway/UNEP expert conference on biodiversity. Directorate for Nature Management and Norwegian Institute for Nature Research, Trondheim.

Janzen, D. H. 1998. How to grow a wildland: the gardenification of nature. Ins. Sci. Appl. 17: 269–276.

Junk, W. J. 1997. The Central Amazon Floodplain. Ecology of a pulsing system. Springer-Verlag, Berlin.

Kato, M., Inoue, T. Hamid, A. A. Nagamitsu, T. Merdek, M. B. Nona, A. R. Itino, T., Yamane, S., Yumoto, T. 1995. Seasonality and vertical structure of light-attracted insect communities in a Dipterocarp forest in Sarawak. Res. Popul. Ecol. 37: 59–79.

Kitching, R. L. 1987. A preliminary account of the metazoan food webs in phytotelmata from Sulawesi. Malay. Nat. J. 41: 1–12.

Kitching, R. L. In press. Container Habitats and Foodwebs. Kluwer, Amsterdam.

Kitching, R. L., Bergelson, J. M., Lowman, M. D., McIntyre, S. & Carruthers, G. 1993. The biodiversity of arthropods from Australian rain forest canopies: general introduction, methods, sites and ordinal results. Aust. J. Ecol. 18: 181–191.

Kitching, R. L., Mitchell, H., Morse, G. & Thebaud, C. 1997. Determinants of species richness in assemblages of canopy arthropods in rain forests. Pp. 131–150. In: Stork, N. E., Adis, J. & Didham, R. K. (eds), Canopy arthropods. Chapman & Hall, London.

Knight, W. J. & Holloway, J. D. 1990. Insects and the Rain Forests of South East Asia (Wallacea). Royal Entomological Society of London, London.

Lawton, J. H. 1991. Species richness, population abundances, and body sizes in insect communities: tropical versus temperate comparisons. Pp. 209–225. In: Price, P. W., Lewinsohn, T. M., Fernandes, G. W. & Benson, W. W. (eds), Plant-animal interactions: evolutionary ecology in tropical and temperate regions. Wiley, New York.

Lawton, J. H., Bignell, D. E., Bolton, B., Bloemers, G. F., Eggleton, P., Hammond, P. M., Hodda, M., Holt, R. D., Larsen, T. B., Mawdsley, N. A., Stork, N. E., Srivastava, D. S. & Watt, A. D. 1998. Biodiversity inventories, indicator taxa and effects of habitat modification in tropical forest. Nature 391: 72–76.

Leigh, E. G., Rand, A. S. & Windsor, D. M. 1996. The ecology of a tropical forest: seasonal rhythms and long-term changes. Second edition. Smithsonian Institution Press, Washington.

Lieth, H. & Werger, M. J. A. 1989. Tropical rain forest ecosystems. Biogeographical and ecological studies. Elsevier, Amsterdam.

Longino, J. T. & Colwell, R. C. 1997. Biodiversity assessment using structured inventory: capturing the ant fauna of a tropical rain forest. Ecol. Appl. 7: 1263–1277.

Lounibos, L. P. 1981. Habitat segregation among African treehole mosquitoes. Ecol. Entomol. 6: 129–154.

Lourenço, W. R. 1988. Synopsis de la faune scorpione de la région de Manaus, Etat d' Amazonas, Brésil, avec description de deux nouvelles espèces. Amazonia 10: 327–337.

Louton, J., Gelhaus, J. & Bouchard, R. 1996. The aquatic macrofauna of water-filled bamboo (Poaceae: Bambusoideae: *Guadua*) internodes in a Peruvian lowland tropical forest. Biotropica 28: 228–242.

Lowman, M. D. 1995. Herbivory as a canopy process in rain forest trees. Pp. 431–455. In: Lowman, M. D. & Nadkarni, N. M. (eds), Forest canopies. Academic Press, San Diego.

Lowman, M. D., Foster, R., Wittman, P. & Rinker, H. B. 1998. Herbivory and insect loads on epiphytes, vines and host trees in the rain forest canopy of French Guiana. Pp. 116–128. In: Hallé, F. (ed), Biologie d'une canopée de forêt equatoriale – III. Rapport de la mission d'exploration scientifique de la Canopée de Guyane, Octobre–Décembre 1996. Pro-Natura International & Opération Canopée, Paris.

Lowman, M. D. & Moffett, M. 1993. The ecology of tropical rain forest canopies. T.R.E.E. 8: 104–107.

Lowman, M. D. & Wittman, P. K. 1996. Forest canopies: methods, hypotheses, and future directions. Annu. Rev. Ecol. Syst. 27: 55–81.

Malcolm, J. R. 1997. Insect biomass in Amazonian forest fragments. Pp. 510–533. In: Stork, N. E., Adis, J. & Didham, R. K. (eds), Canopy arthropods. Chapman & Hall, London.

Martius, C. 1994. Diversity and ecology of termites in Amazonian forests. Pedobiologia 38: 407–428.

May, R. M. 1994. Past efforts and future prospects towards understanding how many species there are. Pp. 71–84. In: Sollbring, O. T., van Emden, H. M. & van Oordt, P. G. W. J. (eds), Biodiversity and global change. C.A.B. International, Wallingford.

McClure, H. E. 1978. Some arthropods of the dipterocarp forest canopy in Malaya. Malay. Nat. J. 32: 31–51.

Missa, O. 1999. Diversity and spatial heterogeneity of a weevil fauna living in the canopy of a tropical lowland rain forest in Papua New Guinea. Thèse, Université Libre de Bruxelles, Bruxelles.

Moffett, M. W. & Lowman, M. D. 1995. Canopy access techniques. Pp. 3–26. In: Lowman, M. D. & Nadkarni, N. M. (eds), Forest canopies. Academic Press, San Diego.

Morais, J. W., Adis, J., Manhert, V., Berti-Filho, E. 1997. Abundance and phenology of Pseudoscorpiones (Arachnida) from a mixedwater inundation forest in Central Amazonia, Brazil. Rev. Suisse Zool. 104: 475–483.

Moran, C. V. & Southwood, T. R. E. 1982. The guild composition of arthropod communities in trees. J. Anim. Ecol. 51: 289–306.

Moran, C. V., Hoffmann, J. H. Impson, F. A. C. & Jenkins, J. F. G. 1994. Herbivorous insect species in the tree canopy of a relict South African forest. Ecol. Entomol. 19: 147–154.

Morse, D. R., Stork, N. E. & Lawton, J. H. 1988. Species number, species abundance and body length relationships of arboreal beetles in Bornean lowland rain forest trees. Ecol. Entomol. 13: 25–37.

Nadkarni, N. M. & Longino, J. T. 1990. Invertebrates in canopy and ground organic matter in a Neotropical montane forest, Costa Rica. Biotropica 22: 286–289.

Ng, R. 1978. The vertical distribution of aerial insects in Pasoh Forest Reserve. Malay. Nat. J. 30: 299–305.

Nicolai, V. 1989. Thermal properties and fauna of the bark of trees in two different African ecosystems. Oecologia 80: 421–430.

Novotny, V. & Basset, Y. 2000. Ecological characteristics of rare species in communities of tropical insect herbivores: pondering the mystery of singletons. Oikos 89: 564–572.

Novotny, V., Basset, Y., Miller, S. E., Allison, A., Samuelson, G. A. & Orsak, L. J. 1997. The diversity of tropical insect herbivores: an approach to collaborative international research in Papua New Guinea. Pp. 112–125. In: Lee, B. H., Choe, J. C. & Han, H. Y. (eds), Proceedings of the International Conference on Taxonomy and Biodiversity Conservation in the East Asia. Korean Institute for Biodiversity Research of Chonbuk National University, Chonju.

Noyes, J. S. 1989. The diversity of Hymenoptera in the tropics with special reference to Parasitica in Sulawesi. Ecol. Entomol. 14: 197–207.

Ødegaard, F. 1999. Host specificity as a parameter in estimates of arthropod species richness. PhD thesis, Norvegian University of Science and Technology, Trondheim.

Palacios-Vargas, J. G., Castaño Meneses, G. & Gómez-Anaya, J. A. 1998. Collembola from the canopy of a Mexican tropical deciduous forest. Pan-Pac. Ent. 74: 47–54.

Paoletti, M. G., Taylor, R. A. J., Stinner, B. R., Stinner, D. H., & Benzing, D. H. 1991. Diversity of soil fauna in the canopy and forest floor of a Venezuelan cloud forest. J. Trop. Ecol. 7: 373–383.

Penny, A. D. & Arias, J. R. 1982. Insects of an Amazon forest. New York, Columbia University Press.

Rees, C. J. C. 1983. Microclimate and the flying Hemiptera fauna of a primary lowland rain forest in Sulawesi. Pp. 121–138. In: Sutton, S. L. Whitmore, T. C. & Chadwick, A. C. (eds), Tropical rain forest: ecology and management. Oxford, Blackwell.

Riede, K. 1997. Bioacoustic monitoring of insect communities in a Bornean rain-forest canopy. Pp. 442–452. In: Stork, N. E., Adis, J. & Didham, R. K. (eds), Canopy arthropods. Chapman & Hall, London.

Roberts, H. R. 1973. Arboreal Orthoptera in the rain forests of Costa Rica collected with insecticide: a report on the grasshoppers (Acrididae), including new species. Proc. Acad. Nat. Sci. Phil. 125: 46–66.

Rodgers, D. J. & Kiching, R. L. 1998. Vertical stratification of rain forest collembolan (Collembola: Insecta) assemblages: description of ecological patterns and hypotheses concerning their generation. Ecography 21: 392–400.

Roubik, D. W. 1993. Tropical pollinators in the canopy and understory: field data and theory for stratum 'preferences'. J. Ins. Behav. 6: 659–673.

Russell-Smith, A. & Stork, N. E. 1994. Abundance and diversity of spiders from the canopy of tropical rain forests with particular reference to Sulawesi, Indonesia. J. Trop. Ecol. 10: 545–558.

Russell-Smith, A. & Stork, N. E. 1995. Composition of spider communities in the canopies of rain forest trees in Borneo. J. Trop. Ecol. 11: 223–235.

Schowalter, T. D. 1994. Invertebrate community structure and herbivory in a tropical rain forest canopy in Puerto Rico following hurricane Hugo. Biotropica 26: 312–319.

Shmida, A. & Wilson, M. V. 1985. Biological determinants of species diversity. J. Biogeogr. 12: 1–20.

Smythe, N. 1982. The seasonal abundance of night-flying insects in a Neotropical forest. Pp. 309–318. In: Leigh, E. G. jr., Rand, A. S. & Windsor, D. M. (eds), The ecology of a tropical forest: seasonal rhythms and long-term changes. Smithsonian Institution Press, Washington.

Stoner, K. E. 1996. Prevalence and intensity of intestinal parasites in mantled howling monkeys (*Alouatta palliata*) in northeastern Costa Rica: implications for conservation biology. Cons. Biol. 10: 539–546.

Stork, N. E. 1988. Insect diversity: facts, fiction and speculation. Biol. J. Linn. Soc. 35: 321–337.

Stork, N. E. 1991. The composition of the arthropod fauna of Bornean lowland rain forest trees. J. Trop. Ecol. 7: 161–180.

Stork, N. E. 1994. Inventories of biodiversity: more than a question of numbers. Pp. 81–100. In: Forey, P. I., Humphries, C. J. & Vane-Wright, R. I. (eds), Systematics and conservation evaluation. Clarendon Press, Oxford.

Stork, N. E. 1996. Tropical forest dynamics: the faunal components. Pp. 1–20. In: Edwards, D. S., Booth, W. E. & Choy, S. C. (eds), Tropical rain forest research – current issues. Kluwer Academic Publishers, Amsterdam.

Stork, N. E., Adis, J. & Didham, R. K. 1997. Canopy arthropods. Chapman & Hall, London.

Stork, N. E. & Brendell, M. J. D. 1990. Variation in the insect fauna of Sulawesi trees with season, altitude and forest type. Pp. 173–190. In: Knight, W. J. & Holloway, J. D. (eds), Insects and the rain forests of South East Asia (Wallacea). The Royal Entomological Society of London, London.

Stork, N. E. & Brendell, M. J. D. 1993. Arthropod abundance in lowland rain forest of Seram. Pp. 115–130. In: Edwards, I. D., MacDonald, A. A. & Proctor, J. (eds), Natural History of Seram, Maluku, Indonesia. Intercept, Andover.

Stuntz, S., Simon, U. & Zotz, G. 1999. Assessing potential influence of vascular epiphytes on arthropod diversity in tropical tree crowns: hypotheses, approaches, and preliminary data. Selbyana 20: 276–283.

Sutton, S. L. 2001. Alice grows up: canopy science in transition from Wonderland to reality. Plant Ecol. 153: 13–21 (this volume).

Sutton, S. L. & Hudson, P. J. 1980. The vertical distribution of small flying insects in the lowland rain forest Zaire. Zool. J. Linnol. Soc. 68: 111–123.

Sutton, S. L., Ash, C. P. & Grundy, A. 1983a. The vertical distribution of flying insects in the lowland rain forest of Panama, Papua New Guinea and Brunei. Zool. J. Linn. Soc. 78: 287–297.

Sutton, S. L., Whitmore, T. C. & Chadwick, A. C. 1983b. Tropical rain forest: ecology and management. Oxford, Blackwell.

Tobin, J. E. 1991. A neotropical rain forest canopy, ant community: some ecological considerations. Pp. 536–538. In: Huxley, C. R. & Cutler, D. F. (eds), Ant-Plant Interactions. University Press, Oxford.

Wagner, T. 1997. The beetle fauna of different tree species in forest of Rwanda and East Zaire. Pp. 169–183. In: Stork, N. E., Adis, J. & Didham, R. K. (eds), Canopy arthropods. Chapman & Hall, London.

Wagner, T. 1998. Influence of tree species and forest type on the chrysomelid community in the canopy of an Ugandan tropical forest. Mus. Reg. Sci. Nat. Torino 1998: 253–269.

Walter, D. E. & Behan-Pelletier, V. 1999. Mites in forest canopies: filling the size distribution shortfall. Annu. Rev. Entomol. 44: 1–19.

Walter, D. E. & O'Dowd, D. J. 1995. Life on the forest phylloplane: hairs, little houses, and myriad mites. Pp. 325–351. In: Lowman, M. D. & Nadkarni, N. M. (eds), Forest canopies. Academic Press, San Diego.

Walter, D. E., O'Dowd, D. & Barnes, V. 1994. The forgotten arthropods: foliar mites in the forest canopy. Mem. Qld Mus. 36: 221–226.

Walter, D. E., Seeman, O., Rodgers, D. & Kitching, R. L. 1998. Mites in the mist: how unique is a rain forest canopy-knockdown fauna? Aust. J. Ecol. 23: 501–508.

Watanabe, H. & Ruaysoongnern, S. 1989. Estimation of arboreal arthropod density in a dry evergreen forest in Northeastern Thailand. J. Trop. Ecol. 5: 151–158.

Watt, A. D., Stork, N. E., McBeath, C. & Lawson, G. L. 1997. Impact of forest management on insect abundance and damage in a lowland tropical forest in southern Cameroon. J. Appl. Ecol. 34: 985–998.

Wells, S. M., Pyle, R. M. & Collins, N. M. 1983. The IUCN invertebrate red data book. International Union for Conservation of Nature and Natural Resources, Gland, Switzerland.

Wilson, E. O. 1987. The arboreal ant fauna of Peruvian Amazon Forests: a first assessment. Biotropica 19: 245–251.

Wolda, H. 1979. Abundance and diversity of Homoptera in the canopy of a tropical forest. Ecol. Entomol. 4: 181–190.

Wolda, H., O'Brien, C. W. & Stockwell, H. P. 1998. Weevil diversity and seasonality in tropical Panama as deduced from light-trap catches (Coleoptera: Curculionoidea). Smiths. Contr. Zool. 590: 1–79.

Wright, J. S. 1995. The canopy crane. p. 15. In: Lowman, M. D. & Nadkarni, N. M. (eds), Forest canopies. Academic Press, San Diego.

Wunderle, I. 1992. Die Baum- und Bodenbewohnenden Oribatiden (Acari) im Tief-landregenwald von Panguana, Peru. Amazonia 12: 119–142.

Young, A. M. 1986. Occurrence of Diptera on tree-trunk mosses in a Costa Rican tropical rain forest. Pan-Pac. Ent. 62: 203–208.

Zug, G. R. & Zug, P. B. 1979. The marine toad, *Bufo marinus*: A natural history resumé of native populations. Smiths. Contr. Zool. 284: 1–58.

Appendix I

Mass–collecting of canopy invertebrates examined in this contribution, listed by alphabetical order of the key reference. Collecting methods: EXT = litter extraction; FOG = fogging; HDC = hand collecting, beating and related; LIT = light traps; OTT = other traps; SEV = several methods. Numbers refers to the codes used in the ordinations.

Key reference	Locality	Focal taxa	Method	No. ind.
1. Adis et al. 1984	Manaus, Brazil	Arthropoda	FOG	9639
2. Adis et al. 1997a	Manaus, Brazil	Arthropoda	FOG	3685
3. Adis 1997	Manaus, Brazil	Arthropoda	FOG	23689
4. Allison et al. 1997*	Wau, Papua New Guinea	Arthropoda	FOG	45464
5. Amedegnato 1997*	Ampiacu, Peru	Acridoidea	HDC	9177
6. Amedegnato 1997*	Yubineto, Peru	Acridoidea	HDC	1156
7. Barrios 1997	Barro Colorado Is, Panama	Curculionidae	LIT	11843
8. H. Barrios*	Parque Metropolitano, Panama	Insecta	OTT	77430
9. Basset 1991	Mt Glorious, Australia	Arthropoda	SEV	51600
10. Basset et al. 1992	Kribi, Cameroon	Arthropoda	HDC	2271
11. Basset et al. 1996b	Wau, Papua New Guinea	Arthropoda	SEV	78500
12. Basset et al. 1999	Mabura Hill, Guyana	Insect herb.	HDC	1659
13. Broadhead & Wolda 1985	Barro Colorado Is, Panama	Psocoptera	LIT	14393
14. Casson & Hodkinson 1991	Dumoga-Bone NP, Sulawesi	Hemiptera	SEV	25000
15. Chey et al. 1998	Brumas, Sabah	Arthropoda	FOG	1640
16. Corbet 1961	Mpanga, Uganda	Diptera	LIT	2010
17. Davies et al. 1997	Henri Pittier NP, Venezuela	Coleoptera	FOG	6132
18. Davis & Sutton 1998	Danum Valley, Sabah	Scarabaeidae	OTT	2378
19. Dejean et al. 1992	Kribi, Cameroon	Formicidae	HDC	na
20. Dejean et al. 1998	Paracou, French Guiana	Formicidae	HDC	na
21. DeVries et al. 1997	Jatun Sacha, Ecuador	Nymphalidae	OTT	6690
22. Dial & Roughgarden 1995	Bishley Watershed, Puerto Rico	Arthropoda	SEV	21200
23. Erwin 1983	Manaus, Brazil	Arthropoda	FOG	24350
24. Erwin 1989	Pakitza, Peru	Arthropoda	FOG	82391
25. Erwin 1995	Tambopata, Peru	Insecta	FOG	42641
26. Erwin & Scott 1980	Pipeline Road, Panama	Coleoptera	FOG	7712
27. Farrell & Erwin 1988	Tambopata, Peru	Chrysomelidae	FOG	2864
28. Fisk 1983	Barro Colorado Is, Panama	Blattodea	SEV	2542
29. Floren & Linsenmair 1997*	Kinabalu NP, Sabah	Arthropoda	FOG	155000
30. Gagné 1979	Mauna Loa, Hawaii	Arthropoda	FOG	na
31. Galindo et al. 1956	La Victoria, Panama	Culicidae	HDC	na
32. Garcia 1999*	Parque Metropolitano, Panama	Coleoptera	HDC	3585
33. Garrison & Willing 1996	El Verde, Puerto Rico	Invertebrates	OTT	4506
34. Guilbert 1997	Pindai, New Caledonia	Invertebrates	FOG	110173
35. Guilbert 1997	Riviere Bleue, New Caledonia	Invertebrates	FOG	63930
36. Hammond 1990	Dumoga-Bone NP, Sulawesi	Coleoptera	SEV	19000
37. Hill & Cermak 1997	Paluma Range, Australia	Arthropoda	OTT	553
38. Hingston 1932	Moraballi Creek, Guyana	Arthropoda	OTT	88
39. Höfer et al. 1994	Manaus, Brazil	Araneae	FOG	235
40. Holloway 1987	Gunung Mulu NP, Sarawak	Lepidoptera	LIT	6066
41. Holloway 1987	Gunung Mulu NP, Sarawak	Lepidoptera	LIT	4834
42. Janzen 1988b	Santa Rosa NP, Costa Rica	Lepidoptera	HDC	54000

Appendix I continued.

Key reference	Locality	Focal taxa	Method	No. ind.
43. Kato et al. 1995	Lambir Hills NP, Sarawak	Insecta	LIT	1023008
44. Kitching et al. 1993	Green Mountains, Australia	Arthropoda	FOG	22984
45. Kitching et al. 1993	Cape Tribulation, Australia	Arthropoda	FOG	9967
46. Longino & Colwell 1997	La Selva, Costa Rica	Formicidae	SEV	6000
47. Lowman et al. 1998	Paracou, French Guiana	Arthropoda	SEV	1941
48. Malcolm 1997	Manaus, Brazil	Insecta	OTT	59795
49. McClure 1978	Genting Sempah, Malaysia	Arthropoda	OTT	na
50. Missa 1999*	Baiteta, Papua New Guinea	Arthropoda	SEV	234743
51. Moran et al. 1994	Pondoland Centre, South Africa	Insect herb.	HDC	na
52. Moran & Southwood 1982	The Cape, South Africa	Arthropoda	FOG	6847
53. Nadkarni & Longino 1990	Monteverde Forest, Costa Rica	Invertebrates	EXT	na
54. Ng 1978	Pasoh Forest Reserve, Malaysia	Insecta	OTT	37198
55. Noyes 1989	Dumoga-Bone NP, Sulawesi	Hymenoptera	SEV	59100
56. Ødegaard 1999*	Parque Metropolitano, Panama	Beetle herb.	HDC	33746
57. Palacios-Vargas et al. 1998	Chamela, Mexico	Arthropoda	FOG	1098248
58. Paoletti et al. 1991	Henri Pittier NP, Venezuela	Invertebrates	EXT	4960
59. Penny & Arias 1982	Manaus, Brazil	Insecta	LIT	na
60. Rees 1983	Morolawi, Sulawesi	Insecta	LIT	3500
61. Roberts 1973	Barro Colorado Is, Panama	Acridoidea	FOG	na
62. Roubik 1993	Parque Metropolitano, Panama	Apoidea	OTT	2356
63. Russell-Smith & Stork 1994	Dumoga-Bone NP, Sulawesi	Araneae	FOG	1649
64. Russell-Smith & Stork 1995	Bukit Sulang, Brunei	Araneae	FOG	945
65. Schowalter 1994	El Verde, Puerto Rico	Invertebrates	HDC	9143
66. Smythe 1982	Barro Colorado Is, Panama	Insecta	LIT	60000
67. Stork 1991	Bukit Sulang, Brunei	Arthropoda	FOG	23874
68. Stork & Brendell 1990	Dumoga-Bone NP, Sulawesi	Arthropoda	FOG	9158
69. Stork & Brendell 1993	Manusela NP, Seram	Arthropoda	FOG	12006
70. Stuntz et al. 1999	Barro Colorado Is, Panama	Arthropoda	SEV	36875
71. Sutton et al. 1983a	San Blas, Panama	Insecta	LIT	4194
72. Sutton et al. 1983a	Buso, Papua New Guinea	Insecta	LIT	7013
73. Sutton et al. 1983a	Temburong, Brunei	Insecta	LIT	8225
74. Sutton et al. 1983a	Labi Safari, Brunei	Insecta	LIT	25151
75. Sutton & Hudson 1980	Scierie, Zaire	Insecta	SEV	32474
76. Sutton & Hudson 1980	Weko, Zaire	Insecta	SEV	33580
77. Tobin 1991	Pakitza, Peru	Formicidae	FOG	19710
78. Wagner 1997	East Rwanda	Coleoptera	FOG	23711
79. Wagner 1997	Kivu, Zaire	Coleoptera	FOG	6279
80. Wagner 1998*	Budongo, Uganda	Coleoptera	FOG	29736
81. Walter et al. 1994*	SE Queensland, Australia	Acari	HDC	15000
82. Watanabe & Ruaysoongnern 1989	Nam Phrom, Thailand	Invertebrates	FOG	na
83. Watt et al. 1997*	Mbalmayo, Cameroon	Arthropoda	FOG	87000
84. Wilson 1987	Tambopata, Peru	Formicidae	FOG	100000
85. Wolda 1979	Barro Colorado Is, Panama	Homoptera	SEV	6363
86. Wolda et al. 1998	Barro Colorado Is, Panama	Curculionoidea	LIT	95333
87. Wolda et al. 1998*	Fortuna, Panama	Insecta	LIT	29467
88. H. Wolda*	Barro Colorado Is, Panama	Insecta	LIT	1247264
89. Wunderle 1992	Panguana, Peru	Oribatei	EXT	17382

* = supplemented by pers. comm. of the author(s).
na = not available, data not expressed in total no. of individuals collected.
herb. = herbivores

Plant Ecology **153:** 109–120, 2001.

Arboreal tropical forest vertebrates: current knowledge and research trends

Roland Kays[1] & Allen Allison[2]
[1]*New York State Museum, CEC 3140, Albany, NY 12230 (E-mail: rkays@mail.nysed.gov);* [2]*Bishop Museum, 1525 Bernice Street, Honolulu, HI 96817-2704, USA (E-mail: allison@hawaii.edu)*

Key words: Amphibians, Birds, Canopy, Mammals, Primates, Reptiles

Abstract

We review the ecology and specialized methods required for studying arboreal mammals, birds, reptiles and amphibians, and use faunal checklists from 12 tropical wet forest sites and an analysis of all articles published during the past ten years in 14 major journals to assess current knowledge and general research trends for these groups. The percentage of arboreal vertebrates was remarkably similar at the different sites (76.2 ± 3.9%). Birds were the most arboreal group and amphibians and reptiles the least. The review of journals showed that primates were overwhelmingly the most studied group (336 papers), followed by bats (105), passeriform birds (73) and rodents (55). Judging by their portion of the arboreal vertebrate community and the number of papers surveyed, birds and amphibians and reptiles are vastly understudied compared to mammals, but this is largely due to the great number of primate studies. The number of publications on arboreal vertebrates has remained relatively stable over the last 10 years for all taxa except primates, which have seen a growth in publications. Canopy vertebrates from Brazil had by far the most publications (120), followed by Madagascar (61), Costa Rica (55) and Indonesia (42). We conclude by highlighting the priorities we see for future studies on tropical canopy vertebrates.

Introduction

Many tropical canopy vertebrates (TCV) are charismatic creatures that hold the attention of the general public with bright colours, beautiful songs and complex behaviours. Monkeys, hornbills, and red-eyed tree frogs are quintessential species that loom large in the popular conception of all things tropical. TCV are also a key to the conservation of tropical forest canopies. Because of their public popularity, and special habitat requirements, TCV often serve as umbrella or flagship species in conservation programs. However, despite the importance and popularity of certain flashy species, most TCV remain little known to science. Many tropical canopy animals are difficult to study and have therefore received much less scientific attention than their temperate or terrestrial counterparts.

This lack of information, combined with their incredible diversity, has limited the scientific discourse on TCV as a group. This is unfortunate, since the ecological roles and conservation threats to TCV are great. However, certain components of the TCV community have been studied, some quite extensively. Adventurous zoologists have climbed into the canopy or used a variety of remote methods to census, mark, and observe canopy vertebrates. These data are slowly accumulating to form the groundwork for an understanding of TCV communities.

In this paper we will review TCV on a general level. The diversity of TCV combined with the lack of information on many taxa, precludes the detailed analyses this group deserves. However, by maintaining a broad scope we can introduce TCV as a group, and highlight some general principles of their organization. First we outline the three major components of the TCV communities and discuss the special efforts needed to census and study each one. Second, we highlight the importance of TCV by describing their portion of the whole vertebrate community across 12 tropical sites in the New World, Africa, South East

Table 1. Number of vertebrate species from 12 tropical rain forest study sites.

Name	Location	Area km^2	Bird species	Mammal species	Amphibian and reptile species	References
New World						
La Selva	North East Costa Rica, 10°N 83°W	ca. 15	366	117	134	Blake et al. 1990; Guyer 1990; Wilson 1990; Voss & Emmons 1996
BCI	Central Panama, 9°N 79°W	ca. 15	387	115	133	Rand & Myers 1990; Voss & Emmons 1996; D. Robinson pers. com.
Cocha Cashu	South Eastern Peru, 11°S 71°W	10+	455	139	129	Janson, C.H. 1990; Robinson & Terborgh 1990; Rodriguez & Cadle 1990; Voss & Emmons 1996; D. Robinson pers. com.
Manaus	Northern Brazil 2°S 60°W	ca. 10	319	105	134	Reis & Peracchi 1987; Malcolm 1990; Voss & Emmons 1996; Zimmerman & Rodrigues 1990
Africa						
Kibale	Western Uganda 0° N 30° E	766	305	87	83	Vonesh 1998; C. Chapman pers com; J. Kerbis pers com.
South East Asia						
Temengor	N. mainland Malaysia 5.47° N, 101.36° E	1489	154	101	72	Davison 1995; Diong et al. 1995; Francis 1995; Kiew et al. 1995; Lim et al. 1995; Ratnam et al. 1995
Belalong Forest	Brunei: 4.60° N, 115.18° E	ca. 144	184	84	91	Charles 1996; Das 1995; Mann 1996
Ulu Endau	S. mainland Malaysia 4.67° N, 103.60° E	ca. 400	195	61	53	Davison 1987; Davison & Akbar 1987; Davidson & Kiew 1987; Kiew 1987
Danum Valley	Sabah, 4.58° N, 117°.48 E	438	275	104	125	Mobilik & Marsh 1993
Australia/ New Guinea						
Finnegan Uplands	N.E. Australia 15.63° S, 146.50° E	281	82	37	57	Williams et al. 1996
Cooktown Lowlands	N.E. Australia, 15.40° S, 145.00° E	44	134	38	79	Williams et al. 1996
Lakekamu	Papua New Guinea 7.74° S, 146.50° E	ca. 15	190	77	74	Allison et al. 1998; Beehler et al. 1995; Mack & Igag 1998; Wright et al. 1998

Asia, and Australia. Third, we survey the literature from the last ten years to identify the strengths and weaknesses in our understanding of TCV. Finally, we will conclude by discussing priorities in TCV research reflected by our results.

The arboreal vertebrate community

In the strict sense TCV would apply only to those species of birds, mammals, amphibians and reptiles that are primarily active in the crowns of rain forest trees. However, our knowledge is insufficient for making such a determination for many species. We have therefore broadened our working definition of TCV to consist of all species that are primarily arboreal, including branch and trunk specialists. Although this approach has certain limitations, it does facilitate making broad comparisons between the world's TCV communities and helps reveal the emergent features of these communities.

Birds are the most diverse and conspicuous component of the TCV community. They play nearly every role within the canopy including pollination, seed dispersal, seed predation, animal predation, and herbivory. Because of their diurnal habits and breeding vocalizations, nearly all bird species can be detected from the ground, permitting accurate census of tropical canopy bird communities without special canopy access efforts. However, collecting more than simple presence/absence data for a canopy bird can require significant effort. Catching and marking canopy birds requires canopy mist netting (Munn 1991). Observing the behaviour and ecology of any but the largest canopy birds requires collecting data from in the trees via ropes, a canopy platform or walkway. However, even tree-borne observers will find it difficult to collect data unless a bird's movement is restricted to a limited area for nesting or foraging.

Like birds, mammals have radiated to fill most canopy niches including pollination, seed predation and dispersal, animal predation, and herbivory. Herbivory by mammals is probably more significant than other vertebrates. Together, sloths and howler monkeys, both canopy herbivores, make up 28.3% and 71.0% of the total mammalian biomass of rain forests in Surinam and Panama, respectively (Eisenberg & Thorington 1973). For our purposes, canopy mammals can be broken into three main groups: large, small, and flying.

A study of large arboreal mammals faces challenges similar to those encountered by ornithologists: the species are not too difficult to detect, but capturing them requires special efforts such as large tree traps or dart guns (e.g., Kays 1999). Many species of large diurnal mammals, especially primates, have been habituated by ground based observers. Armed only with binoculars, these scientists have collected incredible details on the ecology and behaviour of canopy mammals (Terborgh 1983; Smuts et al. 1987; Rowe 1996). Nocturnal, arboreal, large mammals are more difficult to study, but detailed data can be collected with the aid of telemetry and strong headlamps. Clearly, large birds and mammals are the easiest canopy vertebrates to census and study.

A number of small rodents and marsupials are partly or completely arboreal where they play the same ecological roles their terrestrial relatives do, including seed dispersal and predation. While some can be captured on the ground, to even detect many of these species one must put traps into the tree canopy. In Manaus, Malcolm caught a total of 15 species of small mammals with 5010 trap-nights of ground and tree trapping. His study illustrated the importance of tree trapping as four of these species were caught only in canopy and three more were caught significantly more in the canopy (Malcolm 1991). Unfortunately, this tree trapping is rarely done. Of 10 Neotropical sites judged to have the best sampled mammalian faunas by Voss & Emmons (1996), only one had sufficient tree trapping, two had some, and seven had none. Detailed ecological data for most species in this group is practically non-existent and will require significant effort, including regular long-term tree trapping (e.g., Malcolm 1995).

Bats are the most diverse tropical canopy mammal group and play important roles in the forest as seed dispersers and predators. All tropical rain forest bats are arboreal in that they either fly through the canopy or roosting in the vegetation. Mist nets or Harp traps allow the capture of many bat species, and these can be hoisted into the canopy to capture high flying bats. Some bat species can only be captured with these high nets. For example, In French Guiana, Simmons and Voss (1998) caught eight bat species (10.3% of all species) only in high nets (in 2722 net meter hours) despite extensive ground netting (24 957 net-m-hrs). Unfortunately, canopy nets or Harp traps are rarely used in tropical field work (Voss & Emmons 1996). Furthermore, some bat species can detect and avoid mist nets. For these species, other techniques such as intensive roost hunting (Simmons & Voss 1998) and acoustic monitoring (Kalko & Handley 2001) are

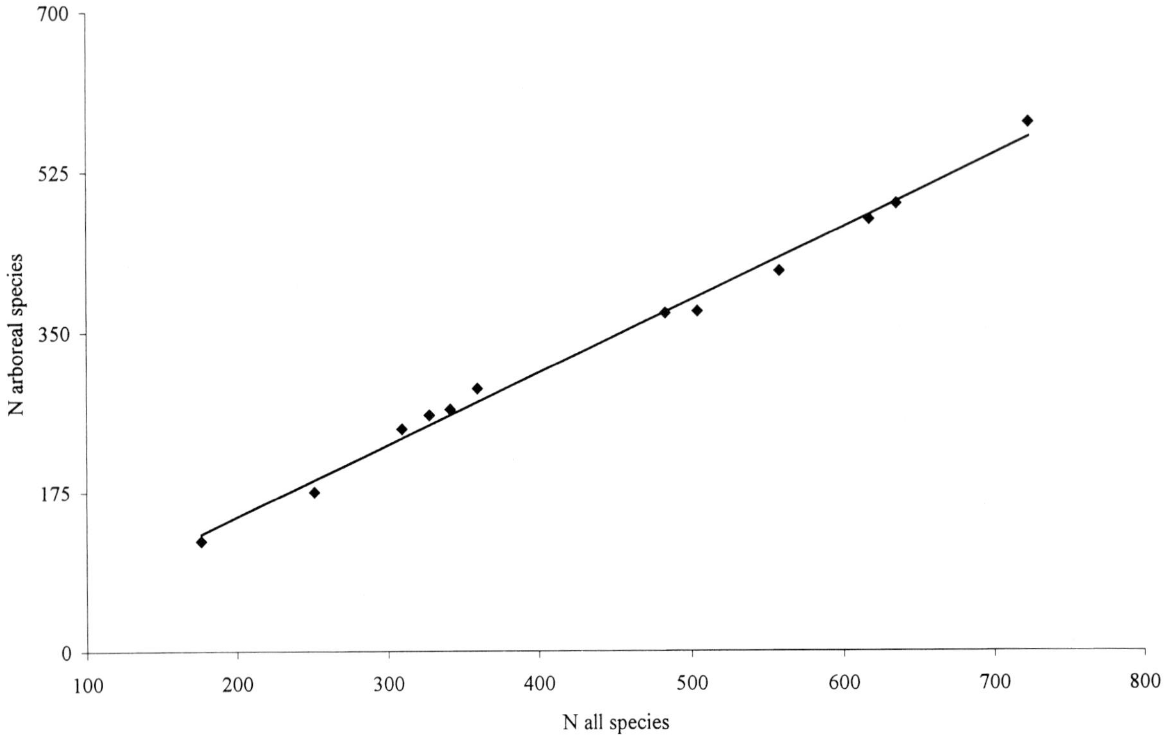

Figure 1. The relationship between total vertebrate species and arboreal vertebrate species from 12 well studied tropical sites. $r^2 = 0.99$. The average percent of vertebrate taxa that was arboreal was 76.2 ± 3.9.

needed. The equipment and identification keys for acoustic monitoring are in the early stages of development, but show great promise for the future of bat biology. Because of their nocturnal nature, small size, and great mobility, bats are almost impossible to observe away from their roosts. Therefore, technological advances such as acoustic bat detectors, radio telemetry and night vision scopes will be a key to discovering details about their behaviour and ecology.

Among the amphibians and reptiles, only frogs, salamanders, lizards and snakes are arboreal. Most major lineages within these groups have arboreal representatives and some, such as the rhacophorid frogs, are almost exclusively arboreal (Duellman & Treub, 1986). Some species of arboreal amphibians and reptiles are relatively easy to detect because they descend to the ground to breed or occur also in low vegetation, or, in the case of frogs can be located by their vocalizations. However, outside the breeding season, most arboreal reptiles and amphibians have cryptic habits and are difficult to study. For the same reasons it is likely that many species have yet to be discovered and described. For example the Peruvian lizard *Tropidurus flaviceps* (Guichenot) lives only in the tree canopy and has only been collected by following loggers and catching animals off recently fallen trees (Duellman 1978). The behavioural and ecological details of these canopy specialists will only be discovered by patient tree-borne observers.

Methods

Faunal lists

We attempted to locate all sites in the wet tropics that were ca. 10–20 km^2 in area and had homogeoeous habitat and a complete list of the non-fish vertebrates. If a number of sites were located close together or were contiguous, we selected only one site or in one case a lowland and upland site. Lists for reptiles, amphibians, small mammals and bats were especially hard to find, and only five sites met our criteria, so we broadened our selection to include seven additional larger area sites. Migrant species were included if they were consistent members of the fauna. Species that were tied to large bodies of water (e.g., many ducks) and never recorded in the forest as well as

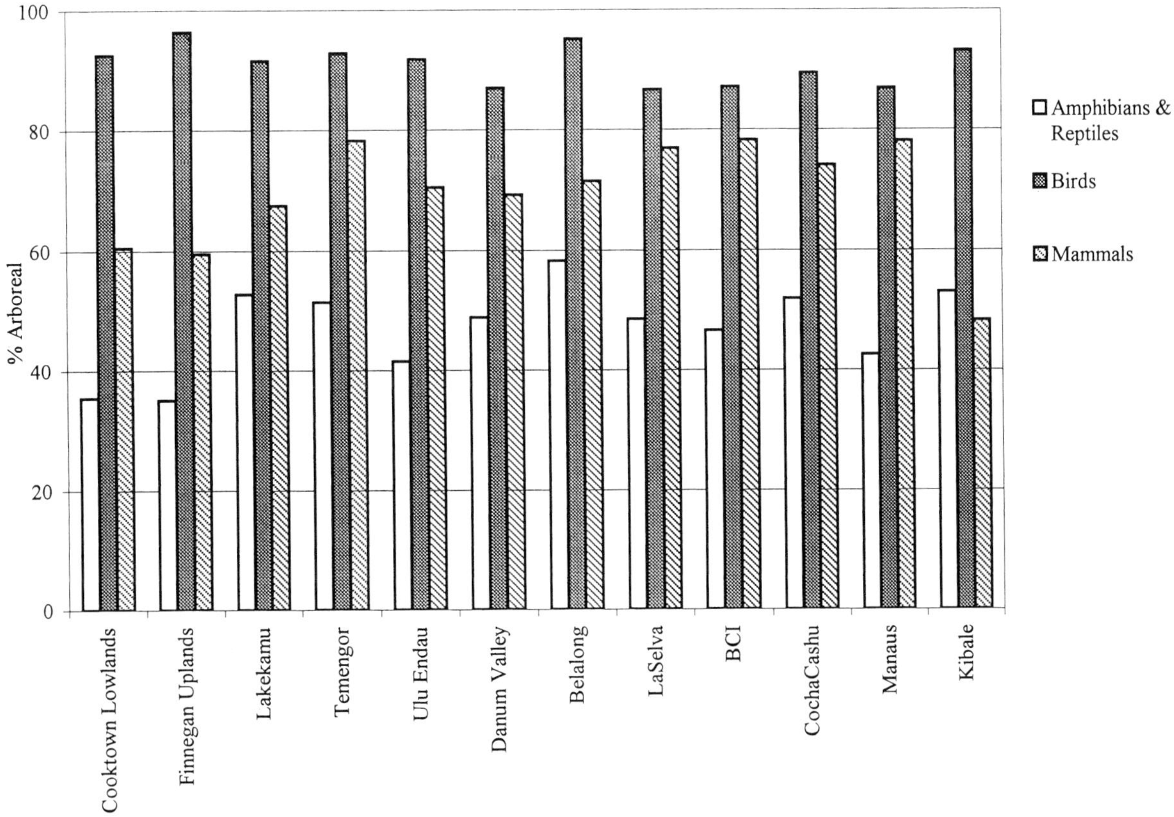

Figure 2. The fraction of mammals, birds, and amphibians and reptiles classified as arboreal from 12 tropical rain forest study sites.

species dependent on humans (e.g., *Rattus rattus* Linnaeus) were not included. For each site we used all available information including publications about the site, regional field guides, expedition reports and personal communication to determine which species were arboreal. Species were considered arboreal if they regularly climbed or perched in trees. This was not an exclusive classification, and many species were rated as both arboreal and terrestrial. We included faunal lists from 12 sites in our review (Table 1).

Literature survey

To assess research trends, we systematically surveyed the last ten years of literature for papers on TCV. Journals from 1988 to 1998 were reviewed. Volumes from 1998 were not complete at the time of our survey. Although it would be preferable to survey all major journals that include articles on arboreal vertebrates, for practical reasons we selected 14 journals (Table 2). Literature surveys are inherently biased by the journals selected. To reduce this bias we included about a third of the major journals published in each major field: ornithology, herpetology, mammalogy, primatology, and general ecology. The number of articles

Table 2. Number of research articles in 14 journals on tropical arboreal vertebrates from 1988–1998.

Journal	Number of articles
African Journal of Ecology	16
Biotropica	74
Condor	90
Emu	23
Folia Primatologica	131
Herpetologica	6
Ibis	12
International Journal of Primatology	161
Journal of Herpetology	5
Journal of Mammalogy	47
Journal of Tropical Ecology	57
Mammalia	108
Tropical Ecology	4
Wildlife Research	18

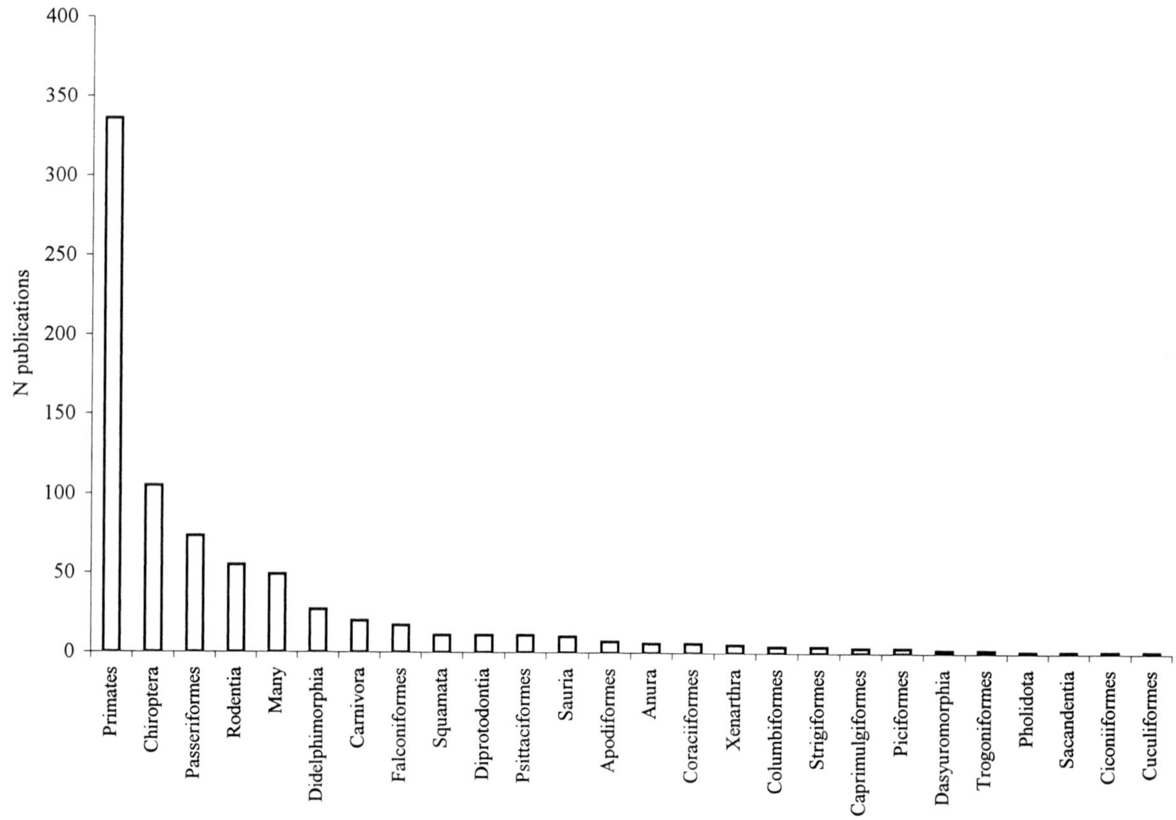

Figure 3. The number of articles published on arboreal species from 25 vertebrate orders in a survey of 14 journals from 1988–1998. Papers that considered 4 or more orders were classified as 'Many'.

per field in our sample from a given discipline was therefore roughly proportional to the contribution that that discipline makes to overall publishing activity. We included a paper in our survey if it contained new data on a tropical arboreal vertebrate and collected data with original field work. All biological topics were included, although no review papers were considered. We included research done in any tropical forest, not just tropical rain forests. We noted for each paper whether it involved an arboreal vertebrate and, if so, which taxa, continents, and countries were involved.

Results

Faunal lists

The total number of vertebrate species recorded from the different sites ranged from 176 to 723 with the Neotropical sites being the richest (Table 1). However, the overall percentage of arboreal vertebrates at the different sites is remarkably constant (76.1 ± 3.8%) (Figure 1). Within the different vertebrate classes, birds were predictably the most arboreal, followed at all but one site by mammals and amphibians and reptiles, respectively (Figure 2). At Kibale, our only African site, the percentage of arboreal amphibians and reptiles slightly exceeded the percentage of arboreal mammals, probably because of the small bat fauna.

Literature survey

We identified 752 articles that presented new results on tropical forest arboreal vertebrates. Not surprisingly, the studies were distributed unevenly across taxa. Mammals were the focus of most (73%) articles, followed by birds (23%), and reptiles and amphibians (4%). A more detailed look at the taxonomic foci of these articles (Figure 3) shows that Primates were the most studied canopy vertebrate order, represented in 336 (44.7%) publications. Bats, Passeriformes, and rodents were the second, third, and forth most studied orders with 105, 73, and 55 publications, respectively.

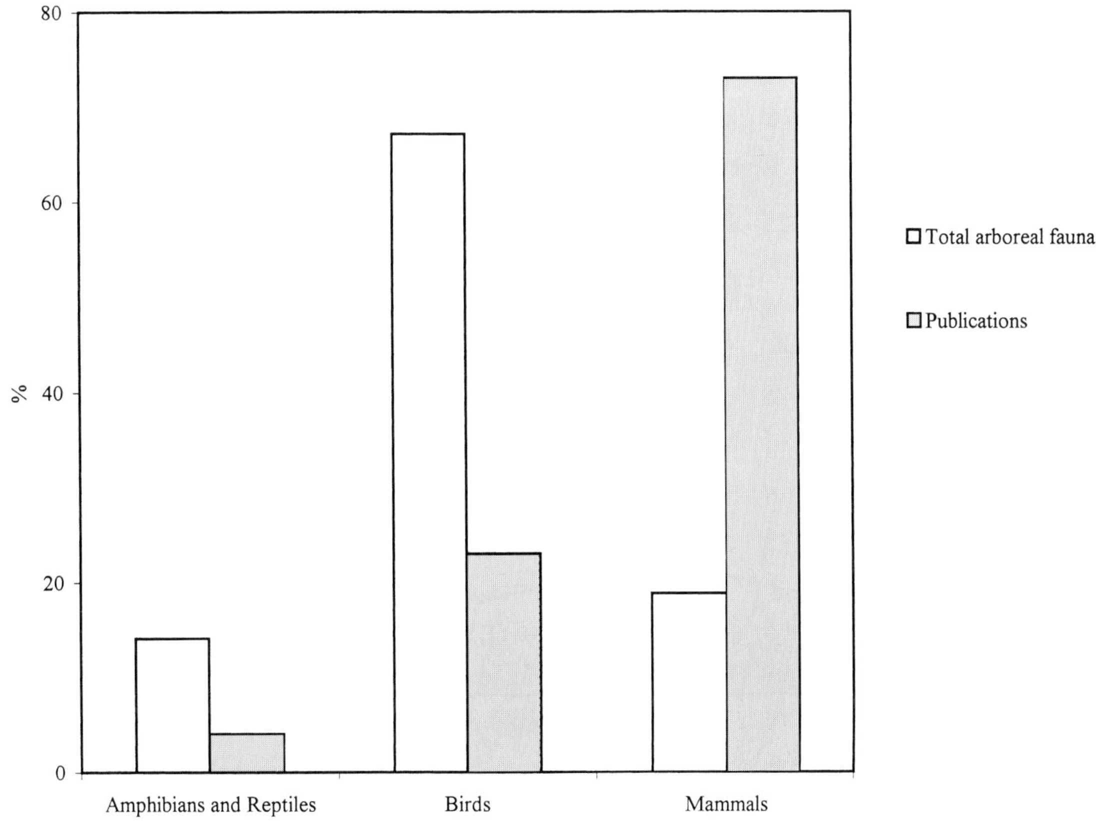

Figure 4. The average fraction of the tropical canopy vertebrate community (TCV) made up by mammals, birds, and amphibians and reptiles compared to their fraction of the papers published on TCV.

More than four orders were included in 49 papers ('Many' in Figure 3). Most of these studies were faunal surveys, especially of bird community. The tail on Figure 3 is long, indicating that most orders of canopy vertebrates are very rarely studied.

To put these results in perspective, we compared them to the average portion of TCV communities made up by each group in our site comparison (Figure 4). Mammals have been studied 3.3 times more than their contribution to the total TCV community while birds and reptiles and amphibians have both been understudied, 0.33 and 0.20 their fraction of the TCV community, respectively. Most of this mammal bias is due to studies with primates, which represent 62% of all publications on arboreal mammals.

Overall, publications on arboreal vertebrates have increased slightly in number over the last ten years (Figure 5). Geographically, this trend is due largely to an increase in papers published on African vertebrates from 1990–1998, with little or no rise in papers from other areas. Because canopy science is a new area of study (Nadkarni 2001) it is encouraging to see a growth of publications on TCV, however slight. Unfortunately, a closer look at the taxonomic distribution of these articles over the years reveals a more discouraging trend (Figure 6). Only papers on primates have grown in number over the last ten years, while publications focusing on birds, reptiles and amphibians, or non-primate mammals have remained stagnant. In sum, the non-primate articles published on tropical arboreal vertebrates in the last ten years have remained relatively level in the last ten years (Figure 7) with an unexplained dip around 1990. The low number of articles from 1998 could be the result of the incomplete volumes available to us at the time of our survey (although this dip is not evident in primates (Figure 6)).

The number of arboreal vertebrate studies published from different tropical countries varied widely (Figure 7). Brazil was represented with, by far, the most studies (120), followed by Madagascar (61), Costa Rica (55), Indonesia (43), Australia (35), Mex-

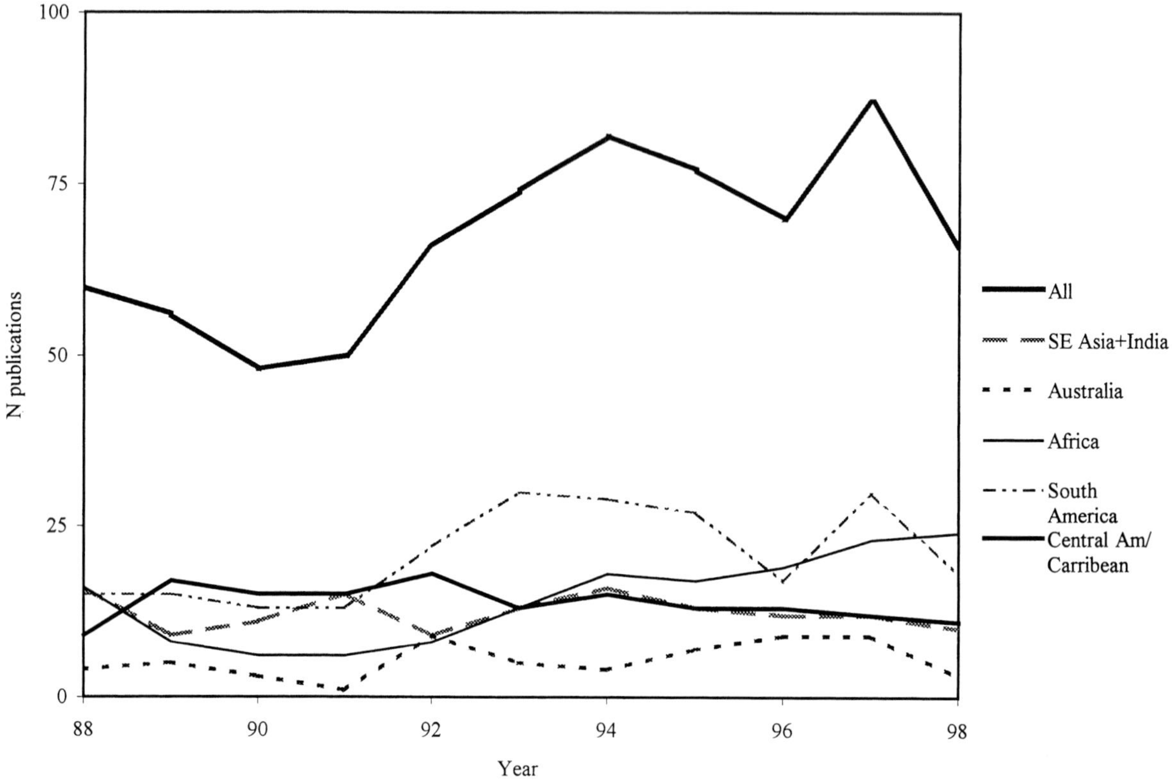

Figure 5. The number of articles published on all tropical canopy vertebrates from 1988–1998 and their geographic location.

ico (29), Venezuela (27), French Guiana (26) and India (24). While the number of studies seems to be generally related to country area, the arboreal vertebrates of some small areas were well studied, including Costa Rica (55 studies), Panama (21), Puerto Rico (16) and Hawaii (15). Some areas known for their megadiversity were understudied including the Philippines (8 studies), Colombia (9), and Zaire (14).

Discussion

Our data demonstrate that throughout the tropics more than three out of four vertebrate species (excluding fish) are fully or partly arboreal, and that across sites the percentage of these species is nearly constant. This is a rather surprising result given the differences in geological and evolutionary history and taxonomic composition of the fauna at the different sites. It suggests that a consistent and dominant fraction of the available habitat is in the trees, regardless of the diversity or complexity of the forest. However, different forests and taxa may reach this convergence in different ways, and the general nature of our comparisons may mask important differences between sites and groups. For example, Erard (1989) reported that the number of arthropod eating birds in French Guiana and Gabon was similar, as was the proportion of this guild that was arboreal, but that nearly twice as many species foraged above 20 m in Gabon than at the French Guiana site. Furthermore, arboreal communities of mammals (Emmons 1995) and amphibians and reptiles (Cadle & Green 1993) converged in different ways across tropical sites, with mammals showing much more similarity. We are planning to examine the terrestrial/arboreal relationship in a future study as we refine the information in our databases.

In addition, our comparisons are only as good as the data from the various sites. It is clear from Figure 4 that compared to mammals, birds, amphibians and reptiles are understudied and deserve more attention. However, despite their high ranking in Figure 3, rodents and bats are very difficult to study, and even complete presence/absence data are rare. This is highlighted by the difficulty we encountered in finding complete mammal lists for any African sites. The taxonomy of African bats is so unsure at this time that a

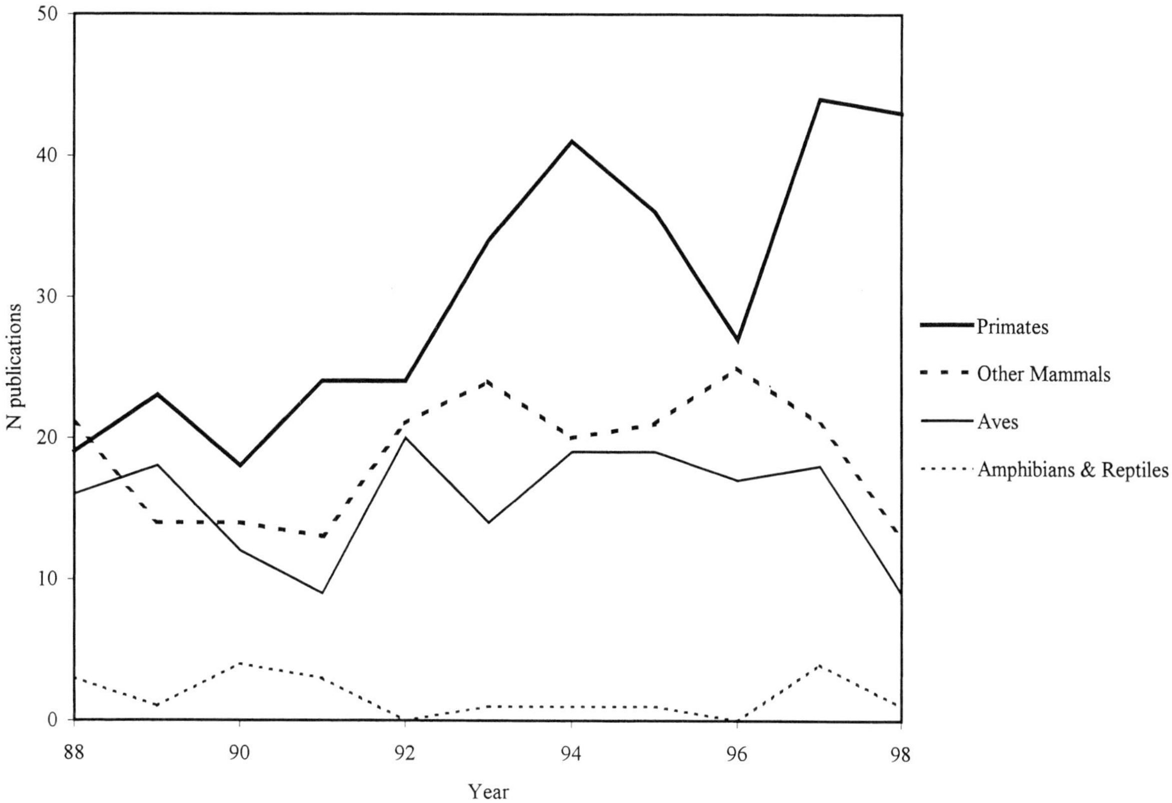

Figure 6. The number of articles published on four groups of tropical canopy vertebrates from 1988–1998.

species list could change by 20%, depending on the taxonomy used (J. Fahr pers. comm.). In addition, relatively few of our sites have been sampled with Harp traps or canopy nets; such trapping is essential for recording all bat species at a site.

Canopy birds, amphibians and reptiles deserve more attention by the scientific community. Because of the ease of censusing singing birds, avifaunal lists are probably accurate, but studies of their natural history are rare. Amphibians and reptiles, on the other hand, need more survey and natural history work. The paucity of studies on arboreal amphibians and reptiles was alarming. The best immediate solution may be more work behind loggers to collect species off recently fallen trees. Clearly, more practical methods for studying canopy amphibians and reptiles are a high priority.

All Taxa Biodiversity Inventories (ATBI) have been suggested for various tropical habitats (Yoon 1993), but none have yet begun in earnest. While the task is enormous, the rewards would be fantastic for the results on a complete TCV community as a whole. An All Vertebrate Biodiversity Inventory may be a more realistic goal that would provide valuable TCV data, and could act as a stepping stone for the eventual completion of an ATBI. Whatever the context, an all vertebrate inventory, including specialized efforts for all canopy species, is badly needed. Because of the varied amounts of canopy sampling used to compile the faunal lists used in this paper, we can not make any but the most general comparisons. However, the potential for addressing important questions with more accurate and complete lists is great. For example, comparison of complete vertebrate fauna from a number of different sites could address broad questions such as the relationship between rainfall amounts and seasonality with different components of vertebrate biodiversity, the effect of dispersal ability on convergence of different groups of tropical vertebrates, or, relationships between the diversity of vertebrate predators and prey.

In addition to more survey work on amphibians and reptiles and non-primate mammals, natural history data for most non-primate TCV's are needed. Primatologists are in an envious position because the distribution and natural history of so many primates has

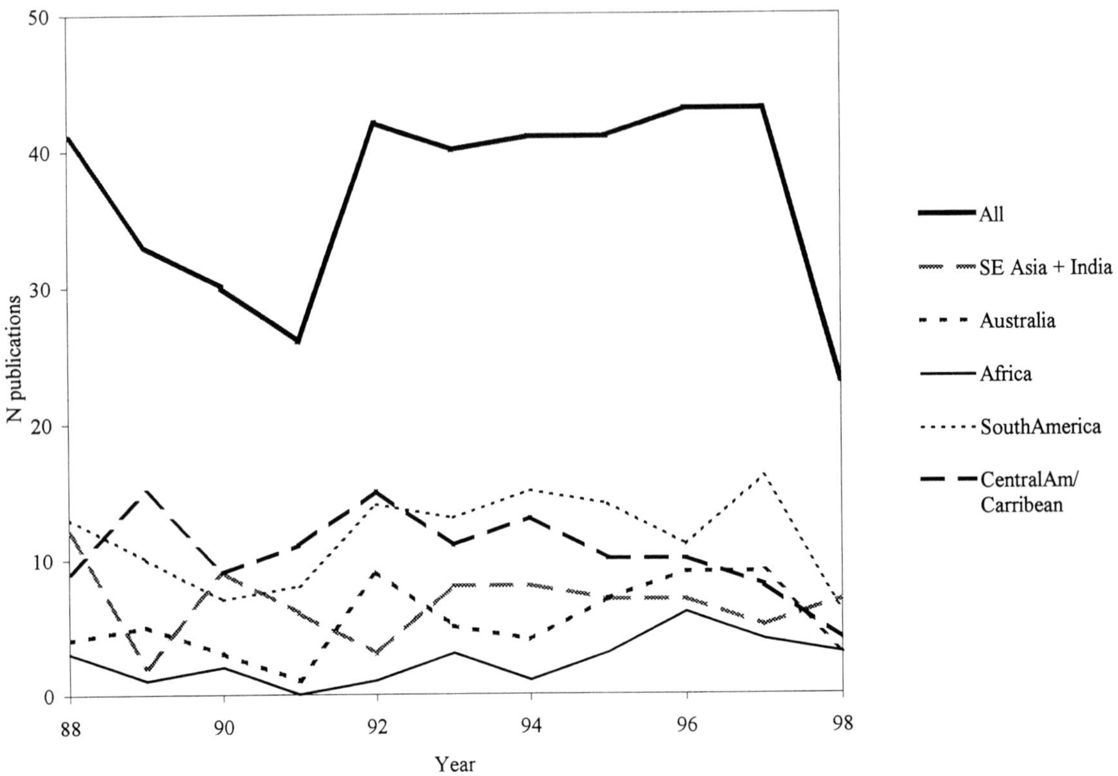

Figure 7. The number of articles published on all non-primate tropical canopy vertebrates from 1988–1998 and their geographic location.

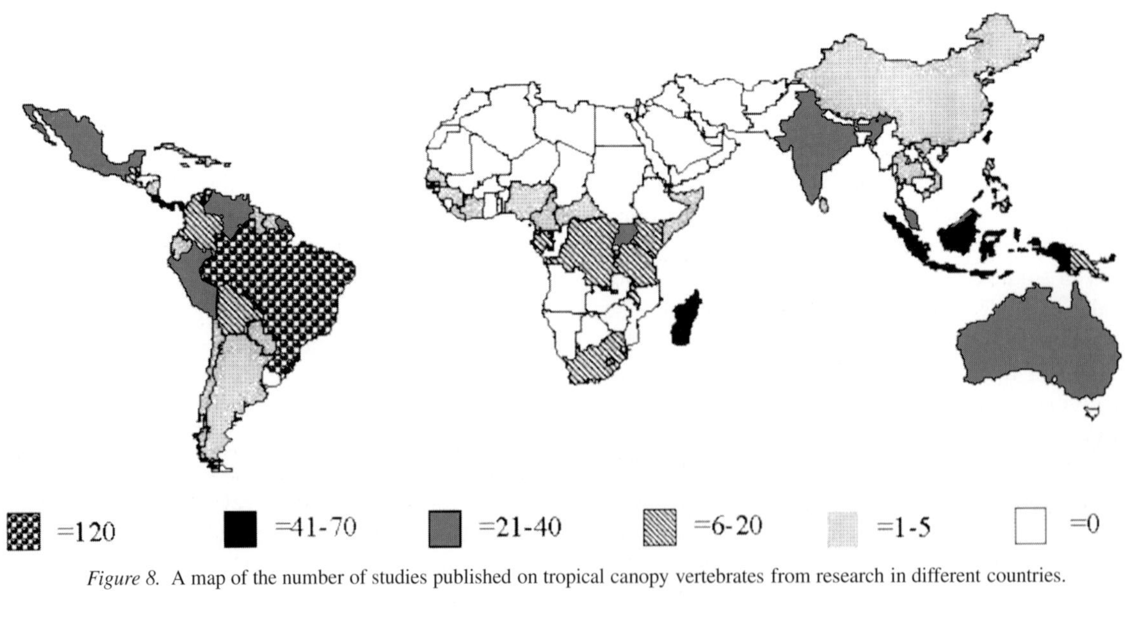

Figure 8. A map of the number of studies published on tropical canopy vertebrates from research in different countries.

been well studied (Rowe 1996). Perhaps the next step in field primatology will be manipulative experimentation to address underlying ecological and evolutionary principals (Janson 1996). Any manipulation must be careful and sensitive to the well being of the subjects, but many possibilities exist within these bounds. Finally, future studies on TCV will be more powerful if they integrate their work with other canopy scientists. Communication between specialists in canopy structure, botany, entomology, and vertebrate zoology will help all groups to integrate data collection and presentation (see Nadkarni 2001). Such interdisciplinary cooperation will benefit each discipline, and lead to a more comprehensive understanding of the tropical forest canopy (Kalko & Handley 2001; Théry 2001).

Acknowledgements

We thank E. Bennett, T. Brooks, T. Butinsky, C. Chapman, R. Drews, J. Fahr, E. Kalko, J. Kerbis, K. Mackinnon, C. Moritz, D. Robinson, T. Smith, J. Vonesh, and S. Williams for help in selecting sites and obtaining vertebrate census dates from the sites. We thank Y. Basset and A. Davis for inviting us to participate in this conference and the European Science Foundation for its support of tropical canopy science. This material is based, in part, upon work supported under a National Science Foundation Graduate Research Fellowship.

References

Allison, A., Bickford, D., Richards, S. & Torr, G. 1998. Herpetofauna. Pp. 58–62, 156–172. In: Mack, A. L. (ed.), A biological assessment of the Lakekamu Basin, Papua New Guinea. RAP Working Papers 9 Conservation International, Washington, DC.

Beehler, B. M., Sengo, J. B., Filardi, C. & Merg, K. 1995. Documenting the lowland rain forest avifauna in Papua New Guinea – effects of patchy distribution, survey effort and methodology. Emu 95 (3): 149–161.

Blake, J. G., Stiles, F. G. & Louiselle, B. A. 1990. Birds of La Selva Biological Station: habitat use, trophic composition, and migrants. Pp. 161–182. In: Gentry, A. H. (ed.), Four Neotropical rain forests Yale University Press, New Haven.

Cadle, J. E. & Greene, H. W. 1993. Phylogenetic patterns, biogeography, and the ecological structure of Neotropical snake assemblages. Pp. 281–293. In: Rickleffs, R. E. & Schluter, D. (eds), Species diversity in ecological communities: historical and geographical perspectives. University of Chicago Press, Chicago.

Charles, J. K. 1996. Small mammal diversity in riparian and dipterocarp habitats in Nelalong forest, Brunei Darussalam. Pp. 175–182. In: Edwards, D. S., Booth, W. E. & Choy, S. C. (eds), Tropical rain forest research-current issues: Proceedings of the conference held in Bandar Seri Begawan, April 1993. Monographiae Biologicae v. 74, Kluwer Academic Publishers, Dordrecht.

Das, I. 1995. Amphibians and reptiles recorded at Batu Apoi, a lowland dipterocarp forest in Brunei Darussalam. Raffles Bull. Zoo. 43 (1): 157–180.

Davison, G. W. H. 1987. The birds of Ulu Endau, Johore, Malaysia, with special reference to birds of heath and fan palm forests. Malayan Nat. J. 41 (2–3): 425–433.

Davison, G. W. H. 1995. The birds of Temengor Forest Reserve, Hulu Perak, Malaysia. Malayan Nat. J. l 48 (3–4): 371–386.

Davison, G. W. H. & Akbar, Z. 1987. The abundance of bats in selected forest types at Ulu Endau, Johore, Malaysia. Malayan Nat. J. 41 (2–3): 425–433.

Davison, G. W. H. & Kiew, B. H. 1987. Mammals of Ulu Endau, Johore, Malaysia. Malayan Nat. J. 41: 435–439.

Diong, C. H., Kiew, B. H. & Lim, B. L. 1995. An annotated checklist of the lizard fauna in the Temengor Forest Reserve, Hulu Perak, Malaysia. Malayan Nat. J. 48 (3–4): 353–356.

Duellman, W. E. 1978. The biology of equatorial herpetofauna in Amazonian Ecuador. Misc. Pub., Univ. Kansas Mus. Nat. Hist. 65: 1–352.

Duellman, W. E. & Trueb, L. 1986. Biology of amphibians. McGraw Hill, New York.

Eisenberg, J. F. & Thorington, R. W. 1973. A preliminary analysis of a Neotropical mammal fauna. Biotropica 5: 150–161.

Emmons, L. H. 1995. Mammals of rain forest canopies. Pp. 199–223. In: Lowman, M. D. & Nadkarni, N. M. (eds), Forest Canopies Academic Press, New York.

Erard, C. 1989. Bird community structure in two rain forests: Africa (Gabon) and South America (French Giana – a comparison. Pp. 89–122. In: Harmelin-Vivien, M. L. & Bourlière, F. (eds), Vertebrates in complex tropical ecosystems. Ecological studies, vol. 69, Springer–Verlag, New York.

Francis, C. M. 1995. The diversity of bats in Temengor Forest Reserve, Hulu Perak, Malaysia. Malayan Nat. J. 48 (3–4): 403–408.

Guyer, C. 1990. The herpetofauna of La Selva, Costa Rica. Pp. 371–385. In: Gentry, A. H. (ed.), Four Neotropical rain forests. Yale University Press, New Haven.

Janson, C. H. 1990. Ecological structure of the nonflying mammal community at Cocha Cashu Biological Station, Manu National Park, Peru. Pp. 314–338. In: Gentry, A. H. (ed.), Four Neotropical rain forests. Yale University Press, New Haven.

Janson, C. H. 1996. Toward an experimental socioecology of primates. Pp. 309–325. In: Norconk, M. A., Rosenberger, A. L. & Garber P. A. (eds), Adaptive radiations of neotropical primates. Plenum Press, New York.

Kalko, E. & Handley, C. 2001. Neotropical bats in the canopy: diversity, community structure, and implications for conservation. Plant Ecol. 153: 319–333 (this volume).

Kays, R. 1999. A hoistable arboreal mammal trap. Wildlife Soc. Bull. 27: 298–300.

Kiew, B. H. 1987. An annotated checklist of the herpetofauna of Ulu Endau, Johore, Malaysia. Malayan Nat. J. 41 (2–3): 413–425.

Kiew, B. H., Diong, C. H. & Lim, B. L. 1995. An annotated checklist of the amphibian fauna in the Temengor Forest Reserve, Hulu Perak, Malaysia. Malayan Nat. J. 48 (3–4): 347–351.

Lim, B. L., Ratnam, L. & Hussein, N. A. 1995. Snakes examined from the Sungai Singgor area of Temengor, Hulu Perak, Malaysia. Malayan Nat. J. 48 (3–4): 357–364.

Mack, A. L. & Igag, P. 1998. Birds. Pp. 62–67, 173–180. In: Mack, A. L. (ed.), A biological assessment of the Lakekamu

Basin, Papua New Guinea. RAP Working Papers 9 Conservation International, Washington, DC.
Malcolm, J. R. 1990. Estimation of mammalian densities in continuous forest north of Manaus. Pp. 339–357. In: Gentry, A. H. (ed.), Four Neotropical rain forests. Yale University Press, New Haven.
Malcolm, J. R. 1991. Comparative abundances of Neotropical small mammals by trap height. J. Mam. 72: 188–192.
Malcolm, J. R. 1995. Forest structure and the abundance and diversity of Neotropical small mammals. Pp. 179–197. In: Lowman, M. D. & Nadkarni, N. M. (eds), Forest canopies. Academic Press, New York.
Mann, C. F. 1996. The avifauna of the Belalong forest, Brunei Darussalam. Pp. 101–116 In Edwards, D. S., Booth, W. E. & Choy, S. C. (eds), Tropical rain forest research-current issues: proceedings of the conference held in Bandar Seri Begawan, April 1993. Monographiae Biologicae, Vol. 74. Kluwer Academic Publishers, Dordrecht.
Mobilik, P. & Marsh, C. 1993. Danum Valley Conservation Area: a checklist of vertebrates. Pusat Luar Lembah Danum (Danum Valley Field Centre), Kota Kinabalu, Sabah, Malaysia.
Munn, C. 1991. Tropical canopy netting and shooting lines over tall treees. J. Field Ornith. 62: 454–463.
Nadkarni, N. 2001. Enhancement of forest canopy research, education, and conservation in the new millennium. Plant Ecol. 153: 361–367 (this volume).
Rand, A. S. & Myers, C. W. 1990. The herpetofauna of Barro Colorado Island, Panama: an ecological summary. Pp. 386–409. In: Gentry, A. H. (ed.), Four Neotropical rain forests. Yale University Press, New Haven.
Ratnam, L., Liat, L. B. & Hussein, N. A. 1995. Mammals of the Sungai Singgor area in Temengor Forest Reserve, Hulu Perak, Malaysia. Malayan Nat. J. 48 (3–4): 409–423.
Reis, N. R. dos, & Peracchi, A. L. 1987. Quirópteros da regiao de Manaus, Amazonas, Brasil (Mammalia, Chiroptera). Boletim do Museu Paraense Emilio Goeldi 3: 161–182.
Robinson, S. K. & Terborgh, J. 1990. Bird communities of the Cocha Cashu Biological Station in Amazonian Peru. Pp. 199–216. In: Gentry, A. H. (ed.), Four Neotropical rain forests. Yale University Press, New Haven.
Rodriguez, L. B. & Cadle, J. E. 1990. A preliminary overview of the herpetofauna of Cocha Cashu, Manu National Park, Peru. Pp. 410–425. In Gentry, A. H. (ed.), Four Neotropical rain forests. Yale University Press, New Haven.
Rowe, N. 1996. The pictorial guide to the living primates. Pogonias Press, East Hampton, New York.
Simmons, N. & Voss, R. S. 1998. The mammals of Paracou, French Guiana: a Neotropical lowland rain forest fauna. Part 1. Bats. Bull. Am. Mus. Nat. Hist. 237: 1–219.
Smuts, B. B., Cheney, D. L., Seyfarth, R. M., Wrangham, R. W. & Struhsaker, T. T. 1987. Primate societies. University of Chicago Press, Chicago.
Terborgh, J. 1983. Five new world primates: a study in comparative ecology. Princeton University Press, Princeton, New Jersey.
Théry, M. 2001. Forest light and its influence on habitat selection. Plant Ecol. 153: 251–261 (this volume).
Vonesh, J. R. 1998. The amphibians and reptiles of Kibale Forest, Uganda: herpetofaunal survey and ecological study of the forest floor litter community. Unpublished MSc. thesis, University of Florida.
Voss, R. & Emmons, L. H. 1996. Mammalian diversity in Neotropical lowland rain forests: a preliminary assessment. Bull. Am. Mus. Nat. Hist. 230: 1–115.
Williams, S. E., Pearson, R. G. & Walsh, P. J. 1996. Distributions and biodiversity of the terrestrial vertebrates of Australia's wet tropics: a review of current knowledge. Pacific Cons. Bio. 2: 327–362.
Wilson, D. E. 1990. Mammals of La Selva, Costa Rica. Pp. 273–286. In: Gentry, A. H. (ed.), Four Neotropical rain forests. Yale University Press, New Haven.
Wright, D. D., Balagawi, S., Raga, T. T. & Ipu, R. T. 1998. Mammals. Pp. 67–71, 181–183. In: Mack, A. L. (ed.), A biological assessment of the Lakekamu Basin, Papua New Guinea. RAP Working Papers 9 Conservation International, Washington, DC.
Yoon, C. K. 1993. Counting creatures great and small. Science 260: 620–622.
Zimmerman, B. L. & Rodrigues, M. T. 1990. Frogs, snakes, and lizards of the INPA-WWF reserves near Manaus, Brazil. Pp. 426–454. In: Gentry, A. H. (ed.), Four Neotropical rain forests. Yale University Press, New Haven.

Plant Ecology **153:** 121–132, 2001.

Vertical stratification of figs and fig-eaters in a Bornean lowland rain forest: how is the canopy different?

Mike Shanahan[1,2] & Stephen G. Compton[1]
[1]*Centre for Biodiversity & Conservation, School of Biology, University of Leeds, Leeds, LS2 9JT, UK (E-mail mikeshanahan@yahoo.com);* [2]*Institute of Biodiversity and Environmental Conservation, UNIMAS, 93400 Kota Samarahan, Sarawak, Malaysia*

Key words: Borneo, *Ficus*, Frugivory, Fruit design characters, Keystone resource, Seed dispersal guilds

Abstract

Fig trees (*Ficus* spp; Moraceae) are a common constituent of many tropical forests, where they produce figs that are eaten by a wide range of bird and mammal species. In our Bornean field site six *Ficus* seed dispersal guilds can be recognised, differentially attracting subsets of the frugivore community. Guild membership appears to be determined by figs' size, colour, crop size and height above ground, and frugivores' size, sensory and locomotory physiology and foraging height. Vertical stratification therefore appears to be an important determinant of fig and frugivore partitioning. The guild structure observed is discussed with respect to implications for seed dispersal and the differences between the canopy and understorey. Regarding figs eaten primarily by birds, larger fruit and crops can be found in the canopy where they are exposed to larger assemblages of potential frugivores than those presented in the understorey.

Introduction

Having been beyond our reach for so many years, we are finally beginning to gain almost routine access to the canopies of tropical forests. Needless to say, in spite of recent advances, much of the detail of ecological interactions in the canopy remains unknown. One of the most studied interactions, the mutualism involving frugivory and seed dispersal of fleshy fruited plants by vertebrate frugivores, is the subject of ongoing research at our lowland forest site in Malaysian Borneo. In this paper we use fig trees (*Ficus* spp; Moraceae) and their putative dispersers to explore how their interaction in the canopy differs from that occurring elsewhere in the forest.

With over 800 species exhibiting a variety of growth forms that includes shrubs, trees, climbers, epiphytes and hemi-epiphytic stranglers, *Ficus* is arguably the world's most diverse woody plant genus (Corner 1988; Berg 1990). The characteristic *Ficus* inflorescence (the fig, or syconium) is remarkably uniform in structure. Differences do occur, however, in terms of the way that figs are packaged and presented. Crops range from tens to millions of red, green, brown or black figs which can be geocarpic (on ground level runners), cauliflorous (growing directly from the stem or trunk) or produced in the leaf axils (Corner 1988). Furthermore, two breeding systems, monoecy and gynodioecy occur among *Ficus* species, with half the individuals of gynodioecious species producing figs that contain no, or very few, seeds.

Ficus species are perhaps best known for their relationship with pollinating wasps (Hymenoptera, Agaonidae) which, with one or two exceptions (see Michaloud et al. 1996), are species-specific (Wiebes 1979). In monoecious *Ficus* species, the need to maintain a supply of pollinator wasps means that ripe figs can be found year-round. Consequently, these figs have been described as 'keystone resources' in tropical forests, potentially sustaining frugivores through lean periods of low fruit availability (Leighton & Leighton 1983; Terborgh 1986; Lambert & Marshall 1991; but see Gautier-Hion & Michaloud 1989; Borges 1993; Patel 1996 for evidence suggesting figs have lesser dietary importance in Indian and Afrotropical forests than they do in the Neotropics or South-East Asia). Particularly high calcium levels in figs add further importance to their role in the diets of tropical frugivores (O'Brien et al. 1998), and the extirpation of such keystone resources has been predicted to precipitate a cascade of further extinctions (Terborgh 1986).

The diversity and abundance of *Ficus* species is reflected in that of the vertebrates recorded eating figs. These include fishes, lizards, tortoises, birds, civets, bears, treeshrews, primates, fruit bats, bears, civets, rodents and deer (Ridley 1930; Baker & Baker 1936; Van der Pijl 1957, 1982; McClure 1966; Hnatiuk 1978; Lambert 1989a, 1990; Mondolfi 1989; Traveset 1990; Heydon & Bulloh 1997; Shanahan & Compton in press). Not surprisingly, figs have been considered an important dietary component for more animal species than the fruit of any other tropical genus (Janzen 1979).

Any animal eating figs has the potential to act as a seed disperser, in that they may move seeds away from their maternal tree. However, depending on plants' germination and establishment requirements, frugivores vary quantitatively and qualitatively in their effectiveness as seed dispersers (Schupp 1993). Furthermore, some frugivores may be true seed predators, destroying the seeds they ingest (Janzen 1971). Features which influence the type of frugivore attracted to a fruit crop might be expected to be adaptive, facilitating feeding by those animals that disperse seeds most effectively. This is particularly obvious in the case of the many hemi-epiphytic (strangling) *Ficus* species, for which only seeds deposited high in the canopy have a chance of successful development (Laman 1995). Studies of fruit characteristics suggest the existence of fruit syndromes, co-varying character complexes that have evolved in response to selection to favour feeding by particular groups of potential dispersers such as birds, fruit bats or primates (Ridley 1930; Van der Pijl 1957, 1982; Turcek 1963; Janson 1983; Gautier-Hion et al. 1985; Wheelwright & Janson 1985; O'Dowd & Gill 1986; Fischer & Chapman 1993, but see Tamboia et al. 1996). Ecological guilds may be defined as groups of organisms sharing, and potentially competing for, a common resource (Root 1967). Since frugivorous vertebrates differ in their choice of fruit, one can recognise guilds of species sharing certain types of fruit resource. Similarly, the fruit-producing plant species can be considered as sharing (and potentially competing for) dispersal resources, in the form of their attendant frugivores (Wheelwright 1985). Guilds of trees attracting similar dispersers can therefore also be recognised. To avoid confusion in this paper the term guild will be used to refer to a group of *Ficus* species sharing potential dispersers, whilst the groups of frugivores attracted are termed assemblages.

Whilst the hemi-epiphytic figs are renowned for attracting diverse frugivore assemblages (e.g. McClure 1966; Lambert 1989a), not all *Ficus* species have this growth form. The diversity of *Ficus* and its attendant frugivores makes this an ideal subject for the study of seed dispersal guilds and of the adaptive significance of fruit characters (putative fruit syndromes). One potentially powerful determinant of guild membership is vertical stratification. Tropical forests can be envisaged as being broadly organised into as many as five vegetational strata, ranging from the lowest shrubs to the emergent layer (Richards 1952). Figs can be found throughout these strata (Ridley 1930). Similarly, vertical stratification of frugivores is also a feature of tropical forests. For example, in Malaysian forests, birds, fruit bats, primates and squirrels all appear to have habitat preferences that partition their environment in the vertical dimension (Harrison 1962; Wong 1986; Francis 1994; MacKinnon et al. 1996).

Studies of seed dispersal are often non-quantitative, or concentrate exclusively on either plants or animals (in the case of the latter, commonly on single taxonomic groups rather than frugivore communities as a whole; Fleming 1979). In their concluding remarks, Kalko et al. (1996) emphasised the need for community-scale studies of fig frugivory aimed at understanding relationships between fruit characters and seed dispersers. This paper, a quantitative description of vertical stratification of fruit and frugivores in a Malaysian rain forest, attempts to describe such patterns and how they vary with distance from the forest floor.

Methods

Study site

Lambir Hills National Park (4°2′ N, 113°50′ E), Sarawak, East Malaysia comprises 6952 ha of mixed dipterocarp forest and kerangas (heath forest). Ashton & Hall (1992) provide a description of the park's forest structure. Altitude varies from sea level to 465 m and annual rainfall exceeds 5000 mm (Inoue & Hamid 1994). The park boasts the highest tree diversity of any site yet described, with 1175 species of tree over

1 cm dbh (diameter at breast height) identified in a 52 ha plot (LaFrankie et al. 1995). It is also home to one of the world's richest *Ficus* communities, with 80 recorded species and varieties that exhibit the full range of growth forms and modes of fruit production reported for the genus (Harrison 1998).

Despite its floristic diversity the park is currently depauperate in larger frugivorous animal species, a probable consequence of the park's small size, proximity to disturbance and the illegal hunting that occurs within its boundaries. In particular, primates, civets, flying foxes and hornbills appear to occur at lower densities than in previously studied Malesian forests (McClure 1966; Medway & Wells 1971; Leighton 1982; Lambert 1990). Lambir's tallest trees reach 70 m in height, but access to the canopy is facilitated by two tree towers (35 and 50 m high), tree ladder systems and 300 metres of walkway which were constructed as part of the Canopy Biology Program in Sarawak (Inoue et al. 1995; Yumoto & Inoue 1995). Fieldwork was conducted between March and September 1998.

Ficus fruiting characteristics

Ficus species were identified by reference to botanical keys (Corner 1965; Kochummen 1983; Primack 1983) and to R. Harrison's specimens, previously matched to those in herbaria in Kuching and Singapore. Regular phenological censuses of over 300 *Ficus* individuals allowed trees that were producing crops to be identified. A wide range of species was observed, rather than a few intensively, and consequently sample sizes per *Ficus* species are low. For gynodioecious species, only female (seed-producing) individuals were considered.

To characterise the attributes of the figs, between 10 and 40 were collected from ripe crops, either directly from the plant or as freshly fallen fruit. An index of fig size was calculated as the product of fig length (basal to apical dimension) and diameter. Fig placement was recorded as cauliflorous, axillary, or geocarpic (see definitions in the Introduction). Crop size was either counted directly or estimated by counting the number of figs on typical branches and then extrapolating to total numbers per tree. Following the forest stratification classification of Yamakura (1992), to which Lambir's forest corresponds, vertical crop placement was classed as either; 1: shrub layer (0–2.5 m), 2: understorey (2.5–12.5 m), 3: subcanopy (12.5–27.5 m), 4: canopy (27.5–42.5 m) or 5: emergent (>42.5 m). A crop-height index was calculated as the percentage of the crop in each of these height classes, multiplied by the class number (one to five). These figures were then summed to give a range from 100 (all fruit in the shrub layer) to 500 (all fruit in the emergent layer). Although a proportion of the figs produced by some geocarpic *Ficus* species may be subterranean, they were considered to have a height of zero metres, rather than a negative value.

Frugivory

Fig trees with ripe crops were selected for the observation of frugivory on the basis of their visibility, the criterion being that at least half of the fruit crop could be seen. Observations of frugivory were made during one to three hour sessions divided into a series of intermittent five minute scan samples. During each of these, the identities and numbers of each frugivore species observed eating figs were recorded. Crops in the canopy were viewed from adjacent ridges or from tree towers, ladders or canopy walkways. Each crop was observed during both day and night with observation sessions usually beginning at 6 a.m. and 7 p.m., respectively; and continuing until each crop had been observed for 10 hours. Diurnal observations were made using 8 × 32 binoculars or a 25 × telescope which aided species identifications at fig crops away from the walkways and towers. Nocturnal observations were made by spotlighting and the use of a Retron night-vision scope. Frugivore identifications were aided by reference to the field guides of Payne et al. (1985) and MacKinnon & Phillipps (1993) for mammals and birds, respectively. Nomenclature used here follows that used in these field guides.

Data analysis

To differentiate *Ficus* dispersal guilds, the DECORANA computer package (Hill 1979) was used to perform detrended correspondence analysis of frugivore species' proportional contributions to total feeding visits at each fig crop. All other statistical tests were performed with the SYSTAT computer program (SPSS 1997), with log transformations of data when necessary to meet requirements of parametric analyses.

Table 1. Fruit characters of *Ficus* species in Lambir Hills National Park. Derivation of crop height index explained in text. Nomenclature follows Corner (1965).

Ficus species	Growth form	Fig diameter (mm) mean ± S.D.	Fig placement	Colour when ripe	Crop height index	Log crop size
F. acamptophylla[1]	tree	11.54 ± 2.03	axillary	orange-red	175	2.48
F. annulata[2]	hemi-epiphyte	25.10 ± 4.10	axillary	green-yellow	300	2.48
F. aurata[1]	tree	10.84 ± 0.39	axillary	orange	200	2.98
F. benjamina[2]	hemi-epiphyte	9.01 ± 1.23	axillary	orange	375	5.59
F. brunneo-aurata[1]	tree	13.20 ± 1.70	axillary	orange	200	2.60
F. callophylla[2]	hemi-epiphyte	10.67 ± 0.52	axillary	orange-red	400	4.00
F. cereicarpa[1]	tree	41.8 ± 2.62	cauliflorous	brown-yellow	150	1.89
F. consociata[2]	hemi-epiphyte	12.04 ± 1.28	axillary	orange-red	400	3.90
F. deltoidea[1]	shrub	9.17 ± 1.44	axillary	orange-red	100	1.14
F. drupaceae[2]	hemi-epiphyte	18.48 ± 1.95	axillary	orange-red to purple	390	4.00
F. dubia[2]	hemi-epiphyte	22.11 ± 1.6	axillary	red to purple	400	3.90
F. fulva[1]	tree	12.78 ± 0.62	axillary	orange-red	200	3.34
F. grossinervis[1]	climber	11.33 ± 0.77	axillary	red	400	4.08
F. kerkhovenii[2]	hemi-epiphyte	10.25 ± 1.17	axillary	orange	400	4.78
F. megalia[1]	tree	16.17 ± 0.76	geocarpic	red	100	0.85
F. obscura[1]	tree	7.66 ± 0.52	axillary	orange	100	3.49
F. pellucido-punctata[2]	hemi-epiphyte	8.48 ± 1.08	axillary	orange-red	375	4.18
F. pisocarpa[2]	hemi-epiphyte	11.91 ± 0.79	axillary	orange	400	4.83
F. punctata[1]	climber	48.95 ± 6.01	axillary	red	315	2.15
F. sarwakensis[1]	climber	20.8 ± 1.80	axillary	red to purple	300	2.48
F. schwarzii[1]	tree	24.10 ± 1.52	cauliflorous	green-yellow	200	1.43
F. sinuata[1]	climber	4.58 ± 0.70	axillary	orange	200	2.95
F. stolonifera[1]	tree	13.25 ± 1.91	geocarpic	red	100	1.42
F. stupenda[2]	hemi-epiphyte	28.75 ± 2.71	axillary	orange-red	400	3.39
F. subcordata[2]	hemi-epiphyte	30.20 ± 2.37	axillary	orange-red to purple	450	4.00
F. subgelderi[2]	hemi-epiphyte	10.98 ± 2.36	axillary	orange-red	394	4.02
F. subulata[1]	climber	8.03 ± 1.27	axillary	orange-red	220	3.48
F. sumatrana[2]	hemi-epiphyte	13.16 ± 2.06	axillary	orange	400	3.73
F. sundaica[2]	hemi-epiphyte	13.78 ± 0.87	axillary	orange-red	375	4.08
F. treubii[1]	tree	14.87 ± 1.04	geocarpic	pink-brown	100	2.35
F. uncinata[1]	tree	21.91 ± 2.14	geocarpic	red	100	1.48
F. uniglandulosa[1]	tree	7.95 ± 0.55	axillary	orange-red	200	3.04
F. urnigera[1]	climber	7.60 ± 0.44	axillary	orange	200	2.48
F. xylophylla[2]	hemi-epiphyte	23.54 ± 2.42	axillary	orange-red	400	3.51

[1](Gyno-)dioecious species.
[2]Monoecious species.

Results

Fruit characters

Thirty-four *Ficus* species were studied with respect to their fruit characters. This suite of *Ficus* species included the whole range of diversity in terms of growth form and fig size, colour and vertical placement (Table 1). Figs were present in all vertical strata in the forest, from the ground level and subterranean runners of the geocarpic species to those of large hemi-epiphytes presented at over 50 m in the emergent layer. Fig size also varied considerably, from the 4.6 mm diameter ripe fruits of *Ficus sinuata* Thunb. to those of *F. punctata* Thunb. which reached more than 50 mm in diameter.

Table 2. Vertebrates observed eating figs in Lambir Hills National Park. n = number of species observed from each family.

Class	Family		n
Mammals	Tupaiidae	Treeshrews	3
	Pteropodidae	Fruit Bats	1
	Cercopithecidae	Monkeys	2
	Sciuridae	Squirrels	6
	Muridae	Rats	1
	Viverridae	Civets	1
	Tragulidae	Mouse-Deer	1
Birds	Columbidae	Pigeons	6
	Psittacidae	Parrots	2
	Bucerotidae	Hornbills	6
	Capitonidae	Barbets	4
	Eurylaimidae	Broadbills	1
	Chloropseidae	Leafbirds	3
	Pycnonotidae	Bulbuls	11
	Oriolidae	Orioles	1
	Corvidae	Crows	2
	Timaliidae	Babblers	1
	Muscicapidae	Flycatchers	1
	Sturnidae	Starlings	1
	Dicaeidae	Flowerpeckers	4

Frugivory and dispersal guilds

Observations of frugivory were made at 29 *Ficus* species and 58 vertebrate species from 20 families were recorded eating figs (Table 2). These comprised 15 mammal species (7 families) and 43 bird species (13 families). Thus, the majority (74.1%) of species recorded eating figs were birds. Comparisons of frugivore assemblages associated with different *Ficus* species indicated the presence of six seed dispersal guilds that we have given names reflecting the frugivores they most frequently attracted (Table 3). The three most distinct guilds, comprising a total of eight *Ficus* species, were those attracting only terrestrial mammals, fruit bats or primates, respectively.

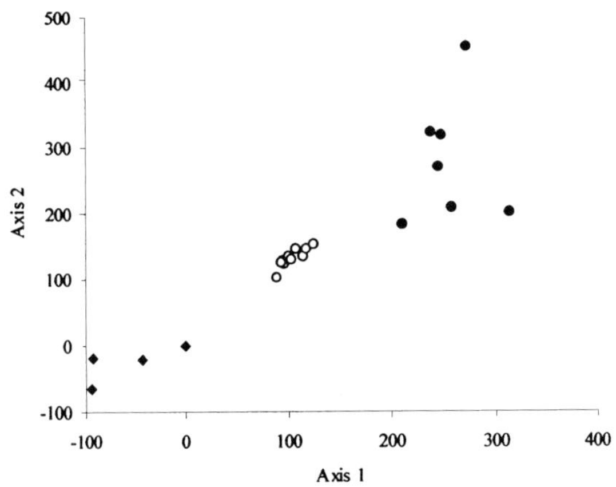

Figure 1. Scattergram of DECORANA axes one and two for 21 *Ficus* species, based on proportional visitation by avian and mammalian frugivores. Filled circles = 'bulbul figs' guild, open circles = 'mixed bird figs' guild, diamonds = 'pigeon figs' guild.

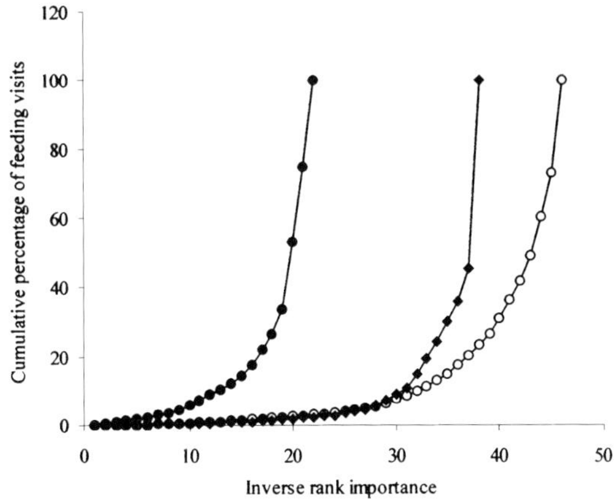

Figure 2. Relative contributions to total observed visits by frugivore species for 21 *Ficus* species attracting both avian and mammalian frugivores. Filled circles = 'bulbul figs', open circles = 'mixed bird figs', diamonds = 'pigeon figs'.

The distinction between the remaining 21 *Ficus* species, which attracted a diverse range of birds and arboreal mammals, is more quantitative. Detrended correspondence analysis of proportional contributions to total feeding visits by individual vertebrate frugivore species separated these *Ficus* species into three further guilds (Figure 1). In each of the three guilds, feeding visits were dominated by a small proportion of the total species attracted (Figure 2). This pattern of dominance is also evident at the familial level. The most commonly observed avian frugivores, bulbuls, barbets and pigeons, hail from just three of the 13 bird families recorded: Pycnonotidae, Capitonidae and Columbidae, yet they account for approximately 80% of all observed feeding visits for each of the three guilds (Table 4). Whilst the combined contributions of these families are similar across all three guilds, the relative importance of each varies considerably (Figure 3). The 'bulbul' guild included gynodioecious small trees and climbers presenting their figs largely in the understorey. The 'mixed birds' and 'pigeon' guilds comprised monoecious hemi-epiphytes bearing figs in

Table 3. Ficus seed dispersal guilds in Lambir Hills National Park. Guild names reflect the dominant frugivores rather than the only frugivores observed. N.B. Guilds defined by patterns of frugivore visitation not by fruit characters.

Dispersal guild	Frugivores attracted	*n Ficus* spp.	Fig fruit characters
Ground mammal figs	Terrestrial mammals only	5	produced on or near the ground, variable size/colour
Fruit bat figs	Fruit bats only	2	green, odorous figs
Primate figs	Primates only	1	large (5 cm plus diameter), red figs with tough exocarp
Bulbul figs	Understorey birds & arboreal mammals	7	orange-red figs in understorey, axillary
Mixed bird figs	Canopy birds & arboreal mammals	10	orange-red figs in canopy, mean diameter < 20 mm, axillary
Pigeon figs	Canopy birds & arboreal mammals	4	orange-red figs in canopy, mean diameter > 20 mm, axillary

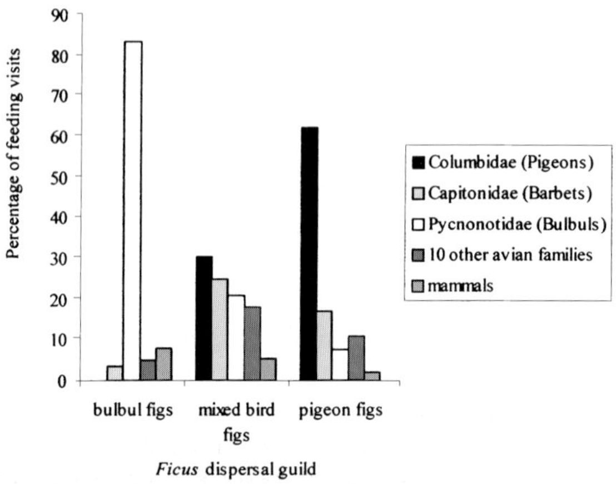

Figure 3. Frugivore families' proportional contributions to total feeding visits observed. Filled circles = 'bulbul figs', open circles = 'mixed bird figs', diamonds = 'pigeon figs'.

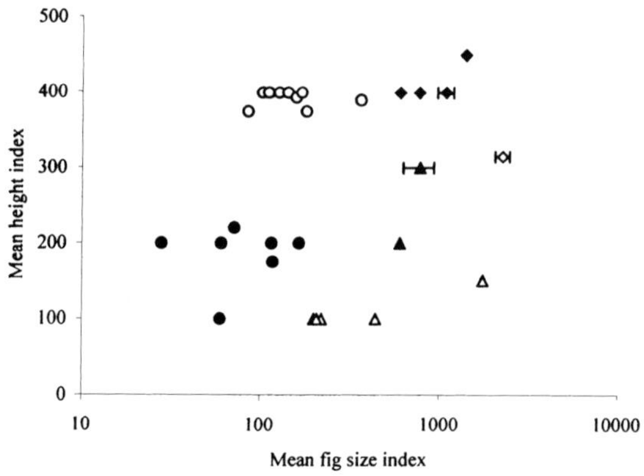

Figure 4. Vertical position and size of figs in Lambir Hills National Park. Fig size index was calculated as the product of fig length (basal to apical dimension) and diameter. Height index was calculated as the percentage of the crop in each of five height classes, multiplied by the class number (see Methods). Open triangles = 'ground mammal figs' guild, filled triangles = 'fruit bat figs' guild, open diamonds = 'primate figs' guild, filled circles = 'bulbul figs', open circles = 'mixed bird' guild, filled diamonds = 'pigeon figs' guild. Standard error bars for fig size are included only when the value exceeds 100. Within each of the six guilds individuals were distributed across the study site making spatial autocorrelation of observations of frugivory unlikely, except in the vertical dimension.

the canopy. *Treron* pigeons dominated feeding visits at the latter, whilst for the 'mixed birds' guild, the bird families' relative contributions were more even.

The six guilds that were identified reflect both qualitative and quantitative differences in frugivore assemblages between fig species, as the composition of the assemblages varied and the frugivores also made different relative contributions to overall fruit removal.

When related to the assemblages of frugivores attracted to fig crops, differences in fruit characteristics largely reflect classically recognised fruit syndromes; birds eating orange or red figs, fruit bats eating green figs and the largest, tough-skinned, red figs being eaten by primates alone (Table 3). We hypothesise that the more fine-grained division of the figs which are fed upon by both birds and arboreal mammals (Figure 1) reflects differences in the size of figs and whether or not they are presented in the canopy (Figure 4). Each of the these three guilds differed significantly in fig size (non parametric analysis of variance, $z = 3.41$, $p < 0.01$, testing the prediction that fig size increases between 'bulbul figs', 'mixed bird figs' and 'pigeon figs'.

Differences between the canopy and understorey

Highlighting the three guilds attracting both birds and arboreal mammals allows us to illustrate specific differences between figs and frugivores in the understorey (one guild) and in the canopy (two guilds). In defining these guilds, we have already demonstrated that differences exist in both the composition of fru-

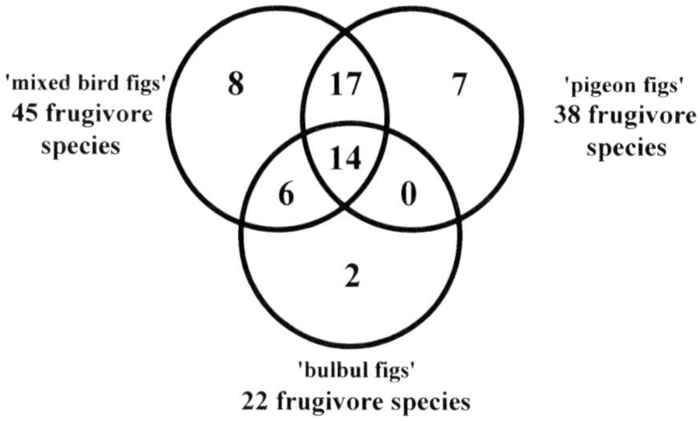

Figure 5. The extent of overlap in frugivore attraction between *Ficus* guilds attracting both avian and mammalian frugivores.

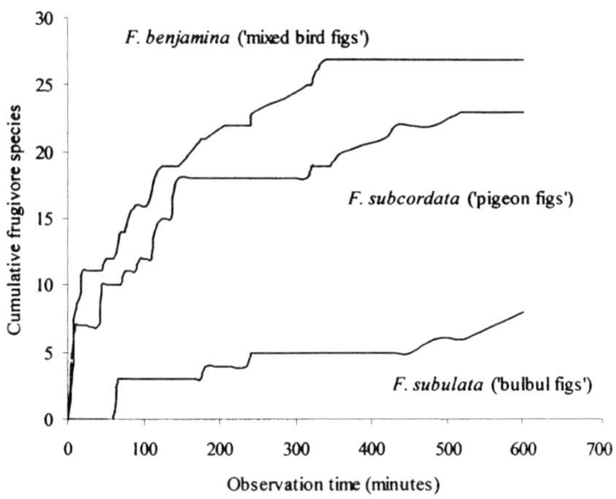

Figure 6. Species discovery curves for frugivores observed at randomly selected fig tree species from each of the guilds for which frugivory is dominated by birds (guilds indicated in parentheses after *Ficus* species' names).

givore assemblages and the relative contributions to overall feeding made by individual frugivore species.

Frugivore assemblages available to these three *Ficus* guilds were generally species-rich (22–45 species, Figure 5). However, the canopy fruiting species appear to have considerably more potential dispersers than do the understorey species. Only two of the 22 species recorded feeding on figs in the understorey were exclusively understorey foragers. In contrast, 32 out of 52 canopy feeding species were never observed to descend to feed on understorey figs. Thus, those species recorded eating the understorey figs largely represent a subset of those recorded feeding in the canopy, rather than a discrete unit of specialist understorey foragers.

The greater species richness of canopy feeding frugivores is reflected in the number of species associated with individual tree species in the two canopy guilds. These attracted larger coteries of potential dispersers than the understorey *Ficus* species of the 'bulbul figs'

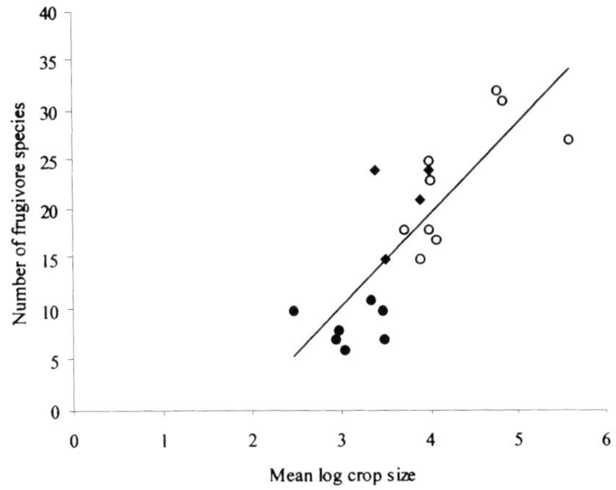

Figure 7. The relationship between mean crop size and number of frugivore species observed feeding in crops 21 *Ficus* species. Filled circles = 'bulbul figs', open circles = 'mixed bird figs', diamonds = 'pigeon figs'.

guild ($U = 0$, $p < 0.01$, Mann–Whitney U test). This pattern is evident when species discovery curves of frugivores observed feeding at figs in each of the three guilds are compared (Figure 6). However, crop height does not necessarily determine assemblage size. Multiple regression analysis allows confounding effects of correlated fruit characteristics to be controlled for (Janson et al. 1986). Such a model, explaining 82.8% of variation in assemblage size ($F = 25.75$, $p < 0.01$, $r^2 = 0.828$), revealed assemblage size to be strongly related to crop size ($\beta = 7.19$, $t = 4.42$, $p < 0.01$) but not to fig height ($\beta = 11.91$, $t = 1.49$, n.s.) or fig size ($\beta = 4.44$, $t = 1.69$, n.s.). Univariate regression shows that variation in crop size accounts for 66% of variation in assemblage size ($r^2 = 0.66$, $F = 34.95$, $p < 0.01$; Figure 7.).

We have already demonstrated that for frugivorous birds and arboreal mammals there are larger figs in the canopy than can be found in the understorey. Figure 8 illustrates the relationship between fig size and mean body length of avian visitors (data from MacKinnon & Phillipps 1993), a value derived after controlling for disproportional visitation rates of each frugivore. As well as attracting richer assemblages of potential dispersers, the canopy-fruiting *Ficus* species drew proportionately more of the larger frugivores (Mann–Whitney $U = 3$, $p < 0.01$). Multiple regression of frugivore size against fig size, crop height and crop size produces a significant model ($F = 25.97$, $p < 0.01$) explaining 83% of variability in frugivore size ($r^2 = 0.83$). However, whilst fig size ($\beta = 4.406$, $t = 4.304$, $p < 0.01$) is a significant factor, crop

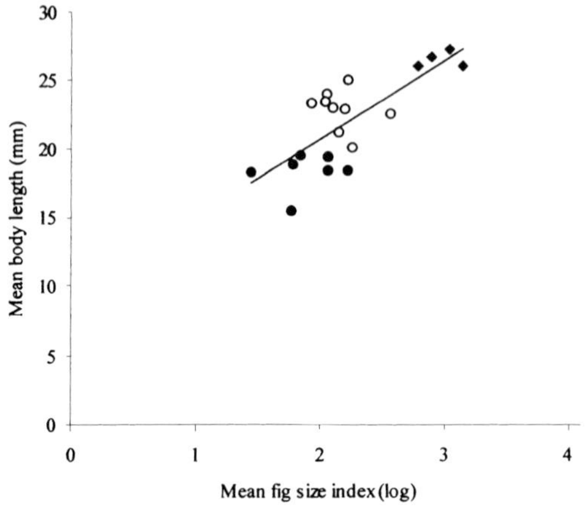

Figure 8. Relationship between mean fig size and mean avian frugivore size for 21 *Ficus* species attracting both avian and mammalian frugivores. Filled circles = 'bulbul figs', open circles = 'mixed bird figs', diamonds = 'pigeon figs'.

height (β = 5.465 t = 1.756, n.s.) and crop size (β = 0.956, t = 1.504, n.s.) are not. Univariate regression of frugivore size on fig size also reveals a highly significant relationship ($r^2 = 0.58$, $F = 33.42$, $p < 0.01$, Figure 8).

Discussion

Our results illustrate the great diversity of fruiting strategies displayed by *Ficus* species living sympatrically at our field site. Similarly, we highlight the diversity of frugivores that eat figs in Lambir, although observations of pteropodid bats, primates and civets were less frequent than might be predicted from the literature (Chivers 1980; Marshall 1985; Rabinowitz 1991). While the range of frugivore species observed eating figs in Lambir lends credence to reports of their dietary importance, our data show that all *Ficus* species are not equal in terms of their suitability for particular frugivores. This variability within *Ficus* allows discrete guilds of *Ficus* species to attract different subsets of the total frugivore community as potential seed dispersers. Meanwhile, competition for dispersers between *Ficus* species and members of other genera is reduced by the fig species' year-round fruiting phenology which contrasts that of non-*Ficus* species in Malesian forests (Medway 1972; Corlett 1984; Yumoto & Inoue 1995; Sakai et al. 1999).

The *Ficus* dispersal guild structure appears to be mediated by the interaction of fruit traits (colour, height, size), with aspects of the vertebrates' ecology, size, sensory capabilities, locomotory ability and behaviour. For example, birds are known to have good colour vision (Hartwig 1993) and, in agreement with earlier studies (e.g., Ridley 1930; Van der Pijl 1982; Turcek 1963; Gautier-Hion et al. 1989), the figs they ate were all red or orange in colour, in stark contrast to the green foliage amongst which they were presented. However, the consumption of a given fruit by one or more frugivore species does not necessarily imply that the fruit is adapted to those frugivores (see Janzen 1985). Investigations of fruit characters in *Ficus* are complicated by the fact that figs act initially as inflorescences, before becoming dispersal units. Selection pressures generated by the pollinators of the trees cannot, therefore, be ignored and we urge caution in the interpretation of the adaptive significance of fig structure, especially as formal comparative analyses utilising the phylogenies of the plants have not yet been conducted. Furthermore, differences in habitat (e.g. closed canopy or forest edge) may also influence the composition of frugivore assemblages recorded at a given crop. Such factors are under consideration in the ongoing studies into fig-frugivore relationships at Lambir.

Vertical stratification of figs and frugivores is clearly an important factor structuring the variety of dispersal guilds in Lambir. Spatial (i.e., vertical) constraints to frugivore mobility limit the potential disperser assemblages available to species producing fruit in a given forest stratum. Thus figs of the 'ground mammal figs' guild present their fruit at or near ground level and are not fed upon by the volant or aboreal frugivores observed at the other guilds (although one might expect ground birds, such as pheasants, to feed on these geocarpic figs). The two *Ficus* species in the 'bat figs' guild presented their fruit in the understorey and subcanopy. Francis (1990, 1994) reports that in Peninsular Malaysia fruit bat abundance is highest in subcanopy, but greater in the canopy than the understorey. In spite of the difficulties posed by nocturnal observations, fruit bats are noisy feeders and are easily detected (M. Shanahan pers. obs.). Our lack of observations of large canopy feeding fruit bats suggest that local populations are artificially low, probably due to hunting.

How is the canopy different?

Considering only the three guilds of fig species that attract both birds and arboreal mammals, it is possible to identify a number of differences between those

Table 4. Contributions of dominant frugivore families to total feeding visits at the three *Ficus* guilds attracting both avian and mammalian frugivores.

Dispersal guild	Percentage of all feeding visits accounted for by Columbidae, Capitonidae and Pycnonotidae: Total	Mean ± S.D.	N trees
Bulbul figs	86.9	79.6 ± 12.8	9 (7 spp.)
Mixed bird figs	76.3	79.0 ± 8.9	17 (10 spp.)
Pigeon figs	86.8	82 ± 9.3	7 (4 spp.)

trees producing their fruit in the canopy and those doing so in the understorey. Canopy figs are produced in a greater range of sizes (size index range; 85.1–1387.2) than those of the understorey (size index range; 28.2–117.9) and part of this greater range is generated by there being some larger figs in the canopy. Fig crops are also considerably larger in the canopy. These patterns of larger fruit and crops in the canopy reflect trends recorded elsewhere in the tropics (see Fleming et al. 1987). Regarding the frugivores eating figs, canopy-fruiting fig trees have the potential to attract more dispersers and, in the case of birds, larger bodied species than the understorey figs. This reflects the higher avian frugivore biomass in the canopy and that the understorey of Malaysian forests may be particularly depauperate in frugivorous birds (Wong 1986; Fleming et al. 1987). Few fig-eating birds were recorded only from the understorey and the frugivore species that eat understorey figs are essentially a subset of those found feeding in the canopy. Perhaps reflecting the lower diversity of dispersers in the understorey, the canopy fruiting species can be divided into two subguilds, whereas the understorey figs all attract superficially similar frugivore assemblages.

These patterns have implications for the seed dispersal service that the different fig guilds experience. By attracting species-rich assemblages, canopy fruiting species may be expected to achieve a relatively greater homogeneity of seed rain, a function of the diversity of sizes, habits, feeding styles and mobility of the frugivores that are attracted. However, our data show that even for canopy fruiting *Ficus*, frugivory is dominated by relatively few vertebrate species. Laman (1996) showed that, per fruiting episode, there is only a 1% chance a single hemi-epiphytic fig seed reaching a 'safe' site for germination at a distance of 60 m from its source. Even if no frugivore foraging in the canopy is particularly adept at fig seed dispersal, production of vast crops and attraction of large frugivore assemblages may function as a bet-hedging trait, maximising the likelihood of a single seed being dispersed.

Frugivorous birds found in the canopy are on average larger than those in the understorey (Wong 1986). Compared to the canopy-fruiting figs, the attraction of smaller birds by *Ficus* species in the 'bulbul figs' guild is therefore not unsurprising. However, among the canopy fruiting species, larger fruited species ('pigeon figs' guild) tended to attract proportionally more larger-bodied birds than the smaller-fruited species of the 'mixed bird figs' guild. Such a relationship between fruit and frugivore size has previously been reported for birds and neotropical fruit bats (Diamond 1973; Lambert 1989a; Kalko et al. 1996) and may be a universal phenomenon. Larger frugivores need to eat more than smaller ones and so may be more reliable fruit removers. They can also travel further and faster and have longer gut transit times for fig seeds (see Milton & May 1976; Lambert 1989b), so should be more likely to move seeds away from the maternal tree. Furthermore, large bodied birds appear to be either restricted to the canopy or not to descend with any regularity into the lower forest strata. For example, the largest avian frugivores in this study (hornbills, pigeons and crows) were never observed feeding on the understorey 'bulbul' figs (M. Shanahan pers. obs.). This is important to canopy-germinating hemi-epiphytes as smaller, vertically transient birds will 'waste' any seeds that they defaecate below the canopy. These functions of large body size do not apply only to the large figs of the 'pigeon figs' guild as, although they contribute relatively more to feeding visits at these species, larger birds than those in the understorey also fed on the smaller figs of the 'mixed birds' guild. An additional advantage to the latter guild of attracting large birds is that they are more likely to ingest seeds because they can consume figs whole rather than pecking at or mashing the pulp, which can scatter the seeds at their source. Once ingested the fate of fig seeds fate depends on the bird's digestive powers. Although larger birds generally retain seeds longer than small birds, the seeds of figs are sufficiently small to pass intact through the guts of most. Hornbills, *Ducula* pigeons and mynahs are among the largest avian frugivores in Asian forests and all pass fig seeds intact (Crome 1975; Midya & Brahmachary 1991; Whitney et al. 1998). Even *Treron* pigeons, previously thought to destroy all fig seeds (Cowles & Goodwin 1959) have been demonstrated to pass a proportion of those ingested in a germinable state (Lambert 1989b).

The existence of two putative dispersal guilds within the canopy may be driven by the greater diversity of canopy frugivores and so may be explicable as a function of resource partitioning of fruit by the frugivores, of dispersers by the figs, or both. In their review of vertebrate frugivore ecology, Fleming et al. (1987) stated that dietary overlap among paleotropical frugivores was especially high in the fig trees' family, Moraceae. However, their statement was based on comparisons of frugivory between plant genera, not species. When this question is examined at a finer scale, dietary overlap is seen to be less pronounced, the reduction being generated by the variety of ways in which figs are presented and packaged.

Interspecific competition among *Ficus* species has already been circumvented to a degree by their species-specific pollination system and the variety of their growth habits. These characteristics have been used to explain the high species packing of *Ficus* in tropical forests (Janzen 1979). *Ficus* species fruit year round (although fertile, seed bearing figs may show seasonality), which not only contributes to their putative keystone status, but also means that competition with other plants for dispersers may be reduced for much of the year. Intra-generic competition may potentially be relatively more important in *Ficus*, but any competition for dispersers should be weakened by the diversity of fruiting strategies they display.

Concluding remarks

Figs continue to be a fascinating aspect of tropical ecology. Whilst demonstrating their diversity and that of the animals that eat them, we have shown that a system of fruit and/or disperser partitioning exists, based on fruit characters and vertical stratification. The canopy is particularly rich in frugivorous birds and arboreal mammals that are able to feed upon larger crops and larger figs than their counterparts in the understorey. As a consequence, canopy-fruiting fig species experience a markedly different seed dispersal service to that of other *Ficus* species.

Acknowledgements

A number of individuals and organisations have helped make this paper possible. Research was conducted with the permission of the Sarawak Forest Department (FDS), for which special thanks due to the director of research, Abang Hj. Hamid. Work in Sarawak was facilitated by the granting of a research fellowship from the Institute of Biodiversity and Environmental Conservation - thanks to Dr. Stuart Davies and Prof. Ismail Ghazally at UNIMAS. Field identifications of *Ficus* individuals were made by Rhett Harrison of Kyoto University during his doctoral research. Use of tree towers and walkways in Lambir was possible through the ongoing work of the Canopy Biology Program in Sarawak. D. Hoare, J Wong, J. Moore and two anonymous reviewers made useful comments on an earlier draft of the paper. The study was funded by States of Jersey Education Committee and Bat Conservation International grants to MS.

References

Ashton, P. S. & Hall, P. 1992. Comparisons of structure among mixed dipterocarp forests of north-western Borneo. J. Ecol. 80: 459–481.

Baker, J. R. & Baker, Z. 1936. The seasons in a tropical rain forests. Part 3. The fruit bats. J. Linn. Soc. Lond. (Zool). 40: 123–141.

Berg, C. C. 1990. Classification and distribution of *Ficus*. Exper. 45: 605–611.

Borges, R. M. 1993. Figs, Malabar giant squirrels and fruit shortages within two tropical Indian forests. Biotropica 25: 183–190.

Chivers, D. J (ed.) 1980. Malayan Forest Primates. Ten years study in tropical rain forest. Plenum Press, New York.

Corlett, R. T. 1984. The phenology of *Ficus benjamina* and *Ficus microcarpa* in Singapore. J. Sing. Nat. Acad. Sci. 13: 30–31.

Corner, E. J. H. 1965. Check list of *Ficus* in Asia and Australasia, with keys to identification. Gard. Bull. (Singapore) 21: 1–186.

Corner, E. J. H. 1988. Wayside Trees of Malaya. United Selangor Press, Kuala Lumpur, Malaysia.

Cowles, G. S. & Goodwin, D. 1959. Seed digestion by the fruit eating pigeon *Treron*. Ibis 101: 253–254.

Crome, F. H. J. 1975. The ecology of fruit pigeons in tropical north Queensland. Austr. Wildl. Res. 2: 155–185.

Diamond, J. M. 1973. Distributional ecology of New Guinea birds. Science 179: 767.

Fischer, K. E. & Chapman, C. A. 1993. Frugivores and fruit syndromes: differences in patterns at the genus and species level. Oikos 66: 472–482.

Fleming, T. H. 1979. Do tropical frugivores compete for food? Amer. Zool. 19: 1157–1172.

Fleming, T. H., Breitwisch, R. & Whitesides, G. H. 1987. Patterns of tropical vertebrate frugivore foraging diversity. Annu. Rev. Ecol. Syst. 18: 91–109.

Francis, C. M. 1990. Trophic structure of bat communities in the understorey of lowland dipterocarp rain forest in Malaysia. J. Trop. Ecol. 6: 421–431.

Francis, C. M. 1994. Vertical stratification of fruit bats (Pteropodidae) in lowland dipterocarp rain forest in Malaysia. J. Trop. Ecol. 10: 523–530.

Gautier-Hion, A., Hecketsweiler, P., Sourd, C., Tiollay, J. M., Rousilhon, C., Quris, R., Decoux, J. P., Dubost, G., Duplantier, J. M., Erard, C., Emmons, L., Feer, F. & Moungazi, A. 1985.

Fruit characters as a basis of fruit choice and seed dispersal in a tropical forest vertebrate community. Oecol. 65: 324–337.

Gautier-Hion, A. & Michaloud, P. 1989. Figs: are they keystone resources for vertebrates throughout the tropics? A test in Gabon. Ecology 70: 1826–1833.

Harrison, J. L. 1962. The distribution of feeding habits among animals in a tropical rain forest. J. Anim. Ecol. 31: 53–63.

Harrison, R.D. 1998. Guide to the *Ficus* flora of Lambir Hills. National Park, Sarawak. Unpublished report. 34 pp.

Hartwig, H.-G. 1993. The central nervous system of birds: A study of functional morphology. Pp. 1–119. In:. Farner, D. S., King, J. R. & Parkes, K. C. (eds), Avian biology, Vol IX. Academic Press, London.

Heydon, M. J. & Bulloh, P. 1997. Mousedeer densities in a tropical rain forest: the impact of selective logging. J. App. Ecol. 34: 484–496.

Hill. M. O. 1979. DECORANA – a Fortran program for detrended correspondence analysis and reciprocal averaging. Cornell University, Ithaca, New York.

Hnatiuk, S. H. 1978. Plant dispersal by the Aldabran giant tortoise *Geochelone gigantea* (Schweigger). Oecologia 36: 345–350.

Inoue, T. & Hamid, A. A. (eds.). 1994. Plant reproductive systems and animal seasonal dynamics: Long-term study of dipterocarp forests in Sarawak. Center for Ecological Research, Kyoto University, Japan.

Inoue, T., Yumoto, T., Hamid, A. A., Lee, H. S. & Ogino, K. 1995. Construction of a canopy observation system in a tropical rain forest of Sarawak. Selbyana 16: 24–35.

Janson C. H. 1983. Adaptation of fruit morphology to dispersal agents in a neotropical forest. Science 219: 187–189.

Janson, C. H., Stiles, E. W., & White, D. W. 1986. Selection on plant fruiting traits by brown capuchin monkeys: a multivariate approach. Pp. 82–92. In: Fleming, T. H. & Estrada, A. (eds), Frugivores and seed dispersal. Junk, Dordrecht.

Janzen, D. H. 1971. Seed predation by animals. Annu. Rev. Ecol. Syst. 2: 465–492.

Janzen, D. H. 1979. How to be a fig. Annu. Rev. Ecol. Syst. 10: 13–51.

Janzen, D. H. 1985. On ecological fitting. Oikos 45: 308–310.

Kalko, E. K. V., Herre, E. A. & Handley, C. O. J. 1996. Relation of fig fruit characteristics to fruit-eating bats in the New and Old World tropics. J. Biogeogr. 23: 565–576.

Kochummen, K. M. 1983. Moraceae. Pp. 135–162. In: Ng. F. S. P. (ed), Tree Flora of Malaya, Vol. 3. Longman, Kuala Lumpur.

LaFrankie, J. V., Tan, S. & Ashton, P. S. 1995. Species List for the 52-ha Forest Dynamics Research Plot Lambir Hills National Park, Sarawak, Malaysia. Miscellaneous Internal Report (2/9/95). Center for Tropical Forest Science.

Laman, T. G. 1995. *Ficus stupenda* germination and seedling establishment in a Bornean rain forest canopy. Ecology 76: 2617–2626.

Laman, T. G. 1996. *Ficus* seed shadows in a Bornean rain forest. Oecologia 107: 347–355.

Lambert, F. R. L. 1989a. Fig-eating by birds in a Malaysian lowland forest. J. Trop. Ecol. 5: 401–412.

Lambert, F. R. L. 1989b. Pigeons as seed predators and dispersers of figs in a Malaysian lowland forest. Ibis 131: 521–527.

Lambert, F. R. 1990. Some notes on fig-eating by arboreal mammals in Malaysia. Primates 31: 453–458.

Lambert, F. R. & Marshall, A. G. 1991. Keystone characteristics of a bird-dispersed *Ficus* in a Malaysian lowland rain forest. J. Ecol. 79: 793–809.

Leighton, M. 1982. Fruit resources and patterns of feeding, spacing and grouping among sympatric Bornean hornbills (Bucerotidae). PhD thesis. University of California.

Leighton, M. & Leighton, D. R. 1983. Vertebrate responses to fruiting seasonality within a Bornean rain forest. Pp. 181–196. In: Sutton, S. L., Whitmore, T. C. & Chadwick, A. C. (eds), Tropical rain forest: ecology and management. Blackwell, Oxford.

MacKinnon, J. & Phillipps, K. 1993. A Field Guide to the Birds of Borneo, Sumatra, Java and Bali. Oxford University Press Inc., New York.

MacKinnon, K., Hatta, G., Halim, H. & Mangalik, A. 1996. The ecology of Kalimantan. Indonesian Borneo. Oxford University Press, Oxford.

Marshall, A. G. 1985. Old World phytophagous bats (Megachiroptera) and their food plants: a survey. Zool. J. Linn. Soc. 83: 351–369.

McClure, H. E. 1966. Flowering, fruiting and animals in the canopy of a tropical rain forest. Malay. Forester 29: 192–203.

Medway, L. 1972. Phenology of a tropical forest in Malaysia. Biol. J. Linn. Soc. 4: 117–146.

Medway, L. & Wells, D. R. 1971. Diversity and density of birds and mammals at Kuala Lompat, Pahang. Malay. Nat. J. 24: 238–247.

Michaloud, G., Carriere, S. & Koobi, M. 1996. Exceptions to the one:one relationship between African fig trees and their fig wasp pollinators: possible evolutionary scenarios. J. Biogeogr. 23: 513–520.

Midya, S. & Brahmachary, R. L. 1991. The effect of birds upon the germination of banyan (*Ficus bengalensis*) seeds. J. Trop. Ecol. 7: 537–538.

Milton, K. & May, M. L. 1976. Body weight, diet and home range area in primates. Nature 259: 459–462.

Mondolfi, E. 1989. Notes on the distribution, habitat, food habits, status and conservation of spectacled bears (*Tremarctos ornatus* Cuvier) in Venezuela. Mammalia 53: 525–544.

O'Brien, T. G., Kinnaird, M. F., Dierenfeld, E. S., Conklin-Brittain, N. L., Wrangham, R. W. & Silver, S. C. 1998. What's so special about figs? Nature 392: 668.

O'Dowd, D. J. & Gill, A. M. 1986. Seed dispersal syndromes in Australian *Acacia*. Pp. 87–121. In: Murray, D. R. (ed.), Seed Dispersal. Academic Press, Sydney.

Patel, A. 1996. Variation in a mutualism: phenology and the maintenance gynodioecy in two Indian fig species. J. Ecol. 84: 667–680.

Payne, J., Francis, C. M. & Phillipps, K. 1985. A Field Guide to the Mammals of Borneo. The Sabah Society & WWF Malaysia, Kota Kinabalu.

Primack, R. B. 1983. Forester's guide to the Moraceae of Sarawak. Forest Department, Sarawak.

Rabinowitz, A. R. 1991. Behaviour and movements of sympatric civet species in Huai Kha Khaeng Wildlife Sanctuary, Thailand. J. Zool. 223: 281–298.

Richards, P. W. 1952. The Tropical Rain Forest. An ecological study. Cambridge University Press, Cambridge.

Ridley, H. N. 1930. The Dispersal of Plants Throughout the World. L. Reeve & Co., Kent.

Root, R. B. 1967. The niche exploitation pattern of the blue-headed flycatcher. Ecol. Mono. 37: 317–350.

Sakai, S., Momose, K., Yumoto, T., Nagamitsu, T., Nagamasu, H., Hamid, A. A. & Nakashizuka, T. 1999. Plant reproductive phenology over four years including an episode of general flowering in a lowland dipterocarp forest, Sarawak, Malaysia. Amer. J. Bot. 86: 1414–1436.

Schupp, E. W. 1993. Quantity, quality and the effectiveness of seed dispersal by animals. Vegetatio 107/108: 15–29.

Shanahan, M. & Compton, S. G. in press. Fig-eating by Bornean treeshrews (*Tupaia* spp.): evidence for a role as seed dispersers. Biotropica.

SPSS (1997). SYSTAT 7.0 for Windows. SPSS, Inc.

Tamboia, T., Cipollini, M. L. & Levey, D. J. 1996. An evaluation of vertebrate seed dispersal syndromes in four species of black nightshade (*Solanum* sect. *Solanum*). Oecologia 107: 522–532.

Terborgh, J. 1986. Keystone plant resources in the tropical forest. Pp. 330–344. In: Soule, M. E. (ed.), Conservation biology, the science of scarcity and diversity. Sinauer, Sunderland, MA.

Traveset, A. 1990. *Ctenosaura similis* Gray (Iguanidae) as a seed disperser in a Central American deciduous forest. Am. Midl. Nat. 123: 402–404.

Turcek, F. J. 1963. Colour preferences in fruit and seed-eating birds. Pp. 285–292. In: Proceedings XIII International Ornithi. Congress 1962. Ithaca, New York.

van der Pijl, L. 1957. The dispersal of plants by bats (Chiropterophily). Acta Bot. Neerl. 6: 291–315.

van der Pijl, L. 1982. Principles of dispersal in higher plants. Springer-Verlag, Berlin.

Wheelwright, N. T. 1985. Competition for dispersers, and the timing of flowering and fruiting in a guild of tropical trees. Oikos 44: 465–477.

Wheelwright, N. T. & Janson, C. H. 1985. Colours of fruit displays of bird-dispersed plants in two tropical forests. Am. Nat. 126: 777–799.

Whitney, K. D., Fogiel, M. K., Lamperti, A. M., Holbrook, K. M., Stauffer, D. J., Hardesty, B. D., Parker, V. T. & Smith, T. B. 1998. Seed dispersal by *Ceratogymna* hornbills in the Dja reserve, Cameroon. J. Trop. Ecol. 14: 351–371.

Wiebes, J. T. 1979. Co-evolution of figs and their insect pollinators. Annu. Rev. Ecol. Syst. 10: 1–12.

Wong, M. 1986. Trophic organisation of understorey birds in a Malaysian dipterocarp forest. Auk 103: 100–116.

Yamakura, T. 1992. Structure and species diversity. Pp. 53–59. In: Shidei, T. & Kira, T. (eds), Considering of tropical forests. Jinbunshoin, Kyoto.

Yumoto, T. & Inoue, T. 1995. Canopy Biology Program in Sarawak – Long term study of plant phenology and plant-animal interactions in a mixed dipterocarp forest. Tropics 4: 307–315.

Plant Ecology **153:** 133–152, 2001.

Understorey versus canopy: patterns of vertical stratification and diversity among Lepidoptera in a Bornean rain forest

Christian H. Schulze[1,2], K. Eduard Linsenmair[1] & Konrad Fiedler[2]
[1]*Lehrstuhl für Tierökologie und Tropenbiologie, Theodor-Boveri-Institut für Biowissenschaften, Am Hubland, D-97074 Würzburg, Germany (E-mail: christian.schulze@uni-bayreuth.de);* [2]*Department of Animal Ecology I, University of Bayreuth, D-95440 Bayreuth, Germany*

Key words: Butterflies, Fruit-feeders, Guild composition, Moths, Nectar-feeders, Species composition

Abstract

We studied the vertical distribution of Lepidoptera from a canopy walkway within a dipterocarp rain forest at Kinabalu Park (Borneo) using three different methods: (1) Bait traps to survey fruit-feeding nymphalid butterflies, (2) standardized counts for predominantly flower-visiting butterflies and their potential predators, aerial-hawking birds, and (3) attraction by blacklight for hawk- and tiger moths. There was a distinct decrease in the abundance of fruit-feeding nymphalids towards the canopy, probably due to a reduced and less predictable availability of rotting fruits in higher strata. These constraints might also be responsible for a higher abundance variation in the canopy, and a significant shift in size from larger species in the understorey to smaller ones in the canopy. Changes of microclimate and the conspicuous increase of insectivorous aerial-hawking birds from ground to canopy layer may be responsible for the prominent change in species composition of fruit-feeding nymphalids between 20 and 30 m. Nectar-feeding Lepidoptera showed a reversed abundance pattern. One main factor contributing to the much higher abundance of flower-visiting butterflies and moth taxa in the canopy, such as Sphingidae and some Arctiinae, might be the increase of nectar resources available in upper vegetation layers. A distinctly higher diversity in hawkmoths was also found in the canopy. A higher abundance of insectivorous aerial-hawking birds in the canopy might contribute to the shift in body design of fruit-feeding nymphalids from more slender bodies at lower vegetation layers to stouter ones (i.e. species which are stronger on the wing) in the canopy. Larval resources could play an additional role in specialisation on but a small part of the vertical gradient. This may explain stratification pattern of the nymphalid subfamilies Morphinae and Satyrinae. Monocotyledoneous larval food plants of both taxa, whose flight activity is largely restricted to the understorey, occur mostly in lower vegetation layers. Our observations on a wide taxonomic and ecological range of butterflies and moths indicate that tropical forest canopies hold a distinct and unique Lepidoptera fauna, whose species richness and abundance patterns differ from lower strata. However, the notion of tropical forest canopies as peaks of terrestrial diversity does not hold uniformly for all taxa or guilds.

Introduction

The vertical structure of tropical rain forests can be described as a number of distinct, though intergrading, vegetation layers (e.g., Pomeroy & Service 1986; Whitmore 1984), which gradually modulate certain biotic (like floristic composition, leaf area, biomass density, species diversity etc., e.g., Terborgh 1985; Parker 1995) and abiotic parameters (like temperature, wind speed, insolation etc., e.g., Kiltie 1993; Parker 1995; Smith et al. 1992; Yoda 1974) along the vertical gradient ranging from ground level to the upper canopy. This results in a high diversity of different microhabitats which may contain different assemblages or communities of organisms adapted to these special conditions of the various forest strata.

Vertical stratification of animals in temperate and tropical forest ecosystems is documented for a wide range of taxa such as mammals (e.g., Francis 1994; Taylor & Lowman 1996), birds (e.g., Pearson 1977; Terborgh 1980), and insects (e.g., Kato et al. 1995;

Longino & Nadkarni 1990; McKey 1991; Sutton et al. 1983). The analysis of vertical distribution patterns and niche breadths of animals along such vertical gradients could provide one of the keys to the understanding of processes underlying species composition in animal communities of multi-layered forest habitats. Moreover, the analysis of stratification patterns and processes will foster our understanding of the enormous species richness of various taxa in such ecosystems like tropical moist forests. In particular, it remains to be critically tested for a broader range of taxa and guilds to what extent canopy communities are really richer in species, as is now widely accepted, because some studies have shown that much of the insect fauna seems to be restricted to the upper canopy (e.g., Rees 1983; Sutton 1983, 1989; although soil faunas may have been severely underrated so far: Groombridge 1992).

Herbivorous animals comprise a very large proportion of species in most terrestrial ecosystems, including tropical forests (e.g., Stork 1987). Therefore, we selected a taxon belonging to this guild to study distribution and diversity of species along the vertical gradient: the order Lepidoptera, whose larvae are almost exclusively herbivores (Scoble 1992). Data on vertical stratification of butterflies and moths are available from various tropical regions (e.g., Australia: Hill et al. 1992; Neotropics: Beccaloni 1997, DeVries 1988, DeVries et al. 1997, Medina et al. 1996, Papageorgis 1975; Africa: Corbet 1961, Jackson 1961). Therefore, a combination of detailed case studies and between-region comparisons may facilitate detecting general patterns in vertical stratification of tropical Lepidoptera.

For the present study we selected Lepidoptera taxa which are taxonomically well known from Borneo (the butterfly families Papilionidae, Pieridae, and Nymphalidae; the hawkmoths (Sphingidae) and tiger moths (Arctiidae: Arctiinae)). This selection has two advantages which are otherwise rarely found in any other species-rich tropical insect group. (1) Almost all species can be reliably identified. (2) For many species, information on adult and larval feeding resources or other biological attributes is available (Fiedler 1998). The present study has four central objectives.

1. Using the guilds of fruit-feeding and flower-visiting Lepidoptera as targets, we investigate differences in various community characteristics like diversity, species composition and abundance patterns, between vegetation layers along a vertical gradient from ground level to the canopy in a Bornean evergreen rain forest. These two guilds should respond in a different way to the contrasting distribution patterns of their resources (rotting fruits: most abundant at ground level; nectar: within closed rain forest most abundant in the canopy layers).
2. We analyze this data set separately for different nymphalid subfamilies to assess how strongly vertical distribution patterns are affected by taxonomic relatedness, and thus presumably by phylogenetic constraints (e.g. body design, conservative larval host plant affiliations).
3. We study the size and body design of fruit-feeding nymphalid butterflies to investigate if butterfly morphology reflects stratum-specific selective forces such as predation and resource availability.
4 Extending our studies to certain nocturnal taxa, we search for general patterns of vertical stratification which may indicate parallel selection pressures, or phylogenetic constraints, imposed on various Lepidoptera in tropical rain forests due to certain life history characteristics (e.g. availability of larval host plants or adult resources).

Methods

Study area

The study area is located at the southeastern border of the Mount Kinabalu National Park (Sabah, East Malaysia) in the vicinity of Poring Hot Springs (6°03′ N 116°42′ E) at an altitude between 600 and 700 m a.s.l. Descriptions of the geology, topography, and vegetational zonation of the Kinabalu area can be found elsewhere (e.g., Beaman & Beaman 1990; Kitayama 1992; Lee Tain Choi 1996; Wong & Phillipps 1996). Our study site at Poring was situated on a steep slope in a dipterocarp rain forest (designated as 'hill forest': Kitayama 1992). The study site at Poring was selected because it was possible to achieve easy access to the canopy layers of the forest using available canopy walkways. The survey of fruit-feeding and flower-visiting butterflies as well as the observations on the vertical distribution of insectivorous aerial-hawking birds were conducted from a walkway at 700 m a.s.l., light trapping was done from a walkway at 600 m a.s.l. The highest platform of both walkways is situated ca. 45 m above ground, roughly 10m lower than the maximum height of the forest canopy at these sites.

Bait trapping of nymphalid butterflies

Bait trapping was conducted with cylindrical gauze-traps (design: Dailey & Ehrlich 1995) suspended from the walkway system. Trap positions ranged from ground level to the upper canopy (ca. 50 m above ground) to survey the vertical distribution of the guild of fruit-feeding butterflies. In South East Asia, all fruit-feeding butterflies belong to the nymphalid subfamilies Satyrinae, Morphinae, Nymphalinae and Charaxinae (Corbet & Pendlebury 1992). Nymphalids were attracted with baits consisting of mashed, rotting bananas. Two traps each – separated by a distance of ca. 10 m – were installed at 0, 10, 20, 30, 40, and 50 m above ground level. Bait trapping was done on 22 days between 25th August and 22nd September 1993, and on 29 days between 20th February and 26th April 1994. Traps were baited on the day before they were checked for butterflies. On every sampling day, all traps were run concurrently and were dealt with twice, once in late morning (between 9:30 and 10:30 h) and again in early afternoon (between 14:00 and 15:00 h). The butterflies were killed to measure certain biometrical parameters later on and to avoid pseudoreplicates, which could bias the data (Beck & Schulze, 2000). For analysis only days without rain during the checks of the traps were used.

Observation of flower-visiting butterflies and insectivorous aerial-hawking birds

Standardized counts were conducted from the walkway to quantify abundance of predominantly flower-visiting species and their potential predators, insectivorous aerial-hawking birds. Birds species were categorized as 'insectivorous aerial-hawking' when this behaviour was directly observed (e.g. hunting insects from a perch) or is mentioned in the literature (Baker 1922, 1924, 1926, 1927, Hails & Jarvis 1987, Henry 1971, Holmes & Nash 1990, King et al. 1980, MacKinnon & Phillipps 1993). Observations were done from three fixed positions on the walkway – 10, 30 and >45 m above ground level – between 10:00 and 14:00 h. From these observation positions, the following vertical intervals were scanned for butterflies and birds: 0–10, 10–20, 20–30, 30–40, 40–50 and >50 m. From every observation position the two neighbouring vegetation layers were scanned alternately for a time period of 20 min by eye and using binoculars. Then, the observer changed to the next position and started to check the next pair of vegetation layers. Birds have been identified to species level, butterflies of the families Papilionidae, Pieridae and Nymphalidae at least to genus level. For every taxon the maximum number of individuals observed at the same time during the 20-min observation-unit was noted to avoid pseudreplicates by counting the identical individuals more than once. However, it is possible that an individual was considered for two strata when it changed to a neighbouring one during the observation period. The only way to standardize the counts was to record every butterfly and bird which could be seen from the walkway. For this reason, the abundance of birds and butterflies might be underestimated within the canopy layers with a higher density of vegetation compared to the less dense vegetation of the understorey and upper canopy layers (see Discussion). A total of 29 observation units (9 between 8–22 September 1993 and 20 between 22 February-24 April 1994) are available for analysis.

Flower-visiting butterflies comprise a wide variety of species belonging to several different taxa. Analyses were restricted to the families Papilionidae, Pieridae and Nymphalidae because only these taxa could be counted adequately in the field to a sufficient degree, while observation of Lycaenidae and Hesperiidae from a distance would have been difficult due to their small size.

Light trapping

To study moths of the canopy and understorey, we attracted them to a gauze tower illuminated by a 15 W blacklight tube between 18:30 and 21:00 h. This weak light stimulus was chosen to minimize attraction of moths from neighbouring vegetation layers (for effective catch radius of much stronger light sources: e.g., Baker & Sadovy 1978; Bowden 1982). The light source was installed either at ground level, or on a platform in the canopy, 45 m above ground. Understorey and canopy samples were alternately taken on subsequent nights whenever permitted by weather conditions. Light trapping was conducted in 1993 (August–September, $n = 4$ sampling nights at ground and in canopy, respectively), 1994 (February–April, $n = 4$), and 1997 (February–April, $n = 8$; August–September, $n = 5$). All attracted individuals of the focal moth taxa were collected manually. In total, the results reported here were assembled during 21 light trapping units each in the canopy and understorey.

We present results only for the family Sphingidae and the arctiid subfamily Arctiinae. While adult hawkmoths – typical income breeders with functional mouth parts, which feed as adults (Tammaru & Haukioja 1996) – are predominantly flower-visiting insects (e.g., Nilsson 1998, Pittaway 1993), the Arctiinae also contain a large proportion of species with a rudimentary, non-functional or completely reduced proboscis. Therefore, many Arctiinae species are capital breeders, which means that the adults depend entirely on resources derived from the larval period (Tammaru & Haukioja 1996) and are independent of the availability of any adult food resources like nectar or rotting fruits.

Identification

For identification of butterflies, we used standard reference publications for the region (Corbet & Pendlebury 1992; Otsuka 1988). For taxonomy and nomenclature, we followed the classification used in the checklist of the butterfly species of the Kinabalu Park (Häuser et al. 1997). Hogenes & Treadaway (1998) and Holloway (1976, 1987, 1988) were used for determination, taxonomy and nomenclature of Sphingidae and Arctiinae. Birds were identified after MacKinnon & Phillipps (1993).

Quantification of α- and β-diversity

A large number of methods to quantitatively express the 'diversity' of communities, or guilds, have been developed in the literature (e.g., Magurran 1988; Hayek & Buzas 1997). We use here three measurements: Fisher's α, Hurlbert rarefaction, and asymptotic extrapolation of species richness based on the Michaelis-Menten model. Fisher's α is an index which quantifies species diversity while being largely independent of sample size (Fisher et al. 1943). Fisher's α and the underlying log-series model have been widely used in biodiversity research and have a number of methodological advantages (Hayek & Buzas 1997). Calculation of Fisher's α was done with the computer program 'divers' (Krebs 1989). Ninety-five percent confidence limits can be calculated using the variance estimate given by Hayek & Buzas (1997).

Hurlbert rarefaction species accumulation curves (e.g., Achtziger et al. 1992) and their 95% confidence intervals were calculated by the program 'rarefact' (Krebs 1989). Such rarefaction analyses yield estimates of expected species numbers (together with a confidence interval) as a function of sample size. Expected species richness can then be compared between samples of different size, thus controlling for the bias produced by variation in sampling success.

A third way of assessing biodiversity is through extrapolation from real sample data to an asymptote, which is thought to estimate 'true' species richness of a community. Colwell & Coddington (1994) describe several methods which allow such extrapolations. This is especially useful for highly diverse habitats like tropical rain forests, where recorded species numbers are usually much lower than the asymptote of the species accumulation curves (e.g., Lamas et al. 1991). Computations of estimated species totals were done here with the computer program 'EstimateS' (Colwell 1997) using the 'MMMeans' method. We selected this method because several studies have empirically demonstrated its usefulness as good predictor for the total richness of a selected sampling site (Süssenbach & Fiedler 1999; León-Cortés et al. 1998, Peterson & Slade 1998). For calculation, the sample order (days of observation) is randomized 50 times to produce 50 species accumulation curves, from which an average species accumulation curve over the units of observation is obtained. The asymptotic value of a fitted hyperbolic Michaelis–Menten function then gives an estimation of total species number (see Colwell & Coddington 1994).

Much as with α-diversity, there are a plethora of ways to quantitatively compare the similarity of species assemblages (β-diversity) (review: Wolda 1981). We here selected the Dice index (M) which is based on presence/absence of species and minimizes bias resulting from different sample sizes (e.g., Wolda 1981). From the index values obtained, a similarity dendrogram was constructed using single-linkage clustering.

Biometrical measurements

To analyse morphological patterns of collected fruit-feeding nymphalids, we measured with a vernier calliper the forewing length (distance from wing base to apex), body length (distance from frons to tip of abdomen) and maximum thorax width of all pinned and set specimens to the nearest 0.5 mm. Forewing length (see Loder et al. 1998) served as a measure of the size of fruit-feeding nymphalids. Body length and thorax width were used to calculate an index for body design, namely the ratio of body length to thorax width (Chai & Srygley 1990). Slenderly built species reach

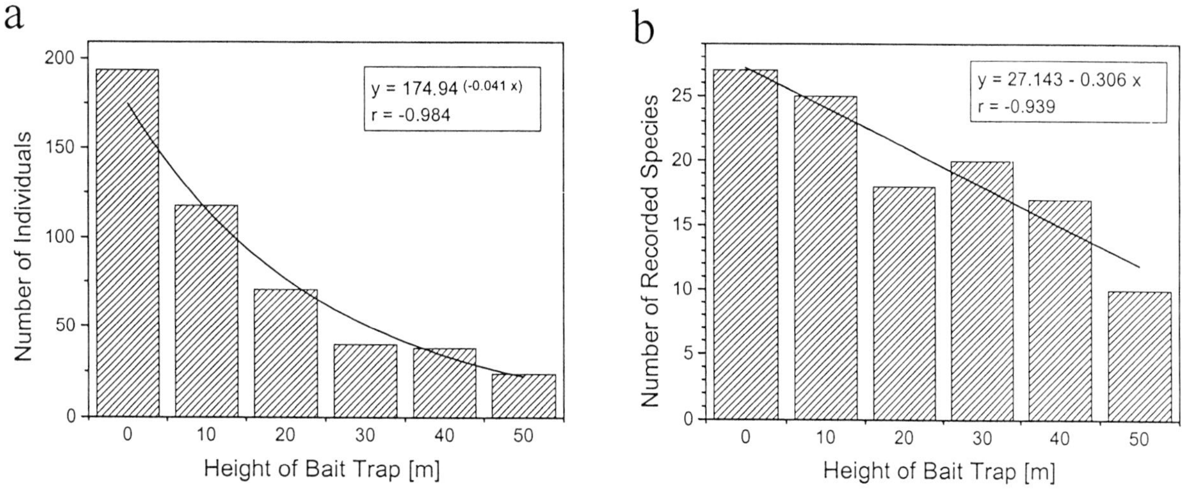

Figure 1. Number of (a) individuals and (b) species of fruit-feeding nymphalids sampled in bait traps at different heights above ground level. For both data sets a regression function and its correlation coefficient were added to describe the decrease of abundance and species records with increasing height.

higher values, and low values characterize stoutly built species.

Statistical analyses were performed using Statistica 5.1 (StatSoft 1997) and WinSTAT 3.1. Where sample data were not normally distributed, elementary non-parametric procedures (χ^2-test, Spearman's rank correlation: Sachs 1992) were used to analyse the data.

Results

Abundance, diversity, and similarity of fruit-feeding butterfly species assemblages in different vegetation layers

In the bait traps we caught a total of 485 fruit-feeding nymphalid butterflies representing 53 species (see Appendix 1 for species list and capture data). Abundance of these nymphalids decreased significantly (Pearson correlation on log abundance data: $r = -0.984$, $N = 6$, $p<0.001$) from ground level towards the upper canopy layer (Figure 1a). Concomitantly, the number of recorded species also fell significantly with trap height (Pearson correlation: $r = -0.950$, $N = 6$, $p<0.005$). However, whereas abundance decreased exponentially, the decline in species number was roughly linear (Figure 1b).

At the subfamily level, patterns of abundance and observed species numbers varied. A significant negative relationship between trap height and abundance, as well as number of recorded species, was found among Satyrinae, Morphinae and Nymphalinae. Charaxinae butterflies, in contrast, were equally rare along the vertical gradient (see Table 1). The number of trapped individuals decreased particularly strongly from ground to upper canopy level in the subfamilies Satyrinae (74.7% of all 166 baited individuals in the understorey at 0 and 10 m) and Morphinae (understorey: 87.9% of 33 individuals). In the canopy layers, only the Nymphalinae reached higher abundances (Table 1).

The number of recorded species (Figure 1b) is not a very meaningful measure of species diversity if sample size effects are not controlled for. To compare species diversity of different strata, we calculated Fisher's α and the species richness estimator MM-Means (Figure 2). Additionally, the estimated number of species was determined for a standardized sample size of 35 individuals (roughly equivalent to the sample size obtained at 40 m height) using Hurlbert's rarefaction (Figure 2). In contrast to absolute numbers of observed species, these three diversity estimates yield notably similar patterns. The highest scores for α, rarified expected species number, and extrapolated 'total' species richness of fruit-feeding nymphalids were obtained at 30 m above ground level in the lower canopy. Inspection of the confidence intervals, however, reveals that this peak in diversity is not statistically significant with regard to Fisher's α, while the rarefaction data indicate a significant increase of species richness at 30 m compared to the neighbouring lower and higher trap heights (Figure 2).

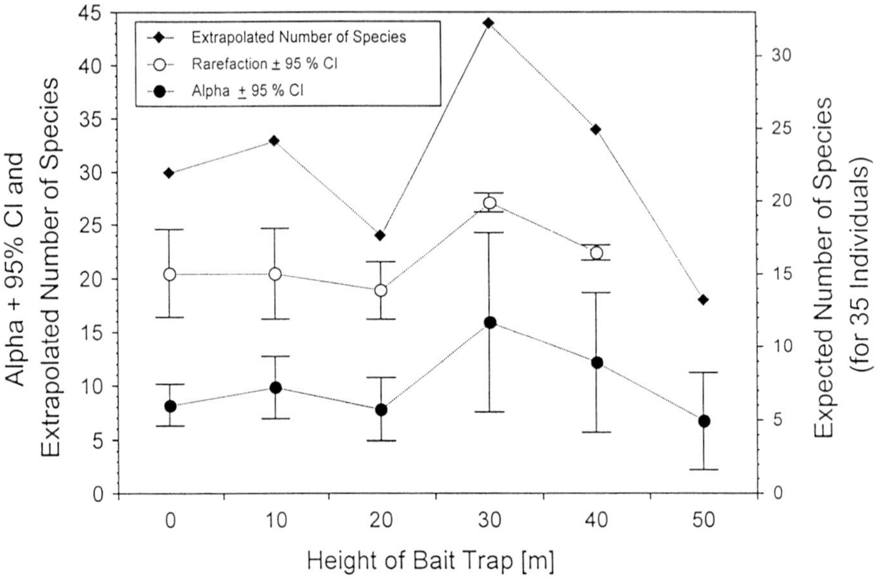

Figure 2. Diversity of fruit-feeding nymphalids at different bait trap heights quantified by three measures. Given are Fisher's $\alpha \pm 95\%$ confidence interval (left ordinate), the extrapolated species totals estimated by a Michaelis–Menten model (left ordinate), and the rarified species numbers (according to Hurlbert's method, $\pm 95\%$ confidence interval, right ordinate) for a standardized sample size of 35 individuals (for description of methods see text).

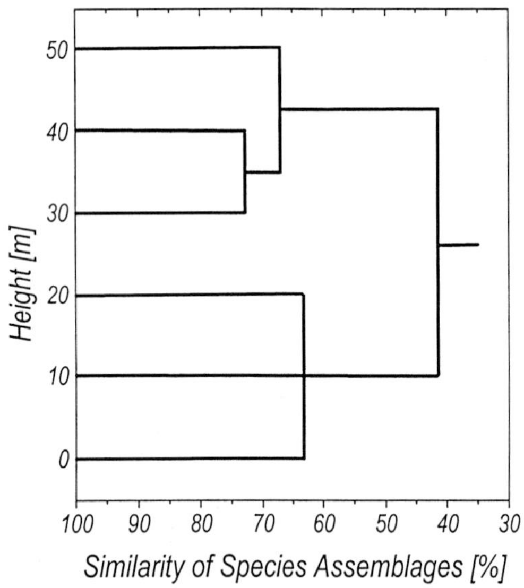

Figure 3. Dendrogram (single-linkage cluster) of similarity among species assemblages of fruit-feeding Nymphalidae between neighbouring trap heights, based on Dice indices.

The similarity between species sets at different trap heights was analyzed by calculating the Dice indices for all possible pairwise combinations of sampled strata. As expected, the highest overlap of species assemblages was invariably found between neighbouring trap heights (Table 2). A cluster analyses using the Dice similarity indices revealed a prominent frontier between 20 and 30 m above ground (Figure 3). These trap heights had the lowest species overlap ($D_{20\text{ m}/30\text{ m}} = 0.44$ compared to Dice indices between 0.67 and 0.90 calculated for all other neighbouring species assemblages) and, therefore, highest species turnover. This pattern recurred on subfamily level within the Nymphalinae ($D_{20\text{m}/30\text{m}} = 0.44$ compared to Dice indices between 0.70 and 1.00 for all other pairs of neighbouring trap heights) (Table 2).

In a study of vertical stratification of fruit-feeding nymphalids in Costa Rica, DeVries (1988) provided detailed sampling data but did not calculate rarified estimates of the diversity of fruit-feeders in different strata. We re-worked his data by calculating Hurlbert rarefaction curves aiming to compare his data with our Bornean samples. Since in this study he just compared the two units 'understorey' and 'canopy', we had to combine our data for the three lower and the three upper trap heights, respectively. Taking into account the high similarity of the species assemblages within these strata in our Bornean samples (see Figure 3 and Table 2), this procedure seems justifiable. Remarkably, DeVries (1988) reported a reverse abundance pattern for his study site in Costa Rica. He trapped more than twice as many fruit-feeding nymphalids in his canopy traps than in the understorey (Figure 4). However, at

Table 1. Abundance and recorded species numbers of fruit-feeding nymphalids on subfamily level and their relationship to bait trap height (r_s = Spearman rank correlation coefficient, p = significance level).

Subfamily	Height of bait trap						Total	r_s	p
	0 m	10 m	20 m	30 m	40 m	50 m			
				Abundance (%)					
Satyrinae	57.2	17.5	12.7	7.8	3.0	1.8	166	−1.0	<0.001
Morphinae	60.6	27.3	6.1	3.0	3.0	0.0	33	−0.986	<0.001
Nymphalinae	31.4	28.5	15.5	7.1	12.1	5.4	239	−0.943	<0.01
Charaxinae	8.5	25.5	23.4	19.1	6.4	17.0	47	−0.314	n.s.
				Number of recorded species					
Satyrinae	10	6	4	6	5	3	15	−0.853	<0.05
Morphinae	6	4	1	1	1	0	8	−0.941	<0.01
Nymphalinae	10	11	10	9	9	5	23	−0.883	<0.01
Charaxinae	1	4	3	4	2	2	7	0.000	n.s

Table 2. The Dice indices (M) for all butterfly assemblage pairs of the six trap heights separately given for family Nymphalidae (upper half) and subfamily Nymphalinae (lower half). Values for assemblages of neighbouring trap heights are printed in bold.

Trap height	0 m	20 m	30 m	40 m	50 m		
0 m		**0.71**	0.44	0.30	0.22	0.40	
10 m	**0.70**		**0.72**	0.35	0.27	0.30	
20 m	0.40	**0.70**		**0.44**	0.44	0.60	Family
30 m	0.33	0.22	**0.44**		**0.67**	0.90	Nymphalidae
40 m	0.20	0.10	0.40	**0.78**		**0.90**	
50 m	0.40	0.00	0.60	0.80	**1.00**		
			Subfamily Nymphalinae				

both sites the rarefaction species accumulation curves of understorey and canopy species assemblages did not differ significantly. At the Costa Rican site the curves for understorey and canopy samples are nearly congruent (Figure 4). So, although abundance patterns of fruit-feeding nymphalids differ markedly between these two study sites (Costa Rica: most individuals in the canopy; Borneo: highest abundance in the understorey), species diversity is roughly equal between both strata, and also rather similar between both sites.

To investigate the temporal abundance variation of understorey versus canopy layers, we again pooled the data for the species assemblages of fruit-feeding nymphalids from 0–20 m and 30–50 m. To test if population fluctuations among fruit-feeders were similar between understorey and canopy, we performed a Pearson correlation analysis between daily samples. The data pooled for both field periods show a significant positive relationship between daily understorey and canopy catch of fruit-feeding nymphalids ($r = 0.401$, $N = 51$, $p<0.01$). However, when conducting this analysis separately for both field periods, the same relationship emerged only for the year 1993 ($r = 0.517$, $N = 22$, $p<0.01$), whereas for the second field period in 1994 the relationship remained positive, yet did not reach significance ($r = 0.284$, $N = 29$, $p = 0.068$). Variation of abundance was quantified using the relative coefficient of variation which is independent of the number of individuals per sample. The average number of individuals trapped per day in canopy strata was significantly lower as indicated by the 95% confidence intervals (Figure 5). However,

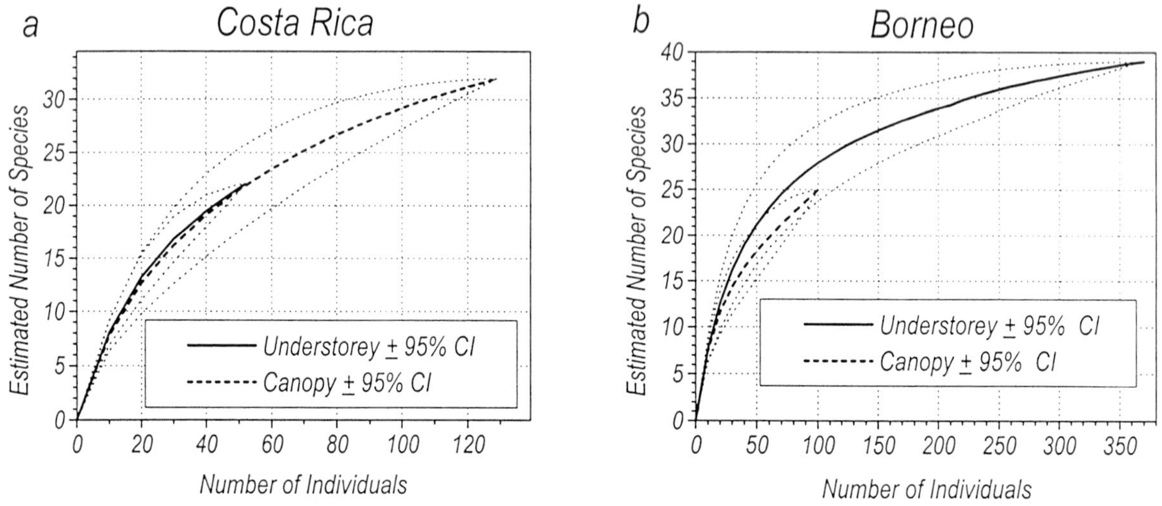

Figure 4. Rarefaction species accumulation curves and their 95% confidence areas for understorey and canopy species assemblages of fruit-feeding nymphalids at a study site in Costa Rica (a: data from DeVries 1988) and Borneo (b: Poring, own data). For comparison, our Poring data are pooled for understorey (0, 10 and 20 m) and canopy traps (30, 40 and 50 m).

in both field sampling periods the relative variation of nymphalid abundance was about 5 times higher in the canopy than in the understorey (Figure 5).

Size and body design of fruit-feeding nymphalids at different trap heights

To avoid a bias resulting by sexual dimorphism, the size of fruit-feeders was analysed for all trap heights separately for boths sexes. In both sexes a significant negative relationship between the mean forewing length of caught individuals and trap height was found (Pearson correlation for males: $r = -0.793$, $N = 6$, $p<0.01$; for females: $r = -0.899$, $N = 6$, $p<0.01$; Figure 6a). Data on forewing length for all species and, in sexually dimorph species, for males and females, are given in Appendix 1. While large Morphinae such as *Zeuxidia* species, *Thauria aliris* (Westwood 1858) and *Amathuxidia amythaon* (Doubleday 1847), Nymphalinae like *Lexias dirtea* (Fabricius 1793), *Bassarona dunya* (Doubleday 1848) and *Kallima limborgi* (Moore 1879), and Charaxinae such as *Prothoe franck* (Godart 1824) and *Agatasa calydonia* (Hewitson 1854), characterise the fruit-feeding butterfly community in the understorey, such extremely large representatives (with forewing lengths ≥45 mm) were a small minority (3 out of 62 (=4.8%) individuals trapped at 40 or 50 m) in the canopy samples (see Appendix 1).

The negative relationship between size and trap height may be biased because of taxonomic differences in sample composition. For example, the large Morphinae are restricted to the ground layer. To control for such taxonomic bias, we performed this same analysis separately for the larger subfamilies Satyrinae and Nymphalinae. Because of small sample sizes, no statistical analysis was possible for the subfamilies Morphinae and Charaxinae. On subfamily level, a similar negative relationship emerged between mean forewing length of individuals and trap height for female (Pearson correlation: $r = -0.946$, $N = 6$, $p<0.01$) and male Nymphalinae ($r = -0.917$, $N = 6$, $p<0.01$). However, in Satyrinae this relationship did not recur in males ($r = -0.005$, $N = 6$, $p = 0.496$) and females ($r = 0.357$, $N = 6$, $p = 0.244$).

Body design was quantified by using the ratio of body length to maximum thorax width. Using this body design index, stouter built species have lower values while slender ones are characterized by high ratios. The index values for all species (and sexes, if appropriate) are given in Appendix 1. We observed a significant negative relationship between mean ratio of body length to thorax width of individuals of fruit-feeding nymphalids and bait trap height in both sexes (Pearson correlation for males: $r = -0.893$, $N = 6$, $p<0.01$; for females: $r = -0.918$, $N = 6$, $p<0.01$). Species with slender-built bodies were more prevalent in lower vegetation layers, while stouter ones dominated in the canopy (Figure 6b). For reasons already mentioned above, the analysis was also performed separately for the subfamilies Nymphalinae and Satyrinae. The relationship found for all fruit-

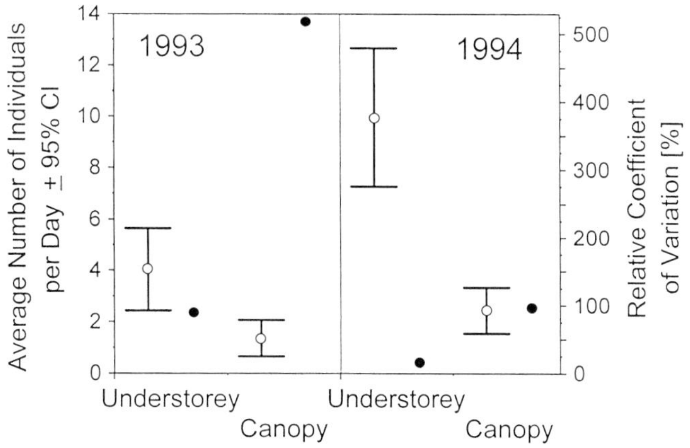

Figure 5. Average number of fruit-feeding Nymphalidae individuals trapped per day (open circles, ±95% confidence interval, left ordinate, $N_{1993} = 22$ trap days, $N_{1994} = 29$ trap days) in understorey (data pooled for 0, 10 and 20 m) and canopy layers (30, 40 and 50 m) in the field sampling periods 1993 and 1994. Additionally, the relative coefficients of variation (filled circles, right ordinate) are given as a measure of relative abundance variation.

feeding Nymphalidae recurred on subfamily level in Nymphalinae (Pearson correlation for males: $r = -0.819$, $N = 6$, $p<0.05$; for females: $r = -0.926$, $N = 6$, $p<0.01$), but not in Satyrinae (males: $r = 0.390$, $N = 6$, $p = 0.222$; for females: $r = -0.035$, $N = 6$, $p = 0.473$).

Abundance of flower-visiting butterflies and bird predators in different vegetation layers

The average number of flower-visiting butterfly individuals counted per 20 min observation unit was significantly higher in the canopy strata between 30 and 50 m, than in the understorey layers between 0 and 20 m (see 95% confidence intervals, Figure 7a).

Parallel to this survey of flower-visiting butterflies, the prevalence of a group of potential butterfly predators – insectivorous aerial-hawking birds – was also noted. This guild of birds increased significantly from ground to canopy (Figure 7b). The only statistically significant difference in abundance of insectivorous aerial-hawking birds between two neighbouring observation layers was found between the strata 10–20 m versus 20–30 m (Figure 7b).

Abundance and diversity of hawkmoths and tiger moths in understorey and canopy

Hawkmoths were attracted by a 15 W blacklight tube in much larger numbers in the canopy, as compared to ground level in a closed forest (see Appendix 2). Moreover, a comparison of their diversity by Hurlbert rarefaction curves shows a significantly higher species richness for the canopy (see Figure 8a). All species noted in the understorey were recorded in just one or two individuals except *Daphnusa ocellaris* Walker, 1856, a species with reduced mouthparts (C. H. Schulze pers. obs.) which seems to be the only specialist understorey hawkmoth at Poring (39 individuals in understorey versus 3 in canopy).

Within the moth subfamily Arctiinae (see Appendix 3) we detected no significant difference in diversity, as revealed by Hurlbert species accumulation curves (Figure 8b). However, the total number of tiger moths attracted to the light source was 28.7% higher in the canopy than in the understorey (test against null hypothesis of equal distribution: $\chi^2_{1df} = 11.76$, $p<0.001$).

Most species of the arctiid subfamily Arctiinae do not feed as adults and are thus capital breeders, while representatives of a few genera (*Amerila*, *Nyctemera*, *Utheteisa*: Common 1990, C. H. Schulze pers. obs.) have retained a functional proboscis and visit flowers for nectar-feeding (at least partial income breeders). The abundance of representatives of these two different life history strategies differed distinctly between understorey and canopy ($\chi^2_{1df} = 23.47$, $p<0.001$). Nectar-feeders clearly dominated in the canopy with more than 90% of all tiger moth individuals attracted by light, whereas in the understorey the occurrence of nectar-feeders and non-feeders was almost balanced (Figure 9).

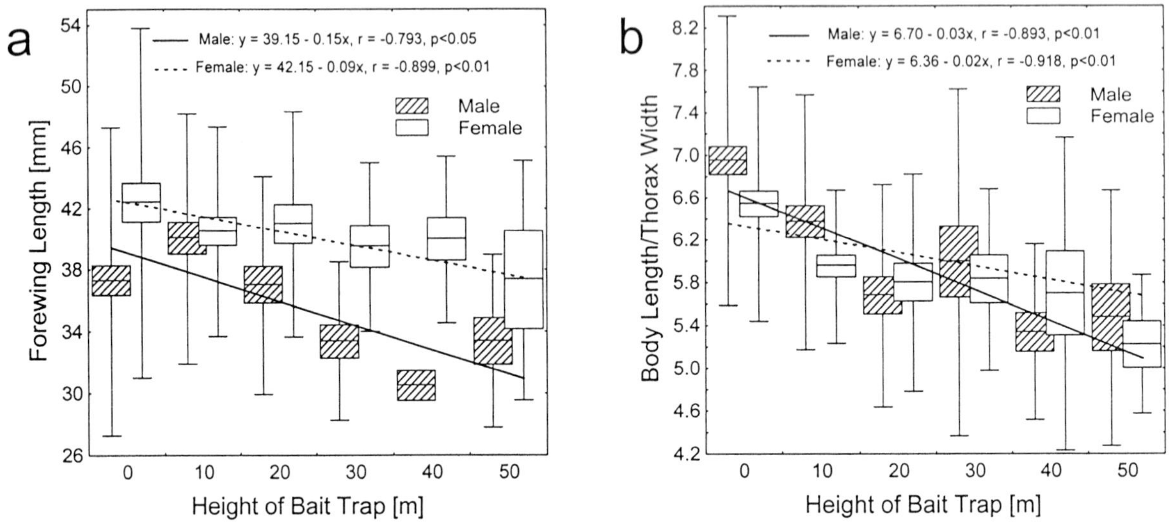

Figure 6. Relationship between morphological parameters of male and female fruit-feeding nymphalids – (a) size (measured as forewing length) and (b) body design (ratio of body length and thorax width) – and height of bait traps (for detailed description of calculation procedure see text). The box-whisker-plots present means as well as their standard errors (box) and standard deviations (whisker). Additionally, regression function, Pearson correlation coefficient and the level of significance are given.

Discussion

Abundance of guilds – understorey versus canopy

The comparison of fruit-feeding nymphalids and flower-visiting butterflies revealed different distribution patterns along a vertical gradient. The numbers of individuals of flower-visiting butterflies increased towards the canopy, whereas fruit-feeding nymphalids showed a decrease in abundance. In both cases the presence of food resources appears to be an important factor that may explain this observation. While the availability of flowers was highest in the canopy during the whole observation period (C.H. Schulze pers. obs.), rotting fruits should be expected to be commonest on the ground.

The abundance data of flower-visiting butterflies – and also of insectivorous aerial-hawking birds – could be biased by differential visibility in the six observation strata due to differences of vegetation density. However, this should not affect the documented pattern in principle. Especially, the extremely low abundances noted in the two understorey layers (0–20 m) cannot be the result of a high vegetational density. The understorey vegetation of this primary forest site was not dense and allowed to reliably recognize butterflies and birds over a distance similar to the upper canopy layers.

If the availability of rotting fruits is less predictable at upper vegetation layers, this should affect the niche breadth of fruit-feeders and maybe their abundance in the canopy. It could be expected that canopy fruit-feeders have to forage in a larger number of different strata than understorey species to find enough food. In accordance with the expectation of a lower abundance, we observed a higher proportion of rare species (<5 recorded individuals; e.g., 48% at trap height 0 m vs. 88% at 40 m; cf., data in Appendix 1) in the canopy. Additionally, the vertical extension of activity ranges of fruit-feeding nymphalid species – measured as number of different trap heights with positive records – increased from the ground layer towards the canopy (Schulze, unpubl. data). Moreover, abundance variation among fruit-feeding canopy nymphalids was five times higher than within the understorey community. The magnitude of daily canopy and understorey samples was positively correlated, indicating that temporal variation in abundance of fruit-feeding nymphalids was largely governed by the same (probably climatic) factors. Hence, differences between understorey and canopy guilds cannot be attributed to seasonal factors. Rather, the higher relative variation in the tree crowns indicates that stochastic factors become more important in the composition of the fruit-feeder community, where the availability of the food resource is less predictable.

Studies on vertical stratification of fruit-feeding nymphalids also exist from the Neotropics. Remarkably, DeVries (1988) found a different abundance

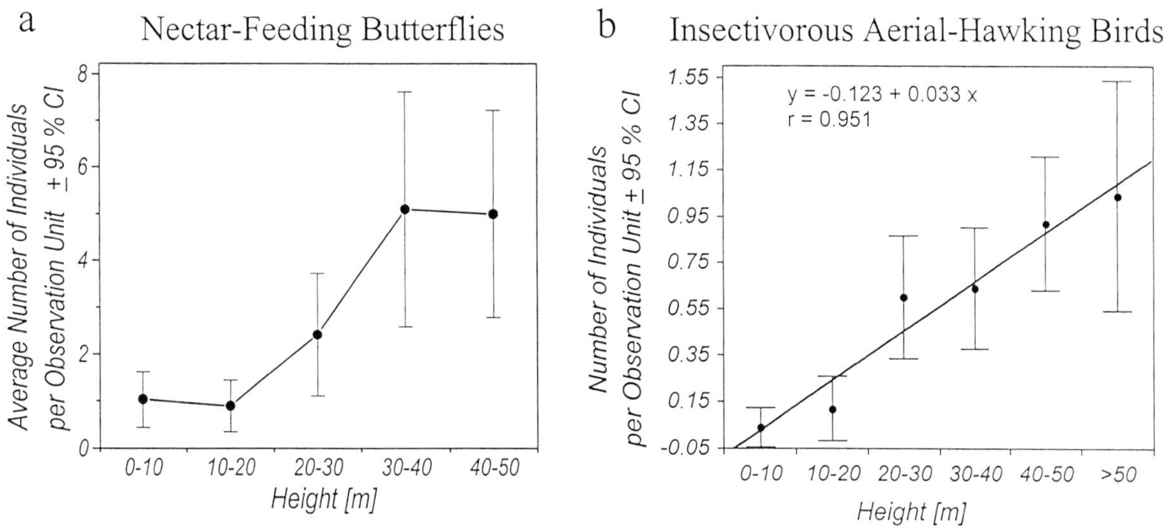

Figure 7. Average numbers of predominantly nectar-feeding butterfly individuals (a) and insectivorous aerial-hawking birds (b) observed per 20 min observation unit (±95% confidence intervals, $N = 25$ observation units) at different observation strata. Additionally, a regression function and its correlation coefficient are added in (b) to describe the increase in abundance of insectivorous aerial-hawking birds from ground to canopy level.

pattern for one Costa Rican study site, where he caught much more fruit-feeders in the canopy. However, a second more recent study from a Neotropical primary forest in the vicinity of the Jatun Sacha in eastern Ecuador (DeVries et al. 1997) revealed results very similar to our own survey at Poring. At Jatun Sacha, fruit-feeding nymphalids were again much less abundant in the canopy (DeVries et al. 1997). If epiphytes (and in particular the funnel-shaped Neotropical bromeliads) act as 'traps', which retain a significant portion of rotting fruits in the canopy, this could increase availability and predictability of this resource and, therefore, abundance of fruit-feeders in epiphyte-rich forests. Such a potential influence of ephiphyte abundance on community structure of fruit-feeding nymphalids in higher forest strata deserves further study.

We observed a higher abundance of nectar-feeding butterflies as well as of moths (Sphingidae, most Arctiinae) in canopy layers. Hawkmoths were nearly six times more abundant in the canopy than in the understorey (for more detailed abundance data see Schulze & Fiedler 1997a), and nearly twice as many tiger moths were attracted in the canopy than in the understorey. Excluding those tiger moth species with vestigial mouthparts demonstrates even more strongly the preponderance of nectar-feeding Arctiinae in the canopy as compared to the forest understorey. These observations match closely the data of Kato et al. (1995) who studied the vertical distribution of nocturnal flying insects at a lowland dipterocarp forest in Sarawak (Malaysia) using light traps and also found the peak in abundance for the guild of nectar collectors to occur at the canopy level.

Unfortunately, for most South East Asian Sphingidae and Arctiinae no precise information on adult behaviour and resource use is available. Most hawkmoth species can safely be regarded as nectar-feeders, because the adults have a well developed proboscis and, accordingly, should at least occasionally visit flowers. A group within the Sphingidae with reduced and non-functional probscis are the Smerinthinae (Lemaire & Minet 1999). The only representative of this subfamily in our samples is *Daphnusa ocellaris*, the only 'true' understorey hawkmoth. A further study on stratification of hawkmoths was published by Corbet (1961). He sampled hawkmoths in a swamp forest in Uganda using light traps in 5 different heights above ground from understorey to canopy level. Although the total sample is very small ($n = 56$), it is notable that 64.3% of the individuals were caught in the two highest traps between 30 and 40 m.

Among the Arctiinae, it is more difficult to assess which species may visit flowers. We categorized species as flower-visiting when their proboscis is well developed as in the genera *Nyctemera*, *Utetheisa* and *Amerila* (nectaring behaviour has been confirmed for other representatives of these genera, e.g., Common

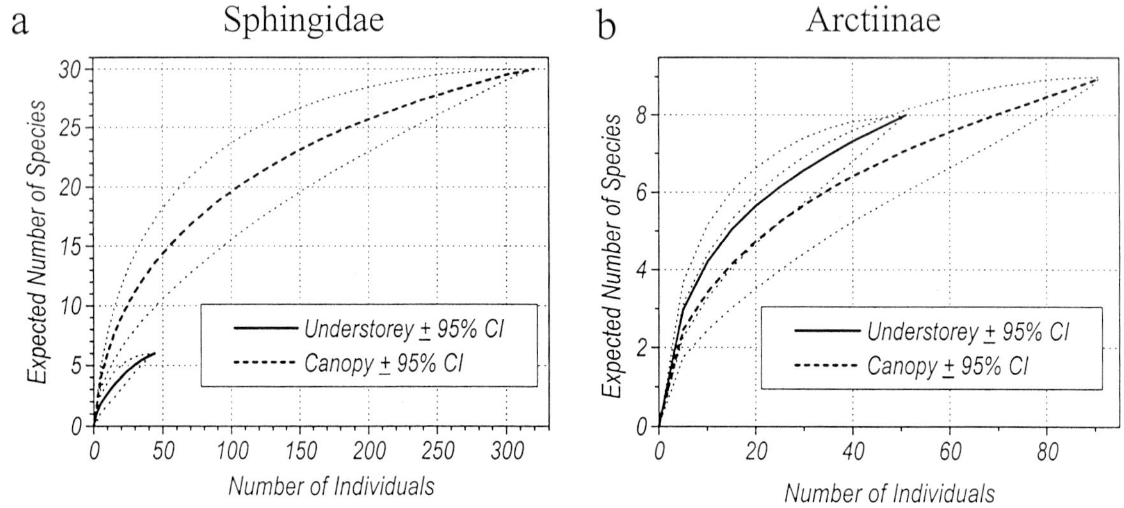

Figure 8. Hurlbert rarefaction curves ±95% confidence areas for understorey and canopy species assemblages of (a) hawkmoths (Sphingidae) and (b) tiger moths (Arctiinae) at Poring. Analysis of hawkmoth assemblages based on 21 light trapping samples in each vegetation layer, of Arctiinae in 13 samples.

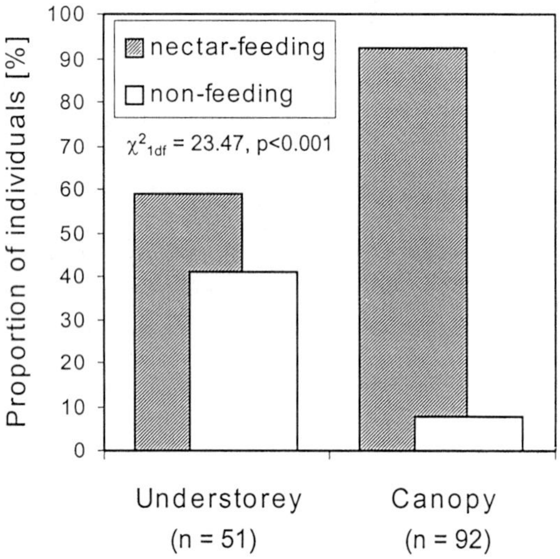

Figure 9. Proportion of nectar-feeding and non-feeding Arctiinae individuals in understorey and canopy sampled during 13 nights of light trapping in each vegetation layer. n = number of collected individuals (set as 100% for each layer). Additionally, the result of a χ^2 test against equal distribution of nectar- and non-feeding Arctiinae individuals between understorey and canopy is given.

1990). In contrast, the proboscis is reduced, and sometimes completely absent, in the genera *Spilosoma*, *Creatonotos*, *Baroa* and *Tinoliodes*.

To summarize, our own data from Borneo as well as available surveys from other tropical forest regions all indicate that abundance patterns of adult Lepidoptera along the vertical gradient from ground level to the canopy are strongly influenced by the availability of the specific resources these butterflies and moths require as adults.

Diversity of Lepidoptera – understorey versus canopy

The canopies of tropical forests are often described as the vegetation layer which contains most of the arthropod diversity of such ecosystems. Accordingly, numerous studies have in recent years focussed on the community ecology of canopy arthopods as a key to understand tropical diversity (e.g., Floren & Linsenmair 1994, 2001; Hammond et al 1996; Kitching & Zalucki 1996; Stork 1991). Our studies on different Lepidoptera guilds and/or taxa exemplify that diversity is not necessarily higher in the forest tree crowns.

Among fruit-feeding nymphalids in Borneo, three measures of α-diversity all indicated that species richness did not differ dramatically between the strata once the effects of sample size were controlled for. We observed an elevated species richness only at a trap height of 30m above ground, in the transition zone between understorey and canopy, because here the two otherwise distinct nymphalid faunas overlap to some degree. We additionally evaluated the Costa Rican study of DeVries (1988) using Hurlbert rarefaction. Remarkably, in his study the two rarefaction curves for understorey and canopy are almost congruent (see Figure 4), which indicates not just a similar species richness but also a very similar abundance structure of the guild of fruit-feeding nymphalids in both forest layers. Our Bornean data reveal much the same pattern. Diversity in understorey and canopy samples did not differ significantly if analysed by Hurlbert rarefaction. The pattern for the fruit-feeding butterfly assemblage surveyed at a rain forest site in eastern Ecuador (DeVries et al. 1997) was intermediate: Observed species richness was lower in the canopy than in the understorey, while rarefaction analysis and species accumulation curves revealed that at this site canopy had higher species richness than understorey.

In our moth samples, we found no significant difference in diversity between understorey and canopy for Arctiinae attracted by light, although species composition was clearly different. Hawkmoths, in contrast, were much more diverse in the canopy than in the understorey. Similarly, in the Uganda study by Corbet (1961), it was observed that the highest number of species was also in the tree crowns. However, his sample size was too small to allow for adequate diversity statistics. No comparable study of the vertical stratification of tropical Arctiinae has been published to date.

Morphological trends along the vertical gradient

In ecological studies, size of arthropods is often used to characterize communities (e.g., Janzen 1973; Erwin & Scott 1980; Morse et al. 1988), for comparison between communities (e.g., of temperate versus tropical regions: Lawton 1991; Ricklefs & O'Rourke 1975; Schoener & Janzen 1968), and as an indicator to analyse the influence of certain abiotic and/or biotic factors on the ecology and evolution of insects (e.g., Barbault 1988; Blackburn et al. 1990; Brown & Maurer 1986; Cushman et al. 1993; LaBarbera 1989; Lindström et al. 1994), sometimes in combination with describing latitudinal trends (e.g., Barlow 1994; Hawkins & Lawton 1995; Nylin & Svärd 1991). We analysed size (measured as forewing length: cf., Loder et al. 1998) of fruit-feeding nymphalid butterflies and found a significant decrease from ground level to canopy layer. This observation corresponds to data of DeVries (1988) from Costa Rica. He also showed that fruit-feeding nymphalid species in the canopy are characterized by smaller forewing lengths than in the understorey.

We used the ratio of body length to thorax width as indicator for flight characteristics. Chai (1990) showed that species with lower ratios and, therefore, stouter bodies, are stronger on the wing while slender-built species with higher quotients do not reach high flight speeds, but have a superior maneuverability. Based on this index, Bornean fruit-feeding nymphalids of the canopy layers might be predicted to be more persistent fliers than their counterparts near ground level. Additionally, Chai (1990) documented for Neotropical butterflies that species with stouter bodies are superior in escaping attacks of caged insectivorous birds. Hence, the increasing presence of insectivorous aerial-hawking birds from ground level towards the forest canopy could be at least partly responsible for the observed change in body design of fruit-feeding nymphalids from more slender built understorey species to representatives with compact bodies like some Nymphalinae and Charaxinae. Higher wind speed in the canopy (e.g., Parker 1995) could further select for species which are stronger on the wing. On the other hand the dense understorey vegetation might select for species with slender bodies and, thus, improved maneuverability.

The body design of flower-visiting butterflies was not studied quantitatively but apparently in this guild more slenderly built species (e.g., Danainae species) are more prevalent. Perhaps, the different abundance patterns of fruit-feeders versus nectar-feeders is caused by the fact that predation pressure is acting in a different way as selective force on the often brightly coloured (aposematic) nectar-feeders, as compared to the mostly camouflage-coloured fruit-feeders. It was shown for Neotropical butterflies that relative to distasteful species, palatable ones are characterized by a larger thorax and a shorther abdomen, they fly faster and more successfully evade attacks from aerial predators (e.g., Srygley & Dudley 1993). These different escape and defense strategies of palatable and unpalatable butterflies can lead to different temporal and spatial activity patterns (Chai & Srygley 1990; Srygley & Chai 1990), and may also explain our observations that cryptic large-bodied fruit-feeders were more prevalent in the understorey whereas nectar-feeders with high maneuverability were more abundant in the canopy.

Factors supporting stratification among tropical forest Lepidoptera – a tentative synopsis

The distribution of adult resources, presence of predators, and variation in microclimatic conditions, may all contribute to differentiation of Lepidopteran faunas across the vertical gradient from ground level up into the canopy. For certain species, larval resources may be another important factor leading to specialisation on certain strata within the vertical gradient. In South East Asia, the majority of butterflies are host plant specialists (e.g., Fiedler 1998). Whole subfamilies often share host plant requirements (e.g., Morphinae and Satyrinae exclusively on monocotyledonous food plants: Ackery 1988; Corbet & Pendlebury 1992). In South East Asian rain forests monocotyledones are mainly restricted to the lower forest strata (except of orchids which never serve as food plants for both subfamilies). This may be one reason why the flight activity of Satyrinae and Morphinae is largely restricted to the understorey (Schulze & Fiedler 1998). A very similar restriction of morphine butterflies to the understorey was recorded for the two neotropical tribes Morphini and Brassolini during a bait trapping study in a Costa Rican rain forest (DeVries 1988). Beccaloni (1997) has shown for Neotropical Ithomiinae that their vertically stratified host plants influence the distribution of adult butterflies. Hill et al. (1992) also mentioned that the distribution of adult skippers in a Northern Australian forest may well be explained through the distribution of larval food plants. At our study site at Poring the pierid butterfly *Dercas gobrias* provides another example of a species whose vertical distribution as adult is congruent with the vertical distribution of the food plant, a large woody Fabaceae climber, which is restricted to the canopy (Schulze & Fiedler 1997b). The restriction of most butterflies and moths to certain larval food plant taxa may therefore select for the usage of only restricted parts of the vertical gradient by adults, at least when searching for egg-laying resources.

Burd (1994) suggested that in the butterfly community on Barro Colarado Island (Panama) certain groups of similar-coloured species preferentially co-occur at the same stratum. For example, orange- and yellow-coloured species had a strong association with the upper layers of the forest. However, human vision may not be a satisfactory surrogate for predator vision and, therefore, predation risk. To ascertain whether colour has an aposematic or cryptic effect in a particular vegetation layer, the ultraviolet wing patterns must be taken into account (e.g., Church et al. 1998) because UV vision is an integral part of the visual capabilities of many visually guided butterfly predators, e.g., most birds (e.g., Bennett & Cuthill 1994) and predatory flies (e.g., Hardie 1985). Therefore, it is thus far impossible, in the absence of physiologically appropriate measurements of wing colour patterns, to assess the significance of butterfly colour patterns for vertical stratification under selection through visually hunting predators.

A combination of factors might be responsible for the conspicuous species turnover of, and therefore lowest similarity among, assemblages of fruit-feeding nymphalid species in the zone between 20 and 30 m above ground. In this transition zone the first gaps exist in the forest canopy at our study site at Poring. Although we did not measure climatic parameters, there should be a distinct change of microclimatic factors such as insolation, temperature, and wind speed. This change of abiotic conditions may support different body and wing designs. In addition, in exactly this transition zone the prevalence of insectivorous aerial-hawking birds increased abruptly.

Although our data show that adults of fruit-feeding nymphalid species are stratified when feeding, this may not necessarily correspond to the vertical distribution in other behavioural contexts, e.g. searching for adequate hostplants and egg laying, courtship and resting behaviour. For example, DeVries (1988) noted that males of the Neotropical Morphinae *Morpho cypris* Westwood, 1851, a species trapped only in the understorey, spend much of their time patrolling in the canopy presumably in search for mates. A further example from our study site is *Polyura athamas* (Drury 1773), a Charaxinae species which could be trapped and observed only in upper vegetation layers, but caterpillars (hostplant: *Albizza* spp. and other Fabales) were found at ground level within closed forest (Fiedler & Schulze pers. obs.). Additionally, many tropical butterfly species (mainly males) take up water and nutrients at moist sites on the ground (e.g., Beck et al. 1999), among them a number of 'canopy' species (e.g. Charaxinae, *Dercas gobrias* (Hewitson, 1864): Schulze & Fiedler 1997b).

It cannot be ruled out that phylogenetic constraints may also play a role in vertical stratification of forest Lepidoptera, as well as adaptive responses to selection forces. Such constraints may be manifest in characters like hostplant relationships or certain morphological traits, which are often fixed across large monophyletic clades. For example, Bornean fruit-feeding Satyrinae and Morphinae butterflies, with their relatively uniform phenotypes and a comparatively small set of larval food plants, are by and large restricted to lower vegetation layers, whereas Nymphalinae with a much higher diversity of phenotypes and larval foodplants (Fiedler 1998) are more evenly distributed across all strata. Conversely, nectar-feeding hawkmoths rarely penetrate into the dense forest understorey, with the exception of those specialists in one monophyletic clade (Smerinthinae) which have irreversibly shifted to a non-feeding reproductive strategy. These cases demonstrate that constraints relevant to vertical stratification have become fixed at higher taxonomic levels, and may have become modified, but not fundamentally changed, during species radiation.

Acknowledgements

We thank Datuk Lamri Ali and Frances Liew for granting permission to work at Mt. Kinabalu NP. Special thanks are due to the staff of Sabah Parks at Poring for their kind support of our studies and the supply of accommodation during the stay, as well as to Elke Aug, Daniel Bartsch, Sabine Brammertz, Justinah Francis, Dr Christoph L. Häuser, André Herzig, Jutta Klein, Anette Malsch, Eva Mühlenberg, Roswitha Mühlenberg, Dr Wolfgang A. Nässig, Edda Roddewig for their help in the field. Our field work was partially supported by the Deutsche Forschungsgemeinschaft (Fi 547/1-2, Fi 547/4-1/4-2, Li 150/15:1-3).

References

Achtziger, R., Nigman, U. & Zwölfer, H. 1992. Rarefaction-Methoden und ihre Einsatzmöglichkeiten bei der zooökologischen Zustandsanalyse und Bewertung von Biotopen. Z. Ökologie Naturschutz 1: 89–105.

Ackery, P. R. 1988. Hostplants and classification: a review of nymphalid butterflies. Biol. J. Linn. Soc. 33: 95–203.

Baker, E. C. S. 1922. The fauna of British India, including Ceylon and Burma. Birds Vol. I, 2nd edition. Taylor and Francis, London.

Baker, E. C. S. 1924. The fauna of British India, including Ceylon and Burma. Birds Vol. II, 2nd edition. Taylor and Francis, London.

Baker, E. C. S. 1926. The fauna of British India, including Ceylon and Burma. Birds Vol. III, 2nd edition. Taylor and Francis, London.

Baker, E. C. S. 1927. The fauna of British India, including Ceylon and Burma. Birds Vol. IV, 2nd edition. Taylor and Francis, London.

Baker, R. R. & Sadovy, Y. 1978. The distance and nature of the light-trap response of moths. Nature 276: 818–821.

Barbault, R. 1988. Body size, ecological constraints, and the evolution of life-history strategies. Evol. Biol. 22: 261–286.

Barlow, N. D. 1994. Size distribution of butterfly species and the effect of latitude on species sizes. Oikos 71: 326–332.

Beaman, J. H. & Beaman, R. S. 1990. Diversity and distribution patterns in the flora of Mount Kinabalu. Pp. 147–160. In: Baas, P., Kalkman, K. & Geesink, R. (eds), The plant diversity of Malesia. Kluwer, Academic Publishers, Dordrecht.

Beccaloni, G. W. 1997. Vertical stratification of ithomiine butterfly (Nymphalidae: Ithomiinae) mimicry complexes: the relationship between adult flight height and larval host-plant height. Biol. J. Linn. Soc. 62: 313–341.

Beck, J. & Schulze, C. H. 2000. Diversity of fruit-feeding butterflies (Nymphalidae) along a gradient of tropical rain forest succession in Borneo with some remarks on the problem of 'pseudoreplicates'. Trans. lepid. Soc. Japan 51: 89–98.

Beck, J., Mühlenberg, E. & Fiedler, K. 1999. Mud-puddling behavior in tropical butterflies: in search of proteins or minerals? Oecologia 119: 140–148.

Bennett, A. T. D. & Cuthill, I. C. 1994. Ultraviolet vision in birds: what is its function? Vision Res. 34: 1471–1478.

Blackburn, T. M., Harvey, P. H. & Pagel, M. D. 1990. Species number, population density and body size relationships in natural communities. J. Anim. Ecol. 59: 335–345.

Bowden, J. 1982. An analysis of factors affecting catches of insects in light-traps. Bull. entomol. Res. 72: 535–556.

Brown, J. H. & Maurer, B. A. 1986. Body size, ecological dominance and Cope's rule. Nature 324: 248–250.

Burd, M. 1994. Butterfly wing colour patterns and flying heights in the seasonally wet forest of Barro Colorado Island, Panama. J. Trop. Ecol. 10: 601–610.

Chai, P. 1990. Relationships between visual characteristics of rain forest butterflies and responses of a specialized insectivorous birds. Pp. 31–60. In: Wickstein, M. (ed.), Adaptive coloration in invertebrates. Proceedings of Symposium sponsored by American Society of Zoologists Galveston: Seagrant College Program, Texas A & M University.

Chai, P. & Srygley, R. B. 1990. Predation and the flight, morphology, and temperature of Neotropical rain forest butterflies. Am. Nat. 135: 748–765.

Church, S. C., Bennett, A. T. D., Cuthill, I. C., Hunt, S., Hart, N. S. & Partridge, J. C. 1998. Does lepidopteran larval crypsis extend into the ultraviolet? Naturwissenschaften 85: 189–192.

Colwell, R. K. 1997. Estimates: Statistical estimation of species richness and shared species from samples. Version 5. User's guide and application published at: http://viceroy.eeb.uconn.edu/estimates.

Colwell, R. K. & Coddington J. A. 1994. Estimating terrestrial biodiversity through extrapolation. Phil. Trans. Roy. Soc. Lond. B 345: 101–118.

Common, I. F. B. 1990. Moths of Australia. E. J. Brill, Leiden.

Corbet, A. S. & Pendlebury, H. M. 1992. The Butterflies of the Malay Peninsula. 4th edition, revised by J. N. Eliot. Malayan Nature Society, Kuala Lumpur.

Corbet, P. S. 1961. Entomological studies from a high tower in Mpanga Forest, Uganda. X. Observations on Sphingidae (Lepidoptera). Trans. Roy. Entomol. Soc. Lond 113: 351–352.

Cushman, J. H., Lawton, J. H. & Manly, B. F. J. 1993. Latitudinal patterns in European ant assemblages: variation in species richness and body size. Oecologia 95: 30–37.

Dailey, G. C. & Ehrlich, P. R. 1995. Preservation of biodiversity in small rain forest patches: rapid evaluations using butterfly trapping. Biodiv. Cons. 4: 35–55.

DeVries, P. J. 1988. Stratification of fruit-feeding nymphalid butterflies in a Costa Rican rain forest. J. Res. Lepid. 26: 98–108.

DeVries, P. J., Murray, D. & Lande, R. 1997. Species diversity in vertical, horizontal, and temporal dimensions of a fruit-feeding butterfly community in an Ecuadorian rain forest. Biol. J. Linn. Soc. 62: 343–364.

Erwin, T. L. & Scott, J. C. 1980. Seasonal and size patterns, trophic structure, and richness of Coleoptera in the tropical arboreal ecosystem: The fauna of the tree *Luehea seemannii* Triana and Planch in the Canal Zone of Panama. Coleopt. Bull. 34: 305–322.

Fiedler, K. 1998. Diet breadth and host plant diversity of tropical- versus temperate-zone herbivores: South East Asian and West Palaearctic butterflies as a case study. Ecol. Entomol. 23: 285–297.

Fisher, R. A., Corbet, A. S. & Williams, C. B. 1943. The relation between the number of species and the number of individuals in a random sample of an animal population. J. Anim. Ecol. 12: 42–58.

Floren, A. & Linsenmair, K. E. 1994. Zur Diversität und Wiederbesiedlungsdynamik von Arthropoden auf drei Baumarten in einem Regenwald in Sabah, Malaysia. Andrias 13: 23–28.

Floren, A. & Linsenmair, K. E. 2001. Changes in the structure of arboreal arthropod communities in forests of different disturbance levels. Plant Ecol. 153: 153–167 (this volume).

Francis, C. M. 1994. Vertical stratification of fruit bats (Pteropodidae) in lowland dipterocarp rain forest in Malaysia. J. Trop. Ecol. 10: 523–530.

Groombridge, B. (ed.) 1992. Global biodiversity. Status of the earth's living resources. Chapman and Hall, London.

Hails, C. & Jarvis, F. 1988. Birds of Singapore. Times Edition, Singapore.

Hammond, P. M., Kitching, R. L. & Stork, N. E. 1996. The composition and richness of the tree-crown Coleoptera assemblage in an Australian subtropical forest. Ecotropica 2: 99–108.

Hardie, R. C. 1985. Functional organization of the fly retina. Pp. 1–79. In: Ottoson, D., Autrum, H., Perl, E. R., Schmidt, R. F., Shimazu, H. & Willis, W. D. (eds), Progress in sensory physiology. Vol. 5. Springer-Verlag, Berlin.

Häuser, C. L., Schulze, C. H. & Fiedler, K. 1997. The butterfly species (Insecta: Lepidoptera: Rhopalocera) of Kinabalu Park, Sabah. Raffles Bull. Zool. 45: 281–304.

Hawkins, B. A. & Lawton, J. H. 1995. Latitudinal gradients in butterfly body sizes: is there a general pattern? Oecologia 102: 31–36.

Hayek, L.-A. C. & Buzas, M. A. 1997. Surveying natural populations. Columbia University Press, New York.

Henry, G. M. 1971. A guide to the birds of Ceylon. 2nd edition. Oxford University Press, London.

Hill, C. J., Gillison, A. N. & Jones, R. E. 1992. The spatial distribution of rain forest butterflies at three sites in North Queensland, Australia. J. Trop. Ecol. 8: 37–46.

Hogenes, W. & Treadaway, C. G. 1998. The Sphingidae (Lepidoptera) of the Philippines. Nach. Entomol. Ver. Apollo, N.F., Suppl. 17: 17–132.

Holloway, J. D. 1976. Moths of Borneo with special reference to Mt. Kinabalu. Malayan Nature Society, Kuala Lumpur.

Holloway, J. D. 1987. The Moths of Borneo. Part 3. Malayan Nature Society, Kuala Lumpur.

Holloway, J. D. 1988. The Moths of Borneo. Part 6. Malayan Nature Society, Kuala Lumpur.

Holmes, D. & Nash, S. 1990. The birds of Sumatra and Kalimantan. Oxford University Press, Singapore.

Hurlbert, S. H. 1978. The measurement of niche overlap and some relatives. Ecology 59: 67–77.

Jackson, T. H. E. 1961. Entomological studies from a high tower in Mpanga Forest, Uganda. IX. Observations on Lepidoptera (Rhopalocera). Trans. Roy. Entomol. Soc. Lond. 113: 346–350.

Janzen, D. H. 1973. Sweep samples of tropical foliage insects: description on study sites, with data on species abundance and size distributions. Ecology 54: 659–686.

Kato, M., Inoue, T., Hamid, A. A., Nagamitsu, T., Merdek, M. B., Nona, A. R., Itino, T., Yamane, S. & Yumoto, T. 1995. Seasonality and vertical structure of light-attracted insect communities in a dipterocarp forest in Sarawak. Res. Pop. Ecol. 37: 59–79.

Kiltie, R. A. 1993. New light on forest shade. Trends Ecol. Evol. 8: 39–40.

King, B. F., Dickinson, E. C. & Woodcock, M. W. 1980: A field guide to the birds of South-east Asia. Collins, London.

Kitayama, K. 1992. An altitudinal transect study of the vegetation on Mount Kinabalu, Borneo. Vegetatio 102: 149–171.

Kitching, R. L. & Zalucki, J. 1996. The biodiversity of arthropods from Australian rain forest canopies: some results on the role of the tree species. Pp. 21–28. In: Edwards, D. S., Booth, W. E. & Choy, S. C. (eds), Tropical rain forest research – current issues. Kluwer, Academic Press, Dordrecht.

Krebs, C. J. 1989. Ecological methodology. Harper Collins Publishers, New York.

LaBarbera, M. 1989. Analyzing body size as a factor in ecology and evolution. Annu. Rev. Ecol. Syst. 20: 97–117.

Lamas, G., Robbins, R. K. & Harvey, D. J. 1991. A preliminary survey of the butterfly fauna of Pakitza, Parque nacional del Manu, Peru, with an estimate of its species richness. Publ. Mus. Hist. nat. UNMSM (A) 40: 1–19.

Lawton, J. H. 1991. Species richness, population abundances, and body sizes in insect communities: tropical versus temperate comparisons. Pp. 71–89. In: Price, P. W., Lewinsohn, T. M., Fernandes, G. W. & Benson, W. W. (eds), Plant-animal Interactions: Evolutionary Ecology in Tropical and Temperate Regions. Wiley, New York.

Lee Tain Choi, D. 1996. Geology of Kinabalu. Pp. 19–29. In: Wong, K. M. & Phillipps, A. (eds), Kinabalu : Summit of Borneo. The Sabah Society & Sabah Parks, Kota Kinabalu.

Lemaire, C. & Minet, J. 1999. The Bombycoidea and their relatives. Pp. 321–352. In: Kristensen, N. P. (ed.), Lepidoptera, moths and butterflies. Vol. 1: Evolution, systematics and biogeography. Walter de Gruyter, Berlin.

León-Cortés, J. L., Soberón-Mainero, J. & Llorente-Bousquets, J. 1998. Assessing completeness of Mexican sphinx moths inventories through species accumulation functions. Diversity Distrib. 4: 37–44.

Levins, R. 1968. Evolution in changing environments: some theoretical explorations. Princeton University Press, Princeton, N.J.

Lindström, J., Kaila, L. & Niemelä, P. 1994. Polyphagy and adult body size in geometrid moths. Oecologia 98: 130–132.

Loder, N., Gaston, K. J., Warren, P. H., & Arnold, H. R. 1998. Body size and feeding specificity: macrolepidoptera in Britain. Biol. J. Linn. Soc. 63: 121–139.

Longino, J. T. & Nadkarni, N. 1990. A comparison of ground and leaf litter ants (Hymenoptera: Formicidae) in a tropical montane forest. Psyche 97: 81–94.

MacKinnon, J. & Phillipps, K. 1993. A field guide to the birds of Borneo, Sumatra, Java and Bali. Oxford University Press, Oxford.

Magurran, A. E. 1988. Ecological diversity and its measurement. Croom Helm, London.

McKey, D. 1991. Interactions between ants and plants: comparison of canopy, understory and clearing environments. Pp. 66–73. In: Halle, F. & Pascal, O. (eds), Biologie d'une canopée de forêt équatoriale. Longman, Paris.

Medina, M. C., Robbins, R. K. & Lamas, G. 1996. Vertical stratification of flight by Ithomiinae butterflies (Lepidoptera: Nymphalidae) at Pakitza, Manu National Park, Peru. Pp. 211–216. In: Wilson, D. E. & Sandoval, A. (eds), Manu – the biodiversity of Southeastern Peru. Smithsonian Institution, Washington.

Morse, D. R., Stork, N. E. & Lawton, J. H. 1988. Species numbers, species abundance and body length relationships of arboreal beetles in Bornean lowland rain forest trees. Ecol. Entomol. 13: 25–37.

Nilsson, L. A. 1998. Deep flowers for long tongues. Trends Ecol. Evol. 13: 259–260.

Nylin, S. & Svärd, L. 1991. Latitudinal patterns in the size of European butterflies. Holarctic Ecol. 14: 192–202.

Otsuka, K. 1988. Butterflies of Borneo. Vol. I. Tobishima Corporation, Tokyo.

Papageorgis, C. 1975. Mimicry in Neotropical butterflies. Am. Sci. 63: 522–532.

Parker, G. G. 1995. Structure and microclimate of forest canopies. pp. 73–106. In: Lowman, M. D. & Nadkarni, N. M. (eds), Forest canopies. Academic Press, San Diego.

Pearson, D. L. 1977. Ecological relationships of small antbirds in Amazonian bird communities. Auk 94: 283–292.

Peterson, A. T. & Slade, N. A. 1998. Extrapolating inventory results into biodiversity estimates and the importance of stopping rules. Diversity. Distrib. 4: 95–105.

Pittaway, A. R. 1993. The hawkmoths of the Western Palaearctic. Harley Books, Colchester.

Pomeroy, D. E. & Service, M. W. 1986. Tropical ecology. Longman Scientific & Technical, Harlow.

Rees, C. J. C. 1983. Microclimate and the flying Hemiptera fauna of a primary lowland rain forest in Sulawesi. Pp. 121–136. In: Sutton, S. L., Whitmore, T. C. & Chadwick, A. C. (eds), Tropical rain forest: ecology and management. Blackwell, Oxford.

Ricklefs, R. E. & O'Rourke, K. 1975. Aspect diversity in moths: a temperate-tropical comparison. Evolution 29: 313–324.

Sachs, L. 1992. Angewandte Statistik. 7th edition. Springer-Verlag, Berlin.

Schal, C. 1982. Intraspecific vertical stratification as a mate finding mechanism in tropical cockroaches. Science 215: 1405–1407.

Schoener, T. W. & Janzen, D. H. 1968. Some notes on tropical versus temperate insect size patterns. Am. Nat. 101: 207–224.

Schulze, C. H. & Fiedler, K. 1997a. Patterns of Diversity and Vertical Stratification in Hawkmoths (Lepidoptera: Sphingidae) of a Bornean Rain Forest. Mitt. Dtsch. Ges. allg. angew. Entomol. 11: 767–770.

Schulze, C. H. & Fiedler, K. 1997b. Notes on the biology of *Dercas gobrias* (Hewitson, 1864) (Lepidoptera: Pieridae, Coliadinae). Trans. Lepid. Soc. Japan 48: 25–30.

Schulze, C. H. & Fiedler, K. 1998. Habitat preferences and flight activity of Morphinae butterflies in a Bornean rain forest, with a note on sound production by adult *Zeuxidia* (Lepidoptera: Nymphalidae). Malay. Nat. J. 52: 163–176.

Scoble, M. J. 1992. The Lepidoptera: form, function and diversity. Oxford.

Smith, A. P., Hogan, K. P. & Idol, J. R. 1992. Spatial and temporal patterns of light and canopy structure in a lowland tropical moist forest. Biotropica 24: 503–511.

Srygley, R. B. & Chai, P. 1990. Predation and the elevation of thoracic temperature in brightly colored Neotropical butterflies. Am. Nat. 135: 766–787.

Srygley, R. B. & Dudley, R. 1993. Correlations of the position of center of body mass with butterfly escape tactics. J. Exp. Biol. 174: 155–166.

StatSoft. 1995. STATISTICA for Windows, vol I-V. StatSoft, Tulsa.

Stork, N. E. 1987. Guild structure of arthropods from Bornean rain forest trees. Ecol. Entomol. 12: 69–80.

Stork, N. E. 1991. The composition of the arthropod fauna of Bornean lowland forest trees. J. Trop. Ecol. 7: 161–180.

Süssenbach, D. & Fiedler, K. 1999. Noctuid moths attracted to fruit baits: testing models and methods of estimating species diversity. Nota Lepid. 22: 115–154.

Sutton, S. L. 1983. The spatial distribution of flying insects in tropical rain forests. Pp. 77–91. In: Sutton, S. L., Whitmore, T. C. & Chadwick, A. C. (eds), Tropical rain forest: ecology and management. Blackwell, Oxford.

Sutton, S. L. 1989. The spatial distribution of flying insect. Pp. 427–436. In: Lieth, H. & Werger, M. J. A. (eds), Tropical rain forest ecosystems. Biogeographical and ecological studies. Elsevier, Amsterdam.

Sutton, S. L., Ash, C. P. J. & Grundy, A. 1983. The vertical stratification of flying insects in lowland rain forests of Panama, Papua New-Guinea and Brunei. Zool. J. Linn. Soc. 78: 287–297.

Tammaru, T. & Haukioja, E. 1996. Capital breeders and income breeders among Lepidoptera – consequences to population dynamics. Oikos 77: 561–564.

Taylor, P. H. & Lowman, M. D. 1996. Vertical stratification of small mammal community in a northern hardwood forest. Selbyana 17: 15–21.

Terborgh, J. 1980. Vertical stratification of a neotropical forest bird community. Proceedings of the 17th International Ornithological Congress, Pp. 1005–1012.

Terborgh, J. 1985. The vertical component of plant species diversity in temperate and tropical forests. Am. Nat. 126: 760–776.

Whitmore, T. C. 1984. Tropical rain forests of the Far East. Clarendon, Oxford.

Wolda, H. 1981. Similarity indices, sample size and diversity. Oecologia 50: 296–302.

Wong, K. M. & Phillipps, A. (eds). 1996. Kinabalu: Summit of Borneo. The Sabah Society & Sabah Parks, Kota Kinabalu.

Yoda, K. 1974. Three-dimensional distribution of light intensity in a tropical rain forest of West Malaysia. Jap. J. Ecol. 24: 247–254.

Appendix 1. Recorded species and numbers of individuals collected with fruit-bait traps at six different heights above ground level. Additionally, for each species forewing length and body design index (ratio of body length to maximum thorax width) are given. When sexual dimorphism exists, values are given for males (before oblique) and females (behind oblique) separately. When one of both sexes was absent in the sample, this is indicated by a dash. For calculation procedure see text (Methods).

Subfamily	Species	Number of individuals at different trap heights						FW (mm)	Body design index
		0 m	10 m	20 m	30 m	40 m	50 m		
Satyrinae	*Melanitis leda*	8	8	9	7	1	1	33.3/36.3	6.2/6.6
	Melanitis zitenius	27	16	9	2	2	1	40.6/41.0	6.1/5.9
	Elymnias panthera	–	–	–	1	–	1	33.5	9.5
	Elymnias nesaea	–	1	2	1	1	–	37.0/43.3	10.5/9.8
	Lethe europa	–	–	–	1	–	–	31.0/–	6.0/–
	Lethe darena	–	–	–	1	–	–	–/42.0	–/6.0
	Neorina lowii	9	2	1	–	–	–	54.3	7.3
	Mycalesis patiana	–	–	–	–	1	–	23.0	7.0
	Mycalesis amoena	14	1	–	–	–	–	26.7/30.0	6.6/6.7
	Mycalesis kina	1	–	–	–	–	–	28.0	10.0
	Mycalesis orseis	7	–	–	–	–	–	23.5/27.5	7.8/8.3
	Mycalesis maianeas	5	–	–	–	–	–	23.0/26.0	8.8/8.5
	Erites argentina	2	–	–	–	–	–	26.0	7.3
	Coelites euptychioides	5	–	–	–	–	–	34.0	6.1
	Ragadia makuta	17	1	–	–	–	–	23.5	9.1
Morphinae	*Amathusia phidippus*	–	–	–	–	1	–	–/55.0	–/7.5
	Amathuxidia amythaon	3	3	–	–	–	–	57.3/62.0	9.3/6.3
	Zeuxidia amethystus	8	2	–	–	–	–	48.4/52.4	10.3/6.9
	Zeuxidia doubledayi	6	3	–	–	–	–	51.2/61.7	9.1/7.4
	Zeuxidia aurelius	1	–	–	–	–	–	–/75.0	–/7.6
	Thauria aliris	1	–	–	–	–	–	–/78.0	–/7.0
	Discophora sondaica	–	–	2	–	–	–	–/41.0	–/6.3
	Discophora necho	1	1	–	1	–	–	–/52.0	–/5.7
Nymphalinae	*Rhinopalpa polynice*	–	–	–	–	1	–	–/36.0	–/6.3
	Kallima limborgi	13	3	–	–	–	–	51.6	6.1
	Neptis nata	–	–	–	1	–	–	26.0	8.5
	Moduza procris	–	–	–	–	1	–	34.0/–	5.0/–
	Tanaecia pelea	1	7	1	–	–	–	32.8/36.4	6.3/6.2
	Tanaecia clathrata	–	1	1	–	–	–	29.0/32.0	5.3
	Tanaecia munda	29	29	5	–	–	–	30.9/36.7	6.3/5.8
	Tanaecia iapis	1	2	–	–	–	–	30.0/37.0	4.8/5.7
	Tanaecia sp.	–	2	1	–	–	–		
	Euthalia monina	1	–	9	3	4	2	27.2/35.7	4.9/5.0
	Euthalia aconthea	–	–	–	1	1	–	28.5/–	4.9/–
	Euthalia alpheda	–	–	–	3	4	1	30.8/39.5	4.2/4.6
	Euthalia merta	1	–	–	1	10	4	29.8/36.0	5.3/5.4
	Euthalia mahadeva	–	–	3	2	3	5	29.4/38.5	5.7/5.0
	Euthalia kanda	–	–	1	–	–	–	29.0/–	4.0/–
	Euthalia sp.	–	–	–	–	1	–		
	Dophla evelina	–	9	7	–	–	–	45.7/54.0	5.4/5.3
	Bassarona teuta	–	6	2	1	1	–	35.8/43.4	5.2/6.2
	Bassarona dunya	3	2	–	–	–	–	45.5/55.0	5.8/5.4

Appendix 1. continued

Subfamily	Species	Number of individuals at different trap heights						FW (mm)	Body design
		0 m	10 m	20 m	30 m	40 m	50 m	index	
	Lexias dirtea	19	–	–	–	–	–	41.7/48.6	5.3/5.9
	Lexias sp.	1	–	–	–	–	–		
	Amnosia decora	5	1	–	–	–	–	42.6/45.0	7.5/8.0
	Stibochiona schoenbergi	1	5	6	1	–	–	33.8/36.6	5.8/5.5
	Rohana nakura	–	–	1	–	1	1	21.0/24.0	3.5/6.0
	Herona sumatrana	–	1	–	–	–	–	34.0/–	6.3/–
	Euripus nyctelius	–	–	–	4	2	–	30.8/42.0	7.3/5.5
Charaxinae	*Prothoe franck*	4	6	1	–	–	–	41.1/47.5	6.1/5.1
	Agatasa calydonia	–	4	2	–	–	–	46.8/–	4.9/–
	Polyura athamas	–	–	–	1	–	–	38.0/–	5.5/–
	Charaxes solon	–	–	–	2	1	2	34.5/–	4.9/–
	Charaxes bernardus	–	1	8	4	2	6	41.6/48.3	5.0/4.5
	Charaxes borneensis	–	–	–	2	–	–	41.0/–	5.0/–
	Charaxes durnfordi	–	1	–	–	–	–	–/50.0	–/4.4

Appendix 2. Recorded Sphingidae species and numbers (hawk moths) of individuals attracted by light in forest under storey and canopy between 18:30 and 21:00 h.

Species	Number of individuals	
	Understorey	Canopy
Acherontia lachesis	1	8
Acosmeryx anceus	–	2
Acosmeryx shervillii	2	9
Agrius convolvuli	–	1
Ambulyx clavata	–	1
Ambulyx pryeri	–	4
Ambulyx substrigilis	–	1
Callambulyx rubricosa	–	1
Cechenena aegrota	–	3
Cechenena helops	–	13
Cechenena lineosa	2	163
Daphnis hypothous	–	3
Daphnusa ocellaris	39	3
Elibia dolichus	–	3
Eurypteryx falcata	–	3
Hippotion echelus	–	1
Hippotion rosetta	–	1
Megacorma obliqua	–	7

Appendix 2 continued. Recorded Sphingidae species (hawk moths) and numbers of individuals attracted by light in forest understorey and canopy between 18:30 and 21:00 h.

Species	Number of individuals	
	Understorey	Canopy
Meganoton analis	–	19
Meganoton nyctiphanes	–	1
Meganoton rufescens	–	1
Panacra dohertyi	–	8
Panacra psaltria	–	3
Psilogramma menephron	–	2
Rhagastis albomarginatus	1	1
Rhagastis rubetra	–	2
Theretra clotho	–	2
Theretra rhesus	2	55
Theretra nessus	–	4
Theretra nr. *rhesus*	–	1

Appendix 3. Recorded Arctiinae species (tiger moths) and numbers of individuals attracted by light in forest understorey and canopy between 18:30 and 21:00 h.

Species	Number of Individuals	
	Understorey	Canopy
Amerila astreus	1	3
Baroa siamica	10	1
Creatonotos transiens	–	4
Nyctemera baulus	2	60
Nyctemera coleta	–	1
Nyctemera muelleri	26	5
Spilosoma ericsoni	–	1
Spilosoma griseabrunnea	–	1
Spilosoma groganae	1	–
Spilosoma hosei	6	–
Tinoliodes dehanna	4	–
Utetheisa abraxoides	1	16

Plant Ecology **153:** 153–167, 2001.

The influence of anthropogenic disturbances on the structure of arboreal arthropod communities

Andreas Floren & K. Eduard Linsenmair
Department of Animal Ecology and Tropical Biology, Biozentrum, Am Hubland, D-97074 Wuerzburg, Germany, (E-mail: Floren@biozentrum.uni-wuerzburg.de)

Key words: Canopy fogging, Community structuring mechanisms, Formicidae, Secondary forests, Stochastic/Deterministic, Tropical rain forest

Abstract

In the framework of our research, aimed at understanding the processes structuring tropical arthropod communities, we investigated the changes occurring in tree crown communities of forests of different disturbance levels. These were a mixed dipterocarp primary lowland rain forest in Kinabalu National Park (in Sabah, Malaysia) and, some kilometres away, three forests of regeneration periods 5, 15, and 40 years following a clear-cut. These disturbed forest sites were adjacent to one another and merged into mature forest. From each forest at least ten individuals of one tree species were sampled using the fogging method. In the primary forest relative proportions of some arthropod taxa differed on the ordinal and familial level significantly within trees. The dominance of Formicidae was characteristic as was the almost complete lack of less mobile arthropods such as Lepidoptera larvae. In the five-year- old forest, differences in relative proportions among most taxa had almost disappeared. Formicidae abundances had declined drastically which coincided with an increase of Lepidoptera larvae. With progressing forest succession, arthropod communities increasingly converged on the pattern of primary forest, and total ant abundance as well as diversity increased significantly. Ant communities in the most disturbed forest were of low structural complexity, and to a large degree predictable in species arrangement, but became more and more unpredictable as the complexity of the forest increased. Several species of Coleoptera and non-formicine Hymenoptera occurred in high numbers in the youngest forest, contrasting with the mature forest where all species were typically rare. These changes may indicate a change in the structuring mechanisms from predominantly deterministic processes in disturbed forests to stochastic processes in mature forest.

Introduction

During the last decade the extraordinary high diversity of tropical lowland rain forests has been demonstrated, particularly in relation to arboreal arthropods (references in Stork et al. 1997). This raises the question how such high diversity is maintained - a matter we hardly know anything about at all. In the broad sense, there are two types of models which try to explain the coexistence in species rich guilds, and which are briefly summarised here. Firstly, deterministic equilibrium models assume that community structure and dynamics are to a large degree predictable. Interspecific competition leads to the establishment of very similar communities in habitats which, are in large measure, comparable. It is assumed that these communities have reached a state of species saturation (see, e.g., Giller 1986; Gotelli & Graves 1996). Secondly, stochastic non-equilibrium models emphasise that chance events influence community structure, which leads them to be unpredictable in space and time. Usually, many species with similar niche requirements are assumed to be able to use any one available resource. It is not mainly interspecific competition that determines the success of a species but, rather, the initial conditions at the time of arrival of the species in the community. These models give plausible explanations for the coexistence of many more species at the same resource than predicted by deterministic models (Chesson & Case 1986; Pimm 1991; Huston 1994; Bascompte & Sole 1995; Pahl-Wostl 1997).

To judge the value of both models, we tested the different predictions for arboreal arthropod communities of some tree species in SE-Asian lowland rain forests. For primary forest we had previously found strong indications of the great importance of stochastic events for many taxa, namely Formicidae (Floren & Linsenmair 1997, 1998b, 2000), Coleoptera (Floren & Linsenmair 1998a), parasitic Hymenoptera (Horstmann et al. 1999) and Orthoptera (Floren unpubl. data). On the basis of these results we extended our investigations to forests along an anthropogenic disturbance gradient and asked how arthropod communities changed in these structurally simplified systems which have a strikingly lower floral diversity. In addition, we asked whether these changes also correlate with changes in the structuring mechanisms. Generally, there are few investigations analysing how diversity changes along man-made disturbance gradients (e.g., Roth et al.1994; Daily & Ehrlich 1995; Bignell 1996; Olson & Andriamidana 1996; Eggelton et al. 1998; Lawton et al. 1998; Holloway 1998), and specifically concerning arboreal arthropods and their communities, we know of no other such investigation; but see Davis & Sutton (1998) and Schowalter & Ganio (1999).

Methods

Study site and trees selected for investigation

All investigations were performed in Kinabalu National Park, in Sabah, Malaysia (6°2.75′ N, 116°42.2′ E). The work started in 1992 with the investigation of a primary mixed dipterocarp lowland forest (this site will be referred to as site 'P' in subsequent discussion) at Poring Hot Spring (Figure 1). This was a multilayered forest, in which at least three permanent canopy strata could be distinguished: the lower canopy stratum with trees of 25 m average height, the upper canopy at 40–50 m in height, and emergent, more isolated trees, up to 80 m. In combination with the high floristic diversity, a structurally and microclimatically extremely heterogeneous mosaic of habitats is thus formed. To date, 79 foggings and refoggings were conducted on 13 tree species. The work in the disturbed forests was carried out at substation Sorinsim at the NE-side of Mount Kinabalu, where three forest types of different disturbance levels were chosen for the comparative investigations. They had been cleared for agricultural use 5, 15, and ca. 40 years ago (in the following named SI, SII, SIII) and then left to regenerate. All forests lay adjacent to one another and SIII merged into primary hill forest.

The results from the foggings of ten *Aporusa lagenocarpa* (Euphorbiaceae) were taken as a representative sample for the primary forest. With a height of 20 to a maximum of 30 m this tree is a typical representative of the lower canopy stratum (GBH 79.3 cm ± 15.1). As this tree species could not be found in the disturbed forests, we had to content ourselves with other tree species for these habitats. In the five-year-old, most disturbed forest type (SI), only pioneer trees were growing, of which ten individuals of a still unidentified tree species (belonging to the family Euphorbiaceae) were chosen. This species measured between 6 and 8 meters in height, with leaves arranged almost as a single crown layer (GBH 58.6 cm ± 9.53), producing only little shade. This species was quite rare in SII and was totally absent in SIII. The most common tree species in SII was a Verbenaceae, *Vitex pinnata*, which also occurred regularly in SIII (GBH in SII 109.7 cm ± 26.25, in SIII 135.8 cm ± 49.49) and of which 10 and 11 individual trees respectively were chosen for the comparative analysis. The *Vitex pinnata* trees grew to around 20 meters in height and already showed 'age related' damage in SIII (partly rotten bough and branches, etc.). Very evidently, tree diversity increased with the age of regenerating forest. In SI, and partly also in SII, common grass-, bush- and climbing plant-species became increasingly rare. In SIII these plants were eventually, to large degree, replaced by a thick growth of young tree saplings. A closed canopy had not developed in SI, SII, or in SIII, and in SIII individual trees (Sterculiacae) protruded above the canopy. Within each study area we only chose trees that grew under similar habitat conditions, e.g., with regard to the edaphic situation, the degree of shading, crown cover with lianas, etc. (Floren & Linsenmair 1998b).

Fogging method

Because of the multilayered structure of the primary forest, the crowns of all *Aporusa* trees were covered with a 100 m^2 cotton roof the day before fogging. This enabled us to prevent arthropods immobilised by the insecticide in the higher canopy from falling in the collecting funnels installed beneath the tree,

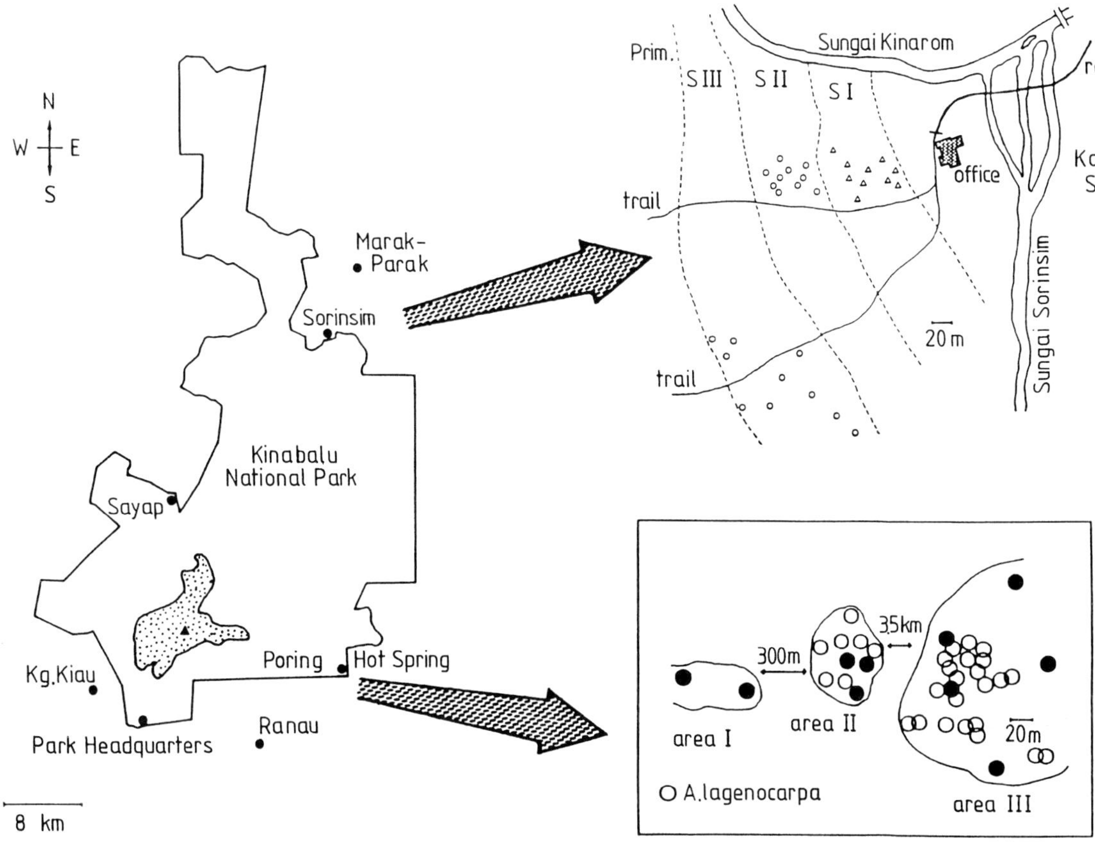

Figure 1. Localisation of research sites. Symbols refer to Sorinsim map: Δ = the pioneer tree species, O = *Vitex pinnata* trees. Distance refers to relative positions of trees within forest types. SIII was about 1 km from SI and SII, the primary forest approximately 3 km from SIII. Sungai = river; Kampong = village. The Poring map shows the distribution of *A. lagenocarpa* trees in the primary forest. Investigation areas were 300 m and 3.5 km away from each other. Dark circles indicate trees fogged.

thereby allowing us to sample arthropod communities in a tree-specific way (i.e., 'tree selective fogging') – an important precondition for our investigation. The vegetation beneath the study tree was cut and 80 to 90% of the crown projection area was covered with funnels: this made it possible to collect the majority of all arthropods killed by the insecticide. At 06:00 on the following morning, up to four trees were fogged for ten minutes. We used natural pyrethrum as insecticide, in 2% concentration, which biodegrades under the influence of light in a few hours into non-toxic components. Depending on the weather conditions, fogging was conducted from the ground or – usually – from within the tree crowns. All arthropods that had dropped into the funnels two hours after fogging were sampled and stored in ethanol. The simpler forest structure in all secondary forests, in which no closed canopy and no distinct crown layers had yet developed, made it possible to perform the fogging without the roof construction. For more details concerning the process of fogging see Floren & Linsenmair (1997), Floren et al. (1998) and Adis et al. (1998).

Taxa considered in the evaluation and dominance classification of Formicidae

Arthropods were sorted to order and Formicidae to morphospecies. To describe the changes between all forests on the ordinal level we refer mainly to those taxa which are usually numerous in the sample trees (see Figure 2) and which had been shown to be effectively captured by fogging (references in Stork et al. 1997). To date, identification of Coleoptera on the morphospecies level refers only to the primary forest (see Floren & Linsenmair 1998a) because the secondary samples are still under evaluation.

Corresponding to our evaluation of the mature forest, we defined an ant species as numerically dominant

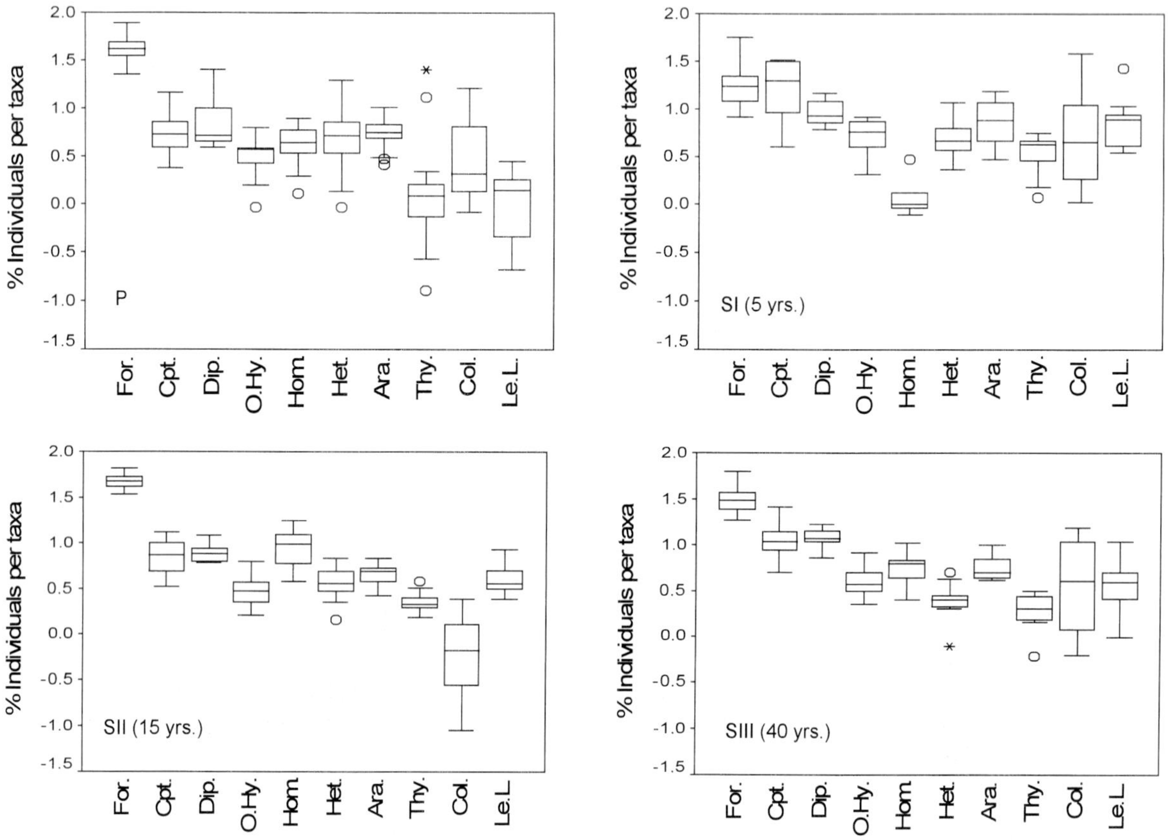

Figure 2. Changes in arboreal arthropod communities of forests at different disturbance levels compared to those of the primary forest. P = Primary forest, SI, SII, SIII = Forests regenerating for 5, 15, and 40 years respectively. The Y-axis is on a log scale. For. = Formicidae; Cpt. = Coleoptera; Dip. = Diptera; O.Hy. = Other Hymenoptera; Hom. = Homoptera; Het. = Heteroptera; Ara. = Arachnida; Thy. = Thysanoptera; Le.L. = Lepidoptera larvae. A circle indicates extreme values between one and three times the box length, an asterisk even larger values.

in a community when it provided most of all ant individuals (Floren & Linsenmair 1997, 2000). Species numbering less than 80% of the dominant species were termed subdominants. A species was considered rare when found with less than five individuals. For many of the rare species, however, we were not able to determine whether they were nesting in a tree and thus forming an integral part of a community. To find out whether an ant species was nesting in a certain tree tuna baits were placed on all branches within reach of the tree trunk. The foraging ants were then followed in order to locate the position of their nests. In such a way, dominant ant species and most of the subdominant species of the relatively small trees were sampled, while only small colonies, for example those which live in the outer crown parts, may not have been recorded completely: this may also be true for some species not attracted by tuna baits (Floren & Linsenmair 2000). As this work is quite labour intensive, surveys have, to date, only been done on the trees in primary forest.

Statistics

Taxa composition within and between forests was compared with a one-way-ANOVA and a post-hoc test (Scheffe's test). For these tests the data were log-transformed. In all cases in which the preconditions for the tests were not fulfilled, a Mann–Whitney U-test was computed. This was the case for the comparison of relative proportions of taxa within SII, and the comparisons of adult Diptera, and Lepidoptera larvae, among forest types. Alpha diversity was characterised by the Shannon–Wiener and Berger–Parker indices (Berger–Parker's index strongly emphasises the importance of numerically dominant species). The differences between trees and between forest types were quantified by Soerensen's index, by the NESS index (Normalized Expected Species Shared) which

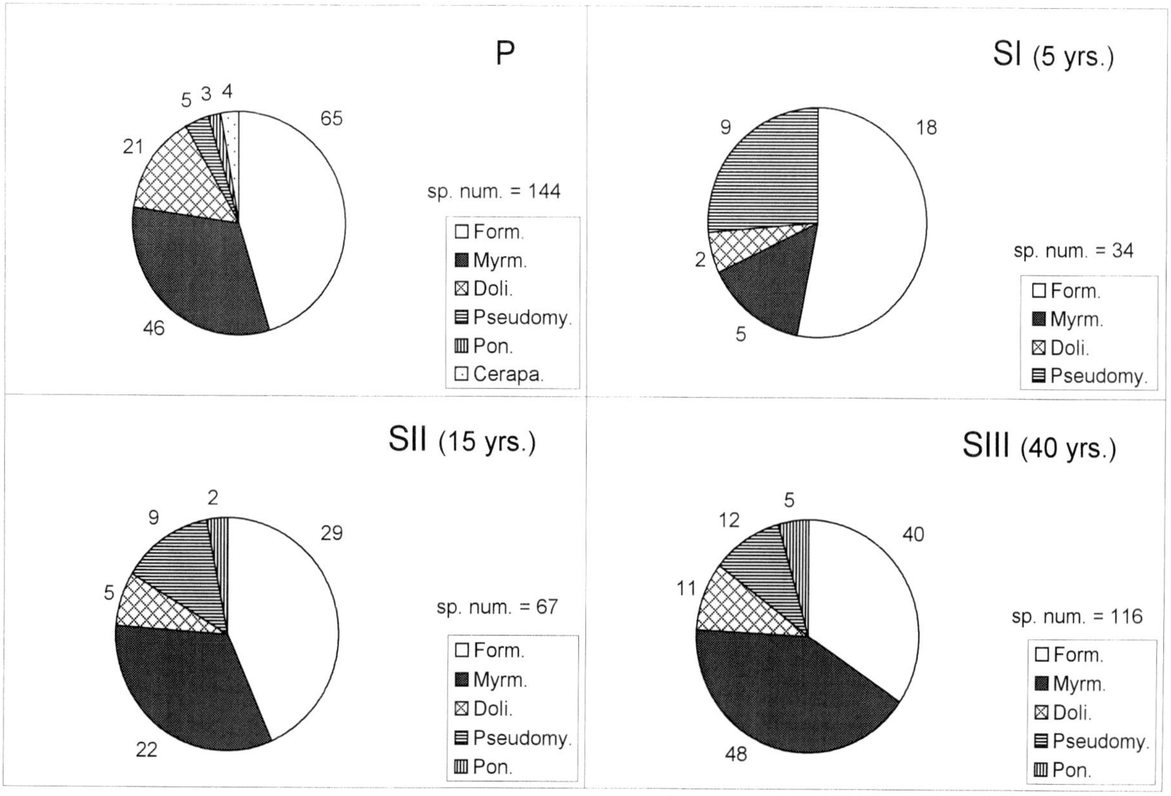

Figure 3. Relative proportion of subfamilies and number of species for each of the investigated forests (abbreviations as in Figure 2). Form. = Formicinae; Myrm. = Myrmicinae; Doli. = Dolichoderinae; Pseudomy. = Pseudomyrmecinae; Pon. = Ponerinae; Cerapa. = Cerapachyinae; Aenic. = Aenictinae).

considers species and their abundances (Grassle & Smith 1976), and by Shinozaki curves (Shinozaki 1963). Similar to the NESS index, the computation of the Shinozaki curve is mathematically derived from the rarefaction statistic but uses qualitative presence/absence data of species (Shinozaki 1963). As every species from a forest type is included in the computation, these curves give a graphical impression of the extent of the beta diversity, and allow assessment of the sampling effort. If the slope of the curve reaches zero, the species of an area are considered to be well represented in the samples. On the contrary, every deviation of the slope from zero indicates that further sampling will result in more species. The introduced parameters are in detail discussed in Magurran (1988) and Floren & Linsenmair (1997).

Table 1. Total number of ant genera and most species rich genera of the forests investigated (abbreviations follow those in Figure 3).

Ant genera	Species number			
	P	SI (5 yr)	SII (15 yr)	SIII (40 yr)
Total number of genera	49	8	14	38
Camponotus (Form.)	45	6	10	12
Polyrhachis (Form.)	33	7	8	16
Crematogaster (Myrm.)	16	3	6	10
Dolichoderus (Doli.)	11	0	0	2
Technomyrmex (Doli.)	6	0	0	6
Tapinoma (Doli.)	8	2	4	3
Tetraponera (Pseudomy.)	5	9	9	12

Results

Arthropod community composition in the forests at different disturbance levels

On the basis of our studies in the primary forest, Figure 2 shows how taxa composition changed with the degree of disturbance. A full description of the composition on the ordinal level is given in Appendix 1.

In the primary forest, relative proportions of taxa varied only slightly among trees, and arthropod communities showed a very characteristic composition (see also Floren & Linsenmair 1997). All trees were dominated by Formicidae, which were significantly more abundant than all other groups ($p < 0.001$). Lepidoptera larvae – here taken as representatives of less mobile arthropods – always remained rare and therefore differed from all other taxa, with the exception of Thysanoptera and Collembola which, in contrast, varied largely in numbers between trees. Coleoptera, Homoptera, Diptera and Arachnida were regularly caught, with around 5% each, from the trees in the mature forest, while numbers of non- formicine Hymenoptera and Heteroptera were usually somewhat lower, although this was not statistically significant. This abundance pattern on the high taxonomic level disappeared fully in the five-year-old forest (SI). During forest succession (SI → SII → SIII) we observed a convergence to the approximate conditions in mature forest (P). In SI only Homoptera differed significantly from all other taxa ($p < 0.001$). In the 15-year-old forest (SII) their proportion was around 5% and remained roughly at the same level in the older forests. Collembola showed the lowest numbers but the largest variance.

With increasing forest age, total arthropod numbers clearly rose (SI = 17,598, SII = 30,592, SIII = 51,864 individuals), mainly as a consequence of Formicidae abundance. Total numbers of arthropods in the primary forest (48,457 individuals) were of comparable size to those of SIII. Besides Formicidae, changes were caused substantially by Coleoptera, Lepidoptera larvae, Homoptera and, to a lesser degree, by non-formicine Hymenoptera (see Appendix 1). Differences in relative proportions of Diptera, Heteroptera, and Arachnida were not distinguishable between forest types. Coleoptera differed significantly between the primary forest, SI ($p < 0.008$), and SIII ($p < 0.001$), with beetles increasing from 6% to more than 10%. There was no significant difference in numbers of beetles between the primary forest and SII ($p = 0.498$). The non-formicine Hymenoptera show a similar result, although only at a 5% significance level. A distinct increase in numbers of Lepidoptera larvae in SI correlated with low ant abundance, from less than 1% in the primary forest to 8.8% ($p < 0.001$). With the increase of Formicidae in SII and SIII their relative proportion dropped again to an average of 4%. However, due to the large variations between trees, this decrease was not significant.

The communities of Formicidae

Formicidae communities of the primary forest were more diverse on all taxonomic levels (subfamily, genus, morphospecies) compared to the disturbed forests (see Figure 3 and Table 1). The greatest loss in species was found in SI, in which most subfamilies showed large reductions in species numbers. Cerapachyinae and Ponerinae were not collected. An exception to the general decrease in ant diversity was observed within the subfamily Pseudomyrmecinae (genus *Tetraponera*). A total of 14 species was found in all disturbed forests (9 species each in SI and SII and 12 species in SIII) while only five species were identified in the sample of the mature forest (P). During the course of forest regeneration, ant communities increased in species richness and also gradually converged, in terms of relative proportions of subfamilies and genera, to the approximate conditions of the primary forest. This is demonstrated in detail in Table 1 for the largest genera: *Camponotus, Polyrhachis* and *Crematogaster*. After 40 years of regeneration 26%, 48% and 62% of the original number of species of these genera were presented once more in samples. In SI, Dolichoderinae were represented by only two species of *Tapinoma*, while *Dolichoderus* and *Technomyrmex* appeared only in SIII – *Dolichoderus* still having fewer species than in the primary forest.

In no forest type were we able to detect mutually exclusive species. Dominant species could usually be found in a subdominant position in other communities (see Table 2). The abundance structure of assemblages in the trees in SI and SII was simple (Figure 4), and in most cases a single species was clearly in a dominant position. In SII, *Camponotus* #282 became the most conspicuous species, dominating eight trees. Two species of *Crematogaster* and *Cladomyrma* #278 were newly found in SII. In SIII, *Camponotus* #282 showed a strong decrease in abundance and many of the former dominants disappeared, and in their place four new dominant species appeared. Of those, only

Table 2. Distribution of dominant ant species on the trees in all forest types. A = Number of trees from which the ant species was collected. B = Number of trees on which the species was dominant.

Dominant ant species	P		SI (5 yr)		SII (15 yr)		SIII (40 yr)	
	A	B	A	B	A	B	A	B
Camponotus #282	–	–	10	4	10	8	7	1
Polyrhachis #273	–	–	5	2	4	–	–	–
Crematogaster #43	–	–	2	2	9	1	–	–
Tetraponera #258	–	–	4	2	7	–	–	–
Camponotus #153	–	–	2	1	–	–	3	1
Paratrechina #108	–	–	3	1	2	–	–	–
Crematogaster #44	–	–	–	–	2	1	–	–
Crematogaster #224	–	–	–	–	2	1	4	1
Cladomyrma #278	–	–	–	–	2	1	2	–
Technomyrmex #94	–	–	–	–	–	–	3	3
Crematogaster #223	–	–	–	–	–	–	1	1
Dolichoderus #23	4	3	–	–	–	–	1	1
Crematogaster #46	–	–	–	–	1	–	4	2
Plagiolepis #154	4	2	–	–	–	–	–	–
Technomyrmex #29	8	1	–	–	–	–	–	–
Crematogaster #155	4	2	–	–	–	–	–	–
Dolichoderus #236	2	1	–	–	–	–	–	–
Camponotus #115	2	2	–	–	–	–	–	–
Crematogaster #83	2	1	–	–	–	–	–	–
Dolichoderus #22	2	1	–	–	–	–	–	–
Camponotus #74	2	1	–	–	–	–	–	–

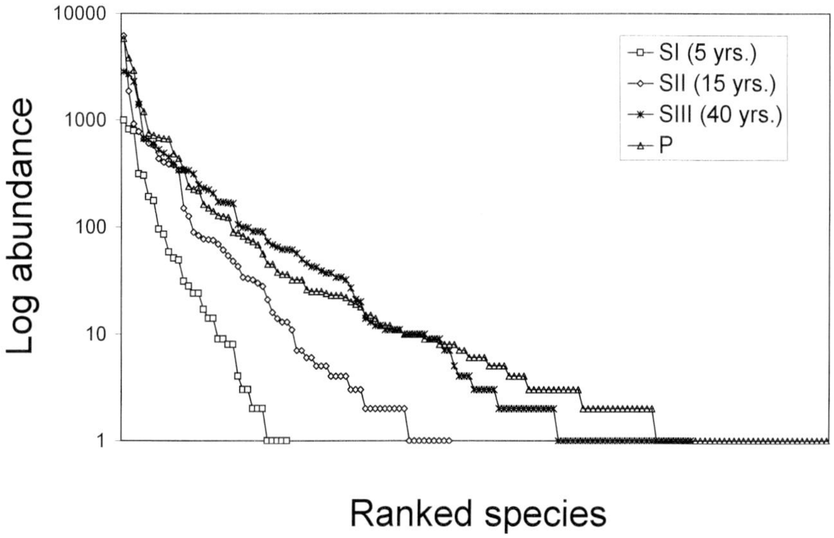

Figure 4. Rank abundance plots for the Formicidae of all forests of different disturbance levels (abbreviations as in Figure 2). The numbers on the x-axis refer to the number of species with less then five individuals collected.

Crematogaster #75 had also been found in a subdominant position (with 16 individuals) in SII. The change of dominant species continued with forest age, and *Dolichoderus* #23 was the only dominant species of primary forest, also found in a disturbed forest type. Dolichoderinae spp., in particular, were mainly collected in the old secondary forest (SIII).

The gradual approximation of ant communities to the conditions of the primary forest is also seen in rank-abundance curves (Figure 4). The steepest curve and therefore the community with the lowest number of species was observed in SI. During forest succession the curves become flatter, caused by an increasing number of total species per community, more species in all abundance classes, and an increasing share of rare species.

Figure 5 shows the structure of the ant communities on the alpha and beta diversity level. With increasing forest regeneration, species numbers (A) as well as total abundances (B) increased significantly. With the alpha correction after Bonferoni for pairwise comparisons (Lehmacher et al. 1991) the exact significance level is reached at $\alpha = 0.0083$. Species numbers differed significantly between forest types ($\chi^2 = 28,819$, $p < 0.001$), the only exception was the mature forest (P) compared to SIII, the oldest secondary forest ($p = 0.791$). The medians of species abundances and minimum/maximum values became larger with forest succession (see B): median SI, 283, range 49 to 1223; median SII, 1161, range 419-2835; median SIII 1479.5, range 424-4670; and median P, 2040, range 524-5010). Kruskal–Wallis ANOVA confirmed significant differences ($\chi^2 = 16,239$, $p < 0.001$), which disappeared when only SII, SIII, and P were compared ($\chi^2 = 0.721$, $p < 0.697$). In spite of alpha correction, the median of SI varied significantly from the older forest types. On the contrary, alpha indices (C, D) did not differ, either within forest type or between the forests. This is a consequence of the similar abundance structure of ant communities on most trees, as well as the dependence of the index value on the sample size and of the distribution of species within a community (Magurran 1988). With increasing species number and abundance of ant communities during forest succession, communities became more complex. This is demonstrated by the mean beta-diversity values of forest types (E, F), which became lower with forest age (SIII and P), corresponding with a low similarity of species composition in ant communities. In contrast, the ant communities in both SI and SII showed great similarity in species composition.

Shinozaki curves (Figure 6) allow direct comparisons between forests through computed numbers of expected species. The fast levelling off of the curves in SI and SII demonstrates that sample size was adequate to represent the respective ant communities, although, it was not suited to give a true picture of ant diversity in SIII and particularly not in the primary forest (P). No estimation can be made for the size of the species pool of the primary forest or how large sample size would have to be extended to reach an asymptote.

Discussion

The changes in communities on the ordinal level

The arboreal arthropod communities of the disturbed forests differed distinctly in composition on the ordinal level in comparison to the primary forest. Here, Formicidae dominated all trees followed by Coleoptera, Diptera, Homoptera, and Arachnida, which each contributed about 5% of all individuals to a community. Lepidoptera larvae were always rare. In the five year- old regenerating forest following clear-cut, relative proportions of most taxa were similar or showed such fluctuations in their abundances that differences were no more significant. Surprisingly, Homoptera which usually occur in high numbers in disturbed forests (Schowalter 1988, 1995) were almost completely lacking.

Aggregations of Homoptera are considered to be the most substantial and permanent source of food (honey dew) for Formicidae (Tobin 1994; Davidson 1997), and can distinctly influence ant species distribution patterns (Hölldobler & Wilson 1990). Defence of continuously available food resources can lead to the establishment of large territories of dominant aggressive and mutually exclusive ant species, so that the impression of a mosaic-like structuring of communities arises (Jackson 1984). Such ant mosaics have mainly been described from plantations and were not found in SI, in which hardly any Homoptera occurred, or in SII and SIII, in which Homoptera reached, on average, about 5% of a community. In contrast to the ant mosaic theory we collected from many trees of every forest type some potentially dominant species, which were integrated in structurally highly different communities, without mutually excluding each other. This result indicates that ant mosaics may only establish under special circumstances, for example in the structurally even more uniform plantations compared to SI.

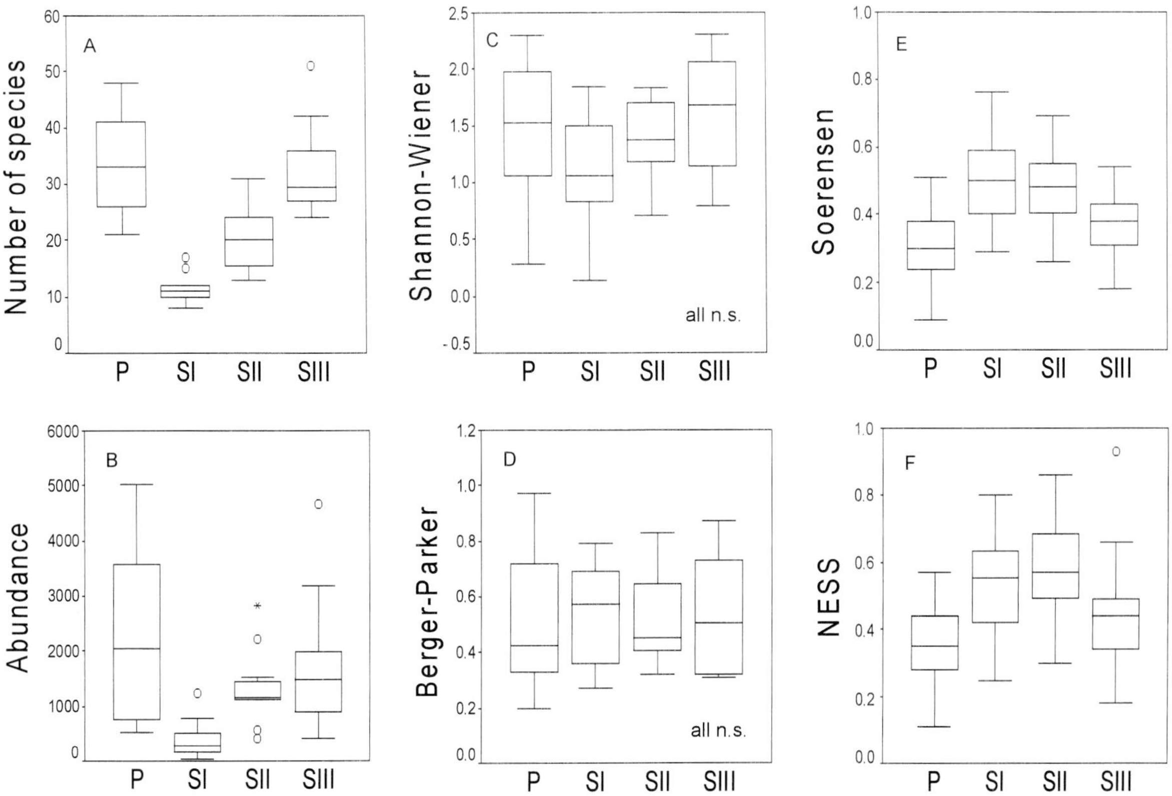

Figure 5. Changes in Formicidae communities of alpha and beta diversity (abbreviations of forest types as in Figure 2). Data are given as Box-Plots. A circle indicates extreme values between one and three times the box length, an asterisk even larger values.

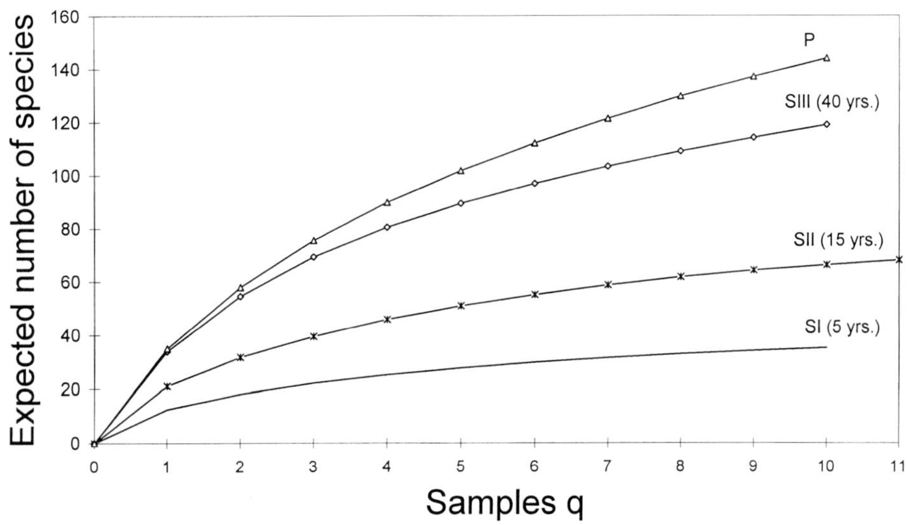

Figure 6. Shinozaki curves for the communities of Formicidae for all forests investigated.

Here, honey dew, provided by Homoptera, is reliably available in large quantities (e.g., Leston 1973; Jackson 1984; Majer 1993). In contrast, ant mosaics were never detected in the primary forest in which these dominant ant species always remained rare (Götzke & Linsenmair 1996; Floren & Linsenmair 1997, 1999, 2000). This is presumably a consequence of the unpredictably dispersed food there, which makes the defence of large territories uneconomic, since these would have to be so large that it would not be possible at all to monitor and defend them effectively (Hölldobler & Lumsden 1980).

A further contrast with the mature forest is the high quantity of Lepidoptera larvae in SI, in which Formicidae had decreased considerably and in whose trees the highly abundant species had entered from the ground (Floren pers. obs.). In these relatively small trees caterpillars sometimes counted for more than 26% of all arthropods in a single community. In contrast, Lepidoptera larvae, like less mobile arthropods in general, were found in very low numbers in the primary forest where all trees were dominated by Formicidae. We suspect that this is due to the high predation pressure of ants on the arthropod fauna (Floren & Linsenmair 1997, 1998a, b), which continuously patrol all parts of a tree and instantly attack most arthropods they encounter. We suggest that the greater numbers of caterpillars in SII and SIII may be due to fewer predatory ant species, exerting a lower predation pressure on the arthropod fauna than in the mature forest. In order to assess this hypothesis, intensive studies of the autecology of ant species, about which only little is known, would be necessary.

The Formicidae communities

Because the analysis of the changes within the ant communities in the disturbed forests refers also directly to the structuring mechanisms, we give first a brief summary of our results of the primary forest, and then discuss the results from the disturbance gradient (the summary refers to results of Floren & Linsenmair 1997, 1998b, 2000).

The primary forest

Ant communities were highly heterogeneous among trees and had neither a uniform abundance structure nor a fixed dominance hierarchy, as demonstrated with behavioural observations at tuna baits (Floren unpubl. data). Neither positive nor negative species associations were detected within all 218 hitherto analysed arboreal ant species. This heterogeneity is maintained during the re-establishment of the ant communities, as demonstrated by refogging individual trees after six months and three years. Within six months after the fogging, ant communities returned to previous levels of abundance and species numbers, but no convergence in the composition of communities at the morphospecies level was detected. This was also true for the ant communities which had established themselves after three years in these trees. The newly established ant communities were just as heterogeneous as the original communities, but showed a completely different species composition. It can not be excluded that the ant communities showed patterns on different functional levels, for example in body size or guild composition (see Gotelli & Graves 1996). Ant species showed a large overlap in their feeding ecology, with most species harvesting every available food item. This finding was corroborated again and again in hundreds of observation hours and bait experiments in the trees. Furthermore, many species were able to use a variety of nest structures such as hollow twigs, stem cavities, carton-, or detritus nests, which indicates a large degree of overlap in this niche dimension as well. Artificial nesting structures (of hollow twigs of different dimensions and detritus bags fixed in the tree crowns) were accepted in more than 50% of cases by ants during a seven month period of time, without our being able to predict which species would settle. Species saturation is even on the basis of 79 fogging samples in the mature forest not reached (Floren unpubl. data) and we can not yet guess at when saturation of ant species might be reached.

Together, these results point to the great importance of chance events which distinctly influence the structure of ant communities in the primary forests. Because one can not assume that ant communities were assembled completely at random, we suspect that community structure is a consequence of a very complex dynamic process (Pahl-Wostl 1997; Floren & Linsenmair 2000).

The disturbance gradient

In the five year-old, most disturbed forest we found only residual ant communities consisting of few ant species. These communities were structurally simple and showed a low degree of beta diversity. During the course of forest succession, Formicidae communities converged on the conditions of the primary forest. Communities developed over time in terms of species richness and abundance, and the proportion of rare species increased continually. This was correlated with an increase in the degree of complexity and in the unpredictability of the ant communities, until they were eventually no longer predictable at all in the primary forest. On the basis of a pre-sorting of all of our 79 foggings in the mature forest, we have identified approximately 300 arboreal species of ants, without the steepness of the Shinozaki curve being altered distinctly (Floren unpubl. data). For single species, we detected preferences with a particular disturbance level with which their maximal abundance correlated. For example, the dominant *Camponotus* #282 in SI (5 yr) and SII (15 yr) decreased during forest succession and could not be found any more in the mature forest.

Similar changes in dominance of individual species were found in different taxa such as Coleoptera and non-formicine Hymenoptera. Among the Coleoptera, individual species of Malachiidae, Anthicidae, and Chrysomelidae occurred in high numbers in the strongly disturbed forests. These species, too, became rarer with increasing forest complexity until they completely disappeared in our primary forest samples. Changes of dominance structure after strong disturbances, accompanied by a drastic decrease in total diversity and the appearance of single highly abundant species, are mainly known from plants (e.g., Leigh et al. 1993; Eggleton et al. 1998) and vertebrate communities (e.g., Lambert 1992; Laurence 1994; Julien & Thiollay 1996). However, such changes have only rarely been analysed within arthropods (e.g., Burbidge et al. 1992; Roth et al. 1994; Daily & Ehrlich 1995; Holloway 1998).

Our results show that strong anthropogenic disturbances, leading to structurally and microclimatically very different habitats, represent very effective filters for many ant species. Above all, from the highly diverse genera (e.g., *Camponotus* or *Polyrhachis*, Formicinae), a great number of species were lacking in the disturbed forests. After 40 years of forest regeneration the number of species of Myrmicinae equalled those of the primary forest, whereas within the Formicinae and Dolichoderinae fewer species were found. This might indicate a change in food availability in the disturbed forest and in any case demonstrates how great the differences between the regenerating and the primary forest still are, even after a – for the human time scale – rather long period of time. Besides the general decrease in ant diversity, only Pseudomyrmecinae (genus *Tetraponera*), showed a distinct increase in species numbers as well as abundances. This might be related to the opening of the forest (P. Ward pers. comm.). To the ten species of *Tetraponera*, which were found in all hitherto evaluated fogging samples from trees of the primary forest, another five species must be added from the disturbed forests. Such an increase has as yet not been recorded for Pseudomyrmecinae. Whether the occurrence of certain species of this genus or its overall diversity may serve as reliable indicators for the extent of forest disturbance is currently being investigated.

General remarks

It was our aim to carry out our research on the same tree species in all forest types. This, however, was impossible because the primary forest tree species did not occur in the disturbed forest, as also the pioneer tree of SI was not found in SII and SIII (despite extensive surveys we were unable to find any tree species occurring in all forest types). This forced us to select three different tree species. Fogging was always carried out on trees growing under similar habitat conditions. Even if some part of the variability of community composition between forest types is a consequence of investigating three different tree species, we are convinced that the big differences we found among forest type communities were mainly caused by the disturbance regime, with tree species-specific effects playing only a minor role. This is indicated by 79 foggings on 13 tree species which were performed in the primary forest, which showed that arthropod community structure was completely independent of the tree species. This result is also supported by foggings of further tree species by other researchers (references in Stork et al. 1997).

Formicidae, in particular, are suited for such an analysis, because – to date – no hints on any tree specific adaptations of the ant fauna could be discovered. In addition to this finding, the following results also support our hypothesis: (1) Ant communities gradually converged on conditions in primary forest with increasing forest regeneration in all aspects of diversity analysed. This occurred in both SII and SIII, in which the same tree species was investigated, as well as the other forest types. (2) Similar changes were also documented for other taxa, namely for phytophagous groups (e.g., Coleoptera, Orthoptera) as well as for predators and parasitoids (e.g., Arachnida, Ichneumonidae). (3) The distribution patterns of species in the disturbed forests were not found in the primary forest (high abundance of individual species and a distinctly reduced number of rare species). (4) New theoretical analyses show that the ant communities of disturbed forests possess – in contrast to those of the primary forest – a deterministic structure which disappears completely in the primary forest (Floren unpubl. data).

These results indicate that extensive anthropogenic disturbances not only reduce the size of the species pool dramatically, but also change the dynamic of the system. During forest regeneration of strongly disturbed forests, deterministic processes are eventually superseded in their significance by stochastic processes (Floren & Linsenmair 2000, Floren unpubl. data). This is a consequence of increasing forest complexity and total diversity. We believe that well founded and scientifically substantiated conclusions about the importance of secondary forests for conservation management will only be reached via the sort of investigation described here.

Acknowledgements

We thank Sabah Parks for permission to work in Kinabalu Park and use the park facilities. We are in particular thankful to Alim Biun, who helped with the choice of investigation areas and study trees. Thanks are also due to Park staff members of substation Sorinsim, who helped during all occurring difficulties and to Andre Kessler who was considerably involved in the field work. For critical comments to the manuscript we thank Andrew Davis and two anonymous referees. The study was funded by grants of the Deutsche Forschungsgemeinschaft (DFG), Li 150/13 1-4.

References

Adis, J., Basset, Y., Floren, A., Hammond, P. M. & Linsenmair, K. E. 1998. Canopy fogging of an overstorey tree - Recommendations for standardisation. Ecotropica 4: 93–97.

Bascompte, J. & Sole, R. V. 1995. Rethinking complexity: modelling spatiotemporal dynamics in ecology. Trends Ecol. Evol. 10: 361–366.

Bignell, N. C. 1996. The diversity, abundance and biomass of termites under differing levels of disturbance in the Mbalmayo forest reserve, southern Cameroon. Phil. Trans. R. Soc. London. B 351: 51–68.

Burbidge, A. H., Leicester, K., McDavitt, S. & Majer, J. D. 1992. Ants as indicators of disturbance at Yanchep National Park, Western Australia. J. R. Soc. West. Austr. 75: 89–95.

Chesson, P. L. & Case, T. J. 1986. Overview: non-equilibrium community theories: chance, variability, history, and coexistence. Pp. 229–239. In: Diamond, J. & Case, T. J. (eds), Community ecology. Harper and Row, New York.

Daily, G. C. & Ehrlich, P. R. 1995. Preservation of biodiversity in small rainforest patches: rapid evaluations using butterfly trapping. Biod. Cons. 4: 35–55.

Davidson, D. W. 1997. The role of resource imbalances in the evolutionary ecology of tropical arboreal ants. Biol. J. Linn. Soc. 61: 153–181.

Davis, A. J. & Sutton, S. L. 1998. The effects of rainforest canopy loss on arboreal dung beetles in Borneo: implications for the measurement of biodiversity in derived tropical ecosystems. Diversity Distrib. 4: 167–173.

Eggleton, P., Bignell, D. E., Sands, W. A., Mawdsley, N. A., Lawton, J. H., Wood, T. G. Dupuy, J. M. & Chazdon, R. L. 1998. Long-term effects of forest regrowth and selective logging on the seed bank of tropical forests in NE Costa Rica. Biotropica 30: 223–237.

Floren, A. & Linsenmair, K. E. 1997. Diversity and recolonisation dynamics of selected arthropod groups on different tree species in a lowland rain forest in Sabah, Malaysia with special reference to Formicidae. Pp. 344–382. In: Stork, N. E., Adis, A. J. & Didham, R. K. (eds), Canopy arthropods. Chapman & Hall, London.

Floren, A. & Linsenmair, K. E. 1998a. Non-equilibrium communities of Coleoptera in trees in a lowland rain forest of Borneo. Ecotropica 4: 55–67.

Floren, A. & Linsenmair, K. E. 1998b. Diversity and recolonisation of arboreal Formicidae and Coleoptera in a lowland rain forest in Sabah, Malaysia. Selbyana 19 (2): 155–161.

Floren, A. & Linsenmair, K. E. 2000. Do ant mosaics exist in pristine lowland rain forest? Oecologia, Oecologia 123: 129–137.

Floren, A., Linsenmair, K. E. & Biun, A. 1998. A very mixed bag. Insect diversity in Sabah's lowland forest. Malayan Naturalist 52 (2): 26–29.

Giller, P. S. 1984. Community structure and the niche. Chapman and Hall, London.

Gotelli, N. J. & Graves, G. R. 1996. Null models in ecology. Smithsonian Institution Press, Washington, London.

Götzke, A. & Linsenmair, K. E. 1996. Does beta-diversity allow reliable discrimination between deterministically and stochastically assembled communities? Ecotropica 2: 79–82.

Grassle, J. F. & Smith, W. 1976. A similarity measure sensitive to the contribution of rare species and its use in investigation of variation in marine benthic communities. Oecologia 25: 13–22.

Horstmann, K., Floren, A. & Linsenmair, K. E. 1999. High species-richness of Ichneumonidae from the canopy of a Malaysian rain forest. Ecotropica 5 (1): 1–12.

Huston, M. A. 1994. Biological diversity. Cambridge University Press, Cambridge.

Hölldobler, B. & Wilson, E. O. 1990. The ants. The Belknap Press of Harvard University Press, Cambridge, Massachusetts.

Hölldobler, B. & Lumsden, C. J. 1980. Territorial strategies in ants. Science 210: 732–739.

Holloway J. D. 1998. The impact of traditional and modern cultivation practices, including forestry, on Lepidoptera diversity in Malaysia and Indonesia. Pp. 567–598. In: Newbery, D. M., Prins, H. H. & Brown, N. D. (eds), Dynamics of tropical communities. Blackwell Science, Oxford.

Jackson, D. A. 1984. Ant distribution patterns in a Cameronian cocoa plantation: investigation of the ant mosaic hypothesis. Oecologia 62: 318–324.

Jullien, M. & Thiollay, J.-M. 1996. Effects of rain forest disturbance and fragmentation: comparative changes of the raptor community along natural and human-made gradients in French Guiana. J. Biogeogr. 23: 7–25.

Lambert, F.R: 1992. The consequences of selective logging for Bornean lowland forest birds. Phil.Trans. R. Soc. London 335: 443–457.

Laurance, W. F. 1994. Rainforest fragmentation and the structure of small mammal communities in tropical Queensland. Biol. Cons. 69: 23–32.

Lawton, J. H., Bignell, D. E., Bolton, B., Bloemers, G. F., Eggelton, P., Hammond, P. M., Hodda, M., Holt, R. D., Larsen, T. B., Mawdsley, N. A., Stork, N. E., Srivastava, D. S. & Watt, A. D. 1998. Biodiversity inventories, indicator taxa and effects of habitat modification in tropical forest. Nature 391: 72–75.

Lehmacher, W., Wassmer, G. & Reitmeir, P. 1991. Procedures for two-sample comparisons with multiple endpoints controlling the experimentwise error rate. Biometrics 47: 511–521.

Leigh, E. G., Wright, S. J., Herre, E. A. & Putz, F. E. 1993. The decline of tree diversity on newly isolated tropical islands: a test of a null hypothesis and some implications. Evol. Ecol. 7: 76–102.

Leston, D. 1973. Leston D (1973) The ant mosaic - tropical tree crops and the limiting of pests and diseases. PANS 19: 311–341.

Magurran, A. E. 1988 Ecological diversity and its measurement. Chapman & Hall, London.

Majer, J. D. 1993. Comparison of the arboreal ant mosaic in Gahna, Brazil, Papua New Guinea and Australia - its structure and influence on arthropod diversity. Pp. 115–141. In: Salle, J. L. & Gauld, J. D. (eds), Hymenoptera and biodiversity. CAB International, Bristol.

Olson, D. M. & Andriamiadana, A. 1996. The effects of selective logging on the leaf litter invertebrate community of a tropical forest in western Madagascar. Primate Report 46: 175–187.

Pahl-Wostl, C. 1997. The dynamic nature of ecosystems. John Wiley & Sons, New York.

Pimm, S. L. 1991. The balance of nature? The University of Chicago Press, Chicago.

Roth, D. S., Perfecto, I. & Rathcke, B. 1994. The effects of management systems on ground- foraging ant diversity in Costa Rica. Ecol. Appl. 4: 423–436.

Schowalter, T. D. 1988. Canopy arthropod community structure and herbivory in old-growth and regenerating forests in western Oregon. Can. J. For. Res. 19: 318–322.

Schowalter, T. D. 1995. Canopy invertebrate community response to disturbance and consequences of herbivory in temperate and tropical forests. Selbyana 16: 41–48.

Schowalter, T. D. & Ganio, L. M. 1999. Invertebrate communities in a tropical rain forest canopy in Puerto Rico following Hurricane Hugo. Ecol. Entomol. 24: 191–201.

Shinozaki, K. 1963. Note on the species-area curve. Proceedings of the 1th Ann. Meet. Ecol. Soc. Japan, Tokyo.

Stork, N. E., Adis, J. & Didham, R. K. (eds) 1997. Canopy arthropods. Chapman & Hall, London.

Tobin, J. E. 1994. Ants as primary consumers: diet and abundance in the Formicidae. Pp. 279–303. In: Hunt, J. H. & Nalepa, C. A. (eds), Nourishment and evolution in insect societies. Westview Press, Oxford.

See Appendix 1 →

Appendix 1. Numbers (N) and relative proportions (%) of major arthropod taxa from fogged trees in the primary forest (P) and the forests at different levels of disturbance (SI, SII, and SIII). Other = comprising all other arthropod groups: Mantodea, Phasmida, Neuroptera, Ephemeroptera, Plecoptera, Trichoptera, Diplopoda and Myriapoda. Topou = local name of the pioneer tree species in SI. Bergil = local name of *Vitex pinnata* (Verbenaceae). A.l. = *Aporusa lagenocarpa* (Euphorbiaceae).

SI 5 yrs	Topou 1 N (%)	Topou 2	Topou 3	Topou 4	Topou 5	Topou 6	Topou 7	Topou 8	Topou 9	Topou 10
Formicidae	$525_{15.44}$	$1223_{56.05}$	$394_{19.81}$	$49_{8.43}$	$146_{8.17}$	$285_{19.08}$	$281_{22.3}$	$219_{13.68}$	$120_{12.06}$	$809_{35.05}$
Coleoptera	$312_{9.17}$	$132_{6.05}$	$308_{15.49}$	$144_{24.78}$	$72_{4.03}$	$440_{29.45}$	$397_{31.51}$	$524_{32.73}$	$324_{32.56}$	$350_{15.16}$
Diptera	$266_{7.82}$	$139_{6.37}$	$122_{6.13}$	$84_{14.46}$	$217_{12.14}$	$139_{9.3}$	$118_{9.37}$	$123_{7.68}$	$126_{12.66}$	$164_{7.11}$
Other Hym.	$85_{2.5}$	$45_{2.06}$	$97_{4.88}$	$23_{3.96}$	$110_{6.16}$	$117_{7.83}$	$90_{7.14}$	$84_{5.25}$	$72_{7.24}$	$189_{8.19}$
Homoptera	$45_{1.32}$	$25_{1.15}$	$20_{1.01}$	$6_{1.03}$	$53_{2.97}$	$12_{0.8}$	$17_{1.35}$	$15_{0.94}$	$9_{0.9}$	23_{1}
Heteroptera	$155_{4.56}$	$51_{2.34}$	$73_{3.67}$	$30_{5.16}$	$208_{11.64}$	$119_{7.97}$	$78_{6.19}$	$75_{4.68}$	$39_{3.92}$	$77_{3.34}$
Arachnida	$394_{11.58}$	$65_{2.98}$	$92_{4.63}$	$89_{15.32}$	$238_{13.32}$	$109_{7.3}$	$70_{5.56}$	$122_{7.62}$	$88_{8.84}$	$84_{3.64}$
Psocoptera	$9_{0.26}$	$18_{0.82}$	$23_{1.16}$	$4_{0.69}$	$5_{0.28}$	$18_{1.2}$	$5_{0.4}$	$15_{0.94}$	$15_{1.51}$	$62_{2.69}$
Thysanoptera	$40_{1.18}$	$33_{1.51}$	$91_{4.58}$	$29_{4.99}$	$74_{4.14}$	$64_{4.28}$	$49_{3.89}$	$89_{5.56}$	$42_{4.22}$	$67_{2.9}$
Collembola	$1300_{38.22}$	$244_{11.18}$	$522_{26.24}$	$25_{4.3}$	$83_{4.64}$	$16_{1.07}$	$15_{1.19}$	$94_{5.87}$	$41_{4.12}$	$43_{1.86}$
Lepidop. Larv.	$118_{3.47}$	$88_{4.03}$	$82_{4.12}$	$50_{8.61}$	$480_{26.86}$	$93_{6.22}$	$103_{8.17}$	$174_{10.87}$	$82_{8.24}$	$174_{7.54}$
Lepidop. Adu.	$11_{0.32}$	$1_{0.05}$	$2_{0.1}$	$2_{0.34}$	$6_{0.34}$	$3_{0.2}$	$5_{0.4}$	$4_{0.25}$	$4_{0.4}$	$5_{0.22}$
Orthoptera	$35_{1.03}$	$13_{0.6}$	$20_{1.01}$	$20_{3.44}$	$31_{1.73}$	$13_{0.87}$	$13_{1.03}$	$15_{0.94}$	$9_{0.9}$	$21_{0.91}$
Blattodea	$56_{1.65}$	$9_{0.41}$	$27_{1.36}$	$3_{0.52}$	$37_{2.07}$	$36_{2.41}$	$1_{0.08}$	$21_{1.31}$	$10_{1.01}$	$34_{1.47}$
Dermaptera	$1_{0.03}$	0_{0}	0_{0}	$9_{1.55}$	0_{0}	$1_{0.07}$	0_{0}	0_{0}	0_{0}	$2_{0.09}$
Other	$38_{1.12}$	$15_{0.69}$	$16_{0.8}$	$8_{1.38}$	$27_{1.51}$	$18_{1.2}$	$11_{0.87}$	$6_{0.37}$	$9_{0.9}$	$18_{0.78}$
Other Larvae	$10_{0.29}$	$16_{0.73}$	$8_{0.4}$	$4_{0.69}$	0_{0}	$5_{0.33}$	$2_{0.16}$	$20_{1.25}$	$2_{0.2}$	0_{0}
Acarina	$1_{0.03}$	$65_{2.98}$	$92_{4.63}$	$2_{0.34}$	0_{0}	$6_{0.4}$	$5_{0.4}$	$1_{0.06}$	$3_{0.3}$	$186_{8.06}$
Sum	3401	2182	1989	581	1787	1494	1260	1601	995	2308

Appendix 1. Continued.

SII 15 yrs	Bergil 1 N (%)	Bergil 2	Bergil 3	Bergil 4	Bergil 5	Bergil 6	Bergil 7	Bergil 8	Bergil 9	Bergil 10	Bergil 11
Formicidae	$419_{43.24}$	$1161_{48.97}$	$1114_{66.23}$	$1544_{47.57}$	$1143_{60.06}$	$567_{34.59}$	$2229_{46.61}$	$1387_{40.12}$	$2835_{57.24}$	$1343_{39.09}$	$1067_{49.54}$
Coleoptera	$47_{4.85}$	$79_{3.33}$	$56_{3.33}$	$159_{4.9}$	$94_{4.94}$	$147_{8.97}$	$353_{7.38}$	$376_{10.88}$	$640_{12.92}$	$434_{12.63}$	$173_{8.03}$
Diptera	$58_{5.99}$	$211_{8.9}$	$128_{7.61}$	$247_{7.61}$	$115_{6.04}$	$106_{6.47}$	$366_{7.65}$	$297_{8.59}$	$296_{5.98}$	$313_{9.11}$	$258_{11.98}$
Other Hym.	$29_{2.99}$	$81_{3.42}$	$38_{2.26}$	$157_{4.84}$	$31_{1.63}$	$37_{2.26}$	$170_{3.55}$	$133_{3.85}$	$128_{2.58}$	$212_{6.17}$	$38_{1.76}$
Homoptera	$94_{9.7}$	$415_{17.5}$	$93_{5.53}$	$518_{15.96}$	$122_{6.41}$	$238_{14.52}$	$336_{7.03}$	$373_{10.79}$	$259_{5.23}$	$348_{10.13}$	$81_{3.76}$
Heteroptera	$62_{6.4}$	$72_{3.04}$	$24_{1.43}$	$138_{4.25}$	$56_{2.94}$	$112_{6.83}$	$170_{3.55}$	$187_{5.41}$	$110_{2.22}$	$112_{3.26}$	$84_{3.9}$
Arachnida	$48_{4.95}$	$127_{5.36}$	$45_{2.68}$	$113_{3.48}$	$72_{3.78}$	$111_{6.77}$	$247_{5.17}$	$168_{4.86}$	$200_{4.04}$	$231_{6.72}$	$81_{3.76}$
Psocoptera	0_{0}	$9_{0.38}$	$1_{0.06}$	$36_{1.11}$	$15_{0.79}$	$1_{0.06}$	$16_{0.33}$	$10_{0.29}$	$11_{0.22}$	$4_{0.12}$	$3_{0.14}$
Thysanoptera	$19_{1.96}$	$36_{1.52}$	$31_{1.84}$	$69_{2.13}$	$40_{2.1}$	$34_{2.07}$	$180_{3.76}$	$112_{3.24}$	$110_{2.22}$	$66_{1.92}$	$61_{2.83}$
Collembola	$11_{1.14}$	$8_{0.34}$	0_{0}	$13_{0.4}$	$25_{1.31}$	$24_{1.46}$	$118_{2.47}$	$40_{1.16}$	$14_{0.28}$	$3_{0.09}$	$5_{0.23}$
Lepidop. Larv.	$81_{8.36}$	$70_{2.95}$	$79_{4.7}$	$89_{2.74}$	$69_{3.63}$	$100_{6.1}$	$185_{3.87}$	$118_{3.41}$	$122_{2.46}$	$180_{5.24}$	$72_{3.34}$
Lepidop. Adu.	$16_{1.65}$	$3_{0.13}$	$4_{0.24}$	$2_{0.06}$	$1_{0.05}$	$7_{0.43}$	$4_{0.08}$	$3_{0.09}$	$13_{0.26}$	$2_{0.06}$	$5_{0.23}$
Orthoptera	$37_{3.82}$	$28_{1.18}$	$33_{1.96}$	$28_{0.86}$	$40_{2.1}$	$72_{4.39}$	$177_{3.7}$	$155_{4.48}$	$116_{2.34}$	$71_{2.07}$	$82_{3.81}$
Blattodea	$13_{1.34}$	$23_{0.97}$	$15_{0.89}$	$8_{0.25}$	$7_{0.37}$	$6_{0.37}$	$16_{0.33}$	$23_{0.67}$	$8_{0.16}$	$9_{0.26}$	$16_{0.74}$
Dermaptera	$13_{1.34}$	$2_{0.08}$	$4_{0.24}$	$26_{0.8}$	$20_{1.05}$	$5_{0.31}$	$34_{0.71}$	$11_{0.32}$	$29_{0.59}$	$33_{0.96}$	$19_{0.88}$
Other	$15_{1.55}$	$40_{1.69}$	$15_{0.89}$	$68_{2.09}$	$31_{1.63}$	$65_{3.97}$	$158_{3.3}$	$37_{1.07}$	$47_{0.95}$	$44_{1.28}$	$28_{1.3}$
Other Larvae	$5_{0.52}$	0_{0}	$2_{0.12}$	$23_{0.71}$	19_{1}	$6_{0.37}$	$8_{0.17}$	$10_{0.29}$	$13_{0.26}$	$28_{0.81}$	0_{0}
Acarina	$2_{0.21}$	$6_{0.25}$	0_{0}	$8_{0.25}$	$3_{0.16}$	$1_{0.06}$	$15_{0.31}$	$17_{0.49}$	$2_{0.04}$	$3_{0.09}$	$81_{3.76}$
Sum	969	2371	1682	3246	1903	1639	4782	3457	4953	3436	2154

Appendix 1 . Continued.

SIII 40 yrs	Bergil 1	Bergil 2	Bergil 3	Bergil 4	Bergil 5	Bergil 6	Bergil 7	Bergil 8	Bergil 9	Bergil 10
	N (%)									
Formicidae	$1258_{32.95}$	$642_{22.29}$	$1638_{35.62}$	$1321_{29.17}$	$2005_{63.19}$	$4670_{37.25}$	$955_{29.06}$	$3257_{43.02}$	$1773_{24.56}$	$424_{18.83}$
Coleoptera	$491_{12.86}$	$232_{8.06}$	$394_{8.57}$	$1181_{26.08}$	$160_{5.04}$	$1716_{13.69}$	$305_{9.28}$	$1170_{15.45}$	$714_{9.89}$	$271_{12.03}$
Diptera	$452_{11.84}$	$389_{13.51}$	$750_{16.31}$	$512_{11.3}$	$340_{10.72}$	$900_{7.18}$	$373_{11.35}$	$640_{8.45}$	$1200_{16.62}$	$317_{14.08}$
Other Hym.	$184_{4.82}$	$226_{7.85}$	$375_{8.16}$	$111_{2.45}$	$71_{2.24}$	$516_{4.12}$	$109_{3.32}$	$240_{3.17}$	$356_{4.93}$	$70_{3.11}$
Homoptera	$244_{6.39}$	$298_{10.35}$	$189_{4.11}$	$276_{6.09}$	$80_{2.52}$	$849_{6.77}$	$143_{4.35}$	$380_{5.02}$	$455_{6.3}$	$161_{7.15}$
Heteroptera	$108_{2.83}$	$142_{4.93}$	$107_{2.33}$	$121_{2.67}$	$25_{0.79}$	$263_{2.1}$	$66_{2.01}$	$204_{2.69}$	$308_{4.27}$	$52_{2.31}$
Arachnida	$167_{4.37}$	$250_{8.68}$	230_{5}	$318_{7.02}$	$141_{4.44}$	$733_{5.85}$	$324_{9.86}$	$309_{4.08}$	$313_{4.34}$	$110_{4.88}$
Psocoptera	$90_{2.36}$	$35_{1.22}$	$29_{0.63}$	$15_{0.33}$	$8_{0.25}$	$24_{0.19}$	$50_{1.52}$	$54_{0.71}$	$41_{0.57}$	$17_{0.75}$
Thysanoptera	$72_{1.89}$	$42_{1.46}$	$145_{3.15}$	$28_{0.62}$	$48_{1.51}$	$256_{2.04}$	$72_{2.19}$	$206_{2.72}$	$144_{1.99}$	$68_{3.02}$
Collembola	$31_{0.81}$	$18_{0.63}$	$162_{3.52}$	$122_{2.69}$	$38_{1.2}$	$565_{4.51}$	$350_{10.65}$	$450_{5.94}$	$1100_{15.24}$	$270_{11.99}$
Lepidop. Larv.	$300_{7.86}$	$310_{10.76}$	$217_{4.72}$	$117_{2.58}$	$31_{0.98}$	$445_{3.55}$	$112_{3.41}$	$151_{1.99}$	$300_{4.16}$	$111_{4.93}$
Lepidop. Adu.	$6_{0.16}$	$2_{0.07}$	$13_{0.28}$	$5_{0.11}$	$1_{0.03}$	$11_{0.09}$	$5_{0.15}$	$10_{0.13}$	$9_{0.12}$	$1_{0.04}$
Orthoptera	$208_{5.45}$	$70_{2.43}$	$65_{1.41}$	$156_{3.44}$	$56_{1.76}$	$707_{5.64}$	$113_{3.44}$	$181_{2.39}$	$169_{2.34}$	$98_{4.35}$
Blattodea	$24_{0.63}$	$42_{1.46}$	$79_{1.72}$	$86_{1.9}$	$18_{0.57}$	$140_{1.12}$	$74_{2.25}$	$88_{1.16}$	$95_{1.32}$	$9_{0.4}$
Dermaptera	$26_{0.68}$	$40_{1.39}$	$27_{0.59}$	$7_{0.15}$	$12_{0.38}$	$166_{1.32}$	$28_{0.85}$	$11_{0.15}$	$24_{0.33}$	$19_{0.84}$
Other	$92_{2.41}$	$14_{0.49}$	$57_{1.24}$	$128_{2.83}$	$89_{2.8}$	$355_{2.83}$	$127_{3.86}$	$94_{1.24}$	$39_{0.54}$	$137_{6.08}$
Other Larvae	$31_{0.81}$	0_{0}	$42_{0.91}$	$23_{0.51}$	0_{0}	$129_{1.03}$	0_{0}	$70_{0.92}$	$17_{0.24}$	0_{0}
Acarina	$34_{0.89}$	$128_{4.44}$	$79_{1.72}$	$2_{0.04}$	$50_{1.58}$	$89_{0.71}$	$80_{2.43}$	$56_{0.74}$	$163_{2.26}$	$75_{3.33}$
Sum	3818	2880	4598	4529	3173	12537	3286	7571	7220	2252

Appendix 1 . Continued.

P	A.I. 5	A.I. 15	A.I. 52	A.I. 57	A.I. 62	A.I. 70	A.I. 71	A.I. 72	A.I. 73	A.I. 74
	N (%)									
Formlcldae	$957_{65.77}$	524_{38}	$3833_{41.29}$	$761_{77.73}$	$1028_{41.07}$	$3052_{51.75}$	$760_{35.8}$	$3052_{46.09}$	$5010_{47.05}$	$3527_{46.61}$
Coleoptera	$59_{4.05}$	$203_{14.72}$	$373_{4.02}$	$23_{2.35}$	$148_{5.91}$	$314_{5.32}$	$144_{6.78}$	$346_{5.23}$	$326_{3.06}$	$286_{3.78}$
Dlptera	$92_{6.32}$	$142_{10.3}$	$361_{3.89}$	$40_{4.09}$	$128_{5.11}$	$261_{4.43}$	$109_{5.13}$	$283_{4.27}$	$533_{5.01}$	$608_{8.03}$
Other Hym.	$23_{1.58}$	$52_{3.77}$	$232_{2.5}$	$9_{0.92}$	$84_{3.36}$	$122_{2.07}$	$81_{3.82}$	$257_{3.88}$	$390_{3.66}$	$215_{2.84}$
Homoptera	$73_{5.02}$	$92_{6.67}$	$390_{4.2}$	$19_{1.94}$	$146_{5.83}$	$351_{5.95}$	$47_{2.21}$	$391_{5.9}$	$836_{7.85}$	$328_{4.33}$
Heteroptera	$39_{2.68}$	$19_{1.38}$	$470_{5.06}$	$9_{0.92}$	$124_{4.95}$	$278_{4.71}$	$108_{5.09}$	$206_{3.11}$	$402_{3.78}$	$529_{6.99}$
Arachnlda	$76_{5.22}$	$104_{7.54}$	$655_{7.06}$	$29_{2.96}$	$177_{7.07}$	$339_{5.75}$	$217_{10.22}$	$398_{6.01}$	$696_{6.54}$	$360_{4.76}$
Psocoptera	$27_{1.86}$	$50_{3.63}$	$29_{0.31}$	$23_{2.35}$	$43_{1.72}$	$215_{3.65}$	$87_{4.1}$	$133_{2.01}$	$180_{1.69}$	$391_{5.17}$
Thysanoptera	$4_{0.27}$	$12_{0.87}$	$2340_{25.21}$	$12_{1.23}$	$324_{12.94}$	59_{1}	$8_{0.38}$	$144_{2.17}$	$151_{1.42}$	$113_{1.49}$
Collembola	$14_{0.96}$	$55_{3.99}$	$157_{1.69}$	$11_{1.12}$	$42_{1.68}$	$463_{7.85}$	$348_{16.39}$	$1071_{16.17}$	$1400_{13.15}$	$389_{5.14}$
Lepldop. Larv.	$22_{1.51}$	$36_{2.61}$	$31_{0.33}$	$3_{0.31}$	$39_{1.56}$	$90_{1.53}$	$46_{2.17}$	$32_{0.48}$	$67_{0.63}$	$64_{0.85}$
Lepldop. Adu.	$2_{0.14}$	$4_{0.29}$	$13_{0.14}$	0_{0}	$1_{0.04}$	$15_{0.25}$	$5_{0.24}$	$5_{0.08}$	$29_{0.27}$	$28_{0.37}$
Orthoptera	$24_{1.65}$	$36_{2.61}$	$141_{1.52}$	$7_{0.72}$	$52_{2.08}$	$87_{1.48}$	$64_{3.01}$	$83_{1.25}$	$203_{1.91}$	$171_{2.26}$
Blattodea	$8_{0.55}$	$6_{0.44}$	$103_{1.11}$	$6_{0.61}$	$15_{0.6}$	$37_{0.63}$	$5_{0.24}$	$24_{0.36}$	$53_{0.5}$	$57_{0.75}$
Dermaptera	$7_{0.48}$	$5_{0.36}$	$17_{0.18}$	$1_{0.1}$	$7_{0.28}$	$26_{0.44}$	$27_{1.27}$	$43_{0.65}$	$46_{0.43}$	$23_{0.3}$
Other	$24_{1.65}$	$9_{0.65}$	$15_{0.16}$	$1_{0.1}$	$10_{0.4}$	$54_{0.92}$	$28_{1.32}$	$32_{0.48}$	$22_{0.21}$	$70_{0.93}$
Other Larvae	$4_{0.27}$	0_{0}	0_{0}	0_{0}	$1_{0.04}$	0_{0}	0_{0}	0_{0}	0_{0}	0_{0}
Acarlna	0_{0}	$13_{0.94}$	$112_{1.21}$	$3_{0.31}$	$134_{5.35}$	$93_{1.58}$	$39_{1.84}$	$96_{1.45}$	$242_{2.27}$	$306_{4.04}$
Sum	1455	1379	9272	979	2503	5898	2123	6622	10648	7567

Plant Ecology **153:** 169–178, 2001.

Seasonal changes in the canopy arthropod fauna in *Rinorea beniensis* in Budongo Forest, Uganda

Thomas Wagner
Zoologisches Forschungsinstitut und Museum Alexander Koenig, Adenauerallee 160, D-53113 Bonn, Germany

Key words: Africa, Fogging, Forest type, Seasonality, Tropical forest

Abstract

Arthropods were collected by insecticidal tree fogging on the understorey tree species *Rinorea beniensis* Engler (Violaceae) in Budongo Forest, a seasonal rain forest in Uganda. Eight trees were fogged in adjacent plots of primary, selectively logged and swamp forest during the wet season and again in the dry season. In all forest types, Psocoptera, parasitoid Hymenoptera, and especially Formicidae and Auchenorrhyncha were more abundant during the dry season, while Ensifera, Heteroptera, Lepidoptera, Diptera, and Coleoptera were more abundant during the wet season. Seasonal change in arthropod composition was very low in the swamp forest located near a stream, where permanent water supply is crucial for the development of the distinct swamp vegetation, with comparatively constant microclimatic conditions all over the year. Seasonal change increased in the primary forest, and shows most significant changes of athropod abundance between dry and wet season in the selectively logged forest which also had the highest heterogeneity in forest structures. Especially small and soft bodied arthropods probably accumulate along a humidity gradient in the dense canopies of *Rinorea* during the dry season, when the forest floor outside the swamp forest is dry. This effect, which is also strongest in the selectively logged forest, leads to a much higher density of canopy dwelling arthropods during the dry season.

Introduction

Samples from all major tropical regions reveal an enormous species richness and diversity of canopy arthropod communities in wet tropical forests. The best way to better understand the mechanisms that generate and maintain such patterns may be a comprehensive overview of a local canopy fauna. In an ongoing project on influence of tree species, management effects, and forest type on canopy arthropods in Central and East African forests, this research focused on Budongo Forest, a seasonal rain forest in Western Uganda. Seasonal change in climatic conditions, especially the availability of water, has not only strong effects on the vegetation (Synnott 1985), but is probably also crucial for dynamics in the assemblage of arthropod communities.

Longterm studies on seasonal fluctuations on abundance in tropical forest arthropods have been carried out on litter arthropods (Levings & Windsor 1985), light captured Homoptera (Wolda 1979, 1992) and other insects (Ricklefs 1975), fogged beetles in Central America (Erwin & Scott 1980) and canopy arthropods in Sulawesi (Stork & Brendell 1990) and Queensland (Frith & Frith 1985; Basset 1991; Kitching & Arthur 1993). Only a few studies concentrate on the seasonal change between distinct seasons, while others are long term studies including, but not focusing on, seasonal change. In this study, the arthropod faunal composition of an understorey tree species, *Rinorea beniensis* Engler, was compared between wet and dry seasons in adjacent plots of primary, selectively logged and swamp forest, and the influence of forest structure, ground water supply, and selective logging on canopy arthropods analyzed.

Methods

Study site

Research was carried out in the Budongo Forest Reserve in Western Uganda around the field station of the Budongo Forest Project (1°45′ N, 31°35′ E, 1200 m a.s.l.; Figure 1). This forest consists of selectively logged compartments next to old primary forest areas and has a well-known history (Eggeling 1947; Plumptre & Reynolds 1994; Plumptre 1996). Dense forest covers an area of 428 km^2, which is divided into compartments for logging purposes. Selective logging started before 1915, increased between 1935 and 1960 when mechanical logging operated in the forest, and was reduced in the early seventies. The dominant forest type in Budongo is *Cynometra* forest, where *Cynometra alexandri* C. H. Wright forms up to 70% of the upper canopy trees, and mixed forest dominated by some *Celtis* species and the mahoganies (*Khaya anthotheca* (Welwitsch) C. DeCandolle and *Entandrophragma* species; Eggeling 1947).

Around the field station an exceptionally large amount of timber (80 m^3 ha^{-1}) has been selectively logged between 1947 and 1952 (Plumptre & Reynolds 1994). This compartment is now with greater gaps in the upper canopy, no distinct canopy storeys, and a dense cover of young shrubs and trees in the understorey (Figure 2). Primary forest occurs in compartment N15, which has been protected as a Nature Reserve since the 1930s. It is a typical mature rain forest, with four distinct canopy storeys. The ironwood *Cynometra alexandri*, and the mahoganies *Khaya anthotheca* and *Entandrophragma utile* (Dawe & Sprague) are the dominant species of the upper canopy between 30 and 40 m (Figure 3). The vegetation of the lowest storey, between 10 and 12 m, is dense, with *Rinorea beniensis* as one of the dominant tree species. Because of the dense canopy strata, the ground is heavily shaded and lacks a shrub layer. The selectively logged forest compartment and the primary forest are separated by a small river. The vegetation on the poorly drained soils next to the river is a distinct swamp forest with the palm *Raphia farinifera* (Gaertner) Hylander as the characteristic plant species. The two-storied canopy is much lower and naturally more open and less homogeneous than in the primary forest on higher elevations (Eggeling 1947). Plant communities and vegetation structure of primary and selectively logged forest outside the swamp are more similiar to each other than is each type to the swamp forest.

The mean annual rainfall in Budongo Forest from 1992 to 1995 was about 1600 mm (A. Plumptre pers. comm.), with a relatively constant precipitation from March to November, and a dry season from late December to February. During this short but strong dry season, about a quarter of tree species lose leaves (Synnott 1985) which leads to a more intensive irradiation of the understorey and ground vegetation, and drying of the soils especially in selectively logged forest compartments.

Data collecting and analyzing

Rinorea beniensis (Violaceae) was selected for the fogging experiments. This is one of the most abundant tree species in the whole forest, and, with a maximum height of 14 m, the dominant tree species of the first canopy layer. Trees were fogged during the wet season in June/July 1995, and during the dry season in January 1997. With the exception of two short rain showers, there was no precipitation for 30–40 days before fogging the trees during the dry season. For both seasons, eight trees of *Rinorea* were fogged in the adjacent compartments of primary, selectively logged and swamp forest (see Figure 1). Different trees were fogged in the same compartments at each season. A group of eight conspecific trees per forest type and season is designated as a ‘collecting unit’. The maximum distance between single fogged trees in the primary and selectively logged forest compartments (swamp forest is situated between both other forest types) is about 4 km. Fogged trees were neither flowering nor fruiting. Canopies of trees selected for fogging isolated from those of other trees as far as possible, i.e., vertical distance to the canopy of the next tree was at least 5 m, horizontally without overlap to other trees.

Before fogging 16 funnel-shaped nylon sheets, 1 m^2 each, were fixed with ropes under the canopy. Trees were fogged for approximately four minutes with a Swingfog SN-50 and a solution of natural pyrethrum with 1% active ingredient, and arthropods were collected for 90 min after fogging. The total height of fogged trees and the height of the lowest branch carrying dense leaves were measured. In the selectively logged forest *Rinorea* were densely covered with leaves even on the lowest branches, while in the primary and swamp forest lowest branches sometimes have sparse leaf cover, and consequentely these branches were neglected in this measuring. The difference between total height of tree and lowest branch with denser leaf cover multiplied by 16 (area of col-

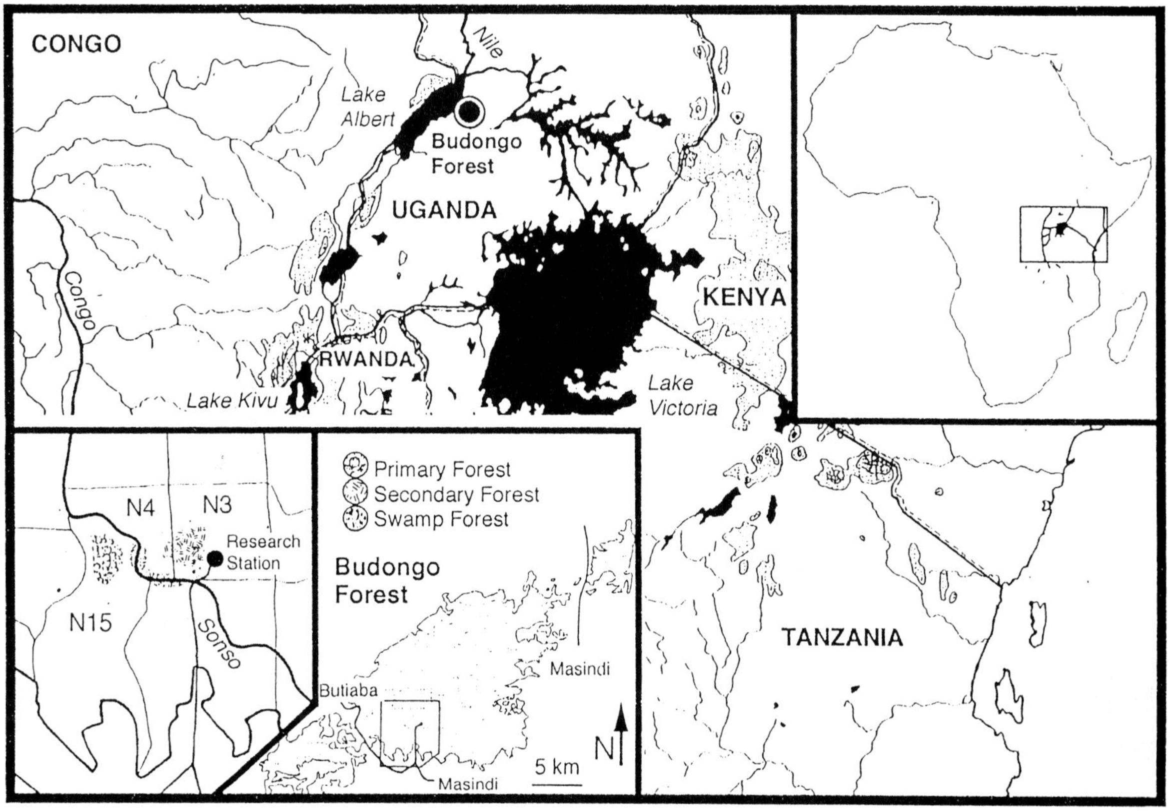

Figure 1. Study site with location of investigated primary, selectively logged (= secondary) and swamp forest plots.

lecting sheets) gives a measurement of the exploited canopy volume. This standardized sampling regime, especially the fogging of many trees of one species only at one locality, allows a clear comparison of the arthropod assemblage between forest types and seasons. Chance effects can be reduced, and the results can be statistically tested.

Material was sorted to groups and numbers of individuals counted. Numbers of individuals of the distinct arthropod groups were transformed to percent data for each tree. Percentages of the collecting units between seasons and forest types were compared using unifactor ANOVA.

Results

On all 48 trees, a total canopy volume of 3728 m^3 was exploited by fogging, and 127 047 arthropods were collected (Table 1). Mean arthropod number for collecting units range between 1822 arthropods per tree, primary forest, wet season, to 3997 arthropods per tree, selectively logged forest, dry season (Table 1, Figure 4). Combining all forest types, the most dominant groups during the wet season were Formicidae (25.1%), Coleoptera (24.0%) and Diptera (11.9%). During the dry season the most dominant groups were again Formicidae (29.9%), and Coleoptera (16.9%), but Auchenorrhyncha were the third most dominant (10.2%; among the Homoptera we collected only few Psyllidae, all others were Auchenorrhyncha). Strongest changes in arthropod composition from wet to dry season were found in the selectively logged forest, with an increase in Auchenorrhyncha from 3.9% to 16.3% ($F = 36.4$, df=15, $p < 0.001$), and a decrease in Coleoptera from 27.6% to 15.3% ($F = 33.7$, df=15, $p < 0.001$; see Figures 5 and 6; Table 2 for statistically significant differences). In the primary forest, such a significant change reaching more than 5% of total arthropod numbers was only found for Formicidae which increased from 23.4% to 32.7% from wet to dry season. These changes mainly occur in taxa with high absolute numbers of individuals, and therefore the above mentioned increase of 9.3% in Formicidae is in accordance with an increase

Table 1. Size, arthropod number and arthropod density (arthropods/m^3 exploited canopy, and arthropods/m^2 collecting sheets) of fogged trees of *Rinorea beniensis*.

Forest type	Primary		Selectively logged		Swamp		Sum/mean
Season	wet	dry	wet	dry	wet	dry	
No. of fogged trees	8	8	8	8	8	8	48
M tree height	9.5–12	8.5–13.5	7.5–10	7–11.5	7.5–11	7.5–12	7.5–13.5
Total m^3 canopy	560	672	768	624	568	536	3728
Total *n* arthropods	14577	22758	19373	31974	19290	19075	127047
Arthropods/tree	1822	2845	2422	3997	2411	2384	2647
	$p < 0.05$, $F = 6.10$		$p < 0.001$, $F = 21.4$		n.s., $F = 0.003$		
Arthropods/m^3	26.0	33.9	25.2	51.2	34.0	35.6	34.3
	n.s., $F = 2.36$		$p < 0.01$, $F = 15.5$		n.s., $F = 0.21$		
Arthropods/m^2	114	178	151	250	151	149	165

Figure 2. Selectively logged forest (compartment N3) along a trail during the dry season with open upper canopy, dense vegetation of seedlings, and layer of fallen leaves on the ground.

of 1.4-times only (see Figures 5 and 6). Despite this strong change, differences are not significant due to the high variation in ant abundance dependent on the distribution of nests. For example, during the dry season percentages of ants on single trees range between 10.4 to 45.9% in the selectively logged and 20.4 to 65.2% in the primary forest.

Relative changes of percentages between seasons were highest in Collembola (Figure 6). Their densities were more than 4-times higher during the dry season in the primary forest ($F = 6.83$, df=15, $p < 0.05$) and selectively logged forest ($F = 5.13$, df=15, $p < 0.05$), while their individual numbers decreased to less than a half in the swamp forest for the same period ($F = 1.10$, df=15, n.s.). Also remarkably high is the greater than fourfold increase of Auchenorrhyncha on selectively logged forest trees during the dry season (see above). In all forest types the increase of Auchenorrhyncha was predominantly caused by nymphs of one cicadellid species. While the increase was insignificant in the primary forest ($F = 2.05$, df=15, n.s.) and only of low significance in the swamp forest ($F = 6.71$, df=15, $p < 0.05$), it was highly significant in the selectively logged forest ($F = 36.4$, df=15, $p < 0.001$). Around 5200 from a total of all Auchenorrhyncha collected during the dry season, and about 4000 individuals from the eight trees from the selectively logged forest only, belong to this single species. This is an unusually high dominance of a single species in a sample from a tropical forest. The significant decrease of Heteroptera from wet to dry season in the selectively logged forest ($F = 16.2$, df=15, $p < 0.01$), and in the swamp forest ($F = 7.67$, df=15, $p < 0.05$) also predominantly depends on

Table 2. ANOVA of percentages of arthropod individuals per tree between seasons (only arthropod groups with significant differences at least in one forest type are shown. df = 15).

Forest type	Primary		Selectively logged		Swamp	
Arthropod group	$p <$	$F =$	$p <$	$F =$	$p <$	$F =$
Acari	**0.05**	**6.54**	n.s.	0.02	n.s.	0.28
Collembola	**0.05**	**6.83**	**0.05**	**5.13**	n.s.	1.01
Ensifera	**0.001**	**153**	**0.001**	**47.9**	**0.01**	**11.9**
Psocoptera	n.s.	0.08	**0.05**	**6.15**	n.s.	0.77
Heteroptera	n.s.	0.12	**0.01**	**16.2**	**0.05**	**7.67**
Auchenorrhyncha	n.s.	2.05	**0.001**	**36.4**	**0.05**	6.71
Coleoptera	**0.05**	**4.65**	**0.001**	**33.7**	n.s.	**0.63**
'Other Hymenoptera'	n.s.	0.27	**0.05**	**7.81**	**0.001**	**33.6**
Diptera	**0.01**	**10.8**	**0.05**	**5.96**	n.s.	0.53

Figure 3. Primary forest (compartment N15) with open understorey and base of an emergent *Khaya anthotheca* tree.

one species only. Fifty three percent of all Heteroptera belong to one species of Miridae, which was found on all fogged trees during the wet season. In the dry season this species again was highly dominant (34% of all Heteroptera), but occurred only on a few trees in higher numbers, while it was absent from most other trees. A very strong decrease in abundance from wet to dry season was also found in Ensifera. They decrease between 2.5-times in the swamp forest ($F = 11.9$, df=15, $p < 0.01$), 3.1-times in the selectively logged forest ($F = 47.0$, df=15, $p < 0.001$) to 3.4-times in the primary forest ($F = 153$, df=15, $p < 0.001$). This pattern is mainly caused by the strong decrease in the number of larvae and nymphs in the dry season.

During the wet season the overall arthropod density was relatively similar in the three forest types investigated, means ranging between 25.2 and 34.0 arthropods per m^3 canopy (Table 1), while during the dry season density ranged between 33.9 and 51.2 arthropods per m^3 canopy. Therefore, arthropod density in general was higher during the dry season, but with significant differences between single forest types. In the selectively logged forest the number of arthropods collected per exploited canopy volume during the dry season was twice as high as during the wet season (+ 203%; $F = 15.5$, df=15, $p < 0.01$), the increase in the primary forest was much lower (+ 30.4%; $F = 2.36$, df=15, n.s.), and density was

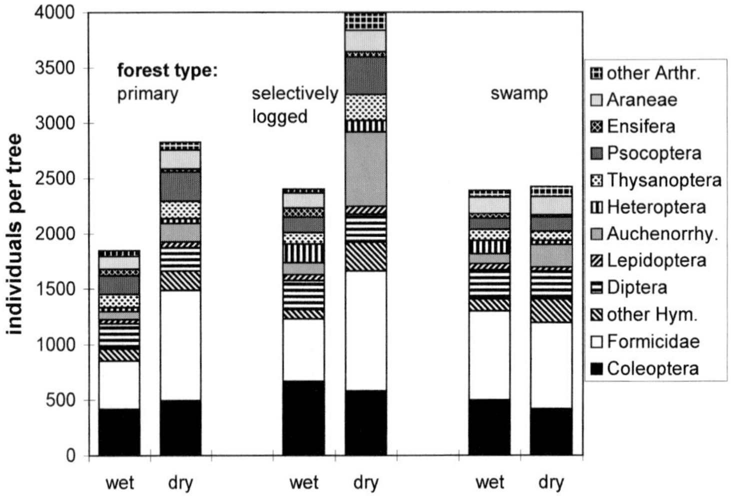

Figure 4. Individual numbers of major arthropod groups, means per collecting unit (each 8 trees per forest type and season).

nearly equal in the swamp forest (+ 4.7%; $F = 0.21$, df=15, n.s.). ANOVA for numbers of arthropod individuals per tree show a significant increase for the dry season in the selectively logged forest ($F = 21.4$, df=15, $p < 0.001$), and primary forest ($F = 6.10$, df=15, $p < 0.05$), and again no difference between seasons in the swamp forest ($F = 0.003$, df=15, n.s.; Table 1).

Discussion

Rainfall and humidity are crucial factors for arthropod abundance in tropical regions as has been shown from the extremely strong fluctuations between seasons in savannas, where long dry seasons lead to a large decrease in insect abundance (Denlinger 1980). Several studies on seasonal fluctuations of arthropods in tropical forests have shown a strong decrease in arthropod density under the influence of seasonal dryness (Janzen & Schoener 1968; Frith & Frith 1985; Kitching & Arthur 1993). At first sight the increase of overall canopy arthropod density from wet to dry season – at least at some sites – in Budongo Forest may be unusual, however in an analysis based on different forest types and arthropod groups the results from Budongo Forest show much concordance with those of other studies.

Influence of forest type

Seasonal changes between the dry and wet season in the arboreal arthropod fauna in general show the same tendencies in the different types of forests investigated, but the extent of the change differs strongly depending on the local forest types. This probably coincides with strong differences affected by the dry season in the different types of forests. Absence of rain for about two months had obviously no negative influence in general on canopy arthropods in the swamp forest. Despite the reduction of water level of the stream during the dry season, soils surrounding the stream were constantly moist. A permanent water supply is one important factor for the development and maintenance of the distinct swamp vegetation, characterized by low leaf abscission during the dry season, and high density of epiphytic mosses and epiphyllic lichens on the trees. This also causes an increase in the structural diversity in plant architecture and consequently the diversity of habitats (see Lawton 1983; Williamson & Lawton 1991), which leads, e.g., to strong differences in the beetle fauna between swamp forest and the other two forest types (Wagner 1998, 1999, 2000). Generally, microclimatic conditions in the swamp forest were comparatively constant all over the year, and seasonal differences in arthropod faunal composition were accordingly low. This result fits well with the fact that populations in an environment with constant climatic conditions like in perhumid lowland rain forests fluctuate less (see Hammond 1990).

The influence of seasonal change is much stronger in the primary and selectively logged forests outside the influence of the stream. Due to the well defined canopy strata in the primary forest, the understorey and ground vegetation is largely shaded even during

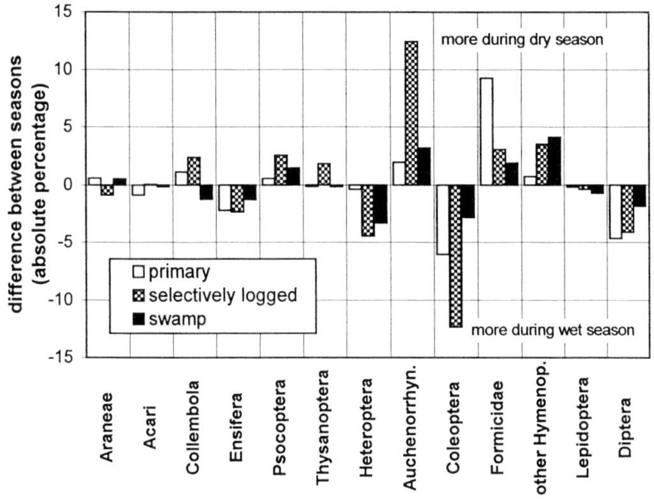

Figure 5. Seasonal change in arthropod composition (only groups with on average more than 1% of individuals are shown). Mean contribution to the community between seasons.

the dry season, keeping surface soil relatively moist. In the selectively logged forest, irradiance is more intensive on the ground especially during the dry season, and the decrease of air and soil moistness on average is higher than in mature forests (Brown & Lugo 1990). On the other hand, the more intensive irradiation in the selectively logged forest understorey leads to higher leaf-density and an increase in the production of young leaves (Brown & Lugo 1990; Aide 1992), and, e.g., to a significant higher abundance of phytophagous Coleoptera and Heteroptera (Wagner 1998, unpubl. data) in comparison to other forest types. In general, primary forests are less productive than selectively logged forests, a fact explained by the comparatively static conditions in primary forests in climax, while successional selectively logged forests show higher structural heterogeneity, higher dynamics, and higher productivity (Brown & Lugo 1990; Chapman & Chapman 1990). *Rinorea* trees in the selectively logged forest show some peculiarities that can be explained by this higher productivity. They are on average of smaller stature, but leaf cover is denser and leaves occur also on the lowest branches.

Studies on several groups of organisms such as birds (Thiollay 1992), rodents (Isabirye-Basuta & Kasenene 1987) and different insect groups (M. Nummelin pers. comm.), show higher fluctuations and differences of the populations in selectively logged forests in comparison to undisturbed forest types. Data from Budongo Forest support this pattern. The composition of the arthropod fauna, and also species composition of the beetle assemblage, is not only most heterogeneous between seasons in the selectively logged forest, but also between single trees within one season (Wagner 1998, 1999). In other words heterogeneity in forest structure correlates with heterogeneity in the tree dwelling arthropod fauna.

Accumulation along a humidity gradient

During the dry season, the canopies of trees with dense foliage are often the most humid habitats, and especially small and soft-bodied insects accumulate presumably along a gradient of humidity; e.g., the flea beetles (Alticinae, Chrysomelidae) significantly increase in the primary and selectively logged forest during the dry season, while the percentage of beetles in general decreases in the dry season. The phytophagous Alticinae are obviously not feeding on *Rinorea* as gut dissection reveal (Wagner 1999), but aggregate in the tree crowns during dry season. This aggregation, sometimes combined with dormancy in comparatively humid sites during the dry season is known from many insect groups (Denlinger 1986), or can be also derived from distribution patterns of understorey arthropods along a gradient of humidity in adjacent forest types (Janzen & Schoener 1968). A group with obviously strong sensitivity to dry conditions are the Collembola (Kitching & Arthur 1993). They reach a high percentage of the arthropod community in soil and in litter of tropical forests, and with exception of few tree crown specialists, they are generally less abundant on trees (Levings & Windsor 1985;

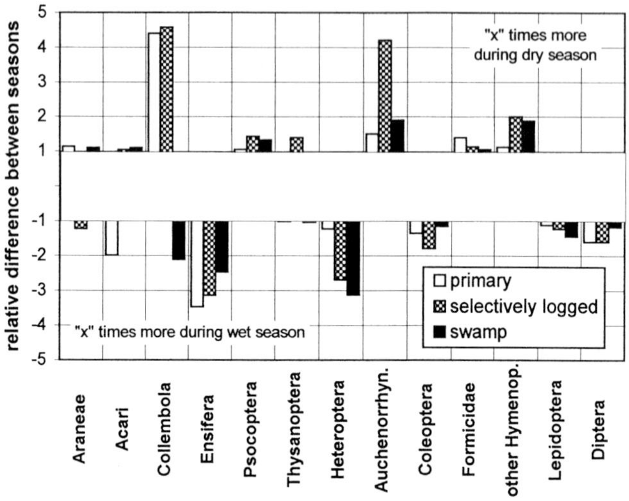

Figure 6. Relative difference in arthropod composition between seasons: decrease or increase of mean percentages.

Stork 1988; Adis 1988), where they additionally show a strong decrease with tree height due to drier conditions in the upper canopy (Kitching & Arthur 1993). The strong increase of Collembola in tree crowns of an understorey tree in the primary and especially in the selectively logged forest during the dry season can be explained by this effect of accumulation in dense crowns of transpiring understorey trees. In the swamp forest this gradient is reversed, because of the permanent high moistness of the soils, and Collembola are less abundant in the comparatively dry canopy.

The decrease of Lepidoptera, Ensifera (Saltatoria), both of which groups occur predominantly as larvae, and Diptera, may be a more general pattern of the dry season, and also coincides with avoidance of dryness in soft bodied insects (Wolda 1978).

Seasonal changes in feeding guilds

Despite the large increase of Alticinae during the dry season, Coleoptera in general decreased. Other phytophagous beetles, especially Eumolpinae and Lagriinae are much rarer during the dry season, which may be explained by insufficient food resources at that time (Wagner 1999, unpubl. data). There is some evidence that the abundance, especially of phytophagous insects, peaked during leaf flush periods (Basset 1991; Aide 1992), because of their preference for young leaves (Price et al. 1980; Raupp 1985; Coley & Aide 1991). In Budongo Forest *Rinorea* trees and also many other tree species, had strong leaf flushes in the late dry season at the end of January, just when the last trees were fogged for this study. It is probable that the number of phytophagous beetles and other insects increased afterwards (see Wolda 1979; Erwin & Scott 1980; Frith & Frith 1985; Kitching & Arthur 1993). Phytophagous suckers are probably less dependent on young foliage which may be confirmed by the extremely high abundance of some Homoptera during the dry season.

The abundance of mycetophagous insects, again predominantly beetles, strongly decreased from wet to dry season. The mould-feeding Latridiidae, averaged about 16%, the most abundant beetle group during the wet season (see also Erwin & Scott 1980), decreased to less than 1% during the dry season when their preferred food is probably rare, due to the pessimal conditions for growing of mould fungus on the trees during the dry season.

Seasonal fluctuations in arthropod communities exist in tropical biomes, but unlike the situation in temperate climates which are strongly influenced by temperature, precipitation is the crucial factor (see Wolda 1978). Consequently, seasonal change is strongest in savannas, and decreases significantly to permanent wet tropical forest. As this study underlines, there can be strong differences even along a transect of only 4 km through different types of forest with different water supply. But arthropod composition in these different forest types shows the same significant changes or at least tendencies which are also concordant to data from comparable studies. Thus, these seasonal changes have many predictable aspects, and reveal a significant deterministic pattern also in a tropical arthropod community.

Acknowledgements

I am very grateful to Andy Plumptre, Jeremy Lindsell and Vernon Reynolds for the invitation and the possibility to work and live in the Sonso Station of Budongo Forest Project, to the Uganda National Council for Science and Technology, and the Uganda Forest Department for acceptance and permission of the field study in Budongo Forest, and to National Geographic Society (NGS), Conservation International (CI – Margot Marsh Fund), Department for International Development (DfID), and Norwegian Agency for Development (NORAD) for supporting The Budongo Forest Project. To Jakisa Deogracius, Ursula Göllner, Michael Schmitt, and Claudia Stuhllemmer for their help collecting the arthropods, to Markus Cousin, Wolfram Freund, Ruth Hasenkamp, Claudia Stuhllemmer and Guido Velten for their valuable help with sorting and counting arthropod material from Budongo Forest, and Patrick Boston, Franz Krapp, Manfred Kraemer and Andy Plumptre for valuable comments. This research was supported by Stifterverband für die Deutsche Wissenschaft and Deutsche Forschungsgemeinschaft.

References

Adis, J. 1988. On the abundancdrivers dre and density of terrestrial arthropods in central Amazonian dryland forests. J. Trop. Ecol. 4: 19–24.

Aide, T. M. 1992. Patterns of leaf development and herbivory in a tropical understory community. Ecology 74: 455–463.

Basset, Y. 1991. The seasonality of arboreal arthropods foraging within an Australian rain forest tree. Ecol. Ent. 16: 265–278.

Brown, S. & Lugo, A. E. 1990. Tropical secondary forests. J. Trop. Ecol. 6: 1–31.

Chapman, C. A. & Chapman, L. J. 1990. Density and growth rates of some tropical dry forest trees: comparisons between successional forest types. Bull. Torrey Bot. Club 117: 226–231.

Coley, P. D. & Aide, T. M. 1991. Comparison of herbivory and plant defenses in temperate and tropical broad-leaved forests. Pp. 25–49. In: Price, P. W., Lewinsohn, T. M., Fernandes, G. W. & Benson, W. W. (eds), Plant-animal interactions. Wiley & Sons, New York.

Denlinger, D. L. 1980. Seasonal and annual variation of insect abundance in the Nairobi National Park, Kenya. Biotropica 12: 100–106.

Denlinger, D. L. 1986. Dormancy in tropical insects. Ann. Rev. Ent. 31: 239–264.

Eggeling, W. J. 1947. Observations on the ecology of the Budongo rain forest, Uganda. J. Ecol. 34: 20–87.

Erwin, T. L. & Scott, J. C. 1980. Seasonal and size patterns, trophic structure, and richness of Coleoptera in the tropical arboreal ecosystem: The fauna of the tree *Luehea seemannii* Triana & Planch in the Canal Zone of Panama. Coleopt. Bull. 34: 305–322.

Frith, C. B. & Frith, D. W. 1985. Seasonality of insect abundance in an Australian upland tropical rain forest. Aust. J. Ecol. 10: 237–248.

Hammond, P. M. 1990. Insect abundance in the Dumoga-Bone National Park, North Sulawesi, with special reference to the beetle fauna of lowland rain forest in the Toraut region. Pp. 197–254. In: Knight, W. J. & Holloway, J. D. (eds), Insects and the rain forest of South East Asia (Wallacea). Roy. Ent. Soc. of London, London.

Isabirye-Basuta, G. & Kasenene, J. M. 1987. Small rodent populations in selectively felled and mature tracts of Kibale Forest, Uganda. Biotropica 19: 260–266.

Janzen, D. H. & Schoener, T. W. 1968. Differences in insect abundance and diversity between wetter and drier sites during a tropical dry season. Ecology 49: 96–110.

Kitching, R. L. & Arthur, M. 1993. The biodiversity of arthropods from Australian rain forest canopies: summary of projects and impact of drought. Selbyana 14: 29–35.

Lawton, J. H. 1983. Plant architecture and the diversity of phytophagous insects. Ann. Rev. Ent. 28: 23–39.

Levings, S. C. & Windsor, D. M. 1985. Litter arthropod populations in a tropical decidous forest: relationship between years and arthropod groups. J. Anim. Ecol. 54: 61–69.

Plumptre, A. J. 1996. Changes following 60 years of selective timber harvesting in the Budongo Forest Reserve, Uganda. Forest. Ecol. Man. 89: 101–113.

Price, P. W., Burton, C. E., Gross, P., McPheron, B. A., Thompson, J. N. & Weis, A. E. 1980. Interactions among three trophic levels: influence of plants on interactions between insect herbivores and natural enemies. Ann. Rev. Ecol. Syst. 11: 41–65.

Raupp, M. J. 1985. Effects of leaf toughness on mandibular wear of the leaf beetle, *Plagiodera versicolora*. Ecol. Ent. 10: 73–79.

Ricklefs, R. E. 1975. Seasonal occurence of night-flying insects on Barro Colorado Island, Panama Canal Zone. J. New York Ent. Soc. 83: 19–32.

Stork, N. E. 1988. Insect diversity: facts, fiction and speculation. Biol. J. Linn. Soc. (London) 35: 321–337.

Stork, N. E. & Brendell, M. J. D. 1990. Variation in the insect fauna of Sulawesi trees with season, altitude and forest type. Pp. 173–190. In: Knight, W. J. & Holloway, J. D. (eds), Insects and the rain forest of South East Asian (Wallacea). Royal Ent. Soc. London, London.

Synnott, T. J. 1985. A checklist of the flora of Budongo Forest Reserve, Uganda, with notes on ecology and phenology. Commonwealth Forestry Institute Occasional Papers, vol. 27. Oxford, UK.

Thiollay, J.-M. 1992. Influence of selective logging on bird species diversity in a Guianan rain forest. Cons. Biol. 6: 47–63.

Wagner, T. 1998. Influence of tree species and forest type on the chrysomelid community in the canopy of an Ugandan tropical forest. Pp. 253–269. In: Biondi, M., Daccordi, M. & D. G. Furth (eds), Proc. IV. Int. Symp. Chrysomelidae; Proc. XX Int. Congr. Ent., Firenze 1996, Torino.

Wagner, T. 1999. Arboreal chrysomelid community structure and faunal overlap between different types of forests in Central Africa. Pp. 247–270. In: Cox, M. L. (ed.), Advances in Chrysomelidae Biology 1. Backhuys Publ., Leiden.

Wagner, T. 2000. Influence of forest type and tree species on the canopy dwelling beetle fauna in Budongo Forest, Uganda. Biotropica, 32: 502–514.

Williamson, M. H. & Lawton, J. H. 1991. Fractal geometry of ecological habitats. Pp. 69–86. In: Bell, S. S., McCoy, E. D.

& Mushinsky, H. R. (eds), Habitat structure – The physical arrangement of objects in space. Chapman & Hall, London.

Wolda, H. 1978. Fluctuations in abundance of tropical insects. Am. Nat. 112: 1017–1045.

Wolda, H. 1979. Abundance and diversity of Homoptera in the canopy of a tropical forest. Ecol. Ent. 4: 181–190.

Wolda, H. 1992. Trends in abundance of tropical forest insects. Oecologia 89: 47–52.

Plant Ecology **153:** 179–191, 2001.

Arthropods in tropical oaks: differences in their spatial distributions within tree crowns

Ulrich Simon* & K. Eduard Linsenmair
*University of Wuerzburg, Biozentrum, Department of Animal Ecology and Tropical Biology, Am Hubland, 97074 Wuerzburg, Germany; (*Present address: Lehrstuhl fuer Landnutzungsplanung und Naturschutz, Faculty of Forest Sciences, TU Muenchen, Am Hochanger 13, 85354 Freising, Germany)*
E-mail: ulrich.simon@lrz.tu-muenchen.de

Key words: Malaysia, *Quercus*, Sabah, Small scale activity, Trap sampling

Abstract

The arthropod activity in the upper and the lower zone of nine emergent oak trees (genus *Quercus*) of a submontane forest in Sabah, Borneo, Malaysia, was investigated by using flight interception traps, yellow colour traps, and arboreal pitfall traps. Number of arthropods did not differ significantly between the upper and the lower region of the tree crowns. Nevertheless, the results reveal significant differences in the abundances of several arthropod groups between the two zones. In the lower part of the oak crowns Homoptera and ants (mostly alates) were sampled in higher numbers, whereas the upper zone was dominated by Diptera and Hymenoptera (ants excluded). Beetles and most other groups showed no significant differences in abundance between the lower and upper zone of the oak trees. Thus, the arthropods of the investigated community are not evenly distributed. The arthropod community in tree crowns, which often has been treated as a homogenous unit, is in fact divided into subunits formed by differently composed arthropod assemblages.

Introduction

The canopy of tropical forests provides habitat for an extremely large number of species. By insecticide fogging of the canopy thousands of arthropod species may be collected from only a few trees (Erwin 1982; Stork 1991; Floren & Linsenmair 1997). However, it is unclear how this huge number of species is able to coexist in tree crowns. Stork et al. (1997, p. 555) mention, that 'in answering these, and related, questions, we need to be aware of the spatial and temporal variation in patterns of arthropod diversity . . . '. On the level of tree-crowns it could be shown that a possible mechanism of coexistence of animal species is spatial separation of species or groups of species. This could be proved in temperate forests (Nielsen 1977: *Fagus sylvatica* (Fagaceae)) and in subtropical trees (Basset 1991: *Argyrodendron actinophyllum* (Sterculiaceae)). In the tropics most studies on canopy arthropods have been carried out by insecticide fogging (Stork et al. 1997). By this method the entire arthropod community of a tree crown is sampled without distinguishing microhabitat preferences. It is true that by a 'restricted canopy fogging' (Basset 1990) one can sample small compartments of the tree crowns and thus describe the spatial distribution within the tree crowns. The advantage of sampling by traps is, however, that it picks out only spot checks of the arthropod community. Repeated trapping allows an assessment of the community composition at a defined site; the impact on community structure can be assumed to be low. To date, such an approach, i.e., the study of distribution patterns of arthropods within tree crowns using traps in different parts of single trees has never been applied in a tropical forest.

A trapping set-up as described below allowed us to investigate the question whether there are within-crown differences in the arthropod community composition in tropical oaks as a first step towards a better understanding of the coexistence of canopy-dwelling species. This study is part of an investigation on the distribution patterns and mechanisms of arthropod species coexistence in oak tree crowns of the genus *Quercus* in both temperate and tropical habitats. In this paper first results of the tropical part of the study are presented on an ordinal classification level. Results of the species composition of different arthropod groups will be published elsewhere.

Area of investigation, studied trees, and methods

Study site

The investigation was carried out at Sayap Ranger Station (6°10′ N, 116°35′ E), located in the north-western area of Mount Kinabalu National Park in the Federal State of Sabah, Republic of Malaysia (Figure 1). The altitude of the study site is ca. 1000 m a.s.l. The Mount Kinabalu National Park is situated in the north-east of the island of Borneo around Mount Kinabalu, the highest peak of South-East Asia (4101 m a.s.l.), covering an area of approximately 75 370 ha (Ismail & Din 1995). The forest is described by Kitayama (1992) as a lower montane forest. Rahman et al. (1995) call it a hill dipterocarp forest. Climatic data for the Sayap area are not available. But own observations and interviews of farmers in the village of Kampung Sayap indicate that there is a rainy season from November to February and a drier period from March/April to October. During the transition periods clouds and fog occur daily with only little rainfall.

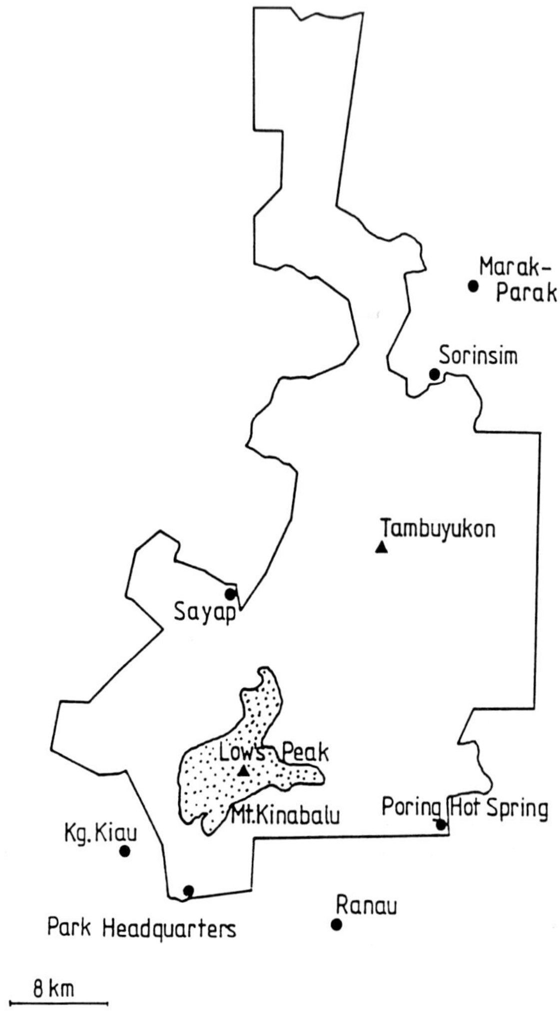

Figure 1. Map of Mount Kinabalu National Park. Sayap is located at the western boundary of the park.

Studied trees

Nine trees of the genus *Quercus* were investigated. All of them were located in the montane forest of the area. All investigated trees were about 45 m tall, emergent trees (see Table 1). The first ramifications occurred within the upper canopy layer of the forest, which was between 17 to 28 m above ground.

All investigated trees shared some special characters like leaves with a sunken midrib and 5-10 secondary nerves, margin of the leaves minutely toothed in the upper half, and acorns of the same shape. After Cockburn (1972) these traits are typical for *Quercus subcericea* A. Camus (1933), a species occurring in lowland and lower montane forest, from 300 to 2600 m a.s.l. But this species is supposed to reach heights of only about 20 m. Thus, to date unequivocal identification of the trees on the species level is impossible due to the lack of inflorescence. However, voucher specimens of sterile branches are deposited at the collection of one of us (U.S.).

Types of traps used, and their distribution within the crowns of the studied trees

Three different types of traps were used in this study.

1. Flight-interception traps (Basset et al. 1997): They were made of gauze with funnels of cotton attached at the bottom and at the upper side. The gauze was one meter long and 50 cm wide. Two gauze pieces were crossed rectangularly. On top a case was attached whereas at the bottom a jar was mounted. This trap type combines the advantages

Table 1. Height of studied trees and diameter at breast height (DBH) (in m), and location of the traps above ground (m) (FIT = flight interception trap; APT = arboreal pitfall trap; YCT = yellow colour trap; FIT-1, APT-1, and YCT-1 = uppermost trap, FIT-2, APT-2, and YCT-2 = lowermost trap). 2 FIT's, 2 YCT's and 2 APT's were installed per tree.

Tree	Tree height	DBH	FIT-1	FIT-2	YCT-1	YCT-2	APT-1	APT-2
B-1	44,0	2.1	35,0	27,0	36,5	30,0	37,0	29,0
B-2	38,0	1.4	27,0	22,5	27,5	22,0	27,5	21,5
B-3	35,0	1.5	26,5	16,5	26,5	17,5	27,0	17,0
B-4	35,0	1.3	23,5	17,0	24,5	19,0	24,5	19,5
B-5	36,0	1.8	28,5	21,0	28,0	22,0	29,0	20,0
B-6	34,0	1.5	28,0	20,5	29,0	22,0	29,0	22,0
B-7	36,0	2.0	27,5	19,5	29,5	21,5	30,0	21,0
B-8	45,0	2.2	37,0	26,5	39,5	28,0	38,5	27,5
B-9	35,0	1.4	24,0	14,5	26,5	16,0	25,0	17,0

of a window-trap with those of a Malaise-trap. It mainly samples flying insects, but also catches ballooning spiders or any other arthropods moving around in its sampling zone.

2. Yellow colour traps: Yellow plastic buckets with a volume of about one liter were tied with a rope to a branch hanging in the air space of the canopy. The aperture width was 12 cm. They work as hyperattractive flowers, and thus preferably sample flying arthropods looking for yellow flowers, but also other arthropods being active in their catchment area.
3. Arboreal pitfall traps: These were modified traps after Weiss (1995) which he used in their original version at spruce trunks in a temperate forest in Germany. They consisted of plastic half-pipes over which a small plastic bag was draped. The open side of the half-pipe was attached to the surface of the bark and fixed by an elastic band. This way, a tight fit of the plastic bag to the surface of the bark and thus a smooth transition of bark-dwelling arthropods into the trap was ensured. The opening width of the half-pipe and thus the length of the sampling edge was about 12 cm.

All sampling jars of the traps were filled with a killing liquid consisting of an aqueous solution of 1% copper sulphate, with a small amount of detergent. This solution kills arthropods very quickly, and is – although an irritant – non-toxic to adult humans.

The relative location of the traps in the crowns was always similar: one flight-interception trap, one yellow colour trap, and one arboreal pitfall trap in the uppermost reachable parts of the tree crowns, and the same set of traps in the zone of the first ramification of the trees. The absolute height of the traps above ground varied due to the height of the studied oak trees and to the zone of first ramification.

Maintaining the traps, and study period

To reach the high canopy of the oaks we used single rope climbing technique (Perry 1978). The study period lasted from 4 April 1997 to 23 October 1997. Traps were emptied twice a month. Damaged traps were repaired during sampling and eventually replaced.

Determination and classification of the sampled arthropods

The specimens of insects were determined to the order level with the exception of Hymenoptera which were divided into 'Formicidae' and 'other Hymenoptera'. Larval stages of Hemimetabola were assigned to each of the respective orders whereas larvae of Holometabola were, whenever possible, determined to order level, and thus treated as separate groups, since their other habitat requirements differ from those of the adults. Larvae of Coleoptera and Lepidoptera were easy to assign, the group 'holometabolan larvae' presumably consisted mainly of Diptera, but this was not unequivocal.

All other arthropods could be allotted to one of the following taxa: Araneae (spiders), Acari (mites), Pseudoscorpiones, Diplopoda, Chilopoda, and Isopoda.

Data analysis

For all statistical analyses we used the program STATISTICA 5.1 (Statsoft, Inc.).

The differences in the efficiency of the traps was tested for each arthropod group separately using the Friedman-test. To find out which trap type showed a significant positive or negative sampling bias, a Wilcoxon-Wilcox-test for a posteriori-testing was used (see Köhler et al. 1996). As a test for differences between abundances in upper and lower parts of the trees, the Wilcoxon-matched-pairs-test was applied. The advantage of this test is that no knowledge about the (normal)-distribution of the data is required. Comparisons of Wilcoxon-tests with t-tests (asymptotic efficiency) showed that the loss of efficiency is low (the tests have 95.5% of the power of the comparable parametric test) (Weber 1986). Thus, a transformation of the data was not necessary.

Results

General results

In total 39,838 arthropods were sampled. The range of the number of individuals per tree was between 3031 (minimum) and 5795 (maximum) (mean = 4426.4 ± 763.6; Table 2). The numbers of arthropods found in the nine tree crowns were not even distributed (χ^2 = 1053, FG = 8, $p<0.0001$). There was no correlation detectable between the installation height above ground of any trap type and the number of arthropod individuals (Spearman-rank-correlation, $R = -0.022$, $p = 0.87$).

The rank-abundance-curve for the determined groups fitted a negative exponential distribution (Figure 2). This pattern was also true for the individual trees, although some deviations were found. For example, ants (Formicidae) were the most abundant group in the entire collection, but in the crowns of B-3 'other Hymenoptera' were the first-ranked, and in B-4, and B-5 Diptera were the most abundant (Table 3).

The high abundance of larvae of holometabolan insects in the traps was a very peculiar result. Larval stages of different taxa were found in high abundance ('holometabolan larvae', 'Coleopteran larvae' and 'Lepidopteran larvae', Table 3). The group 'holometabolan larvae' which apparently mainly consisted of Dipteran larvae was sixth-ranked in this investigation. Coleopteran larval stages then were the tenth frequent group of arthropods. Together with the Lepidoteran larvae (12th-ranked) about 4800 specimens of holometabolan larvae (12% of total) were sampled.

Table 2. Number of all arthropods sampled over the whole period of investigation (March 1997–October 1997) on nine studied trees.

Tree	Number of arthropods
B-1	4493
B-2	3031
B-3	4487
B-4	5015
B-5	4729
B-6	5795
B-7	4244
B-8	4160
B-9	3884
Total	39838

Thysanoptera revealed high individual numbers on most of the investigated trees (mean = 324; SD = 276) with the exception of B-9 (n = 67), and, in contrast, it was the most abundant group of all arthropod groups on B-6 (n = 914).

Efficiency of the traps

The three trap types differed significantly in their sampling efficiency for all arthropod groups (χ^2-test, $p<0.001$), and also for the most abundant ones (χ^2-test, $p < 0.001$) (see Figure 3).

Ants (mainly alates) were predominantly caught in the flight interception traps (FIT) (Wilcoxon–Wilcox-test, p = 0.01); they were underrepresented in the yellow colour traps (YCT).

Although Diptera were mostly sampled in the YCT, and surprisingly in the arboreal pitfall traps (APT), no significant difference between the samples of the three trap types could be found (Friedman-test, p = 0.056). However, there is a tendency that, compared to other flying-insect groups, the FIT were less satisfactory.

Hymenoptera (excluding ants) and Coleoptera were caught by FIT very well, less by YCT and lowest by APT. The differences were highly significant between the samples of FIT compared to the other trap types in both insect groups (Wilcoxon–Wilcox-test, $p < 0.01$); in Hymenoptera (excluding ants) there were also significant differences between APT and YCT with the latter being more effective (WW-

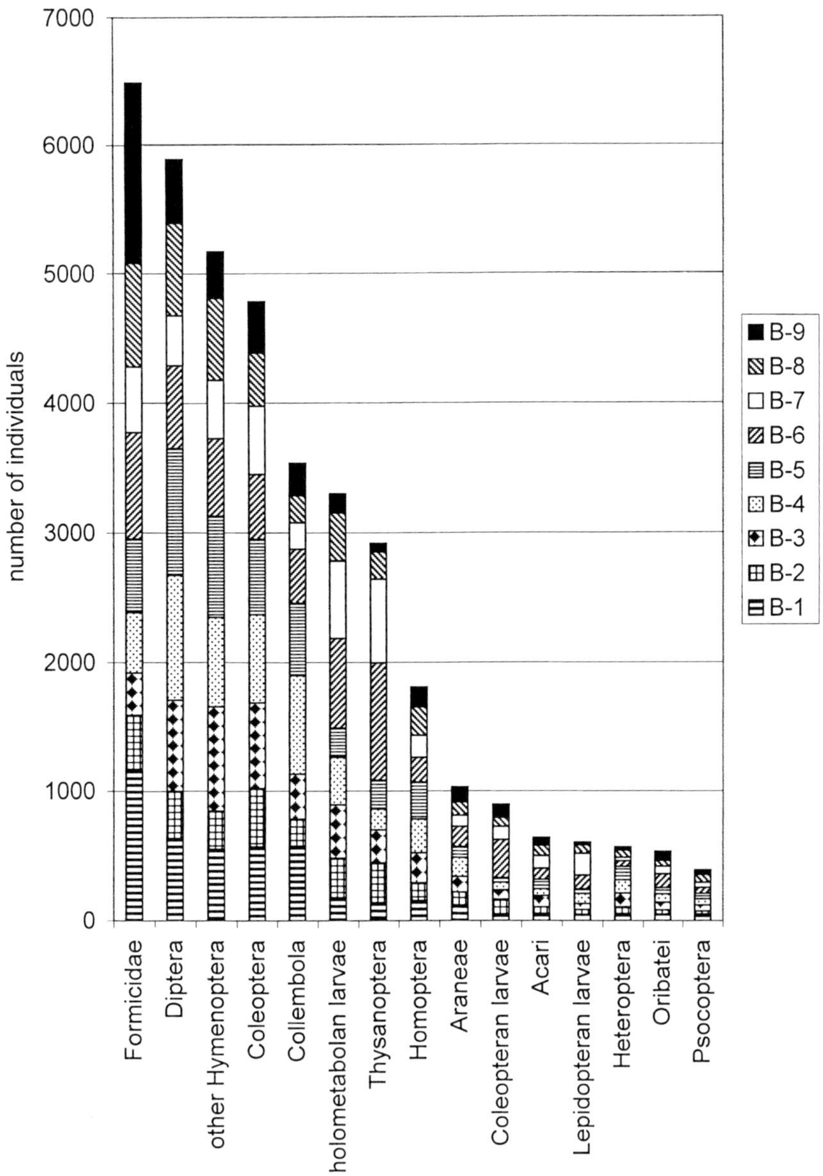

Figure 2. Cumulative numbers of individuals of the most abundant arthropod groups.

test, $p = 0.01$); in Coleoptera no differences between APT and YCT could be found.

The Collembola are, bark-active arthropods and thus occur in high abundance in APT; but surprisingly they were also found in comparably high numbers in the FITs (WW-test, $p > 0.05$). YCTs sampled significantly lower numbers (WW-test, $p = 0.01$).

It was very surprising that the group 'holometabolan larvae' (which mainly consists of Dipteran larvae) was sampled with the highest efficiency in FIT (WW-test, $p = 0.01$). Our only explanation is that litter accumulated in the lower funnel of the traps during each sampling period of 15 days. This litter seems to be attractive to decomposers (of which the group 'holometabolan larvae' presumably consist) and their predators (partly beetle-larvae). All these larvae found in the FIT's were very small, and thus presumably hatched from eggs laid into the leaf-litter material.

For Thysanoptera FITs were the most efficient traps (WW-test, $p = 0.01$). Specimens of these insects were intercepted in high numbers at the gauze,

Table 3. Number of the 15 most abundant arthropod groups sampled in nine oak tree crowns (genus *Quercus*) at Sayap, Kinabalu Park, Sabah, Borneo.

Tree	Formicidae	Diptera	other Hymenoptera	Coleoptera	Collembola	holometabolan larvae	Thysanoptera	Homoptera	Araneae	Coleopteran larvae	Acari	Lepidopteran larvae	Heteroptera	Oribatei	Psocoptera
B-1	1162	635	553	571	575	175	140	154	120	37	53	47	54	48	51
B-2	431	362	292	448	206	308	307	139	101	126	51	37	52	33	22
B-3	328	712	813	670	348	409	252	231	120	75	90	47	107	63	50
B-4	463	965	688	678	769	366	160	258	146	57	51	79	103	63	46
B-5	568	978	782	582	560	232	223	285	83	36	74	35	102	48	42
B-6	824	638	600	503	413	692	914	190	153	291	84	106	43	106	45
B-7	506	387	448	524	205	598	646	174	87	101	98	168	27	59	40
B-8	797	712	634	408	210	370	207	221	101	68	78	62	55	43	60
B-9	1410	501	358	402	255	150	67	158	117	105	59	22	24	70	35
Total	6489	5890	5168	4786	3541	3300	2916	1810	1028	896	638	603	567	533	391

but they were not attracted by the yellow colour of the YCT.

Homoptera are the only group of arthropods which is sampled with highest efficiency in YCT (WW-test, $p = 0.05$). This group seems to be attracted by yellow colour; they were sampled in significantly lower numbers in APTs, which goes along with their more sessile, sap-sucking feeding behaviour.

Differences in the abundance of arthropod groups between upper and lower tree crown zones

The total number of arthropods did not differ significantly between the upper and the lower region of the oaks (Wilcoxon-matched-pairs-test, p=0.95) (Figure 4).

To reveal differences in the activity of arthropod groups between upper and lower crown regions data were first analysed for each trap type separately (Wilcoxon-matched-pairs-test). Each sampling date per tree yielded one pair of data. Given 12 sampling dates on nine trees a maximum of 108 pairwise comparisons were possible. The level of significance was $p = 0.05$; values of $0.1 > p > 0.05$ were considered to show a tendency.

Arboreal pitfall traps

Diptera and Hymenoptera (excluding ants) were the only arthropod groups with significant differences in their abundance between zones (Hymenoptera p=0.005; Diptera p=0.032). Both groups were sampled in lower numbers in lower pitfall traps (Figures 5a and 5b). Beetles and Collembola showed a marginally significant tendency (Coleoptera: p=0.08; Collembola: p=0.0503) to occur in lower numbers in lower pitfall traps. For all other groups no significant differences could be found.

Yellow colour traps

Diptera again were significantly less abundant in lower traps than in upper ones (p=0.02). 'Other Hymenoptera' did not differ in abundance (p=0.66). Ants were found less frequently in upper YCT's than in lower ones (p=0.03). Due to the trapping type most of the sampled ants were winged ones (alates) flying within the tree crowns. Homoptera showed also a significantly lower activity in the upper parts of the tree crowns (p=0.03) (Figure 6). Beetles showed just a very slight tendency towards differences in their occurrence (p=0.08). All other groups did not differ in abundance.

Flight-interception traps

These traps did not conform to the former results in respect to Diptera. No significant difference in abundance was detectable. Ants, on the other hand, were sampled in significantly higher numbers in lower traps than in upper ones (p=0.004). And also 'other Hy-

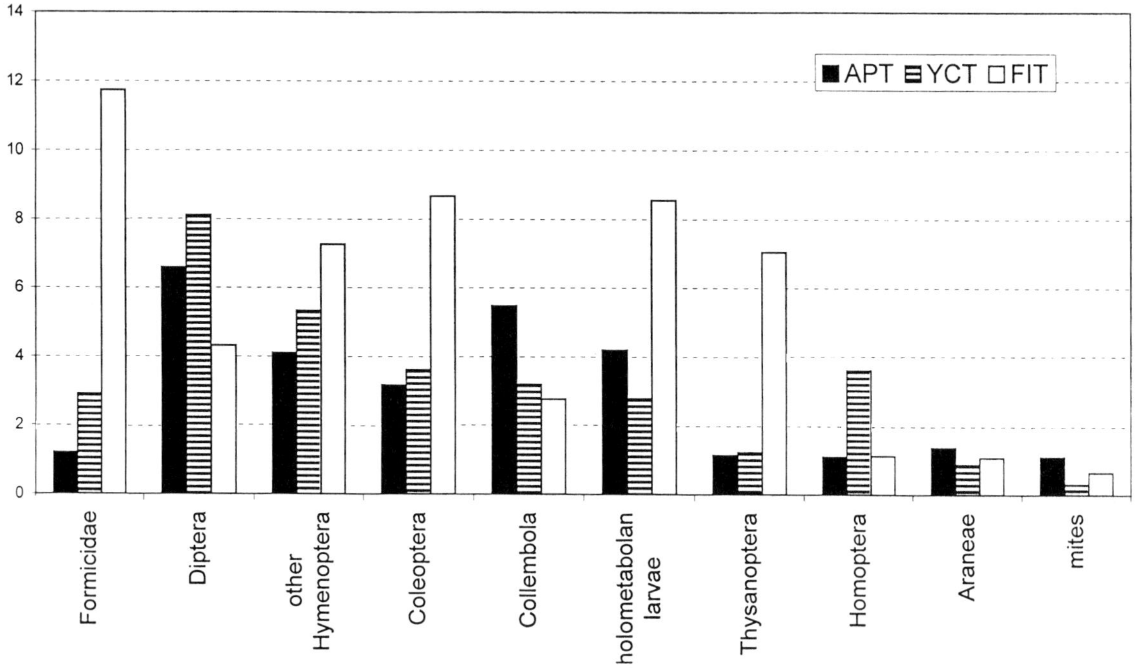

Figure 3. Comparison of the efficiency of the three trap types. APT = arboreal pitfall trap; YCT = yellow colour trap; FIT = flight interception trap. The total number of sampled arthropods equals 100%.

menoptera' were significantly different in numbers between the lower and the upper zone; as with the arboreal pitfall traps higher numbers were detected in the upper traps ($p<0.001$) (Figure 7a). Thysanoptera were sampled with high significance more frequently in upper traps compared to lower traps ($p=0.007$) (Figure 7b). Heteroptera showed a tendency to be more frequent in lower traps ($p=0.064$).

In a second step of analysis the total number of specimens of an arthropod group sampled in one height and in one tree were summed up irrespective of trap type.

Diptera, Psocoptera, 'other Hymenoptera', and Thysanoptera were not equally distributed in the crowns. All these groups occurred in significantly smaller numbers in the lower parts of the tree crowns than in the upper parts (Diptera: $p=0.04$; 'other Hymenoptera': $p=0.03$; Psocoptera: $p=0.04$; Thysanoptera: $p=0.02$). However, ants in this calculation neither were significantly different in their occurrence between upper and lower parts of the crowns nor showed any tendency to do so ($p=0.11$). For all other groups no significant differences in their activity were found.

Discussion

Usefulness of trap types

The traps used in this investigation differed in sampling efficiency. Yellow colour traps and flight interception traps preferentially sample flying animals whereas the arboreal pitfall traps mainly catch bark-active arthropods. Thus, it is necessary to apply not only one trapping method to study the activity of arthropods in tree crowns (see also Basset et al. 1997).

Arboreal pitfall traps are cheap and easy to handle, but are not the best solution for sampling bark-active arthropods. To investigate arthropod activity on the bark, stem-eclectors (Funke 1971; Behre 1989) may be more appropriate, and Floren (1996) sampled a high number of arthropods with this method. On branches, branch eclectors, either those used by Koponen et al. (1996) or the ones described by Simon (1995) are sufficient. Floren (1996) sampled a very low number of arthropods with the latter trap type which might be caused by differences in details of their construction.

Yellow colour traps work very well for Diptera and, new for the tropics, also for Homoptera. On the other hand, they obviously sample few beetles and ants. In summary, YCTs reveal additional information on the

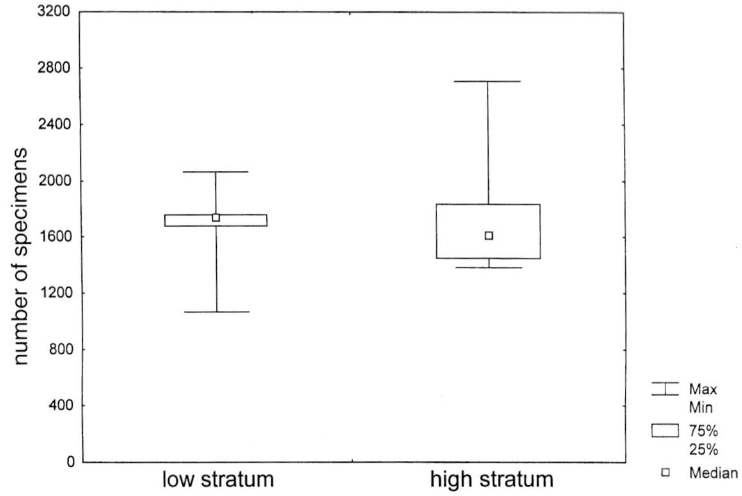

Figure 4. Median and range of arthropod numbers per crown stratum and tree. $N = 9$ trees.

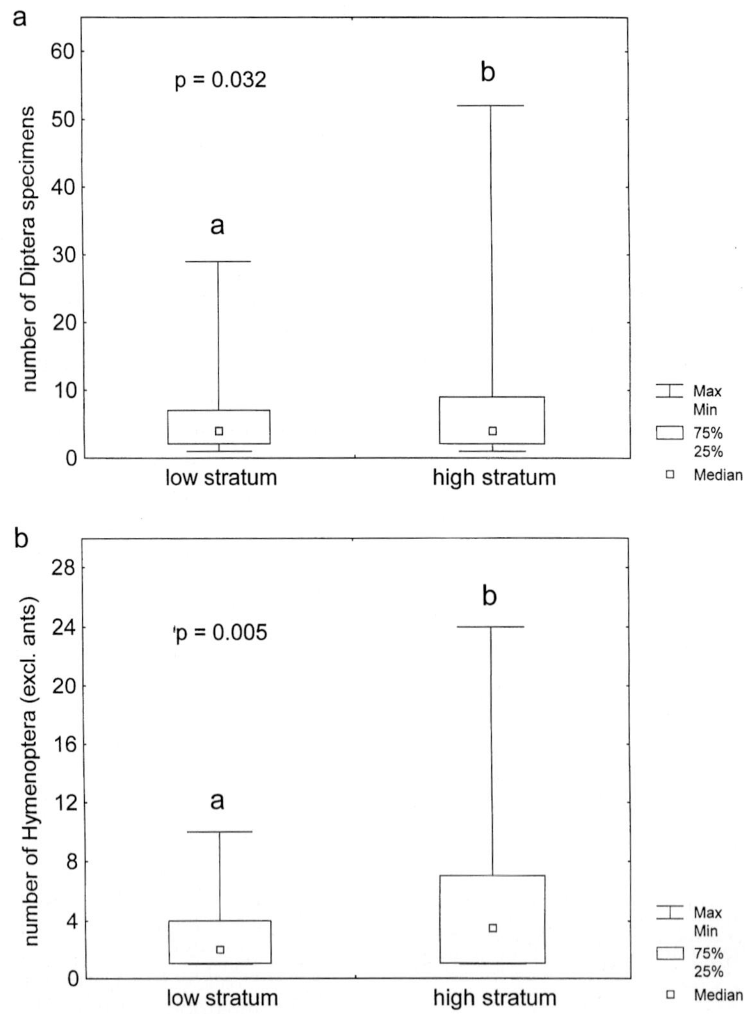

Figure 5. Differences in the numbers of Diptera (5a) and Hymenoptera (excl. ants) (5b) per sampling interval, sampled in upper and lower arboreal pitfall traps ($N = 108$ per height).

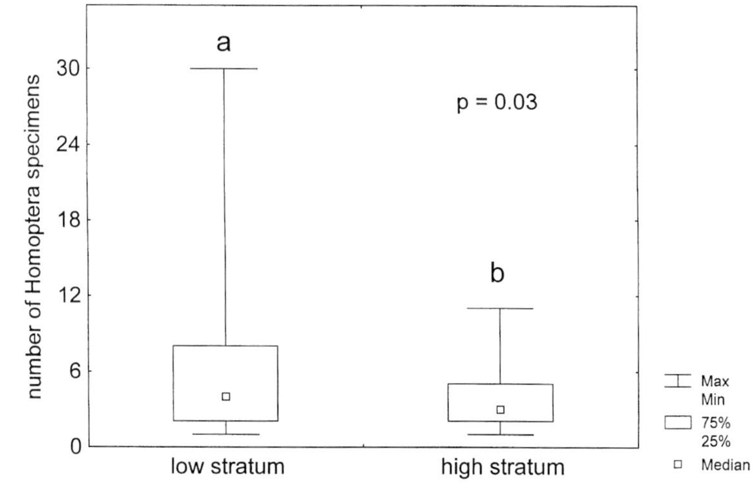

Figure 6. Differences in the numbers of Homoptera per sampling interval, sampled in upper and lower yellow colour traps ($n = 108$ per height).

arthropod fauna in tree crowns, and thus are useful as a complementary trap type in a trapping set-up.

Flight-interception traps were the most effective traps in this investigation. Being non-attractive to specific groups they provide a measure of 'density activity' (see Basset 1991). FITs contributed the most to the total number of sampled arthropods. In this investigation they provided best results for the (flying) ants, the 'other Hymenoptera', and for the beetles. Thus, none of the important arthropod groups was missing; only Heteroptera, which are very abundant in tree crowns of temperate forests (Barsig & Simon 1995; Schubert 1998), were sampled in low numbers; but Heteroptera are, in most cases, not very abundant in tropical canopies (e.g., Floren 1996). Diptera, which are an important insect group in the canopies (see discussion below) were sampled in FITs in lower numbers, compared to YCTs. This might be due to the reaction of these insects after contact with the gauze; e.g., Coleoptera stop moving and, mostly, fall into the sampling jar at the bottom, whereas flies keep on moving after contact, and thus eventually escape (personal observations). But all in all, flight interception traps are a very efficient trap type, and are strongly recommended if an investigation of flight-active arthropods is to be executed.

The trap type introduces a bias towards particular arthropod groups, and arthropod behaviour also influences trapping efficiency. All traps used in this investigation are activity-dependent. If the activity of some arthropods (either single species or larger taxonomic groups) is low they will be sampled in lower proportion compared to their abundance in the habitat. For example, during an investigation of the diel activity of the arthropods in the investigated tree crowns (Simon, unpubl. data), we observed for several days a big grasshopper (Ensifera-Tettigoniidae) sitting in one pleat of the gauze of a flight-interception trap. No such grasshopper was ever found in the samples of this trap. As a conclusion, some groups of arthropods will always be underrepresented in their individual numbers, or will actually not be sampled at all although they live in the investigated tree crowns. Moreover, the effectiveness of the traps is likely to change during the period of investigation. Open traps like the arboreal pitfall traps and yellow colour traps could be attractive to moisture-depending arthropods during dry time-spans; we found a number of minute mosquito-larvae (presumably hatched from eggs laid on the surface of the killing liquid) in one trap showing that water-dependent arthropods eventually choose even low water quality if availability of water is reduced.

Notes on the host-specificity of the arthropod fauna

The determination of the tree species was unfortunately not fully unequivocal. Since host-specificity is high in temperate forests this could be seen as a draw-back of our study. Mawdsley & Stork (1997) showed, at least for tropical canopy beetles, that host-specificity of insects in tropical tree crowns was apparently very low. There was only a significant difference in effective specialisation (May 1990) of about 10% between trees of different families; no signifi-

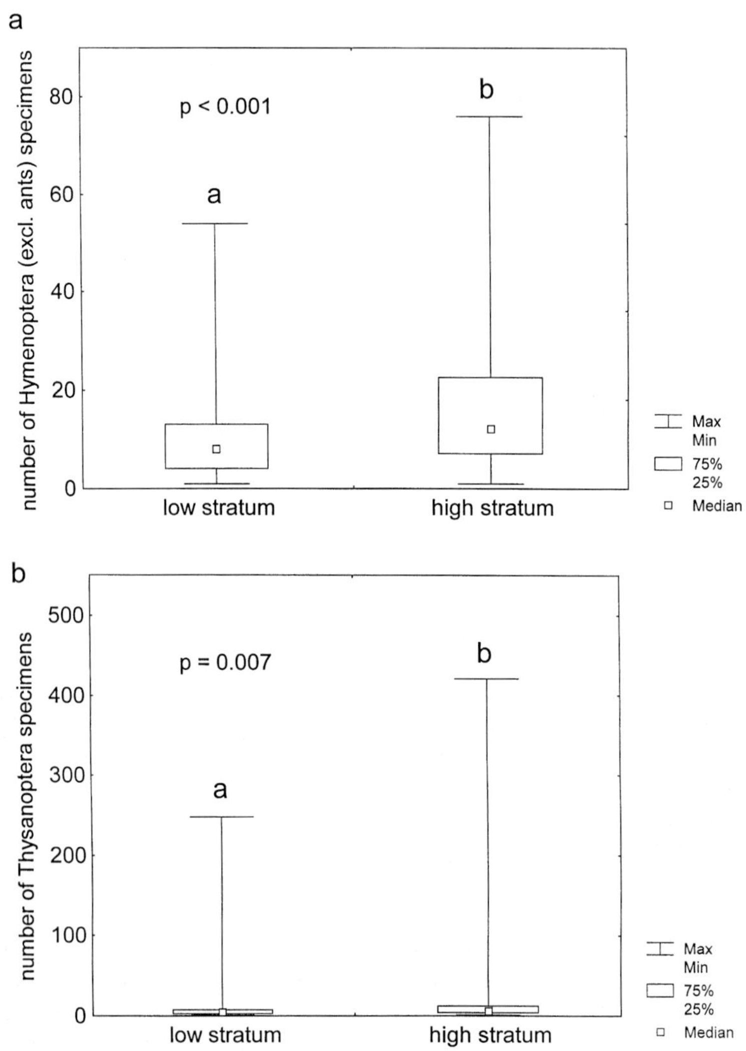

Figure 7. Differences in the number of Hymenoptera (excl. ants) (a) and Thysanoptera (b) per sampling interval, sampled in upper and lower flight interception traps ($N = 108$ per height).

cant differences were found on the genus nor on the species level. Mawdsley & Stork conclude from their results that arboreal arthropod communities may be not so very specific to tree species than compared to temperate trees. Several authors state that there are several factors potentially explaining the host-specificity in tropical tree-dwelling insects (reviewed in Basset 1992b; see also Gauld et al. 1992). Further to this, Mawdsley & Stork 'suggest that plant relatedness may also produce lower levels of host-specificity in tropical as compared with temperate canopy insects'. The differences between two host species which are closely related are said to be very fine in the tropics. According to these authors, host plant diversity is mainly represented at the family level.

However, all these statements must still be doubted: (1) in no study (and in no investigated taxon) species-saturation was observed, (2) in some taxa severe differences in the species composition on neighbour conspecific trees was revealed (Floren & Linsenmair 1997, 1998), (3) the regeneration of communities are not convergent on conspecific tree species over a period of up to three years (Floren & Linsenmair 1997), and (4) the high number of singletons about which more basic knowledge on their natural history is needed to get an idea of the structuring mechanisms of the communities.

For our study we can state that the trees investigated belong to either one species or, at most, to two very near-related species of the genus *Quercus*. The trapping methods used and the level of determination of arthropods are not likely to separate differences in community structures if the trees belong to more than one species.

The composition of the sampled arthropod fauna, and its variability

Arboreal arthropod communities, as revealed in several studies, were found to be not homogenous over several tropical sites. More than this, remarkable differences from site to site were found (see references below).

A remarkable result of this study was the high number of imagines of Diptera. This differs from other studies of our group in Borneo. Floren & Linsenmair (1994) record only a low proportion of Diptera in the communities of tree crowns of trees of the genus *Aporusa* in a lowland forest (sampling technique: insecticide fogging). However, some other authors mention a high percentage of this order in their collections, performed also by canopy fogging (Stork 1991; Stork & Brendell 1990; Kitching et al. 1993). From the results of Stork & Brendell (1990) one could draw the conclusion that the differences may be site-dependent. Moreover, these different results in an insect group, as mobile as flies, may well be a result of the sampling set-up. Basset (1990) also found low numbers of Diptera in crowns of a subtropical tree by using 'restricted canopy fogging' covering branches or twigs with a plastic sheet before the fogging procedure. In the study of Floren & Linsenmair (1994) a cotton roof was stretched out over the top of a fogged tree to avoid sampling of arthropods non-specific for the tree, e.g., from higher strata. Both procedures doubtless disturb many of the highly mobile Diptera, and this may eventually lead to an underestimation of the proportion of flies and mosquitoes (and thus their meaning) in the community of the tree crowns.

The proportion of Collembola in the investigation of Stork & Brendell (1990) was very low (summed under 'other insects'), Basset (1990) did not even count them, whereas Kitching et al. (1993) ranked them as fifth group in their tropical plot but as first group in the subtropical plot. Stork (1991) mentioned Collembola as the ninth ranked group in his study. In our collection Collembola were the fifth ranked group. Collembola on trees surely depend on the occurrence of non-vascular epiphytes like lichens, algae, and bryophytes (Prinzing 1997). Unfortunately, these epiphytes are only seldom recorded or quantified. The high number of Collembola in our study may be due to the high abundance of accumulated humus ('elevated soil', see Nadkarni & Longino 1990; Paoletti et al. 1991). In a study of a student of ours (Unsicker, unpubl. data) the soil-dwelling fauna of lower and higher parts of similar oak trees at the same site was compared. She found between 8 and 20 individuals of Collembola per litre elevated soil which means 8 to 20×10^6 per m^3. There is a large amount of accumulated 'elevated soil' in the big, first ramifications of the studied oak trees. Thus, the number of 3541 sampled Collembola appears to be only a small proportion of the real number of to be expected in this group in the studied trees.

One factor which could have influenced abundance was installation height above ground of the traps. But as mentioned earlier, there was no correlation between the installation height and the arthropod numbers. Maybe tree-intrinsic factors like microclimatic conditions or structural features may in fact influence the arthropod number in tree crowns. But this is unfortunately still speculative here; this should be studied in further investigations.

Spatial distribution of arthropods within the investigated oak tree crowns

The results of this investigation indicate that, on a higher taxonomic level, there are differences in the activity of arthropods within tree crowns (Figures 5–7). Some arthropod groups were found in significantly higher numbers in the lower zone of the tree crowns. The most conspicuous group here were the (mostly winged) ants, which were sampled in significantly higher numbers in lower flight-interception and yellow colour traps. One explanation is that individual-rich nests occur in the lower parts of the trees due to good nesting conditions. In this zone the young queens are supposed to look preferably for nesting sites. The lower zone of the oak trees with its accumulation of humus is presumably an ideal nesting site for some ant species (Nadkarni & Longino 1990; Paoletti et al. 1991). An examination of the sex distribution within the winged ants is one of the necessary next steps of analysis.

On the other hand, Thysanoptera, Diptera, and 'other Hymenoptera' were sampled in higher individual numbers in the upper zone of the tree crowns. For the latter two groups similar results were obtained in a temperate oak forest (Gaspard et al. 1967) but also in a lowland rain forest (Sutton & Hudson 1980). The more open environment in the upper regions of a forest seems to be preferred by both groups. The 'other Hymenoptera' in our study mainly consisted of fig wasps which seem to spread in this upper zone from the trees of their origin to other fig trees. The higher number of Thysanoptera in the upper zone could be related to the occurrence of young leaves. Resources used by the Diptera in the upper zone of the trees are unknown. An analysis of the family composition would give preliminary information of the natural history in regard to the probable proportion of gall-builders or leaf-eaters.

An uneven distribution of arthropods in tree crowns has already been demonstrated several times in other studies for herbivores, either directly through sampling, or indirectly via proving stratum-specific patterns in of herbivory by measuring leaf demage (Nielsen 1977; Lowman 1995). Generally, densities of herbivorous arthropods are predicted to be higher in the upper canopy than in lower since in the upper parts of the crowns it is there where new foliage of high food quality is mainly produced (Coley 1983; Lowman 1985). Foliage attributes influence the distribution pattern of herbivores within in the subtropical tree species *A. actinophyllum* tree crowns, and, in addition, illumination and temperature may also have indirect effects on the overall spatial distribution of arboreal arthropods in *A. actinophyllum* (Basset 1992a). Since new leaves on the studied trees occurred apparently all around the study period the upper parts offer more soft leaf-tissue. This might explain the preference of upper parts by predominantly phytophagous groups as the Thysanoptera in our study.

Basset (1992a) found no significant differences, only trends, in distribution patterns of similar taxa as in our study in lower and upper parts of the crowns; he mentioned that 'arthropod abundance and activity in the outer periphery of the crown may not be extremely different from that in the inner core'. The results of arthropod activity in upper and lower canopy in our study in tropical oaks differ drastically from Basset's results. The differences in the activity of ants, Thysanoptera, Diptera, and 'other Hymenoptera' are highly significant. Reasons for these distinctions may have either been caused by differences in structural features (Southwood et al. 1982), or may be due to climatic differences between the sites studied at Sayap (near the equator) on the one hand and the subtropical site of Basset in Australia on the other.

Acknowledgements

We are grateful to the Director of Sabah Parks, Datuk Ali Lamri, for permission to work in Kinabalu Park, and Mr Francis Liew and Mr Wong for various kinds of support. Ansow Gunsalam of the Botanical Department of Kinabalu Park helped us to find and determine the trees around Sayap Ranger Station. Peninsus Guliong and his family offered great hospitality at Sayap Ranger Station. Many thanks to our Malaysian assistent Soimin Magindol who emptied and maintained the traps over a period of more than four months with great reliability. Furthermore, three of our students, Heinrich Bardorz, Reinhard Bertele, and Sibylle Unsicker, and furthermore Dipl.-Biol. Jan Pfütze helped us to start the project, and with sampling and dismantling the traps. We are also grateful for the annotations of two anonymous reviewers and the editorial comments of Andrew Davis. The fieldwork was supported by travel grants of the Tropical Canopy Programme (1994 – 1998) of the European Science Foundation.

References

Barsig, M. & Simon, U. 1995. Vitalitätsveränderungen von Kiefernnadeln und ihre Auswirkungen auf die Phytophagenfauna. Landschaftsentwicklung und Umweltforschung 98.

Basset, Y. 1990. The arboreal fauna of the rain forest tree *Argyrodendrum actinophyllum* as sampled with restricted canopy fogging: composition of the fauna. Entomologist 109: 173–183.

Basset, Y. 1991. The spatial distribution of herbivory, mines and galls within an Australian rain forest tree. Biotropica 23 (3): 271–281.

Basset, Y. 1992a. Influence of leaf traits on the spatial distribution of arboreal arthropods within an overstorey rain forest tree. Ecol. Entomol. 17: 8–16.

Basset, Y. 1992b. Host-specificity of arboreal and free-living insect herbivores in rain forests. Biol. J. Linn. Soc. 47: 115–133.

Basset, Y., Springate, N. D., Aberlenc, H. P., Delvare, G. 1997. A review of methods for sampling arthropods in tree canopies. in: Stork, N. E., Adis, J. & Didham, R. K. (eds), Canopy arthropods. Chapman & Hall, London.

Behre, G. F. 1989. Freilandökologische Methoden zur Erfassung der Entomofauna (Weiter- und Neuentwicklung von Geräten). Jber. Naturwiss. Ver. Wuppertal 42: Abb. 1–6.

Cockburn, P. F. 1972. Fagaceae. Tree Flora of Malaysia, vol. 1.

Coley, P. D. 1983. Herbivory and defensive characteristics of tree species in a lowland tropical forest. Ecol. Monogr. 53: 209–233.

Erwin, T. L. 1982. Tropical Forests: their richness in Coleoptera and other arthropod species. Colepts. Bull. 36: 74–75.

Floren, A. 1996. Diversität und Wiederbesiedelungsdynamik arborikoler Arthropodengemeinschaften in einem Tieflandregenwald auf Borneo, Sabah, Malaysia. Wissenschaft & Technik Verlag, Berlin, pp. 124. (Also: PhD-thesis University of Wuerzburg.)

Floren, A. & Linsenmair, K. E. 1994. Zur Diversität und Wiederbesiedelungsdynamik von Arthropoden auf drei Baumarten in einem Regenwald in Sabah, Malaysia. Andrias 13: 23–28.

Floren, A. & Linsenmair, K. E. 1997. Diversity and recolonization dynamics of selected arthropod groups on different tree species in a lowland rain forest in Sabah, Malaysia with special reference to Formicidae. Pp. 344–382 in: Stork, N. E., Adis, J. & Didham, R. K. (eds), Canopy arthropods. Chapman & Hall, London.

Floren, A. & Linsenmair, K. E. 1998. Non-equilibrium communities of Coleoptera in trees in a lowland rain forest of Borneo. Ecotropica 4: 55–67.

Funke, W. 1971. Food and energy turnover of leaf-eating insects and their influence on primary production. Ecol. Studies 2: 81–93.

Gaspar, C., Krzelj, S., Verstraeten, C. & Wolf, F. 1967. Insectes récolteés dans des bacs d'eau. Bull. Rech. Agron. Gembloux N. S. III (1): 83–100.

Gauld, I. D., Gaston, K. J. & Janzen, D. H. 1992. Plant allelochemicals, tritrophic interactions and the anomalous diversity of tropical parasitoids: the 'nasty' host hypothesis. Oikos 65: 353–357.

Ismail, G. & Din, L. B. (eds) 1995. A scientifc journey through Borneo, Sayap-Kinabalu Park, Sabah. The Faculty of Resource Science and Technology, Universiti Malaysia Sarawak. Pelanduk Publications, Selangor Darul Ehsan, Malaysia.

Kitayama, K. 1992. An altitudinal transect study of the vegetation on Mount Kinabalu, Borneo. Vegetatio 102: 149–171.

Kitching, R. L., Bergelson, J. M., Lowman, M. D., McIntire, S. & Carruthers, G. 1993. The biodiversity of arthropods from Australian rain forest canopies: general introduction, methods, sites and ordinal results. Austr. J. Ecol. 18: 181–191.

Köhler, W., Schachtel, G. & Voleske, P. 1996. Biostatistik. Springer-Verlag, Berlin.

Koponen, S., Rinne, V. & Clayhills, T. 1996. Arthropods on oak branches in SW Finland, collected by a new trap type. Entomol. Fennica 8: 177–183.

Lowman, M. D. 1985. Temporal and spatial variability in insect grazing of the canopies of five Australian rain forest tree species. Aust. J. Ecol. 10: 7–24.

Lowman, M. D. 1995. Herbivory as a canopy process in rain forest trees. In: Lowman, M. D. & Nadkarni, N. M. (eds), Forest canopies. Academic Press, San Diego.

Mawdsley, N. A. & Stork, N. E. 1997. Host-specificity and the effective specialization of tropical canopy beetles. Pp. 104–130. In: Stork, N. E., Adis, J. & Didham, R. K. (eds), Canopy arthropods. Chapman & Hall, London.

May, R. M. 1990. How many species? Phil. Trans. Royal Soc. London, Series B 330: 293–304.

Nadkarni, N. M. & Longino, J. T. 1990. Macroinvertebrate communities in canopy and forest floor organic matter in a montane forest, Costa Rica. Biotropica 22: 286–289.

Nielsen, B. O. 1977. The distribution pattern of herbivory in a beech canopy. Ecol. Entomol. 2: 293–299.

Paoletti M. G., Taylor, A. J., Stinner, B. R., Stinner, D. H. & Benzing, D. H. 1991. Diversity of soil fauna in the canopy and forest floor of a Venezuelan cloud forest. J. Trop. Ecol. 7: 373–383.

Perry, D. R. 1978. A method of access into the crowns of emergent and canopy trees. Biotropica 10 (2): 155–157.

Prinzing, A. 1997. Spatial and temporal use of microhabitats as a key strategy for the colonization of tree bark by *Entomobrya nivalis* L. (Collembola: Entomobryidae). In: Stork, N. E., Adis, J. & Didham, R. K. (eds), Canopy arthropods. Chapman & Hall London.

Rahman, M. A., Abidin, Z. Z., Nor, B. M. & Abdullah, M. T. 1995: A brief study of bird fauna at Sayap-Kinabalu Park, Sabah. In: Ismail, G. & Din, L. B. (eds), A scientifc journey through Borneo, Sayap-Kinabalu Park, Sabah. The Faculty of Resource Science and Technology, Universiti Malaysia Sarawak. Pelanduk Publications, Selangor Darul Ehsan, Malaysia.

Simon, U. 1995. Untersuchung der Stratozönosen von Spinnen und Weberknechten (Arachn.: Araneae, Opilionida) an der Waldkiefer (Pinus sylvestris L.). PhD-thesis, TU Berlin. W & T–Verlag, Berlin.

Southwood, T. R. E., Moran, V. C. & Kennedy, C. E. J. 1982. The assessment of arboreal insect fauna: comparisons of knockdown sampling and faunal lists. Ecol. Entomol. 7: 331–340.

Stork, N. E. 1991. The composition of the arthropod fauna of Bornean lowland rain forest trees. J. Trop. Ecol. 7: 161–180.

Stork, N. E.; Brendell, M. J. D. 1990. Variation in the insect fauna of Sulawesi trees with season, altitude and forest type. Pp. 173–190; In: Knight, W. J. & Holloway, J. D. (eds), Insects and the rain forest of South East Asia (Wallacea) Roy. Ent. Soc. of London.

Stork, N. E., Adis, J. & Didham, R. K. (eds), 1997: Canopy arthropods. Chapman & Hall, London.

Sutton, S. L. & Hudson, P. J. 1980. The vertical distribution of small flying insects in the lowland rain forest of Zaire. Zool. J. Linn. Soc. 68: 111–123.

Weber, E. 1986. Grundriss der biologischen Statistik. VEB Gustav Fischer Verlag, Jena.

Weiss, I. 1995. Spinnen und Weberknechte auf Baumstämmen im Nationalpark Bayrischer Wald. In: Ruzicka V. (ed.), Proceedings of the 15th European Colloquium of Arachnology Czech Academy of Sciences, Ceske Budejovice, Czech Republic.

Plant Ecology **153:** 193–202, 2001.

Behaviour and ecology of birds in tropical rain forest canopies

H. Winkler & M. Preleuthner
Austrian Academy of Sciences, Konrad Lorenz-Institute for Comparative Ethology, Savoyenstraße 1A, A-1160 Vienna, Austria

Key words: Behavioural flexibility, Ecomorphology, Foraging, Frugivory, Habitat

Abstract

The canopy of rain forests exhibits several distinct features that determine avian adaptations. Light intensity and colour, wind, and temperature are the most important components of the physical environment. Tree architecture poses many challenges for the evolution of avian morphological features, and resource availability varies greatly. The canopy is easily accessible for birds that are equipped to travel over considerable distances. Given these conditions, high behavioural flexibility, and adaptations in behaviour and morphology related to a nomadic life-style, should prevail among canopy birds. The easy access for certain groups of birds also has consequences for the historical processes that shape canopy bird communities. Birds may contribute greatly to forest dynamics and composition, mainly due to their role as seed dispersers. We elaborate on these points with behavioural, morphological, and ecological data collected at our Neotropical study site in southern Venezuela (Surumoni Canopy Crane Project). We discuss how behaviour and morphological features reflect the ecological demands in the canopy. We contrast the specific behavioural and ecological characteristics of canopy birds with those of species found in the other tiers of the rain forest. It is pointed out that much of our knowledge is biased towards the Neotropics, and that better knowledge of the larger canopy species is needed for conservation.

Introduction

The role birds can play in an ecosystem depends on their behavioural, morphological and physical endowment as much as on the characteristics and history of the ecosystems they play their role in. Rain forest canopies form an integral part of the most complex ecological systems on earth (Erwin 1988; Lowman & Moffett 1993). High structural complexity, high species diversity, and pronounced fluctuations in microclimate and resource availability separate them from forest interiors.

The first question regarding the ecological role of birds that needs to be answered is how strongly the canopy affects birds. Among other things, this depends on how intensively birds use the canopy. In this context it should be noted that many forest edge species also inhabit the canopy (Stiles 1980; Cohn-Haft & Sherry 1994). It has been suggested that canopy birds exhibit high inter-individual variance in their diet while at the same time their individual diet spectra narrow (Cohn-Haft & Sherry 1994), and that frugivores and omnivores prevail (Greenberg 1981; Loiselle 1988; Levey & Stiles 1992). Since they are less restricted to certain forest types (Lovejoy 1974), canopy birds should be relatively little affected by disturbances and comprise few threatened species (Cohn-Haft & Sherry 1994; but see Levey & Stiles 1994). Resources in the canopy are patchily distributed in time and space ('boom or bust' resources; Frankie et al. 1974; Greenberg 1981; Loiselle 1988; Levey & Stiles 1994) and therefore one would assume that many canopy birds are well adapted to long distance movements (Karr & James 1975), and one also would infer that local populations show pronounced fluctuations.

The second question is whether and how canopy birds might contribute to rain forest processes. The idea that birds could exert any significant effect on the function of highly productive and complex ecosystems seemed to be rather unrealistic in the days when researchers tried to understand ecosystems mainly on the basis of energy flow and nutrient cycling. However, abundance, energy and nutrient flow alone may say little about the effects an organism may have on the ecosystem it lives in (Strong 1992; Polis & Strong 1996). Canopy birds could influence rain forest dynamics if they constitute important pollinators or seed dispersers, or if they are omnivorous top predators (Stiles 1985; Fagan 1997).

In this paper we intend to elaborate on these notions and related questions by reviewing some of the relevant literature and presenting our own preliminary data. We investigated particularly the birds in a lowland rain forest in southern Venezuela, seeking patterns in their ecomorphology, habitat use and behaviour, and discuss them in relation to the reviewed present knowledge.

Methods

Our study site (3°10′ N; 64°40′ W, elevation 105 m a.s.l.) is located near the mouth of a small black water tributary to the Upper Orinoco (river Surumoni) halfway between the settlements La Esmeralda and Tamatama (Est. Amazonas, Venezuela). At this location, the canopy can be accessed by means of a mobile crane system over an area of approximately 1.5 ha. The Liebherr tower crane 50EC is 42 m tall with a horizontal jib 40 m long and can move on a rail track of 120 m. Two separate personnel baskets with a rated capacity of 200 kg can take off alternately. Operated from inside the gondola by the respective investigator with a remote control they can reach every point within the plot in all dimensions.

The plot represents a primary rain forest and covers a gradient from a forest temporarily flooded by blackwater to *terra firme*, moist evergreen forest. It contains more than 100 tree species, only four of them are deciduous (Anhuf et al. 1999). The dominating species is *Goupia glabra* Aubl. (Celastraceae), a typical pioneer tree usually frequent in secondary forests (Gentry 1993). It forms 30% of the crown cover (Nieder et al., in press). The demographic structure of this species suggests that the population originated from a single event. Other common species are *Dialium guianense* (Aubl.) (Caesalpinaceae), *Ocotea* sp. (Lauraceae) and *Ruizterania trichanthera* (Warm.) (Vochysiaceae). The height of the canopy ranges between 20 and 30 m and there are no emergent trees in the plot. Due to a comparatively low crown density and an indistinctly pronounced stratification, light penetration is high resulting in a relatively dense understorey. Typical representatives of this stratum are species of the families Melastomataceae and the Araceae. The mean annual rainfall amounts to about 2900 mm per year.

Bird observations in the canopy were made from the tower, the jib, and the gondola during the whole day. In addition, we recorded the behaviour of midstory and understorey birds in the same area along trails nearby and at similar sites along the Orinoco in the area between our site and La Esmeralda.

All behavioural data refer to observations made at the first encounter with the bird; to avoid dependent observations no further data were obtained from that individual within an interval of 3 min (Hejl et al. 1990). Based on the suggestion by Remsen & Robinson (1990) we recognised for the present analyses six different foraging techniques or tactical manoeuvres: aerial captures, gleaning (picking up food without extreme body movements or postures), lunge, and the three sub-categories of 'reach': reach down, reach out, and reach up. The latter split seemed justified because the morphological adaptations for these techniques may be quite different (Moermond et al. 1986).

Mist nets were set up from the ground up to the canopy. For the ground and the canopy we usually used four-shelf mist nets of 16 mm black nylon mesh (Denier 110/1 and 210/1), with a length of 6 to 12 m on the ground, and 6 meters in the canopy. Some modified standard 12 m nets with 24 shelves were hung in the subcanopy/understorey. On the ground, a netline of max. 258 m was deployed in transects of 36 to 204 m. Nets were mostly opened 8–9 hours per day (about twice a week depending on weather conditions and availability of the crane).

We selected 13 morphological features which could be readily measured on standard museum skins as well as in the field and for which we had sufficient sample sizes. The data were collected from study skins. They consisted of three measurements related to the flight apparatus (wing length, projection of the primaries from the secondaries, length of the inner tail feathers), three characters on the bill (length from the base to the tip, width at the basis, depth measured at the proximal edge of nasal openings), and seven hind limb characters (tarsus length, width of the tarsus measured parallel and perpendicular to the longitudinal body axis, lengths of the first and third toe, and the lengths of the corresponding claws). They form a representative subset of the characters recommended for ecomorphological studies (Leisler & Winkler 1985, 1991; Leisler et al. 1997).

Since weight data were not available for all species, we used total body length as our measure for size. An analysis of 86 Neotropical species for which we had weight and habitat (canopy, $n = 29$; interior, $n = 34$; both, $n = 23$) data had shown that this character did not vary with stratum (one way ANOVA, $F_{2,83} = 0.154$, $p > 0.85$). So, data were divided by total body length, and log-transformed, and thus represent measures of shape relative to total body length. For multivariate analysis we applied Principal Component Analysis (PCA) based on species means and on correlations, and Canonical Discriminant Analysis. In order to avoid over-interpretation of PCA results, we adopted the following criterion: a component was thought to be relevant if its corresponding eigenvalue was greater than one in 95% of 10 000 bootstrap samples and explained more than 10% of the total variance (Jackson 1993; Gamauf et al. 1998). By applying this rule we avoided discussing 'significant' but biologically uninteresting components. To measure behavioural versatility or diversity we used the Shannon-Wiener diversity index (Pielou 1975) of tactical manoeuvres. To compare groups of species statistically, we took 20 000 bootstrap samples.

About 270 bird species have been observed so far in an area of about 5 ha around the crane. The basis for our analyses are field observations on 182 species. Data were few for 90 species and were complemented with information in the literature (Karr 1971; Greenberg 1981; Gentry 1990; Rosenberg 1990; Stotz et al. 1996). Scientific and vernacular bird names follow Sibley & Monroe (1990).

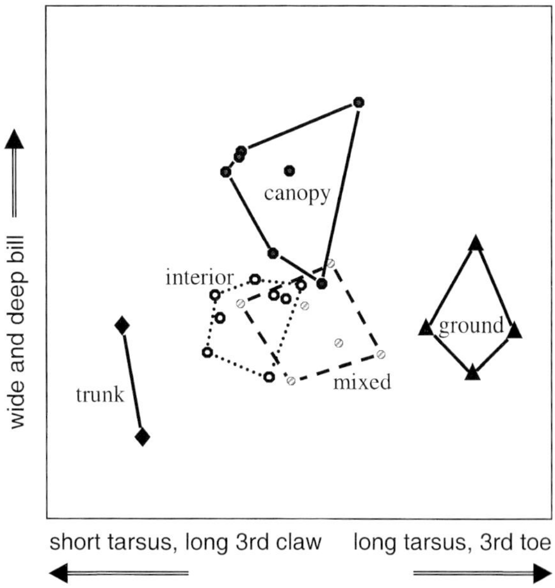

Figure 1. Morphological differentiation of Surumoni bird families according to their stratum use (canopy, canopy and interior, interior, ground, and trunk). In the canonical analysis the first axis correlates mostly with features of the hindlimb, length of the tarsi and the third toe with its claw. The second axis represents width and depth of the bill.

Results and discussion

Characteristics of canopy birds

In early attempts to study avian communities with ecomorphological methods, it has been noted that, in pigeons, canopy feeders are larger and have relatively longer wings than ground foragers (Karr & James 1975). Winkler & Leisler (1994) compared passerine families of the New and Old World, mainly Old World Warblers (Sylviini) and Wood Warblers (Parulini). Ground living species were most distinct, and canopy species showed a tendency towards well developed wings and small size within the families analysed. Strong phylogenetic effects were also noted.

To characterise canopy birds by their morphology, we tried to avoid phylogenetic effects by grouping species within families according to their stratum use (canopy, interior and canopy, interior, ground, and trunk) and taking overall means. Then the data were subject to Canonical Discriminant Analysis. In the ordination, ground and trunk foragers were vastly different from the other groups and formed the poles of the first axis (Figure 1). Features of the hindlimb, length of the tarsi and the third toe with its claw, dominated the first axis, which accounted for 59% of the total variance ($p < 0.01$). Ground foragers had the longest tarsi and third toes (Figure 1). The hindlimbs of trunk foragers were mainly characterised by their relatively short tarsi, only the canopy birds were shorter legged, and by their very long third claws (Figure 1). Birds of the canopy were equivalent to those of the interior and those that use both strata, and were only separated from them on the second axis. According to the analysis, canopy birds differed mainly because of their wide and deep bills. However, the differences were not significant which may be due to the small sample sizes because of the rigorous lumping. These results suggest that morphological differences refer mainly to the way the habitat is used (perching, climbing, hopping and running) and underline the importance of locomotion (see Leisler & Winkler 1991).

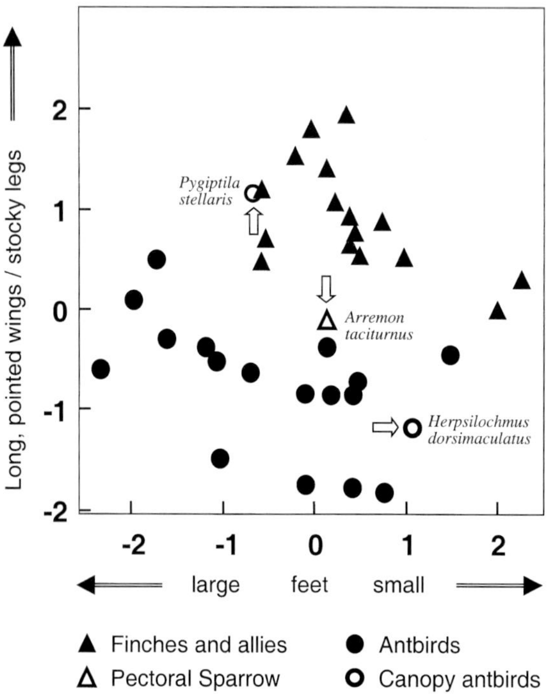

Figure 2. Morphological differentiation of Fringillidae (canopy species) and Thamnophilidae (species of the interior) based on a Principal Component Analysis of 13 characters. The first component describes mainly hindlimb features (length of feet and lateral width of tarsus). The second component reflects the tendency to longer pointed wings in combination with stocky legs characteristic for the canopy dwelling fringillids. The three species depicted as open symbols differ from their relatives in terms of habitat use.

Because adaptations for canopy life vary widely among lineages, we carried out an analysis restricted to the two families Fringillidae and Thamnophilidae. The former was represented by 17 species, 16 of which are typical canopy species, and the latter comprised 18 species with only two canopy species. The first component explained 41% of the total variation ($p < 0.0005$) and was related mainly to hindlimb features (length of first toe and its claw, length of the third claw, and lateral width of the tarsus), the second component explained 24% of the total variance ($p < 0.0005$) and was characterised by parallel variation in wing length and pointedness. The third component accounted for 10% of the variance ($p < 0.005$) and correlated mainly with the width of the bill. The two groups in the corresponding ordination were well separated basically along the second component, indicating that the canopy dwelling fringillids possess long and pointed wings. The understorey Pectoral Sparrow *Arremon taciturnus* (Hermann), however, was placed among the understorey antbirds, and the canopy dwelling Spot-winged Antshrike *Pygiptila stellaris* (Spix) among the finches (Figure 2). Only the Spot-backed Antwren *Herpsilochmus dorsimaculatus* Pelzeln, a bird that forages in the canopy, did ordinate among its understorey relatives (Figure 2).

Behaviour

At the time when ecologists focused on coexistence of species and on niche sizes, behavioural ecologists regarded behavioural stereotypy, or low number of responses with little variation, as a measure of niche size. They hypothesised that the tropics, with their climatic stability, would allow birds to specialise to a great extent (Klopfer 1967). ‘Behavioural stereotypy’ referred both to perceptual and motor stereotypy. Sherry (1990) extended ‘stereotypy’ to diet spectra and stressed the point that evolutionary generalists are necessarily also dietarily broad-niched, while stereotyped behaviour does not inevitably imply stenophagy. Birds could eat a broad array of foods by using a single foraging technique (Sherry 1990). Thus, behavioural stereotypy means that behaviour varies little with substrate or food. In terms of diet breadth, Cohn-Haft & Sherry (1994) stated that canopy birds, in contrast to birds of the forest interior, show little stereotypy. Their analyses of stomach contents corroborated this view. Sherry (1984) found that within the Neotropical tyrannid flycatchers canopy species tend to be more opportunistic than forest-interior species. If it is generally true that the range of behavioural responses will increase with environmental variability (Komers 1997), canopy birds should exhibit a lower degree of stereotypy.

Pearson (1975) enumerated the number of individual birds observed using one of seven foraging techniques in each of seven vertical strata. We reanalysed these data by calculating diversity indices as measure for stereotypy (low diversity implies high stereotypy). We calculated the greatest diversity of foraging methods for the open area separating understorey and midstory, and in the dense midstory. The ground exhibited the lowest diversity and the values for the canopy attained average values. However, with such data the question of species-specific behavioural diversity can be answered indirectly only.

Rosenberg (1990) studied the foraging behaviour of the birds of Amazonian river islands in Peru and presented data on habitat use and foraging techniques. We analysed the diversities of foraging techniques for those species for which there were more than 15 observations available. The foraging technique diversity of the four canopy species (more than 75% of their habitat use confined to the canopy; see also Pearson 1980) was 0.57, and for the birds of the interior the value was 0.35. This difference was significant ($p < 0.01$).

Finally we analysed our own data on five understorey species, five species which were found in the interior and in the canopy, and seven canopy species. Six foraging techniques were considered and analysed as we did with Rosenberg's (1990) data. The respective values for the foraging technique diversities were 0.35, 0.83, and 1.26 ($p < 0.0001$), showing a clear trend towards less stereotypy in the canopy.

Seasonal fluctuations in bird numbers are pronounced even in tropical avifaunas (Karr 1976a; Greenberg 1981; Loiselle 1988; Karr et al. 1996). So far, data suggest the hypothesis that in the high stratum, assumed to be less buffered against environmental fluctuations, avian community composition changes seasonally more than in the lower segments (Karr 1976a; Greenberg 1981; Loiselle 1988; Levey & Stiles 1992). One reason, among other things, is the rather seasonal production of fruit (Terborgh 1986). Yet, certain terrestrial species may vary considerably, too (e.g., *Geotrygon montana*, Columbidae; Bierregaard 1990; Karr et al. 1990). In the Neotropics at least, altidudinal and intratropical migrants are frequently canopy frugivores (Levey & Stiles 1992).

The high mobility of most canopy birds makes them excellent colonisers. The most successful oceanic and habitat island species among small passerines are often edge and canopy species (Morse 1977; Stouffer & Bierregaard 1995).

In frugivores, the group of birds that exerts the most important effects on ecosystem dynamics (see below), there are some differences between the Old and the New World. Nomadic movements of frugivorous birds seem to be more common in Australia and in Southeast Asia than in the Neotropics, but this notion needs further confirmation (Karr 1976a; Fleming et al. 1987). There may also be differences between specialised frugivores that eat large and nutritious fruits (hornbills, Bucerotidae; fruit pigeons, Columbidae; toucans, Ramphastidae) and the more common generalists like barbets (Megalaimidae, Capitonidae) or tanagers (Thraupidae). Neotropical large to medium-sized canopy frugivores tend to occur solitary or in single species flocks, while smaller species join large canopy flocks that include also many omnivores (Thiollay & Jullien 1998).

On all continents, long-distance migrants tend to winter in drier, more open, disturbed and peripheral habitats, and thus overlap strongly ecologically with tropical canopy and edge species (Leisler 1990; Levey & Stiles 1992). Since they are not restricted by the demands of breeding, they usually occur in a wider range of habitats, often use irregularly available food sources, and may feed more actively on a wider range of food types (Sherry 1984; Thiollay 1984; Leisler 1990).

In some areas, migrants temporarily form a dominant portion of the local canopy avifauna. In the Neotropics, there is a strong north-south and west-east gradient of decreasing impact of migrants. In Central America, migrants may completely dominate the canopy avifauna, and a great deal of the fluctuations in bird numbers is due to the seasonal influx of migrants (Stiles 1980; Greenberg 1981; Levey & Stiles 1994). Because migrants stay there for more than five months, they form an integral part of the avifauna, but show little dietary overlap with residents (Levey & Stiles 1994; Poulin & Lefebvre 1996).

Our study site resembles Amazonian sites where only 0.1% of the birds caught with ground-based mistnets were northern migrants (Stotz et al. 1992). The proportion of long-distance migrant species is low (2.6%, without austral migrants 1.9%; $n = 270$ species), in terms of number of individuals the figure is certainly much lower still.

Ecological role

Canopy birds comprise 25–40% of the total number of species in rain forest avifaunas (Orians 1969; Lovejoy 1974; Karr 1976b). Taking only common species into account, Pearson (1980) considered common species (> 0.12 sightings h^{-1} of observation) only and found in his Peru, Bolivia, and New Guinea sites 42–89% of them in the highest strata, while most of the common birds occurred in the mid-strata in Ecuador, Borneo, and Gabon. African species seem to be broader in vertical foraging heights than Neotropical ones (Karr 1976b; Brosset 1996). These differences should be taken cautiously, however, because forest structure and sampling techniques may have differed widely.

At the Surumoni site, species richness compares well with other Neotropical sites (Karr 1980; Bierregaard 1990; Karr et al. 1990). The canopy is used by 69% of the more common species, and 42% used this habitat exclusively. First analyses of mist netting data confirm this pattern. Hence, there is little support for the notion that the lower strata maintain a more diverse avifauna (Karr 1976a).

Bird biomass is generally low and significantly lower than mammal biomass (Terborgh 1986), and there is a negative correlation between the number of frugivores in either group (Pearson 1980). Biomass of arboreal (vs. terrestrial) avian frugivores at Terborgh's study site (Amazonia, southeastern Peru) amounted to 53% of the total 158 kg km^{-2}. Total frugivore biomass in a Panamian site was about 44 kg km^{-2} and 26 kg km^{-2} in Liberia (Karr 1976b). Generally, the biomass of avian frugivores is concentrated in the canopy (Fleming et al. 1987). Biomass of ground foragers (e.g., Casuariidae, Tinamidae, Psophidae, Cracidae) seems to be generally high (Pearson 1975; Stocker & Irvine 1983; Terborgh 1986; Bierregaard 1990), whereas the biomass of arboreal birds increases with stratum height (Pearson 1975).

Birds do not contribute greatly to the energy flow within forest ecosystems. For example, they consumed only 0.17% of the annual net primary aboveground productivity in a North American hardwoods ecosystem (Holmes & Sturges 1975), and Glowacinski & Weiner (1983) found a peak value of 0.5%. According to Leigh (1999), arboreal birds on, respectively, Manu (Peru) and Barro Colorado Island (Panama) consume fruits at a rate of 21 kg dry weight ha^{-1} $year^{-1}$, and between 25 (Manu) and 32 (BCI) kg dry weight ha^{-1} $year^{-1}$ in insects. Compared with mammals, birds are the most important consumers of arboreal insects, while mammals are the dominant arboreal fruit and leaf consumers and are the most important consumers on the ground in all food categories. Recently, it has been stressed that previous models of trophic interactions are not adequate for describing ecosystem dynamics. The role of an organism within an ecosystem may not at all be reflected in its biomass, consumption or energy flow (Polis & Strong 1996). In complex terrestrial ecosystems, top predators and omnivores may have significant impact (Fagan 1997), and vertebrates and especially birds because of their mobility tend to impose spatially density-dependent depletion of invertebrates (Dolman & Sutherland 1997).

Some feeding categories in Pearson's (1975) study showed clear vertical trends. Taking fruits prevailed in the canopy, ant-following was, of course, mainly observed close to the ground, and gleaning was fairly evenly distributed over the strata. Greenberg (1981) noted that, at his Panamian study site, omnivores were the most common species in the canopy. Frugivores dominated the canopy avifauna with 75% in a Costa Rican lowland forest (Loiselle 1988), and in central Amazonia they comprised only 11% of the understorey (Bierregaard 1990). In these sites and in Malaysian dipterocarp forests (Wong 1986), insectivores dominate the understorey.

Considering foraging guilds, the proportion, in terms of numbers of species, of obligate frugivores in the total avifauna in our study was 15%, 42% are insectivores, and 43% omnivores. As in other sites (see Leigh 1999 for an overview), where frugivores comprise between 0.51 and 1.15 kg (wt) ha^{-1} (34–68% of total avian biomass), these figures may not reflect ecological impact. Among the canopy species, frugivores and omnivores account for 74% (30%, 44%) of species, whereas 68% of the interior species comprise insectivores, and 28% are omnivorous ($\chi^2 = 27.1$, df $= 2$, $p < 0.00001$; 133 species). In the latter group only 4% are obligate frugivores, compared to 30% among the canopy species. Similar results were obtained by Thiollay & Jullien (1998) in French Guiana where 45% of canopy species were insectivores compared to 78% in the lower levels.

Seed dispersers and pollinators exert major influences on ecosystem dynamics. The importance of birds as potent seed dispersers is undisputed (Snow 1976, 1981; Howe & Smallwood 1982; Stiles 1985). However, few canopy trees are bird pollinated (Stiles 1985; Bawa 1990; Kress & Beach 1994), and the high mobility of avian pollinators may be of greater significance for epiphytes only (Nadkarni & Matelson 1989; Sazima et al. 1995). A large proportion of canopy trees in the Neotropics is bird dispersed, but understorey plants depend even more on bird dispersal (Karr 1976a; Stiles 1985). In addition, birds become more important as dispersers with altitude and wetness of the forests (Stiles 1985). At a wet montane site in the Choco, Columbia, 83% of the canopy trees and 94% of the trees and small trees of the understorey were categorised as being bird dispersed (Hilty 1981). Bird dispersed canopy trees form the most important category of trees at our study site, with the *Goupia glabra* (Celastraceae) and *Oenocarpus bacaba* (Arecaceae) being the most abundant species (J. Wesenberg pers. comm.).

The canopy and the forest interior do not only differ in the relative frequency of animal dispersed plants, they also differ in several details of frugivore interactions. Canopy fruits are generally produced in larger crops, are larger in size with larger seeds, are more seasonal in occurrence, and persist for shorter periods on the mother plant (Denslow et al. 1986;

Fleming et al. 1987). Generally, fruit availability is extremely variable in both space and time in tropical forests (Fogden 1972; Terborgh 1986). The biomass of avian frugivores in the tropics is concentrated in the canopy (Terborgh 1986). Several authors have noted that canopy frugivores seem to be larger on average (Karr & James 1975; Loiselle 1988; Levey & Stiles 1994). However, this needs to be confirmed by further comparative analyses. The role of avian frugivores in the canopy is most pronounced in the Neotropics and least in Southeast Asia (Karr 1980; Fleming et al. 1987) where in some areas only 10% of the trees are suited for frugivorous birds (Fogden 1972). However, Africa is poorest in plants of high quality for avian frugivores (Snow 1981).

Conclusions

Vertical segregation of species seems to be more prominent in tropical than in temperate bird communities (Marra & Remsen 1997; Shaw & Flick 2000). About 42% of the 182 species analysed in our study site were restricted to the canopy and one can therefore expect that they are adapted to this habitat segment.

The low success of several studies, including this one, to characterise canopy birds according to their morphological features can be interpreted in several ways. Birds may have evolved different solutions to the same ecological demands in the course of phylogeny, phylogenetic constraints preclude convergence, the proper morphological features have not yet been detected, and ecological factors in the canopy may not be that stringent for convergent features to evolve. We believe that the latter is true, and that the apparently great influence of phylogenetic provenance on the morphology of canopy birds does not reflect phylogenetic inertia. Rather, it demonstrates the richness of the environment that allows many different life styles. We suggest that morphological differences are more related to the characteristic methods of foraging for fruits, insects and special resources like epiphytes (Nadkarni & Matelson 1989), than to structural features of the canopy affecting all species in similar fashion. The forest floor and the tree trunks require more specialisation, and much clearer evidence for convergence can be found there. All these findings confirm the notion that the coupling of morphology and ecology is more likely in resource specialists, and that in community-wide analyses only gross associations between ecology and morphology can be found (Wiens 1989). Why, in the Neotropics at least, canopies are occupied by families largely restricted to this habitat, why understorey and canopy avifaunas are so different in their phylogenetic origin (Greenberg 1981; Traylor & Fitzpatrick 1982; Levey & Stiles 1992, 1994), and why sibling species that are syntopic but occupy different strata seem to be absent (compare with African forests, Brosset 1996), clearly all need further study.

Birds can respond to the high environmental variability of the canopy with flexible behaviour in two ways. Either they stay and change their behaviour opportunistically, or their behaviour is stereotyped, and they appear at a certain location only when conditions are favourable. Our analyses show that, on average, canopy birds have a more diverse behavioural repertoire. Further analyses with long-term data should enable clarification of the relationships between environmental predictability and the mobility and behavioural plasticity of birds. In particular, we predict that canopy birds that show little nomadic behaviour would be the most behaviourally versatile ones.

Birds can depress the abundance of free-living herbivorous insects in temperate forests (Holmes et al. 1979; Otvos 1979). Gradwohl & Greenberg's (1982) exclosure experiments proved that deadleaf foraging antbirds can significantly deplete populations of arthropods particular to this microhabitat. Dial & Roughgarden (1995) showed that *Anolis* lizards can heavily influence canopy insect communities. Future exclosure experiments in the canopy should test the hypothesis that omnivorous birds are able to control insect populations (Strong 1992; Polis & Strong 1996).

With respect to seed dispersal, many questions remain unresolved. Avian frugivore biomass may be concentrated in species rarely considered in studies on frugivory (Terborgh 1986), or in a few large and increasingly endangered species (Levey & Stiles 1994). Whether these observations are relevant also for our present understanding of seed dispersal remains to be shown. The role of birds in seed dispersal may be crucial in fragmented landscapes where movements of non-volant dispersers are severely restricted. The role of canopy birds as seed predators is virtually unknown in quantitative terms.

Even this short survey has shown that canopy avifaunas are as different as the forests they live in. Clearly, the greatest difficulties a general assessment of the role of canopy birds is faced with is the lack of sufficient quantitative data, and the bias of research towards the Neotropics. Even within the Neotropics, data from study sites on rich soils (Central America, Andean slopes) prevail. Only if we are able to obtain data from the vast areas hardly covered by quantitative studies will we be in the position to appreciate the role of those birds that live in rain forest canopies. Canopy birds are thought to be less affected by disturbance because they are better equipped to overcome environmental fluctuations (Cohn-Haft & Sherry 1994). This may be true at the global level, and for those species that also frequent edge habitats. But there is also some evidence that the full number of canopy species is only attained after many years (e.g., Robinson & Terborgh 1997). Levey & Stiles (1994) noted that at La Selva canopy and edge species have decreased most in recent years. There are certainly differences between the larger species which require large areas of undisturbed forest and which are dependent on widely scattered resources (e.g., large parrots, toucans, Levey & Stiles 1994; raptors, Gamauf et al. 1998) and the small omnivorous edge and canopy species (e.g., tanagers). Further research should concentrate on these species, determine their resource use patterns, and investigate the effects of forest fragmentation on their movements.

Acknowledgements

The study was supported by the Fonds zur Förderung der wissenschaftlichen Forschung, project P11563-BIO. A long-term study like this depends on the help and encouragement of a large number of people, more than can all be acknowledged here. Many data were contributed by Pia Grubbbauer, Anita Gamauf, and Erwin Nemeth; they were also untiring in their field assistance.

References

Anhuf, D., Winkler, H., Motzer, T. & Szarzynski, J. 1999. Geographical and ecological settings of the Surumoni-Crane-Project (Upper Orinoco, Estado Amazonas, Venezuela). Anz. Sitzber. Österr. Akad. Wiss. (in press).

Bawa, K. S. 1990. Plant-pollinator interactions in tropical rain forests. Annu. Rev. Ecol. Syst. 21: 399–422.

Bierregaard, R. O. Jr. 1990. Species composition and trophic organization of the understory bird community in a central Amazonian terra firme forest. Pp. 217–236. In: Gentry, A. H. (ed.), Four Neotropical rain forests. Yale University Press, New Haven.

Brosset, A. 1996. Role of the sibling species in the dynamics of the forest-bird communities in M'Passa (northeastern Gabon). Pp. 251–289. In: Cody, M. L., Smallwood, J. A. (eds), Long-term studies of vertebrate communities. Academic Press, San Diego.

Cohn-Haft, M. & Sherry, T. W. 1994. Evolution of avian foraging stereotypies in tropical rain forest habitats. J. Ornithol. 135: 481.

Denslow, J. S., Moermond, T. C. & Levey, D. J. 1986. Spatial components of fruit display in understory trees and shrubs. Pp. 37–44. In: Estrada, A. & Fleming, T. H. (eds), Frugivores and seed dispersal. Dr. W. Junk Publishers, Dordrecht.

Dial, R. & Roughgarden, J. 1995. Lizards, leaves, and arthropods: Interaction web for a tropical rain forest canopy. Ecology 76: 1821–1834.

Dolman, P. M. & Sutherland, W. J. 1997. Spatial patterns of depletion imposed by foraging vertebrates: theory, review and meta-analysis. J. Anim. Ecol. 66: 481–494.

Erwin, T. L. 1988. The tropical forest canopy. The heart of biotic diversity. Pp. 123–129. In: Wilson, E. O. (ed.), Biodiversity. National Academic Press, Washington, D.C.

Fagan, W. F. 1997. Omnivory as a stabilizing feature of natural communities. Am. Nat. 150: 554–567.

Fleming, T. H., Breitwisch, R., Whitesides, G. H. 1987. Patterns of tropical vertebrate frugivore diversity. Annu. Rev. Ecol. Syst. 18: 91–109.

Fogden, M. P. L. 1972. The seasonality and population dynamics of equatorial forest birds in Sarawak. Ibis 114: 307–343.

Frankie, G. W., Baker, H. G. & Opler, P. A. 1974. Comparative phenological studies of trees in tropical wet and dry forest in the lowlands of Costa Rica. J. Ecol. 62: 881–919.

Gamauf, A., Preleuthner, M. & Winkler, H. 1998. Philippine birds of prey: interrelations between habitat, morphology, and behavior. Auk 115: 713–726.

Gentry, A. H. 1990 (ed.). Four Neotropical rain forests. Yale University Press, New Haven.

Gentry, A. H. 1993. A Field Guide to the Woody Plants of Northwest South America. Conservation International, Washington D.C.

Glowacinski, Z. & Weiner, J. 1983. Successional trends in the energetics of forest bird communities. Holarctic Ecol. 6: 305–314.

Gradwohl, J. & Greenberg, R. 1982. The effect of a single species of avian predator on the arthropods of aerial leaf litter. Ecology 63: 581–583.

Greenberg, R. 1981. The abundance and seasonality of forest canopy birds on Barro Colorado Island, Panama. Biotropica 13: 241–251.

Hejl, S. J., Verner, J., & Bell, G. W. 1990. Sequential versus initial observations in studies of avian foraging. Pp. 166–173. In: Morrison, M. L., Ralph, C. J., Verner, J., Jehl, J. R. Jr. (eds), Avian foraging: theory, methodology, and applications. Cooper Ornithological Society, Lawrence, Kansas.

Hilty, S. L. 1981. Flowering and fruiting periodicity in a premontane rain forest in Pacific Columbia. Biotropica 12: 292–306.

Holmes, R. T., Schultz, J. C., Nothnagle, P. 1979. Bird predation on forest insects: an exclosure experiment. Science 206: 462–463.

Holmes, R. T., Sturges, F. W. 1975. Bird community dynamics and energetics in a northern hardwoods ecosystem. J. Anim. Ecol. 44: 175–200.

Howe, H. F. & Smallwood, J. 1982. Ecology of seed dispersal. Annu. Rev. Ecol. Syst. 13: 201–228.

Jackson, D. A. 1993. Stopping rules in principal component analysis: a comparison of heuristical and statistical approaches. Ecology 74: 2204–2214.

Jones, S. E. 1977. Coexistence in mixed species antwren flocks. Oikos 29: 366–375.

Karr, J. R. 1971. Structure of avian communities in selected Panama and Illinois habitats. Ecol. Monogr. 41: 207–233.

Karr, J. R. 1976a. Seasonality, resource availability, and community diversity in tropical bird communities. Am. Nat. 110: 973–994.

Karr, J. R. 1976b. Within- and between-habitat avian diversity in African and Neotropical lowland habitats. Ecol. Monogr. 46: 457–481.

Karr, J. R. 1980. Geographical variation in the avifaunas of tropical forest undergrowth. Auk 97: 283–298.

Karr, J. R. & James, F. C. 1975. Eco-morphological configurations and convergent evolution in species and communities. Pp. 258–291. In: Cody, M. L & Diamond, J. M. (eds), Ecology and evolution of communities. Belknap Press of Harvard University Press, Cambridge, Massachusetts.

Karr, J. R., Robinson, S. K., Blake, J. G. & Bierregaard, R. O. Jr. 1990. Birds of four neotropical forests. Pp. 237–251. In: Gentry, A. H. (ed.), Four neotropical rain forests. Yale University Press, New Haven.

Karr, J. R., Schemske, D. W. & Brokaw, N. V. L. 1996. Temporal variation in the understory bird community of a tropical forest. Pp. 441-453. In: Leigh, E. G. Jr., Rand, A. S. & Windsor, D. M. (eds), The ecology of a tropical forest: seasonal rhythms and long-term changes, 2nd ed. Smithsonian Tropical Research Institute, Balboa, Panama.

Klopfer, P. H. 1967. Behavioral stereotypy in birds. Wilson Bull. 79: 290–300.

Komers, P. E. 1997. Behavioural plasticity in variable environments. Can. J. Zool. 75: 161–169.

Kress, J. W. & Beach, J. H. 1994. Flowering plants reproductive systems. Pp. 161–182. In: McDade, L., Bawa, K. S., Hespenheide, H. A. & Hartshorn, G. S. (eds), La Selva, ecology and natural history of a Neotropical rain forest. University of Chicago Press, Chicago.

Leigh, E. G., Jr. 1999. Tropical forest ecology. Oxford University Press, New York.

Leisler, B. 1990. Selection and use of habitat of wintering migrants. Pp. 156–174. In: Gwinner, E. (ed.), Bird migration. Springer-Verlag, Berlin.

Leisler, B. & Winkler, H. 1985. Ecomorphology. Curr. Ornithol. 2: 155–186.

Leisler, B. & Winkler, H. 1991. Ergebnisse und Konzepte ökomorphologischer Untersuchungen an Vögeln. J. Ornithol. 132: 373–425.

Leisler, B., Winkler, H. & Siebenrock, K.-H. 1997. Ökomorphologische Untersuchungen am Beispiel der Webervögel (Ploceidae) und Eisvögel (Alcedinidae). Ann. Ornithol. 21: 17–43.

Levey, D. J. & Stiles, F. G. 1992. Evolutionary precursors of long-distance migration: resource availability and movement patterns in Neotropical landbirds. Am. Nat. 140: 447–476.

Levey, D. J. & Stiles, F. G. 1994. Birds: Ecology, behavior, and taxonomic affinities. Pp. 217–228. In: McDade, L., Bawa, K. S., Hespenheide, H. A. & Hartshorn, G. S. (eds), La Selva, ecology and natural history of a Neotropical rain forest. University of Chicago Press, Chicago.

Loiselle, B. A. 1988. Bird abundance and seasonality in a Costa Rican lowland forest canopy. Condor 90: 761–772.

Lowman, M. D. & Moffett, M. 1993. The ecology of tropical rain forest canopies. Trends Ecol. Evol. 8: 104–107.

Lovejoy, T. E. 1974. Bird diversity and abundance in Amazon forest communities. Living Bird 13: 127–191.

Marra, P. P. & Remsen, J. V., Jr. 1997. Insights into the maintenance of high species diversity in the Neotropics: habitat selection and foraging behavior in understory birds of tropical and temperate forests. Ornithol. Monogr. 48: 445–483.

Moermond, T. C., Denslow, J., Levey, D. J. & Santana, C. E. 1986. The influence of morphology on fruit choice in neotropical birds. Pp. 137-146. In: Estrada, A., Fleming, T. H. (eds), Frugivores and seed dispersal. Dr. W. Junk Publishers, Dordrecht.

Morse, D. H. 1977. The occupation of small islands by passerine birds. Condor 79: 399–412.

Nadkarni, N. M & Matelson, T. J. 1989. Bird use of epiphyte resources in Neotropical trees. Condor 91: 891–907.

Nieder, J., Engwald, S., Klawun, M., & Barthlott, W. Spatial Distribution of Vascular Epiphytes in a Lowland Amazonian Rain Forest (Surumoni Crane Plot) in Southern Venezuela. Plant Ecol. (in press).

Orians, G. H. 1969. The number of bird species in some tropical forests. Ecology 50: 783–801.

Otvos, I. S. 1979. The effects of insectivorous bird activities in forest ecosystems: an evaluation. Pp. 341–374. In: Dickson, J. D., Connor, R. N., Fleet, R. R., Jackson, J. A., Kroll, J. C. (eds), The role of insectivorous birds in forest ecosystems. Academic Press, New York.

Pearson, D. L. 1975. The relation of foliage complexity to ecological diversity of three Amazonian bird communities. Condor 77: 453–466.

Pearson, D. L. 1980. Patterns of foraging ecology for common and rarer bird species in tropical lowland forest communities. Acta XVII Congr. Int. Ornithol.: 974–978.

Pielou, E. C. 1975. Ecological diversity. John Wiley & Sons, New York.

Polis, G. A. & Strong, D. R. 1996. Food web complexity and community dynamics. Am. Nat. 147: 813–846.

Poulin, B. & Lefebvre, G. 1996. Dietary relationships of migrant and resident birds from a humid forest in Central Panama. Auk 113: 277–287.

Remsen, J. V. & Robinson, S. K. 1990. A classification scheme for foraging behavior of birds in terrestrial habitats. Pp. 144–160. In: Morrison, M. L., Ralph, C. J., Verner, J. & Jehl, J. R. Jr (eds), Avian foraging: theory methodology, and applications. Cooper Ornithological Society, Lawrence, Kansas.

Robinson, S. K. & Terborgh, J. 1997. Bird community dynamics along primary successional gradients of an Amazonian whitewater river. Ornithol. Monogr. 48: 641–672.

Rosenberg, G. H. 1990. Habitat specialization and foraging behavior by birds of Amazonian river islands in northeastern Peru. Condor 92: 427–443.

Sazima, I., Buzato, S., & Sazima, M. 1995. The Saw-billed Hermit *Ramphodon naevius* and its flowers in Southeastern Brazil. J. Ornithol. 136: 195–206.

Shaw, D. C. & Flick, C. 2000. Are resident songbirds stratified within the canopy of a coniferous old-growth forest? Selbyana 20.

Sherry, T. W. 1984. Comparative dietary ecology of sympatric, insectivorous Neotropical flycatchers (Tyrannidae). Ecol. Monogr. 54: 313–338.

Sherry, T. W. 1990. When are birds dietarily specialized? Distinguishing ecological from evolutionary approaches. Std. Avian Biol. 13: 337–352.

Sibley, C. G & Monroe, B. L 1990. Distribution and Taxonomy of Birds of the World. Yale University Press, New Haven.

Snow, D. W. 1976. The web of interactions. Collins, London.

Snow, D. W. 1981. Tropical frugivorous birds and their food plants: a world survey. Biotropica 13: 1–14.

Stiles, F. G. 1980. Evolutionary implications of habitat relations between permanent and winter resident landbirds in Costa Rica. Pp. 421–435. In: Keast, A., Morton, E. S. (eds), Migrant birds in the neotropics: ecology, behavior, distribution, and conservation. Smithsonian Institution Press, Washington, D.C.

Stiles, F. G. 1985. On the role of birds in the dynamics of Neotropical forests. Pp. 49–59. In: Diamond, A. W. & Lovejoy, T. (eds),

Conservation of tropical forest birds. ICBP Technical Publication no. 4. International Counsil for Bird Preservation, Cambridge.

Stocker, G. & Irvine, A. 1983. Seed dispersal by cassowaries (*Casuarius casuarius*) in north Queensland's rain forests. Biotropica 15: 170–176.

Stotz, D. F., Bierregaard, R. O., Cohn-Haft, M., Petermann, P., Smith, J., Whittaker, A. & Wilson, S. V. 1992. The status of North American migrants in central Amazonian Brazil. Condor 94: 608–621.

Stotz, D. F., Fitzpatrick, J. W., Parker, T. A. & Moskovits, D. K. 1996. Neotropical birds. Ecology and Conservation. University of Chicago Press, Chicago.

Stouffer, P. & Bierregaard, R. O. Jr. 1995. Use of Amazonian forest fragments by understory insectivorous birds. Ecology 76: 2429–2443.

Strong, D. R. 1992. Are trophic cascades all wet? differentiation and donor-control in speciose ecosystems. Ecology 73: 747–754.

Terborgh J. 1986. Community aspects of frugivory in tropical forests. Pp. 371–384. In: Estrada, A. & Fleming, T. H. (eds), Frugivores and seed dispersal. Dr. W. Junk Publishers, Dordrecht.

Thiollay, J.-M. 1984. Essai d'analyse du succès de la chasse chez les oiseaux insectivores arboricoles d'une fôret neotropicale. Gerfaut 74: 99–125.

Thiollay, J.-M. & Jullien, M. 1998. Flocking behaviour of foraging birds in a neotropical rain forest and the antipredator defence hypothesis. Ibis 140: 382–394.

Traylor, M. A. Jr. & Fitzpatrick, J. W. 1982. A survey of the tyrant flycatchers. Living Bird 19: 7–50.

Wiens, J. A. 1989. The ecology of bird communities, vol. 1. Cambridge University Press, Cambridge.

Winkler, H. & Leisler, B. 1994. Ecomorphological patterns of adaptation and convergence in forest birds. J. Ornithol. 135: 472–473.

Wong, M. 1986. Trophic organization of understory birds in a Malaysian dipterocarp forest. Auk 103: 100–116.

Plant Ecology **153:** 203–214, 2001.

Canopy processes: implications for transpiration, interception and splash induced erosion, ultimately for forest management and water resources

Ian R. Calder
Centre for Land Use and Water Resources Research, University of Newcastle, Porter Building, NE1 7RU, UK (e-mail: i.r.calder@ncl.ac.uk)

Key words: Canopy processes, Dropsize modification, Water resources

Abstract

Some recent experimental and theoretical developments in research related to tropical forest canopies are reviewed. Deuterium tracing studies in India, which rely on the collection of samples of transpired water from leaves in the canopy, have established the importance of 'tree size' as an index of whole tree transpiration and the use of leaf area, basal area and sapwood area for scaling transpiration from the tree to plot scale is discussed.

Experimental studies of the interception loss from tropical trees in Indonesia, India and Sri Lanka have established that conventional interception models of the Rutter type, although adequate for use in temperate climates, are entirely inappropriate for use in the tropics. It is now recognised that the failure of this approach is due to neglect of the process of drop size dependent canopy wetting. The use of rainfall simulators and optical disdrometers in the tropics has demonstrated that to achieve the same degree of canopy wetting upwards of ten times the depth of rainfall may be required for high intensity tropical storms (with large drop sizes) as compared with low intensity, frontal rainfall systems, common in temperate climates. Rainfall simulator studies have also demonstrated that the final degree of wetting is also reduced when vegetation is wetted with drops of large size. These studies have also demonstrated that the drop size of secondary drops falling from vegetation is dependent on the vegetation type and is very much greater for large leafed species such as *Tectona grandis* as compared with species such as *Pinus caribaea* with smaller needle formed leaves.

The combination, in the tropics, of large drop sizes in the primary rainfall and large drop sizes in secondary drops falling from vegetation helps to explain why in absolute and relative terms interception loss from tropical trees is less than that in temperate climates. The recognition that drop size generation is related to vegetation type has important implications for splash induced erosion and the choice of tree species on soils subject to erosion.

Knowledge of forest canopy processes is now leading to a better appreciation of the hydrological, meteorological and water resource impacts of forests. The impact of trees in Amazonia and the Sahel on climate and trees in the Zambezi basin on water resources is discussed. New spatially distributed modelling methods which are being incorporated within Decision Support Systems for forest and water resource management, which also take account of ecological and socio-economic objectives, are also outlined.

Introduction

Improved knowledge of canopy processes is allowing better estimation of both the influence of forests on hydrology and the influence of hydrology (and meteorology) on forest function. Examples of current research relating to these interactions are presented in this journal (Frieberg 2001; Szarznski & Anhuf 2001; Chappell et al., 2001). Some of the earlier developments in forest canopy related hydrological research are reviewed here (see also Bruijnzeel 1990, 1996; Calder 1996a, 1999), particularly as they relate to the estimation of transpiration, interception loss and splash induced erosion. The influence of tropical forest vegetation on climate and water resources is also discussed and the development of new management tools for land use and water resources management is outlined.

Canopy measurements of forest transpiration - deuterium tracing

Tracing methods, using heat or a stable isotope as the tracer, are being increasingly used to estimate tree transpiration. These methods have the advantage that the measurement relates to the whole tree and does not involve the scaling issues which need to be resolved when dealing with say diffusion porometer measurements of stomatal conductance on individual leaves or bunches of leaves.

The deuterium tracing method is essentially a dilution gauging technique for measuring flow rate. It is based originally on the use of the 'total counts' method developed for use with radioactive tracers (Hull 1958). The 'total counts' equation is given by:

$$M = F \int_0^\infty f'(t)\,\mathrm{d}t, \tag{1}$$

where, for a radioactive tracer, M is the total activity of the tracer administered, F is the 'measured flow rate' and $f'(t)$ is the specific activity. The integral of $f'(t)$ dt is equivalent to the 'total counts' recorded by the Geiger counter or other radiation detector. The method is equally applicable for non-radioactive tracers. The equation may then be rewritten as:

$$M = F \int_0^\infty C\,\mathrm{d}t, \tag{2}$$

where M is the total mass of tracer administered, F is the 'measured flow rate' and C is the concentration of the tracer. For operational use this equation is usually recast in the finite difference form:

$$F = \frac{M}{\sum_{i=1}^{i=T} C_i \Delta t_i}, \tag{3}$$

where C_i is the concentration in the ith time increment, Δt_i is the duration of the ith time increment and T is the last time increment in which tracer is present.

Kline et al. (1970) recognised that the equation was equally valid for instantaneous or prolonged administration of the tracer. It was further recognised that the only conditions imposed by the method are:

(1) all of the tracer passes through the tree with the tracer stream during the course of the experiment;

(2) complete mixing of the tracer must take place in the trunk of the tree before the flow streams split to separate branches.

However the validity of the method in situations of fluctuating rather than steady flow has at times been questioned. Kline et al. (1970), recognising that transpiration does not give rise to a steady flow condition, state that the total counts equation (Equation (1)) 'yields flow as if it were steady for the entire period of the experiment'. Unfortunately they did not provide any justification for this statement and practitioners of the method were not able to answer satisfactorily questions such as 'how does the 'measured' flow relate to the actual, fluctuating flow?', and 'do the large diurnal fluctuations away from the steady state ideal introduce a large error in the measurement – if, for example, the period of measurement encompasses more nights than days?'.

Calder (1991) showed that answers could be found to these questions using two methods; one from consideration of the principle of the conservation of mass and by following the reasoning of the total counts method, and the second by considering the process as a convective-dispersive system. The reasoning based on conservation of mass is outlined below.

From the principle of conservation of mass the time integral of the product of flow and concentration out of the tracer system must equal the total mass injected, i.e.:

$$M = \int_0^\infty qC\,\mathrm{d}t, \tag{4}$$

where q is the instantaneous flow rate. Writing Equation (4) in finite difference form gives:

$$M = \sum_{l=1}^{i=T} q_i C_i \Delta t_i, \tag{5}$$

where q_i is now the flow rate in the ith time increment. Substituting for M as expressed in Equation (5) in Equation (3) gives:

$$F = \frac{\sum_{i=1}^{i=T} q_i C_i \Delta t_i}{\sum_{i=1}^{i=T} C_i \Delta t_i}, \tag{6}$$

which shows that the 'measured flow rate', F, as determined by the 'total counts' method is identical to

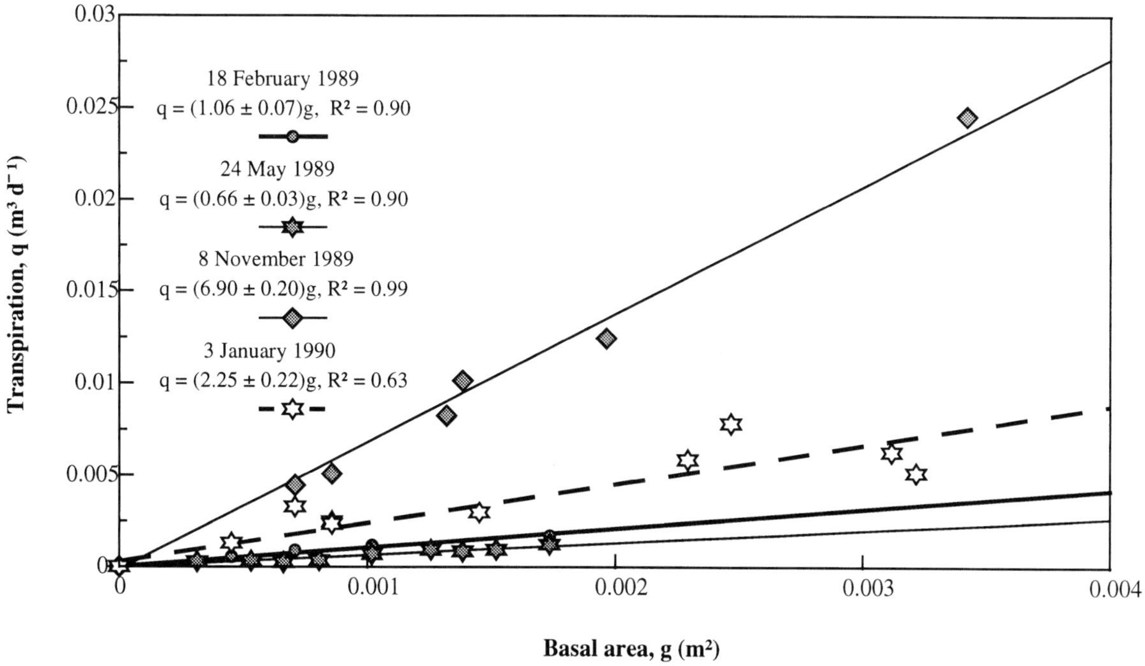

Figure 1. Measured transpiration rates as a function of tree basal area (g) at the Puradal site, Karnataka, southern India, prior to and after the monsoon (July – September) of 1989 (after Calder, 1992b).

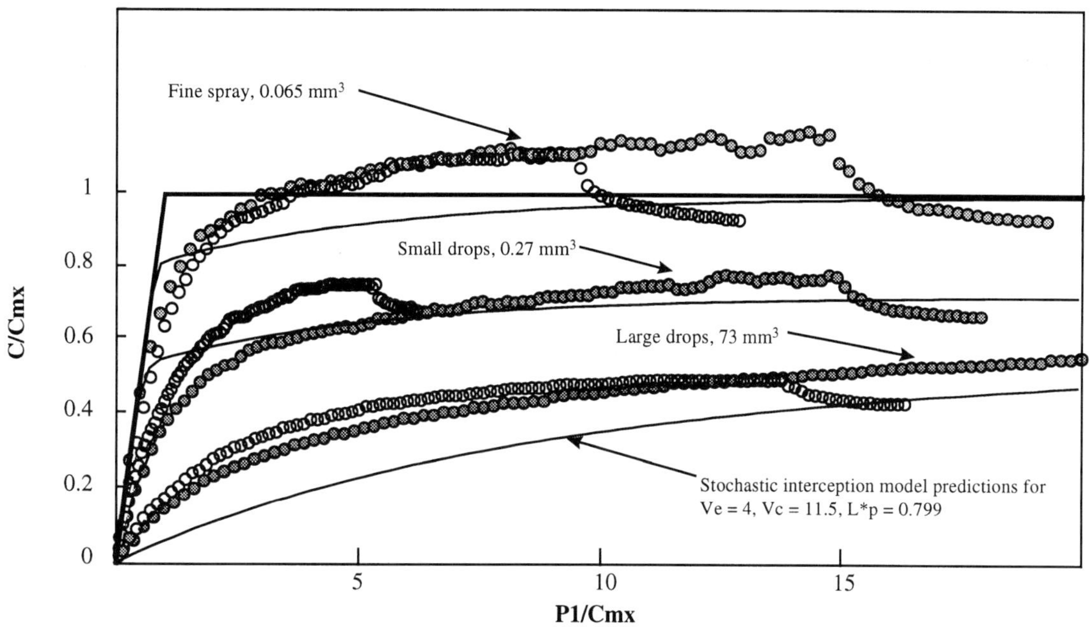

Figure 2. Canopy storage measurements obtained with a rainfall simulator showing that the final degree of canopy saturation is much greater when wetted with drops of smaller size. The storage of water on the sample, C, is shown normalised by the maximum storage obtained with fine drops, C_{mx}, and the depth of simulated rainfall applied, P_1, is also normalised by C_{mx}. The wetting functions predicted by the stochastic interception model for these drop sizes are shown as continuous lines. Also shown, as straight lines, is the wetting function implicit in conventional interception models of Rutter type. (After Calder et al. 1996).

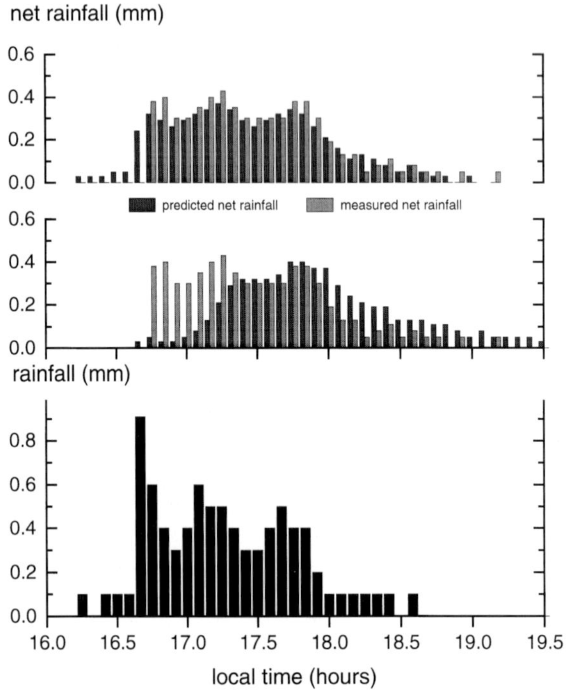

Figure 3. A comparison of the gross (bottom) and net rainfall measured and predicted by the Rutter (middle) and two-layer stochastic models for a storm at a site near Kandy, Sri Lanka on 28 May 1993 (after Hall & Calder 1999).

the weighted mean of the instantaneous flow rates, as denoted by the values of q_i, over the period 1 to T, weighted according to $C_i \Delta t_i$ for each time period. If the sampling time periods, as denoted by Δt_i, are equal then Equation (6) reduces to the simpler form:

$$F = \frac{\sum_{i=1}^{i=T} q_i C_i}{\sum_{i=1}^{i=T} C_i}, \quad (7)$$

which shows that the 'measured flow rate', F, is the weighted mean flow rate over the period 1 to T, where the weighting factor is the concentration C_i.

This reasoning indicates that the methodology is applicable to fluctuating flow and that in fluctuating flow situations it is desirable that the 'spread' of the concentration-time curve is present for a time sufficient to include a number of fluctuations so that uncertainty in the estimate of the mean flow is reduced. The 'width at half height' of the concentration-time curve is of the order of days in tree crops so diurnal fluctuations are effectively 'averaged out' and there is some averaging of the individual daily rates.

Use of the deuterium tracing method in studies of the water use of eucalyptus plantations in India has established the seasonal variability of transpiration, related to soil moisture availability, and perhaps most significantly, for the relatively young plantation trees used in the study (less than 7 years old) a linear, scaling relationship between transpiration rate and basal cross-sectional area (Figure 1).

The choice of scaling relationship for scaling from the tree to plot scale is a subject of current debate amongst forest hydrologists. Leaf area is commonly used as the means for scaling but the studies in India (Calder et al. 1992) showed that higher correlation coefficients were obtained when transpiration rate is regressed against basal area rather than leaf area (although it was recognised that the poorer correlation coefficients may be a reflection of the larger experimental errors associated with the leaf area measurements). Sapwood area is often suggested as the appropriate scalar and this would imply that for young trees, where virtually all the basal area of the tree is composed of sapwood, the transpiration rate would be related to the basal area of the tree, that is transpiration rate is related to the tree diameter to the power of two. For larger trees sapwood area would

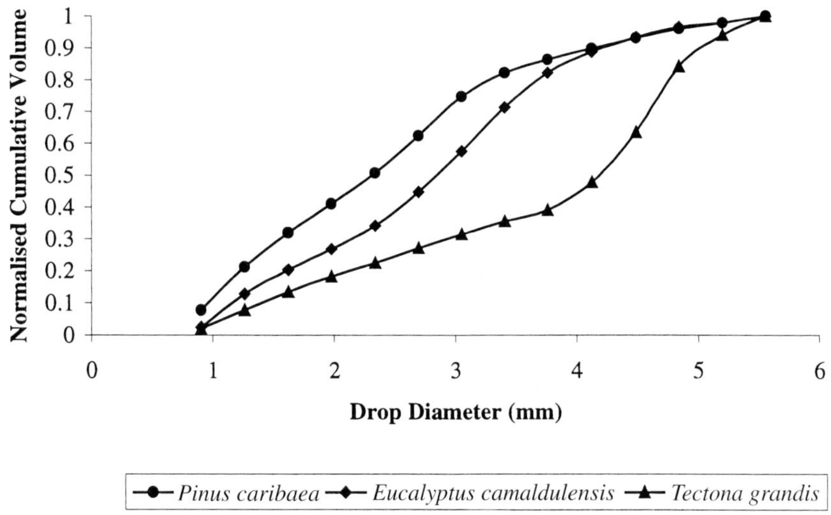

Figure 4. Characteristic net-rainfall drop-size spectra for *Pinus caribaea, Eucalyptus camaldulensis* and *Tectona grandis* (from Hall and Calder 1993).

tend to increase with diameter (power of 1) and we might expect a transition in the transpiration rate with diameter relationship when heartwood is formed. Deuterium and heat pulse tracing studies are suggesting now that the appropriate scalar may be a power of DBH somewhere between 1 and 2 but that the largest trees appear to transpire more than would be expected from consideration of just their sapwood areas.

Interception loss and canopy wetting

The importance of rain drop size in determining canopy wetting and interception losses from forest was first realised from studies carried out in rain forest in Indonesia (Calder et al. 1986). Application of a stochastic interception model, which explicitly took into account drop size, was required to describe the interception process in these conditions.

This model uses the Poisson distribution function to estimate n, the mean number of drops retained per elemental area of the canopy, as a function of m, the mean number of drop strikes per element. It predicts an asymptotic wetting of the canopy, without an explicit drainage function, at rates dependent upon the size of the drops as indicated by the parameter q which is the maximum number of drops which can be retained by a surface element. The basic equation of the model gives the mean number of drops retained per element as:

$$n = q(1 - e^{-m}) + e^{-m} \sum_{i=1}^{r} (i - q)\frac{m^i}{i!}, \tag{8}$$

where r is the truncated value of q. The canopy storage C and rainfall P are related to the model parameters through the relationships:

$$C = n\upsilon L \tag{9}$$

and

$$P = m\upsilon L, \tag{10}$$

where υ is the mean volume of the raindrops and L is the mean number of elemental surfaces per unit ground area.The maximum storage capacity is given by $C_{\max} = q\upsilon L$.

The original, one layer, model has been calibrated and tested in field conditions in Indonesia (Calder et al. 1986) and India (Hall et al. 1992). This model and later developments of the model (Calder 1996b) show that up to ten times as much rain may be required to achieve the same degree of canopy wetting for tropical convective storms, with large drop sizes, than would be necessary for the range of smaller drop sizes usually encountered for frontal rain in the UK. There are also results from studies using rainfall simulators, which show that the final degree of canopy saturation also varies with drop size, being greater for drops of smaller size (Figure 2).

The advantages to be gained by incorporating the drop size dependence of wetting, from both primary raindrops and secondary drops falling from higher levels in the canopy, in interception models were

Figure 5. Understorey regrowth protects the soil from drops falling from the leaves of a teak forest.

demonstrated by Hall et al. (1996a) for a tropical forest site in Sri Lanka.

The performance of the two layer stochastic interception model (Calder 1996b; Calder et al. 1996), which explicitly takes into account drop size, was very much better in describing the initial wetting phase of the storm, Figure 3, and the overall interception loss, than conventional interception models (Rutter et al. 1971).

Canopy processes and splash induced erosion

Contrary to popular belief, forest canopies do not necessarily 'protect' the soil from raindrop impacts. In recent years there has been new understanding of the importance and significance of this effect particularly as it relates to the choice of tree species to minimize erosion. Although not generally recognised, the potential for increased erosion from drops falling from forest canopies was demonstrated almost fifty years ago by Chapman (1948). However the importance of species in determining drop size and erosive impacts has not always been well understood. There have been claims that the drop size spectra of drops falling from vegetation are largely independent of vegetation type, (Brandt, 1989). The logical consequence of this line of thought is that erosivity would be considered to be unrelated to the type of vegetation.

However, new theory derived from observations developed from disdrometer observations of the modified drop size spectra beneath canopies of different tree species (Hall & Calder 1993), suggests a different perspective. These measurements show a well defined, repeatable relationship between the spectrum of drop sizes recorded for a particular tree species, termed the characteristic spectrum (Figure 4).

These characteristic spectra can also be compared usefully with the spectra expected in natural rainfall as predicted by the Marshall-Palmer equation. For *Pinus caribaea*, a rainfall intensity exceeding 50 mm h^{-1} will be required before any diminution of drop sizes will occur. For *Eucalyptus camaldulensis* the 'break-even' intensity is about 200 mm h^{-1} whilst for *Tectona grandis* it would be 3000 mm h^{-1}, an intensity which could never be reached in natural rainfall.

These results also have important implications for splash-induced erosion. If it is assumed that the height of vegetation is sufficient for all drops from the different vegetation types to have reached terminal velocity,

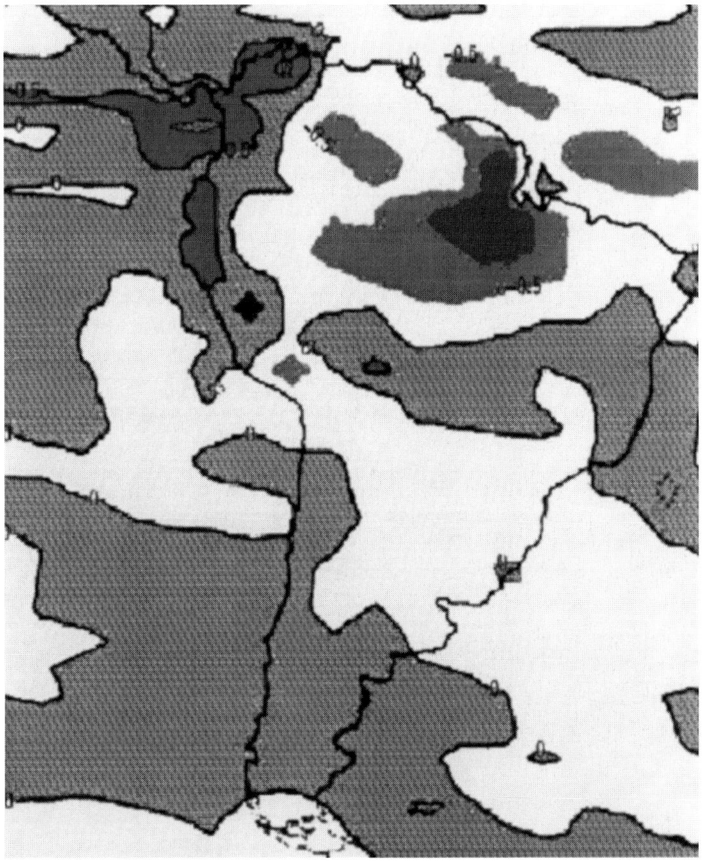

Figure 6. Predictions made by the Hadley Centre GCM of the spatial variation of the annual change in rainfall (mm d^{-1}) over Amazonia resulting from complete removal of the Amazon forest, from Institute of Hydrology 1994.

median volume drops from *Tectona grandis* will have nine times greater kinetic energy than those from *Pinus caribaea*. The high erosivity of drops falling from teak vegetation has been recorded visually following a forest fire in a teak plantation in southern India which destroyed most of the understorey. Regrowth following the fire ameliorated the kinetic energy of the drops falling from the canopy, Figure 5.

Tropical forest impacts on the rainfall climate.

The role of forests and tropical forests in determining the rainfall climate has long been a subject of great controversy. Pereira (1989) has denounced the linkage between forests and increased rainfall as a myth yet there may be some situations where this positive linkage cannot be discounted totally, where the presence of forests does lead to a small increase in rainfall. Theory indicates that the height of trees will slightly increase the orographic effect which will, in turn, lead to a slight increase in the rainfall. Modelling studies using mesoscale climate models have suggested that some of the intercepted water retained by forest canopies and re-evaporated will return as increased rainfall (Blythe et al. 1994) but this result, although indicating an increase in the gross rainfall above the vegetation, would indicate that the overall net rainfall reaching the ground surface would be reduced as a result of the interception by the forest.

Application of Global Circulation Models (GCMs) (Rowntree 1988) indicates that vegetation changes will have a regional impact on climate. Use of these models in Amazonia shows that total removal of the Amazonian rain forest would affect rainfall patterns with reductions in the rainfall, particularly in the drier north-east of the continent by about 0.5 mm per day on average (Figure 6). For the whole of the Amazon basin, rainfall would be reduced by 6% (Institute of Hydrology 1994). In the context of a scenario involv-

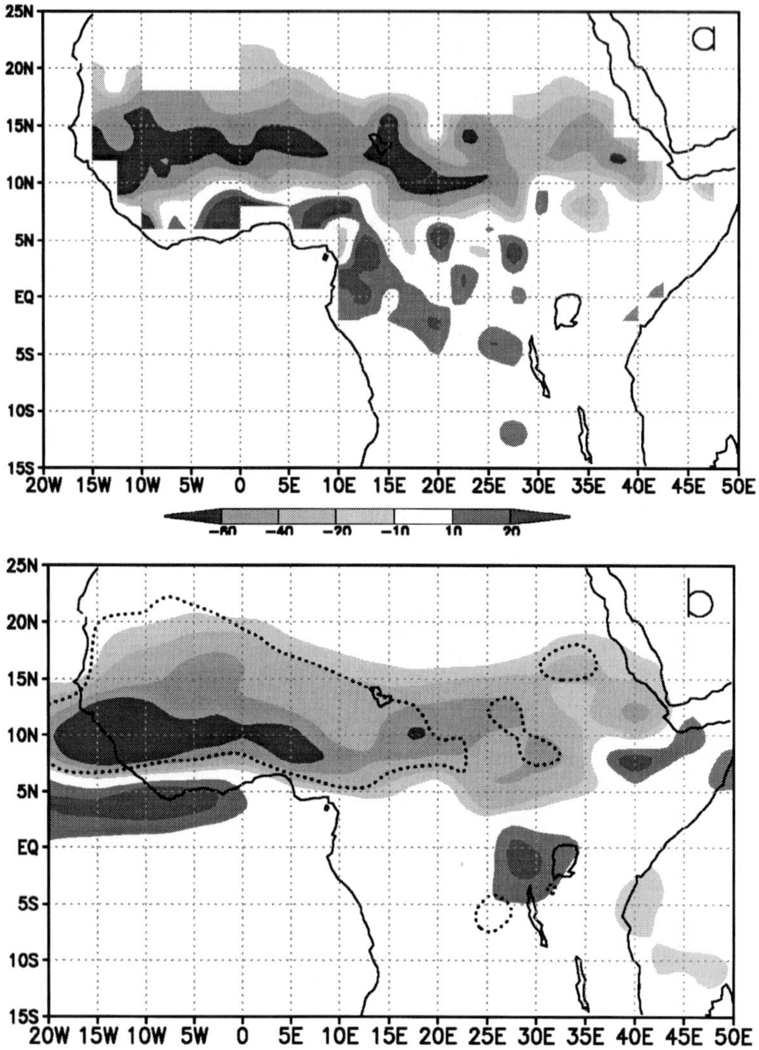

Figure 7. Spatial variability of the predicted change in rainfall (b), for the months July – September over central Africa as a result of the degradation of the Sahel vegetation occurring over the last 30 years, together with the observed change in the rainfall pattern (a) (after Xue 1997).

ing total removal of the Amazonian forest a figure of 6%, although significant, perhaps would not be regarded as a major impact.

Similarly GCM modelling studies for the Sahel (Xue 1997) indicate that past removal of the indigenous bush vegetation will have altered the spatial distribution of rainfall in a manner that bears a close correlation with observed changes in the distribution patterns (Figure 7). Yet in southern India studies of historical rainfall records (Meher-Homji 1980), indicate that annual rainfall over the last 100 years has NOT decreased despite the large scale deforestation that has taken place as the dry deciduous forest has been converted to agriculture, although there is some evidence for a decrease in the number of raindays.

Forests and water resources

An example of how new knowledge of forest evaporation, gained partially through the canopy level tracer studies in India, can be applied at a regional scale for water resource applications is provided by the Lake Malawi water balance study.

Lake Malawi is central to the economy of Malawi; virtually all the country's electricity is generated from hydro-electric schemes on the Shire River which

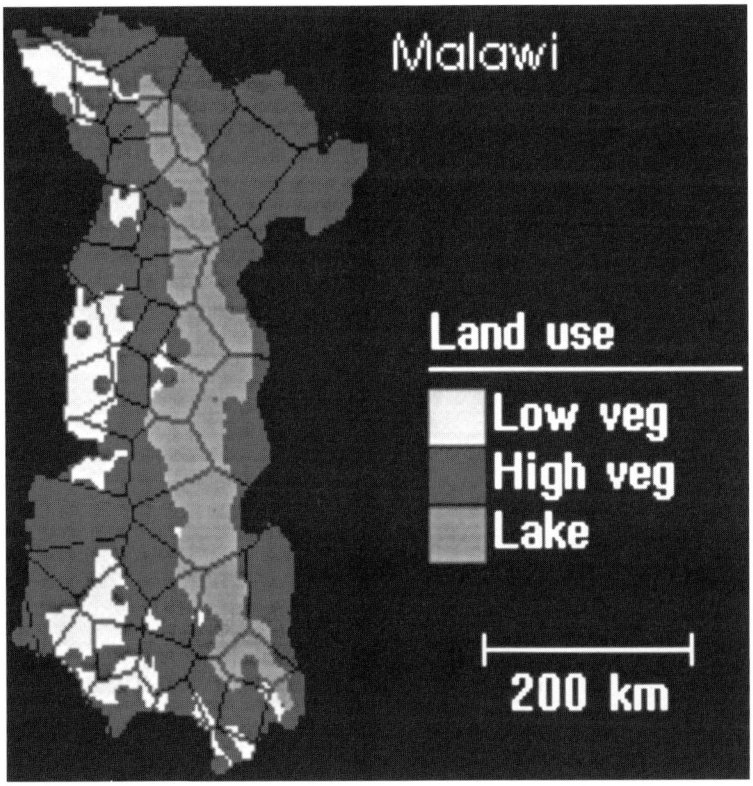

Figure 8. Location of raingauges and vegetation types on the Lake Malawi catchment used with the GIS version of the HYLUC model.

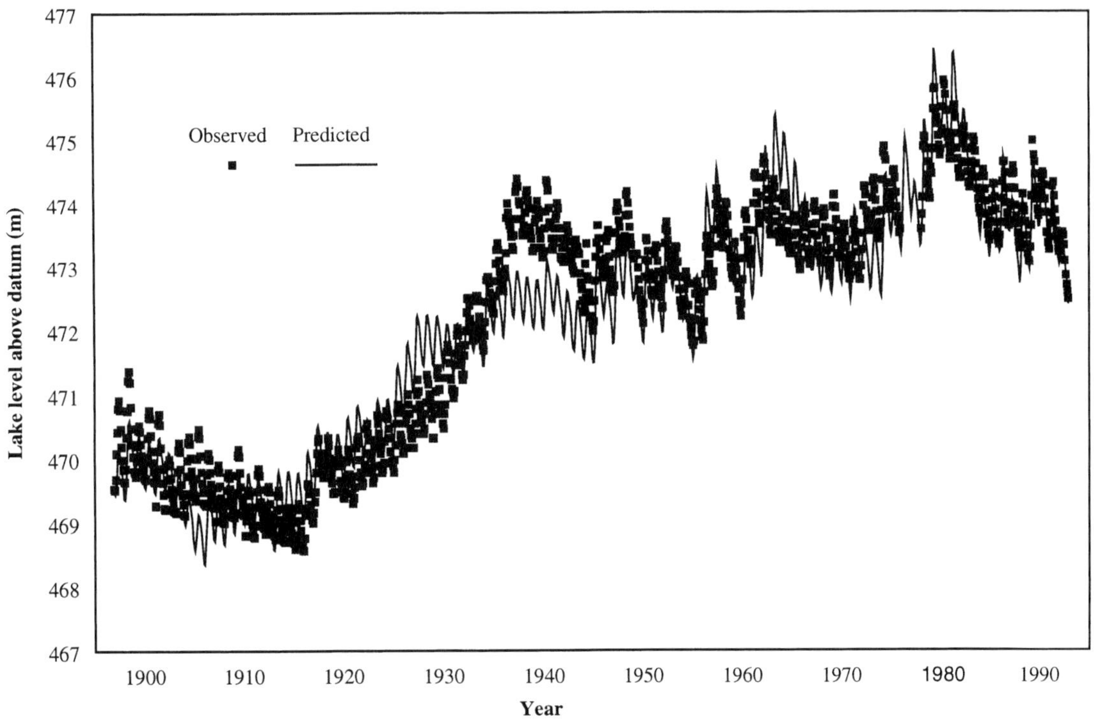

Figure 9. Lake Malawi levels, observed and predicted using a forest cover scenario of 64% till 1967 and declining to 51% by 1990 (after Calder et al. 1995).

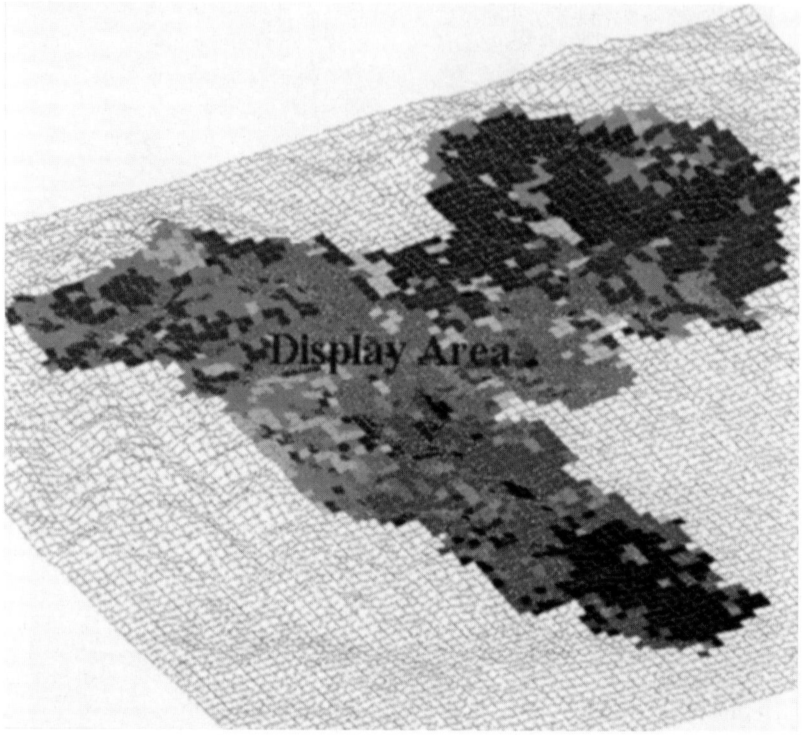

Figure 10. The user interface of the NELUP DSS.

drains the lake; the fishing industry on the lake provides most of the country's protein food supply; the tourism industry is based on the lakeside hotels and the lake also provides an important national and international ship borne transport network. The ecology of the lake is also unique, containing many hundreds of species of cyclid fish and ranks high in the UNESCO classification of World Heritage sites.

The levels of Lake Malawi have undergone very considerable changes over the last century and the cause of these fluctuations is clearly of great interest to the nation. Various explanations have been put forward to account for these changes including tectonic movement, blockage of the lake outlet and linkages with sunspot cycles. To investigate these possible causes and to determine to what extent the changes in level might have been the result of land use changes, the most significant of which has been the clearance of the dry deciduous miombo woodland for rainfed agriculture, a water balance modelling study was carried out (Calder et al. 1995). The evaporative component of the model considers the catchment to be composed of one of three surface types – forest, agricultural land, or water surface.

Using recorded rainfall data for the last hundred years, and assuming that the present relationship between lake level and the flow from the lake (the stage/discharge relationship) applied for the whole of this period, the model was operated to generate predicted lake levels. From 1896 to 1967 most of the major fluctuations in level, both seasonally and annually were well described (apart from a period in the 1930s when the stage/discharge relationship for the lake may have been affected following the prolonged period of no outflow in the early part of the century) by this model by using a value of 64% for the forest coverage of the catchment, (Figures 8 and 9). The overall agreement between prediction and observation indicates that variations in rainfall alone, without any other changes in either evaporative demand or the hydraulic regime of the lake, was sufficient to explain lake level changes during this period.

For the more recent period, model predictions which take into account a decrease in forest cover of 13% over the period 1967 to 1990 (consistent with measurements of the decrease in forest cover for this period) agree well with observations both annually and seasonally. Without this decrease in forest cover it is predicted that the lake level would have been almost

1 m lower than that actually observed before the onset of the southern African drought of 1992. As the country is reliant on the lake for hydropower generation, fisheries, tourism and transport, any further lowering of the lake level would have caused much more serious disruption, implying a significant benefit to water resources by the removal of the miombo woodland.

The overall agreement between prediction and observation indicated that variations in rainfall alone, without invoking any other changes in either evaporative demand or the hydraulic regime of the lake or other esoteric explanations such as sunspot cycles, was sufficient to explain lake level changes during this period.

The model has also been used, in association with the development of a Water Resources Management Strategy for Malawi, to investigate how proposed irrigation developments would affect the water balance of the lake and lake levels (Calder & Bastable 1995).

Increasingly it is being recognised that land use and water resource decision making requires an integrated approach where not only the impacts of land use on water resources but also the impacts on basin economics, socio-economics and ecology are taken into account and evaluated.

Decision Support Systems (DSSs) provide mechanisms not only for testing out the impacts of water resource management strategies on stakeholder interests: equity, environmental, ecological and socio-economic impacts, but can also assist water resource managers by providing a focussing framework for defining stakeholder issues and interrelationships.

Geographic Information System (GIS) based Decision Support Systems provide a mechanism for assisting with this task. An example of one such system is NELUP, developed under the Natural Environment and Economic and Social Research Councils' Land Use Programme (O'Callaghan, 1996). The NELUP DSS was designed to:

1. integrate models covering economics, ecology and hydrology that describe the changes in the spatial pattern of forestry and land use and the impacts of these changes;
2. integrate nationally available data sets within a database which describe the biophysical and economic conditions within a river basin;
3. create an interactive, user-friendly interface to the database and models to allow exploration of future land use scenarios.

Three types of model are incorporated in the NELUP DSS: agro-economic, ecological and hydrological. These models can be used to investigate the characteristics of a region under its present land use and to evaluate how these characteristics will respond to specific land use changes.

The knowledge gained from studies of forest canopy processes provides a significant input to the ultimate decision-making process with regard to the integrated management of land use and water resources.

References

Blyth, E. M., Dolman, A. J. & Noilhan, J. 1994. The effect of forest on mesoscale rainfall: An example from HAPEX-MOBILHY. J. Appl. Met. 33: 445–454.

Brandt, J. 1989. The size distribution of throughfall drops under vegetation canopies. Catena 16: 507–524.

Bruijnzeel, L. A. 1990. Hydrology of moist tropical forests and effects of conversion: a state of knowledge review. Unesco, Paris.

Bruijnzeel, L. A. 1996. Predicting the hydrological impacts of land cover transformation in the humid tropics: the need for integrated research. Pp. 15–55 In: Gash, J. H. C., Nobre, C. A., Roberts, J. M. & Victoria, R. L. (eds), Amazonian deforestation and climate Wiley, Chichester.

Calder, I. R. 1991. Implications and assumptions in using the 'Total Counts' and convection-dispersion equations for tracer flow measurements – with particular reference to transpiration measurements in trees. J. Hydrol. 125: 149–158.

Calder, I. R. 1992. A model of transpiration and growth of Eucalyptus plantation in water-limited conditions. J. Hydrol. 130: 1–15.

Calder, I. R. 1996a. Water use by forests at the plot and catchment scale. Commonwealth For. Rev. 75: 19–30.

Calder, I. R. 1996b. Dependence of rainfall interception on drop size: 1. Development of the two-layer stochastic model. J. Hydrol. 185: 363–378.

Calder, I. R. 1999. The blue revolution – land use and integrated water resources management. Earthscan Publishers.

Calder, I. R. & Bastable, H. G. 1995. Comments on the Malawi government water resources management policy and strategies. Report to ODA. Institute of Hydrology, Wallingford.

Calder, I. R., Hall, R. L., Bastable, H. G., Gunston, H. M., Shela, O., Chirwa, A. & Kafundu, R. 1995. The impact of land use change on water resources in sub-Saharan Africa: a modelling study of Lake Malawi. J. Hydrol. 170: 123–135.

Calder, I. R., Hall, R. L., Rosier, P. T. W., Bastable, H. G. & Prasanna, K. T. 1996. Dependence of rainfall interception on drop size: 2. Experimental determination of the wetting functions and two-layer stochastic model parameters for five tropical tree species. J. Hydrol. 185: 379–388.

Calder, I. R., Swaminath, M. H., Kariyappa, G. S., Srinivasalu, N. V., Srinivasa Murthy, K. V. & Mumtaz, J. 1992. Deuterium tracing for the estimation of transpiration from trees. Part 2. Measurements of transpiration from Eucalyptus plantation, India. J. Hydrol. 130: 37–48.

Calder, I. R., Wright, I. R., Murdiyarso, D. 1986. A study of evaporation from tropical rain forest – West Java. J. Hydrol. 89: 13–33.

Chapman, G. 1948 Size of raindrops and their striking force at the soil surface in a red pine plantation, Eos Trans. AGU 29: 664–670.

Chappell, N. A., Bidin, K. & Tych, W. 2001. Modelling rainfall and canopy controls on net- precipitation beneath selectively-logged tropical forest. Plant Ecol. 153: 215–229 (this volume)

Freiberg, M. 2001. The influence of epiphyte cover on branch temperature in a tropical tree. Plant Ecol. 153: 241–250 (this volume)

Hall, R. L. & Calder, I. R. 1993. Drop size modification by forest canopies - measurements using a disdrometer. J. Geophys. Res. 90: 465–470.

Hall, R. L. & Calder, I. R. 1999. Dependence of rainfall interception on drop size- Sensitivity of the stochastic interception model to the choice of the rainfall intensity – drop size relationship. J. Hydrol. (submitted).

Hall, R. L., Calder, I. R., Gunawardena, E. R. N. & Rosier, P. T. W. 1996. Dependence of rainfall interception on drop size: 3. Implementation and comparative performance of the stochastic model using data from a tropical site in Sri Lanka. J. Hydrol. 185: 389–407.

Hall, R. L., Calder, I. R., Rosier, P. T. W., Swaminath, M. H. & Mumtaz, J. 1992. Measurement and modelling of interception loss from eucalypt plantations in southern India. In: I. R. Calder, R. L. Hall & P. G. Adlard (eds), Proc. Int Conf. on Growth and Water Use of Forest Plantations, Bangalore, 7-11 February 1991. Wiley, Chichester, UK, pp. 270–289.

Hull, D. E. 1958. The total count technique: a new principle in flow measurement. Int. J. Appl. Radiation Isotopes 4: 1–15.

Institute of Hydrology. 1994. Amazonia: forest, pasture and climate – results from ABRACOS. ISBN 0 948540 61 3, 18 pp. Institute of Hydrology , Wallingford.

Kline, J. R., Martin, J. R., Jordan, C. F. & Koranda, J. J. 1970. Measurement of transpiration in tropical trees with tritiated water. Ecology 51: 1068–1073.

Meher-Homji, V. M. 1980. Repercussions of deforestation on precipitation in Western Karnataka, India. Aech. Met. Geogph. Biokl. B28: 385–400.

O'Callaghan, J. R. 1996. Land Use. The interaction of economics, ecology and hydrology. Chapman & Hall, London.

Pereira, H. C. 1989 Policy and practice in the management of tropical watersheds. Westview Press, Colorado, USA.

Rowntree, P. R. 1988. Review of general circulation models as a basis for predicting the effects of vegetation change on climate. Forests, Climate and Hydrology: Regional impacts, E. R. C. Reynolds & F. B. Thompson (eds), Kefford Press, 162–193.

Rutter, A. J., Kershaw, K. A., Robins, P. C. & Morton, A. J. 1971. A predictive model of rainfall interception in forests. I: Derivation of the model from observations in a plantation of Corsician pine. Agric. Meteorol. 9: 367–384.

Shiklomanov, I. A. 1999 Climate change hydrology and water resources: the work of the IPCC, 1988–1994. Pp. 8–20. In: Van Dam, J. C. (ed.), Impacts of climate change and climate variability on hydrological regimes. Cambridge University Press-UNESCO International Hydrology Series.

Szarzynski, J. & Anhuf, D. 2001. Micrometeorological conditions and canopy energy exchanges of a neotropical rain forest (Surumoni-Crane-Project, Venezuela) 153: 231–239 (this volume)

Xue, Y. 1997. Biosphere feedback on regional climate in tropical north Africa. Quart. J. Roy. Meteorol. Soc. 123: 1483–1515.

Plant Ecology **153:** 215–229, 2001.

Modelling rainfall and canopy controls on net-precipitation beneath selectively-logged tropical forest

Nick A Chappell[1], Kawi Bidin[2] & Wlodek Tych[1]
[1]*Centre for Research on Environmental Systems and Statistics, Institute of Environmental and Natural Sciences, Lancaster University, Lancaster LA1 4YQ, UK;* [2]*School of Science and Technology, Universiti Malaysia Sabah, 88999 Kota Kinabalu, Malaysia*

Key words: ENSO, Interception loss, Time-series modelling, Variogram, Wet canopy evaporation

Abstract

Understanding spatio-temporal patterns in rainfall received beneath tropical forest is required for eco- hydrological modelling of soil-water status, river behaviour, soil erosion, nutrient loss and wet-canopy evaporation. As selective-logging of tropical forest leaves a very complex mosaic of canopy types, it is likely to add to the spatio-temporal complexity of this sub-canopy or net precipitation. As a precursor to addressing this problem, the analysis presented here will examine the two dominant biophysical controls on sub-canopy precipitation. These controls are: (a) the spatial and temporal patterns in above-canopy or gross rainfall, and (b) the rate of wet-canopy evaporation associated with each type of canopy structure created by selective-forestry. For this study, over 400 raingauges were installed within a 10 km^2 area of lowland dipterocarp forest affected by selective-forestry some 9-years prior to this work. Gauges were located beneath various canopy types and within large openings. The spatial distribution of gross rainfall (monitored within the openings) was modelled using variography, while the effects of different canopy types on sub-canopy preciptation was analysed by comparing 6-month totals. The temporal distribution of gross rainfall over an 11-year record collected at the same site (Danum Valley Field Centre) was modelled with Data-Based-Mechanistic (DBM) approaches. These DBM approaches were also applied to the rainfall time-series of the two adjacent meteorological stations; all three gauges being contained within a 5000 km^2 region of Eastern Sabah in Malaysian Borneo.

Strong diurnal modulation was apparent within gross rainfall for the inland rainforest site, with a distribution consistent with a dominance of local convective rain cells. A similarly strong cycle coincident with the periodicity of the El Niño-Southern Oscillation (ENSO) was present within all of the region's rainfall records, though marked differences in annual and intra-annual seasonality were apparent. The preliminary variogram modelling indicated that a deterministic drift was present within the local-scale gross rainfall data, probably related to local topographic effects. Notwithstanding the need to remove this drift, the work indicated that spatial models of gross rainfall could be identified and used to interpret similar models of net-precipitation. During the ENSO drought-period monitored, the lowland dipterocarp forest allowed 91% of the gross rainfall to reach the ground as throughfall. These rates were, however, reduced to between 80%–86% beneath representative plots of moderately impacted to creeper-covered, highly damaged patches of forest.

Introduction

Tropical forest canopies modify the spatial and temporal distribution of rainfall before it reaches the ground. A component of the rainfall never reaches the ground, because the water intercepted by leaves, branches and stems evaporates by a process called 'wet-canopy-evaporation' or 'interception-loss'. A further component temporarily stored by the canopy will then drip from canopy surfaces as 'leaf-drip', or be channelled down tree stems as 'stemflow'. The remaining component of rainfall will drop through small gaps

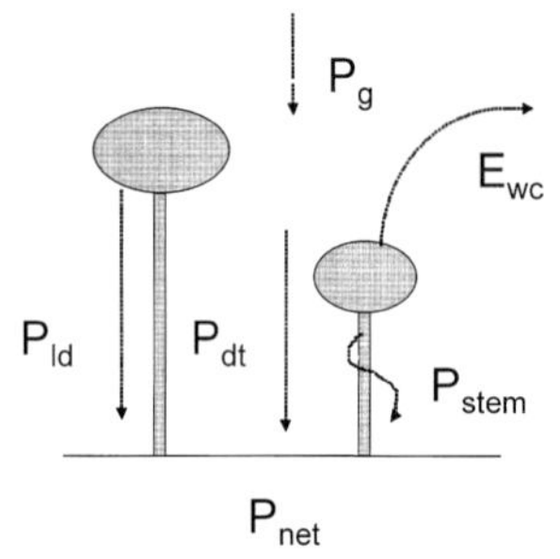

Figure 1. Schematic diagram of the interaction between tropical forest canopies and received rainfall. Symbol P_g is the gross precipitation (i.e., water received above the forest canopy or in large gaps between different tree canopies), P_{net} is the net precipitation (i.e, water received by the forest floor beneath a canopy), P_{ld} is the leaf-drip component of the net precipitation, P_{dt} is the direct throughfall component of the net precipitation (i.e, the raindrops penetrating the canopy that do not hit leaves, branches or stems of trees), P_{stem} is the stemflow component of the net precipitation and E_{wc} is the wet-canopy evaporation.

in the canopy as 'direct-throughfall' (Figure 1). The total volume and spatio-temporal distribution of rainfall reaching the ground, called 'net precipitation' or 'sub-canopy rainfall' (P_{net}), is, therefore, not only dependent on (a) the amount and distribution of the rainfall received by the canopy (Lloyd 1990), but also on (b) the structure of the canopy itself (Herwitz 1985).

Localised removal of trees as part of commercial, selective forestry is common within the rain forests of the Eastern Tropics (Whitmore 1998). Disturbance to the canopy structure as a result of the selective removal of trees will affect the amount and distribution of net-precipitation by changes to (i) the canopy storage and sub-canopy pathways, and (ii) the efficiency of the wet-canopy-evaporation (E_{wc}) process. Quantifying E_{wc} rates is important to understanding the development of small-scale convective rainfall cells and hence to the local cycling of water (Shuttleworth 1988) but may also play a role in macro-scale processes described within Global Circulation Models (Riehl 1954; Henderson-Sellers et al. 1993). Indeed, one theory controversially places the upward flux of water vapour above equatorial Southeast Asia as one of the main driving forces in global circulation (Newell & Gould-Stewart 1981)

Losses or additions of water to catchments as a result of forestry impacts on E_{wc} have equally important implications for local water resources and nutrient losses (Bruijnzeel 1990). As the tropical environment may experience a significant proportion of its rainfall within highly localised and intense convective cells, accurate modelling of soil moisture, river generation or soil erosion requires estimates of water input to a high spatial and temporal resolution (Brooks & Spencer 1995; Chappell et al. 1998, 1999a, b; Calder 2001). As a consequence, provision of detailed data on the spatio-temporal structure of sub- canopy rainfall becomes important to the understanding of patterns of plant water-stress, aquatic ecology and nutrient pathways (Bruijnzeel 1990; Eschenbach et al. 1998; Martin-Smith 1998).

While there are clear reasons for understanding P_{net} (and its controls) within tropical rain forests, its magnitude and spatio-temporal structure are difficult to quantify, particularly within selectively harvested areas. This is because of (a) the relative importance of highly localised convective rainfall cells in the tropics (Riehl 1954), and (b) the 'patchiness' of the different canopy forms produced by selective forestry, giving locally variable E_{wc} rates (Asdak et al. 1998). The dearth of published studies on the spatio-temporal structure of sub-canopy rainfall and its controls within selectively-disturbed forest also adds to the complexity of the problem. As a result, this study focuses on selectively-logged forest, and is the first stage in a more comprehensive analysis of sub-canopy rainfall (Bidin, in preparation). The specfic aim of this initial analysis is to address the key spatio-temporal controls on sub-canopy rainfall of (a) the temporal structure of gross rainfall over the short- and long-term, (b) the spatial structure of gross rainfall over the scale of a typical 'experimental research catchment' (i.e., 0.5–10 km^2), and (c) the time-averaged differences in P_{net} between the different 'canopy types' produced by selective forestry.

Study areas

The rain forest site chosen for study lies close to the Danum Valley Field Centre or 'DVFC' (4°58′ N and 117°48′ E) in the Malaysian state of Sabah, Borneo Island (Figure 2). The area lies within the Ulu Segama Forest Reserve, where the Forestry Upstream Division of Yayasan Sabah manage the forest for commercial timber production and conservation. The selective forestry management system adopted within this forestry concession is typical of that practiced within other concessions of Southeast Asia (Pinard & Putz 1996). The DVFC locality formed the focus of the rainfall analysis, as river- and sediment-generation processes within a series of experimental catchments

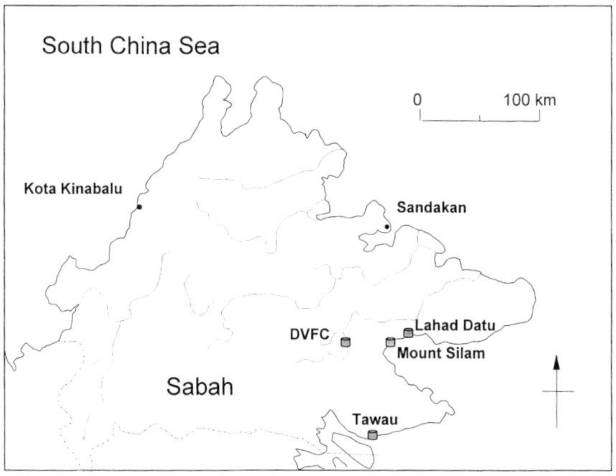

Figure 2. Location of the Danum Valley Field Centre (DVFC), Lahad Datu, Mount Silam and Tawau rainfall stations in Sabah, Malaysian Borneo.

have been studied here since 1986 (Douglas et al. 1992, 1999). Moreover, recent hydrological modelling of these data (Chappell et al. 1998, 1999a) would benefit from a better understanding of the spatio-temporal controls on the rainfall received below the forest canopy.

Regional scale

Daily rainfall has been measured at the DVFC meteorological station since 1986, and over the 13-year monitoring period (1986–1998) an average annual rainfall (*AAR*) of 2663 mm (409 mm σ) has been recorded. The DVFC rainfall study (Bidin in preparation) was undertaken over a 'water-year' 1 May 1997 to 30 April 1998 which covered the two driest calendar-years on record. In 1997 only 1921 mm rainfall or 72% *AAR* was recorded, and only 2130 mm (80% *AAR*) in 1998. The DVFC meteorological station (149 m a.s.l.) is important as it is one of the few stations within the interior of Malaysian Borneo providing records to the Malaysian Meteorological Service (MMS or Perkhidmatan Kajicuaca Malaysia). The nearest long-term rainfall stations to DVFC are Lahad Datu Airport 60 km to the east, and Tawau Airport 83 km to the south southeast. Some rainfall records are also available from Silam Camp (10 m a.s.l.) and Mount Silam (884 m a.s.l.) which are 47 and 43 km directly east of DVFC, respectively (Bruijnzeel et al. 1993). These four stations are all located on the East coast of Sabah (Figure 2). Automatic raingauges have been installed within the DVFC locality and some preliminary analysis has been undertaken by Sherlock (1997). He found that high rainfall intensities are relatively frequent with rates in excess of 4.167 mm/5-min (50 mm h^{-1} equivalent) having a return-period of 23.3 days and in excess of 8.333 mm/5-min (100 mm h^{-1} equivalent) a return-period of 139.6 days.

Local scale

A 10 km^2 area by DVFC was chosen as the focus for local-scale, spatial analysis of above- and below-canopy rainfall. This area has an undulating topography with an altitudinal range from 155 m (Sapat Kalisun river) to 435 m (Bukit Atur) and is covered by 'lowland dipterocarp rain forest' (Newbery et al. 1992) logged during 1988–1989. Timber extraction was by a 'selective', rather than a 'clearfell', logging system where only commercial timber trees with a girth greater than 0.6 m were felled. This type of selective silvicultural system produced a mosaic of 'canopy types'. Bidin (in preparation) has classified these 'patches' of different canopy into:

- Category 1: undisturbed forest canopy,
- Category 2: moderately impacted forest canopy,
- Category 3: very disturbed and climber-covered canopy,
- Category 4: cleared areas now dominated by *Macaranga* spp. pioneer trees,
- Category 5: cleared areas now supporting only shrubs, herbs and sprawling plants, and
- Category 6: no canopy.

These classes were chosen as (i) they could be easily identified by ground survey (with only limited experience), (ii) their expected differences in wet-canopy-evaporation rates (Asdak et al. 1998), and (iii) their importance to foresters in the estimation of future timber yields. Category 1 includes patches of virgin (protected) forest in 'riparian buffer zones' or in areas where slope angles are in excess of 30 degrees (Pinard & Putz 1996). Category 2 forest patches are those areas where the selective removal of commercial timber trees by tractor yarding has not severely damaged the remaining canopy. Category 3 forest is found in areas affected by 'high-lead' timber extraction (Conway 1982) where the remaining trees are severely damaged and have become covered in woody climbers typically of the *Merremia*, *Mezoneuron* and *Uncaria* spp. (Whitmore 1998). This type of high-impact yarding is now banned within the Yayasan Sabah forestry concession. Category 4 areas have been colonised by *Macaranga* spp. pioneer trees and are typically found in the 40 m-wide road corridors where 'Matahari clearance' (Conway 1982) has taken place. Such roadside clearance is undertaken to aid road drying and hence trafficability for haulage lorries. Some of the rotational landslides (ca. 0.5 to 5 ha) developed along these haulage roads have also been colonised by *Macaranga* spp. trees (Chappell et al. 1999b), though others only support category 5 vegetation. Where high-lead logging or Matahari clearance has removed all trees and *Macaranga* spp. pioneer trees have failed to colonise then only the shrubs (notably Zingiberaceae), herbs and sprawlers of category 5 canopy are found. Category 6 areas lack a vegetation canopy and are associated with gravel-surfaced haulage roads or highly compacted dirt roads, 'skid trails' or 'log-landings' (Conway 1982). This study aimed to examine the above- and sub-canopy rainfall associated with forest patches clearly within one of these canopy classes.

Instrumentation and monitoring

For the initial identification of temporal structures in above-canopy rainfall, daily records monitored at the DVFC, Lahad Datu and Tawau meteorological stations were compared. An 11-year time-series (i.e., 1 January 1986 to 31 December 1997) for DVFC was used, while $14\frac{1}{2}$-year (i.e., 1 January 1982 to 23 June 1997) and 17-year (i.e., 1 January 1980 to 31 December 1997) records were used from Lahad Datu and Tawau, respectively. Measurements had been made with UK Meteorological Office Mark II daily storage gauges sited away from obstructions and with the gauge orifice at 0.3 m above the local ground-surface, but lacking any turbulence protection. The diurnal rainfall distribution at the DVFC locality utilised data collected from a single tipping-bucket raingauge (Station name: 'km63') over the 3-year period from 1 January 1995 to 31 December 1997. The 'km63' raingauge, a Casella W5724 tipping-bucket collector, was sited on an outcrop of chert (site photograph in Whitmore 1998, Figure 7.24) and at 20 m from the nearest trees.

For the initial analysis of the spatial structure of above-canopy rainfall, a network of 34 storage gauges located within the 10 km^2 area East of DVFC was used. These gauges formed part of a larger network of 50 gauges for above-canopy rainfall measurement, and a further 345 gauges for sub-canopy measurement. These storage gauges incorporated funnels with a 0.15 m diameter orifice, sited 0.56 m above the ground and having a drop of 0.23 m to the funnel base to prevent 'out-splash'. Evaporation losses from the 5 dm^3 plastic collection bottles were reduced by (a) directing the flow through narrow diameter holes in the funnel base, and (b) painting the gauges white to reduce radiation absorption. The whole gauge was held exactly vertical by attachment to a wooden stake. The accuracy of these non-standard gauges was extensively tested (Bidin in preparation). In one experiment, catch error was estimated by comparing the results of 10 non-standard gauges sited adjacent to two standard (Mark II) gauges for one month and was found to be only 1.04%. Evaporation losses from the project raingauges were similarly negligible at less than 0.6 mm/week (10 December 1997-23 March 1998). The precise location of all of the gauges used for above- and sub-canopy rainfall measurement was surveyed in closed-loop traverses using a Leica TC400 Total Station, a unit incorporating an electronic theodolite and laser-distance-measurement system. Such measurements, and the associated fixed gauge locations, were required for geostatistical modelling of spatial aspects of the rainfall data.

The initial assessment of the effect of forestry-related changes in canopy structure on P_{net} utilised data from all 345 sub-canopy gauges and 12 of the above-canopy gauges. A total of 80 sub-canopy gauges were located in two 'plots' of category '1' canopy. Similarly category '2' and '3' canopy each had 80 gauges in two plots, while category '4' forest was assessed with 85 gauges in two plots and category '5' canopy 20 gauges in a single plot (Table 1). The raingauge plots, each approximately 300 m^2, were located only in forest patches very clearly falling within one of the canopy categories. Within the plots, the gauges were located randomly, giving a 'stratified random design' for the whole network of sub-canopy gauges. The above-canopy rainfall to be associated with that beneath the canopy of each plot was measured with three to eight gauges located in a nearby clearing. These 'clearings' were mostly haulage roads, skid trials and log-landings, which had a minimum diameter of 10–40 m and also 10–40 m away from the plots. All storage gauges were monitored on a variable frequency from an individual storm basis to once every two weeks.

Table 1. Sub-canopy rainfall (and associated gross rainfall) statistics measured beneath five canopy types produced by selective logging of lowland dipterocarp forest in the Ulu Segama Forest Reserve (Sabah) during the approximately 6-month period from 1 May to 27 November 1997.

Canopy category[a]	1	2	3	4	5
Gross rainfall (P_g)					
N^o P_g gauges	3	3	6	8	2
P_g total[b] (mm)	895.0	902.3	903.9	921.9	902.0
σ[c] (mm)	32.6	21.8	30.5	52.4	5.6
CV[d] (%)	4	2	3	6	1
95% confidence interval (mm) 37.6	25.1	24.9	37.0	7.9	
Difference in P_g variance[e]	–	ns[f]	ns	ns	$p < 0.10$
Difference in P_g mean[g]	–	ns	ns	ns	ns
Net rainfall (P_{net})					
N^o P_{net} gauges	80	80	80	85	20
P_{net} total[h] (mm)	815.0	758.7	719.2	795.5	771.6
σ(mm)	118.4	185.2	195.8	143.5	234.3
CV (%)	15	24	27	18	30
95 % confidence interval (mm)	25.9	40.6	42.9	30.5	102.7
Difference in P_{net} variance[i]	–	$p < 0.05$	$p < 0.01$	ns	$p < 0.01$
Difference in P_{net} mean[j]	–	$p < 0.05$	$p < 0.001$	ns	ns
Comparison of gross and net rainfall					
Difference in mean (P_g vs P_{net})[k]	$p < 0.001$	$p < 0.001$	$p < 0.001$	$p < 0.001$	$p < 0.05$
$P_g - P_{net}$ (mm)	80	144	185	126	130
95% confidence interval (mm)[l]	45.1	47.5	49.4	47.4	103.0
Difference in $P_g - P_{net}$ mean[m]	–	$p < 0.10$	$p < 0.01$	ns	ns
$\%P_{net}$[n] (%)	91.1 ± 6.7	84.1 ± 6.9	$79.6. \pm 6.9$	86.3 ± 6.8	85.6 ± 12.1

[a] 1 = undisturbed forest canopy; 2 = moderately impacted forest canopy; 3 = very disturbed and climber-covered canopy; 4 = cleared areas now with *Macaranga* spp. trees; 5 = cleared now with shrubs/herbs/sprawlers; canopy category 6 (no canopy) not shown.
[b] Only raingauges close to the P_{net} plots used within this analysis.
[c] Standard deviation.
[d] Coefficient of variation (or standard deviation normalised by the mean expressed as percent).
[e] Level of significance of difference in variance of P_g measured 'above' category 1 canopy and P_g measured 'above' the other canopies.
[f] Not significant, even at $p < 0.100$ level.
[g] Level of significance of difference in mean of P_g measured 'above' category 1 canopy and P_g measured 'above' the other canopies.
[h] Net preciptation figures exclude the estimated 1 percent stemflow.
[i] Level of significance of difference in variance of P_{net} measured beneath category 1 canopy and P_{net} measured beneath the other canopies.
[j] Level of significance of difference in mean of P_{net} measured beneath category 1 canopy and P_{net} measured beneath the other canopies.
[k] Based on standard errors for both P_g and P_net added in quadrature (Taylor 1982).
[l] Level of significance (2α) of difference in means of P_g and P_{net} measured at each canopy type.
[m] Level of significance of difference in mean of $P_g - P_{net}$ measured at category 1 canopy and $P_g - P_{net}$ measured at the other canopies. (NB. The *T-test* takes account of uncertainty in both P_g and P_{net}).
[n] Percent of the local P_g that is received beneath the canopy. (NB. $\%P_{net}$ is a quotient, so the fractional uncertainties in both P_g and P_{net} are added directly (Taylor 1982).

(A) Temporal structure of above-canopy rainfall

The temporal structure of the rainfall delivered to the canopy is the most significant control on the quantity of water delivered beneath the canopy (Lloyd 1990). This temporal structure might include (i) a diurnal distribution, (ii) an annual seasonality related to phenomena such as large scale reversals of the wind regime, (iii) inter-annual cycles related to phenomena such as the El Niño-Southern Oscillation (ENSO), (iv) any 'longer-term' drift in rainfalls and (v) rainstorm frequency and duration.

The diurnal distribution of above-canopy rainfall was examined by simple graphical and statistical summary of hourly and four-hourly data collected over a three year period (1995–1997) at the 'km63' raingauge in the DVFC locality. In contrast, identification of longer-term, temporal structures (i.e., annual and inter-annual cycles, and drift) within the DVFC daily records were analysed using Data-Based-Mechanistic (DBM) modelling. To aid interpretation, these techniques were also applied to the daily records for the nearest two MMS rainfall stations to DVFC, i.e., Lahad Datu and Tawau airports. The specific DBM modelling tool used in this instance is the Dynamic Harmonic Regression (DHR) model, which is a recursive interpolation, extrapolation and smoothing algorithm for non-stationary time-series (Young 1998; Young et al. 1999). The DHR model identifies three components in the time-series, i.e.,

$$P_{g(t)} = T_t + S_t + e_t, \qquad (1)$$

where $P_{g(t)}$ is the observed (above-canopy) rainfall time-series, T_t is the trend which includes the drift in long-term average rainfall and the inter-annual cycles, S_t is the periodic component related to annual and intra-annual seasonality, and e_t is the white noise. The S_t term is further defined as:

$$S_t = \sum_{i=1}^{R} \{a_{it} \cos(\omega_i t) b_{it} \sin(\omega_i t)\}, \qquad (2)$$

where $a_{i,t}$ and $b_{i,t}$ are the Time-Variable-Parameters (TVPs) of the model, R is the number of seasonal components, and ω_i are the set of frequencies chosen by reference to the spectral properties of the time-series. Optimisation of the TVPs was achieved by first estimating the Noise-Variance- Ratio (NVR) of the TVPs. This is achieved in the frequency domain by fitting the logarithmic pseudo-spectrum of the DHR model to the estimated logarithmic AutoRegressive (AR) spectrum of the observed rainfall series. The order of the AR model is identified via the Akaike Information Criterion. Once NVR parameters are optimised, a single run of two recursive algorithms, the Kalman Filter and Fixed-Interval-Smoothing equations provide estimates of the various components. The estimated trend component is further split into a very slowly changing drift and the inter-annual cyclic component. Since no assumptions are made as to the periodicity of the cycle, it is unlikely that any artefacts are introduced in the procedure. More detailed discussion of the DHR recursive techniques can be found within Young (1998) and Young et al. (1999). All analyses of the hourly and daily time-series were performed in the Matlab® programming environment. Simulation results of rainstorm frequency and duration in the DVFC and Eastern Sabah regions will be presented within Discenza et al. (in preparation).

Results

Plotting the four-hourly rainfall for the three year 'km63' record (Figure 3a) as a grouped histogram shows that there is a tendency for greater rainfall totals within the afternoon. The presence of extreme events within each of the four-hour time-series do, however, obfuscate clear diurnal patterns within the grouped histogram. To overcome this, rainfall occurring within the first hour of every day is totalled, the procedure is then repeated for all 24-h periods, and finally plotted. The resultant diagram (Figure 3b) clearly shows a strong diurnal modulation, with 43% of rainfall occurring between 2:00 to 4:00 p.m. Afternoon rains also characterised the diurnal rainfall distribution monitored about 40 km to the east of DVFC on the summit of Mount Silam during a 38-day sampling period in 1987 (Bruijnzeel et al. 1993). An afternoon peak in rainfall is similarly observed within some other tropical rainforests, for example, at Bukit Tarek, 50 km from the West coast of Peninsular Malaysia (Noguchi et al. 1996) and Reserva Florestal Ducke in the central Amazon (Lloyd 1990). This single peak in the midlate afternoon is consistent with an inland equatorial rainfall regime where rain clouds develop as a result of convectional heating during the day (Watts 1955; Ramage 1964). In regions where the effects of phenomena such as sea breezes or katabatic winds become significant, the diurnal distribution may have different single or multiple times of peak rainfall (Watts 1955; Ramage 1964; Houze et al. 1981).

Figures 4a–c show the spectral frequency of the gross rainfall monitored on a daily basis at the DVFC,

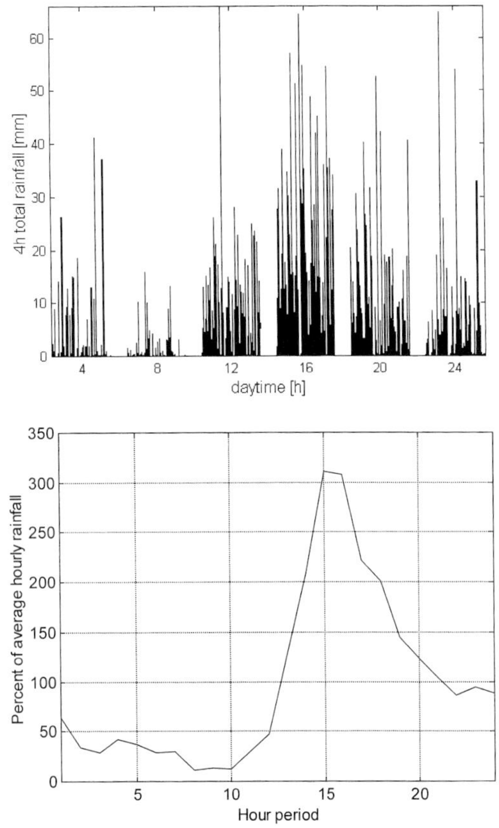

Figure 3. Diurnal distribution of rainfall monitored at the 'km63' tipping-bucket raingauge near to DVFC over a 3-year period shown as (a) a grouped histogram illustrating the time-series of rainfall in 4-hour diurnal periods, and (b) the percent of the average hourly rainfall.

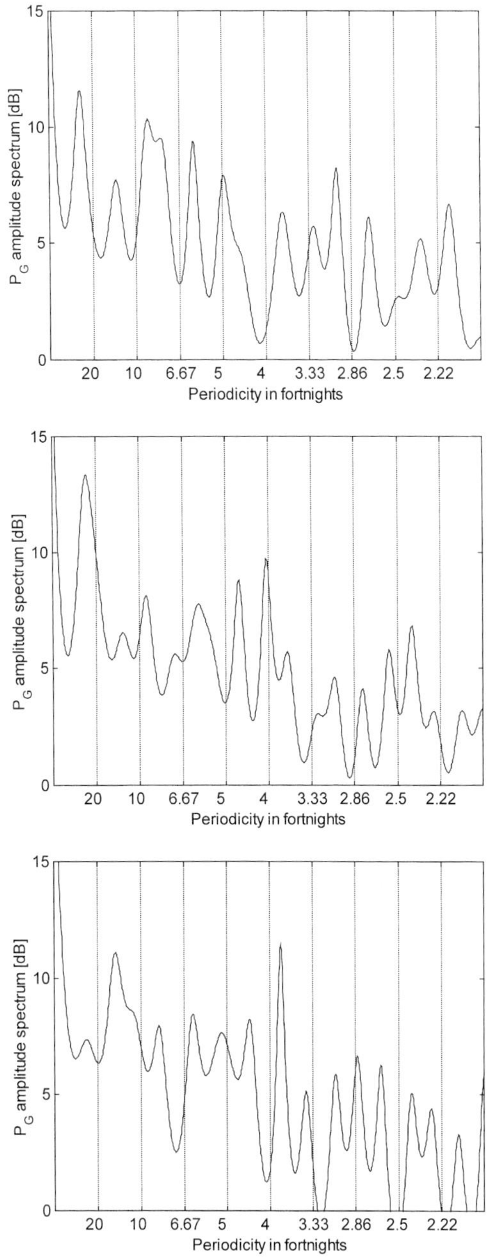

Figure 4. The spectral frequency of gross rainfall monitored at (a) DVFC, (b) Tawau, and (c) Lahad Datu in eastern Sabah.

Tawau and Lahad Datu meteorological stations. For simulation efficiency, the data are averaged over fortnightly periods. The records for Tawau (Figure 4b) show a pronounced spectral peak at 26-fortnightly periods and thereby indicates that there is a marked annual seasonality in rainfall at Tawau. The annual peaks in rainfall tend to occur during the late summer, though this unimodal pattern is disrupted during certain years (i.e., 1984–1985, 1988 and 1995; Figure 5b). Tawau records, therefore, show 'summer monsoon rains' also observable within the records of central Borneo (Kripalani & Kulkarni 1997, 1998). The data for DVFC (Figure 4a) have a annual peak (i.e., 26-fortnights), but it is neither (a) as large as at Tawau, nor (b) markedly larger than the peaks associated with its harmonics. This indicates that while some annual seasonality is present within the DVFC gross rainfall records, it is not as pronounced as that seen at Tawau. Furthermore, examination of a time-series of the seasonal component (Figure 5a) indicates that the most stationary annual feature is a drier period in spring. Indeed, separate analysis of the monthly rainfall totals (1986–1998) supports this observation, with April having the smallest monthly total of only 120 mm (89 mm σ) or 54% of the monthly mean rainfall. The second most significant spectral peak within the DVFC records appears at 8.7-fortnights or every 4 months (Figure 4a). This peak relates to a relatively consistent increased rainfall during the three periods of (a) January–February, (b) July, and (c) October–November (Figure 5a). This weak intra-annual cycle is also just observable within the Lahad Datu records (Figure 5c), though it is the two drier periods that are the most marked (i.e., 12-fortnights spectral peak, Figure 4c) and they are of similar intensity (i.e., no marked spectral peak at 26-fortnights). The phenomena leading to the marked 'summer monsoon rains' at Tawau are, therefore, not as pronounced 81–89 km to the North (at DVFC or Lahad Datu). Adding further to the spatial inhomogeneity at the regional scale, the Lahad Datu and Tawau records exhibit a relatively pronounced peak at about 4-fortnightly periods (Figures 4b and 4c). Such a periodicity coincides with the 30–60-day oscillations in the eastward migration of the circulation cell over the tropics (Madden & Julian 1972; Weickmann et al. 1985). This cycle is not clearly observable within the DVFC records. All these spectral analyses clearly demonstrate the dangers of spatial interpolation of intra- annual rainfall over a meso-scale (i.e., 1000–10 000 km^2) tropical region where rainfall stations are tens of kilometres apart and convective cells are important. Spatial inhomogeneity in the temporal pattern within the 'eastern Sabah' region relate, at least in part, to the effects of land-sea interaction in coastal regions (Johnson & Priegnitz, 1981) and the interactions of different large-scale airflows (e.g., equatorial easterlies, trades and equatorial westerlies) in central and eastern Sabah (Wood 1948; Thompson 1951; Kripalani & Kulkarni 1998).

Figure 6 shows the dominant inter-annual cycles identified by the DHR model in the 11–17-year DVFC, Tawau and Lahad Datu data-series. These cycles have amplitudes of 2-4 mm d^{-1} against the mean daily rainfalls of about 6 mm d^{-1}. Dry periods centred on 1986–1987, 1992 and 1997 are observed within all three data-series. The longer records for Tawau and Lahad Datu also exhibit a drought in 1982-1983. These troughs in the inter-annual rainfall cycles have a periodicity in the 1980s and 1990s of 4.5–5.5 years and coincide with shifts in the El Niño-Southern Oscillation (Wolter & Timlin 1998). While each station experiences these dry periods, intra-regional variations in the intensity and timing are apparent. For example, the 1986–1987 dry spell is much less intense at Tawau than at the other two stations. Similarly, variation in the drought timing can be seen by closer examination of the same drought which is most intense at Tawau in July 1986, at DVFC in December 1986

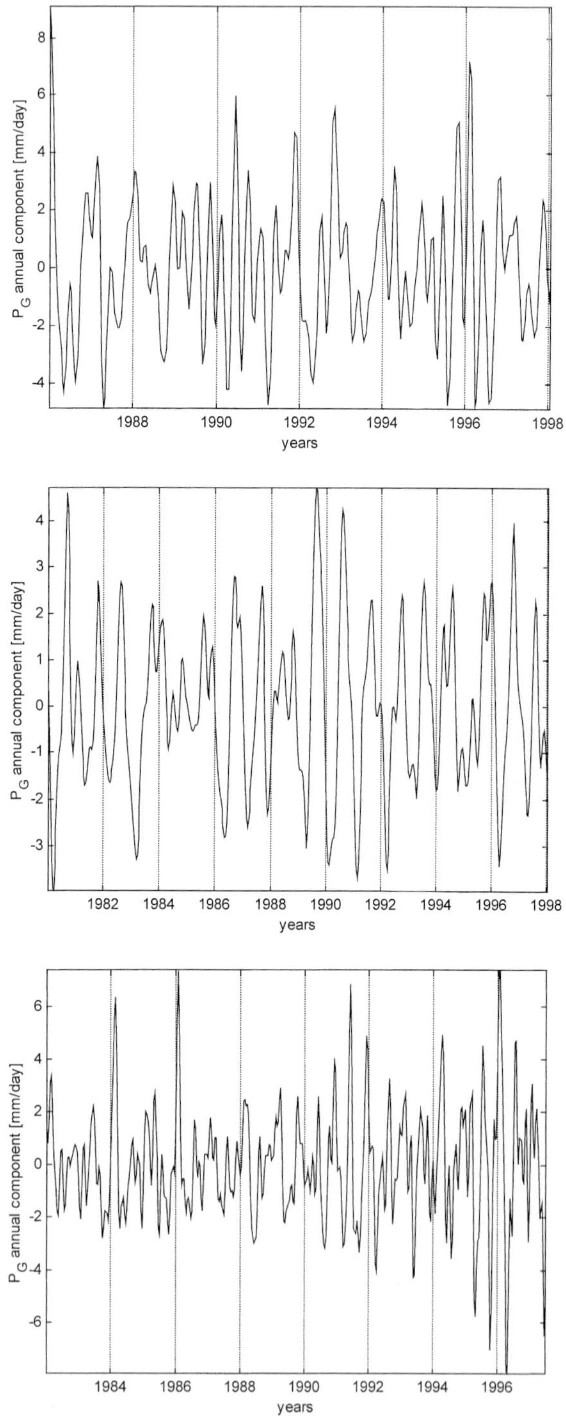

Figure 5. The time-series of the seasonal component (annual and intra-annual) of the rainfall distribution modelled from data collected at (a) the DVFC, (b) the Tawau, and (c) the Lahad Datu meteorological stations.

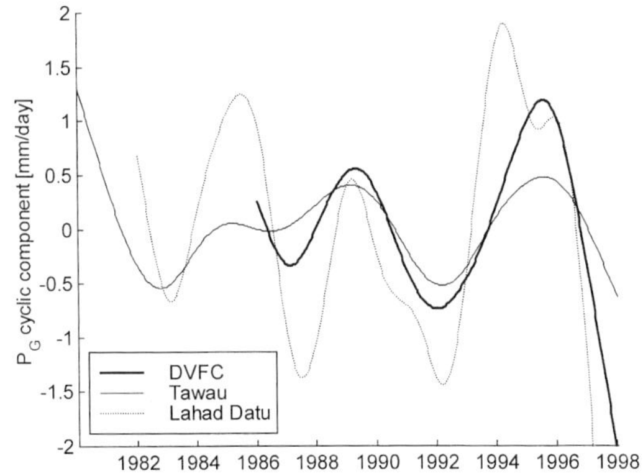

Figure 6. The time-series of the inter-annual cycles in rainfall modelled from data collected at the DVFC, Tawau and Lahad Datu meteorological stations.

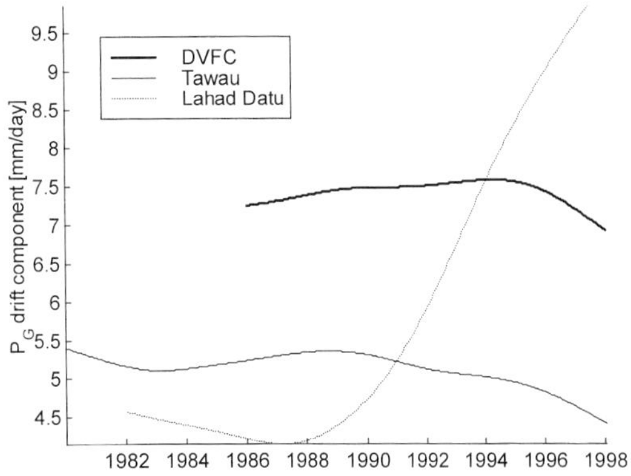

Figure 7. The time-series of 'longer-term' drift in the rainfall modelled from daily data collected at the DVFC, Tawau, and Lahad Datu meteorological stations.

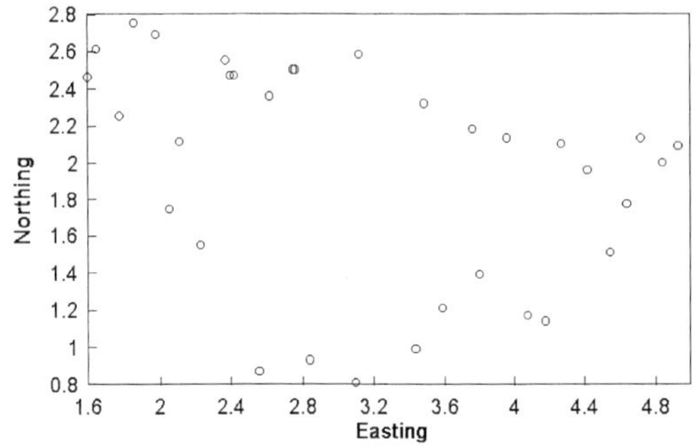

Figure 8. The spatial distribution of the raingauges monitoring gross-precipitation within the 10 km^2 study region near to DVFC. Axes are in kilometres horizontal distance.

and at Lahad Datu in June 1987. Again, the differences in time- series of raingauges a few tens of kilometres apart demonstrate the risks of interpolating temporal phenomena over meso-scale tropical regions.

The presence of drift in the rainfall totals over the 11–17 years of records was identified by the DHR modelling. While such 'longer-term' trends were not observed within the DVFC and Tawau rainfall records, over the period 1989–1997 Lahad Datu received a significant addition of rainfall of 6 mm d^{-1} (Figure 7). High convectional instability in equatorial regions means that the local development of cumulus clouds, and their associated rains, will be very sensitive to even small modifications to the local surface albedo, heat capacity or evapo-transpiration resulting from land-use changes (J.F.R. McIlveen, pers. comm.). As a consequence, the recent urbanisation and plantation development in the Lahad Datu area (Talip 1996) may be having an impact on the local rainfall. As this preliminary analysis only utilises relatively short (i.e., 11–17 year) records, it is, however, unrealistic to relate changes in the 'drift component' to impacts of local land-use change without further analysis, as longer-term cycles and trends within natural climatic phenomena are known to be present within the Borneo data-series (Walsh 1996). Indeed, Tych and Chappell (unpubl. data), again using the DHR model, have identified 18-year cycles within a 102-year rainfall record for Sandakan airport on the northern coast of Sabah. Further, Kripalani & Kulkarni (1997) describe 'epochs' of above- and below-average rainfall within time-series for Southeast Asia which exhibit a periodicity ranging from 10 to 30 years. Possibly, the 'epochs' for Lahad Datu, Tawau and DVFC are (a) out-of- phase, (b) have different intensities, or (c) have different durations.

The DVFC time-series analysis clearly identifies the importance of temporal variations in above-canopy rainfall caused by the presence of mid-late afternoon convective storms and ENSO phenomena in particular. Future analyses of the spatial variations in net precipitation (and wet-canopy-evaporation), therefore, ought to consider spatial patterns during individual convective storms and ENSO droughts.

(B) Spatial structure of above-canopy rainfall

The complex spatial patterns of sub-canopy rainfall will be dependent on the spatial patterns of (a) the rainfall received by the forest canopy and (b) the canopy properties. As the spatial characteristics of the P_g are expected to be less complex than those beneath the canopy, geostatistical (or stochastic) analysis of the P_g is, therefore, a first step in understanding the spatial structure of sub-canopy rainfall (Moliŏvá & Hubert 1994). Within this analysis, P_g data from 34 of the 50 storage gauges within the 10 km^2 DVFC study area (Figure 8) were examined using geostatistical modelling. The relationship between the raingauge catch and horizontal distance between raingauges was analysed by constructing 'experimental semi-variograms', where an estimate of the semi-variance $\hat{\lambda}(h)$ at lag distance (h) is:

$$\hat{\gamma}(h) = \frac{1}{2(n-h)} \sum_{i=1}^{n-h} \{z(i) + z(i+h)\}^2 , \qquad (3)$$

where z is the point rainfall total and n is the sample position (Journel & Huijbregts 1978). The parameters of the most appropriate model structure describing the relationship between the semi-variance in raingauge totals and lag distance were then identified. This analysis was performed on rainfall totals recorded over the 6-month period from 6 May to 27 November 1997 and for a single late afternoon event.

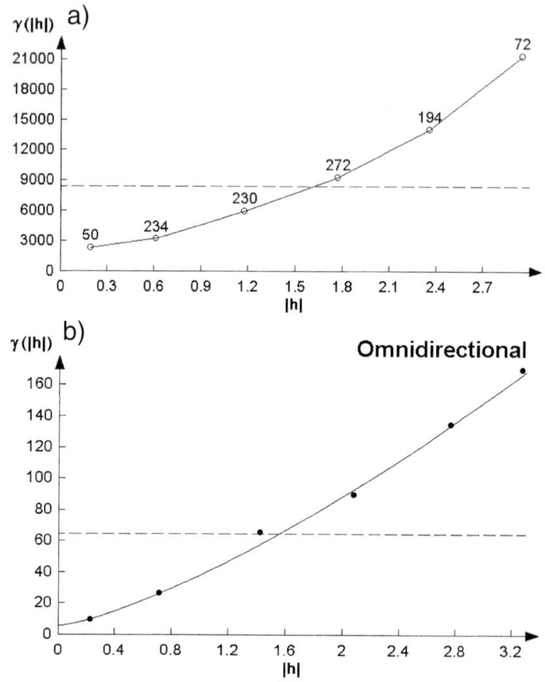

Figure 9. 'Experimental variograms' characterising the spatial variability in gross-precipitation within the 10 km^2 study region near to DVFC for (a) the 6-month period from 6 May to 27 November 1997, and (b) a single 15 mm convective event on the 17 September 1997. The number of raingauges pairs used to derive each value in (a) are shown, while the Power Variogram Model (solid line) fitted to the 'experimental variogram' for the single event is shown in (b). The lag spacing (h) has the units of kilometres, while the estimate of the semi-variance $(\hat{\lambda}(h))$ has the units of mm^2.

Results

The 34 raingauge totals for the 6-month period have a spatial mean of 1009.1 mm, a variance of 8633.0 mm^2 and a spatial coefficient of variation (CV) of 9.2%, while the data for the single event that occurred on the 17 September 1997 had a 14.7 mm mean, 66.7 mm^2 variance, and 55% CV. The clear reduction in the spatial CV from one to N-events is expected, as the patchiness of individual, local convective events will be smoothed by integration with other equally patchy events in the time-series. At the 10 km^2 scale, the 9.2% CV remaining after this integration process may be caused by the presence of local topographic (i.e., altitudinal, aspect or slope) effects on rainfall. Such topographic analysis is ongoing (Bidin in preparation).

The 'experimental variography' undertaken shows a monotonically increasing semi-variance with lag distance. This indicates that as the distance between the raingauges increases, the rainfall totals recorded by pairs of gauges become more dissimilar. This phenomenon is exhibited within the data averaged over 6-months (Figure 9a) and within the data specific to the individual convective storm of the 17 September 1997 (Figure 9b). Local convective rain cells beneath individual cumulus clouds are typically 1–10 km in diameter (Barry & Chorley 1982). As the experimental variogram for the single event does not exhibit a 'sill' towards the largest observed lag distances (ca. 3 km), this indicates that the 10 km^2 area contained either (a) a larger convective cell, (b) a group of small convective rain cells, or (c) a moving small cell. Clearly, at the scale of the 5000 km^2 'eastern Sabah' region, maximum lag distances would exceed the size of individual convective cells and be comparable to that of cell clusters called 'mesoscale convective systems' (Houze et al. 1981). A 'sill' in the variogram (where the maximum extent of spatial correlation has been exceeded) might, therefore, be observable if intensive spatial sampling were to be undertaken at this larger scale. Indeed, other recent studies examining the spatial structure of rainfall in areas where convective cells are important exhibit variogram sills only after 10–40 km lag distances (Borga & Vizzaccaro 1997; Lebel & Le Barbé 1997).

As the rate of change of dissimilarity in raingauge totals in the DVFC area increases with increasing distance and a 'sill' is not reached within the experimental variogram, a (theoretical) Power Variogram Model (PVM) best describes the change in the semi-variance:

$$\gamma(h) = wh^{\alpha}, \quad (4)$$

where w is the positive variance contribution (or scaling term) and α is the power exponent of between 0 and 2. The PVM for the 17 September 1997 event has an exponent of approximately 1.3 (i.e., a concave-upward power relation) and a scaling term of about 35 (Figure 9b). A 'nugget variance' (c_0) of 6 has been added to the PVM as the experimental variogram appears to cross the y-ordinate slightly above zero. Such a small nugget variance (i.e., 9% of total variance) does, however, indicate both the similarity of totals where gauges are only metres apart and little error within the raingauge measurements themselves (Goodrich et al. 1995). In some contrast, the power exponent of a PVM fitted to the 6-month data approaches 2 and may, therefore, indicate that 'deterministic drift' is present within the experimental variogram (Pannatier 1996). Such drift might be caused by local altitudinal enhancement of convection (Goodrich et al. 1995) or upwind/downwind effects of local topography (Lyons & Bonell 1992). An *a priori* parameterisation of these effects would lead to more robust spatial models. The preliminary geostatistical analysis does, however, indicate that as the variography lacks both large 'nugget' and 'hole' effects it is useful in identifying spatial structures in above-canopy rainfall over a 10 km^2 area of rainforest. These spatial models will allow kriging techniques (Journel & Huijbregts 1978) to be used to interpolate rainfall fields accurately from the sparse raingauge networks of other research programmes. These improvements are required for further developments in the modelling of rainfall-runoff and rainfall-sediment behaviour within 'experimental catchments' (Chappell et al. 1998, 1999a). Furthermore, identification of models that define the stochastic and topography-related patterns in above-canopy rainfall are important to the separation of the independent controls on P_{net} of (a) the above-canopy rainfall, and (b) the canopy structure.

(C) Canopy structure and sub-canopy rainfall

Summary data for the period 1 May to 27 November 1997 (211 days) were used to examine the relationship between sub-canopy rainfall and each of the 'types' of canopy left by selective logging of lowland dipterocarp forest (Table 1). Throughfall (i.e., $P_{ld} + P_{dt}$ in Figure 1) within each vegetation category was calculated as a proportion of the local above-canopy precipitation. The remaining component of the sub-canopy rainfall, stemflow, which only amounted to approximately 1% of the above-canopy rainfall (Bidin in preparation) is not examined in detail within this preliminary analysis,

Results

Over the 6-month monitoring period, the 80 raingauges beneath the undisturbed patches of rain forest collected 91.1% of that monitored within nearby canopy gaps (Table 1). Assuming stemflow is 1% of the above-canopy rainfall, the resultant 8% of wet-canopy-evaporation (E_{wc}) equates to an annual depth of 113 mm during the ENSO drought year (i.e., $65.2 \times 365/211$). While this rate of E_{wc} from virgin forest is relatively low compared to many other published rates for undisturbed lowland dipterocarp forests in Southeast Asia (Table 2), similarly low rates can be found in the literature. For example, a rate of 8% E_{wc} has been observed in the undisturbed lowland dipterocarp forest at Pasoh, West Malaysia (UNESCO 1978), and Lloyd et al. (1988) has reported a figure of 9% E_{wc} for undisturbed Amazonian terra firma rain forest. Furthermore, Asdak et al. (1998) has noted a large variability in E_{wc} from different undisturbed tropical rain forests and suggested that more studies, particularly those incorporating intensive sampling strategies, are required to understand such variations within the tropics. Recently published results from Southeast Asia (Table 2) do, however, appear to exhibit a trend. There is a strong and significant positive linear relationship between the annual totals of above-canopy rainfall ($P_{g(yr)}$) and the annual percentage of wet-canopy-evaporation ($\%E_{wc(yr)}$), where:

$$\%E_{wc(yr)} = 0.0088 P_{g(yr)} - 7.5535 \quad (r^2 = 0.91,\ p < 0.05). \quad (5)$$

In other words, the proportion of the rainfall lost to wet-canopy-evaporation increases with increasing annual rainfall. The lower proportion of the E_{wc} for virgin forest at DVFC observed in 1997 compared to that observed in 1989–1990 at a nearby site (Sinun et al. 1992) may be, therefore, associated with the reduced total of gross rainfall recorded during the intense 1997–1998 El Niño year. Such an apparent correlation between E_{wc}-proportion and total P_g probably results from changes in storm frequency through ENSO cycles (W. T. Swank, pers. comm.).

Rainfall beneath disturbed forest patches within the study area was monitored over the same 6-month

Table 2. Recent studies presenting annual rates of wet-canopy-evaporation (E_{wc}) measured in undisturbed lowland dipterocarp forest of Southeast Asia.

Reference	Location	No. gauges (sub-canopy)	Annual		
			Rainfall (mm)	E_{wc} (mm)	E_{wc} (%)
This study	East Sabah	80	1549[a]	113	8
Asdak et al. 1998	Central Kalimantan	50	2199	251	11
Kasran 1989	West Malaysia (central)	4[b]	3722	1005	27
Sinun et al. 1992	East Sabah[c]	40	2824	569	17
Yusop 1996	West Malaysia (central)	4	2723	354	13

[a]Extrapolated from a 211 day total of 895.5 mm.
[b]0.76 m long troughs, rather than raingauges.
[c]Only 5 km from the site used within this study.

period as within the virgin forest patches. First, Table 1 shows that the wet-canopy evaporation (i.e, $P_g - P_{\mathrm{net}}$) from *Macaranga* spp. trees (category 4 cover) and the shrubs/herbs (category 5 cover) was not statistically different ($p < 0.100$) to that penetrating the virgin forest.

Secondly, the data from the 160 gauges located beneath category 2 and 3 vegetation – the moderately and severely impacted patches of tree canopy are significantly different to those from the undisturbed forest. Throughfall beneath the two disturbed forest categories amounted to 84.1% and 79.6% of received rainfall, respectively (Table 1). These are statistically significant reductions in the quantity of rainfall penetrating the disturbed canopies relative to the situation within virgin forest. Again assuming for this preliminary analysis a stemflow of 1% of above-canopy rainfall, this gives apparent rates of wet-canopy-evaporation of 15% and 19%, respectively. There are three preliminary explanations for the apparent increases in E_{wc} following canopy disturbance. First, there is an increased density of leaves on the outer surfaces of the disturbed canopy resulting from the expansion of the woody climbers. This outer surface experiences the highest temperatures and rates of net radiation and, therefore, the potential for high rates of wet-canopy-evaporation. Thus, increasing leaf density in this part of the canopy may have a disproportionate effect on the whole-forest E_{wc}. Second, selective removal of upper canopy trees during silvicultural operations, increases the roughness of the forest surface, which in turn increases the levels of turbulence over the canopy. As increasing atmospheric turbulence over the canopy (i.e., the 'atmospheric conductance' term, Dingman 1994) increases evaporation, then rates of E_{wc} are likely to increase. The third explanation is related to the interaction between the canopy surface and the above- canopy rainfall. Turbulent eddies in the wind above the forest canopy will mean that rainfall often approaches the forest canopy at an acute angle. This may mean that patches of high virgin forest 'intercept' more rainfall at the canopy surface than lower patches of disturbed canopy. In other words, the higher canopies of virgin forest may shelter the lower disturbed canopies (cf., Barry & Chorley 1982, p. 300).

Conclusions and implications

Ten conclusions can be derived from the analysis of the gross rainfall and canopy controls on net precipitation within the studied rainforest.

(1) Forty-three percent of the above-canopy rainfall received by the Ulu Segama rain forest is observed in the mid-afternoon (2–4 p.m.). A similar proportion of the sub-canopy or net rainfall is, therefore, expected between these two hours. A single afternoon or early evening peak in rainfall is indicative of a system dominated by local convective cells (cumulus clouds) and as a result, one that is expected to give a highly heterogeneous spatial rainfall pattern. Clearly, such a situation would demand both a relatively high density of raingauges and an understanding of the spatial structure of the rainfall for estimates of both (a) spatially-averaged rainfall over a typical 'experimental catchment', and (b) rainfall received by 'research plots' distributed throughout at region of a

few square kilometres. The extent of seasonal variations in the diurnal rainfall distribution is currently under investigation (Bidin in preparation).

(2) The dominance of mid-afternoon rains combined with the qualitative observation that the canopies wetted by these rains stay wet until the next day, may have local meteorological implications. First, during calm nights, such canopy water would probably enhance the formation of 'radiation fog' within and just above the upper canopy (Watts 1955; J. F. R. McIlveen, pers. comm.). Second, it may also mean that some water stored on the canopy surfaces is not evaporated from wet surfaces until after perhaps as many as 20 hours (i.e., 3 p.m. to 10 a.m.). The diurnal distribution in wet-canopy-evaporation may, therefore, significantly lag behind that of the above- and sub-canopy rainfall.

(3) Very little annual seasonality is apparent within the daily rainfall records at DVFC. Twelve-month sampling periods of net precipitation (or wet-canopy-evaporation), required in areas with a marked annual seasonality, therefore, may not be necessary for hydrometeorological work in the Ulu Segama region of Sabah. As annual cycles in the rainfall are apparent within the Tawau records, this means that net precipitation or wet-canopy-evaporation studies undertaken in areas only 80 km to the south southeast of DVFC would need at least annual time-series. This also highlights the risks in extrapolation or interpolation of rainfall totals throughout sparsely instrumented tropical regions as small as 5000 km^2.

(4) The presence of marked droughts with a recurrence interval of 4.5 to 5.5 years within the rainfall series for eastern Sabah, that coincide with the periodicity of El Niño-Southern Oscillation (ENSO) phenomena, mean that any short-term studies on for example the soil-moisture-status of rain forest soils, the rates of erosion in rain forest terrain or the associated status of the aquatic biology, should be replicated within both drought and non-drought periods.

(5) The preliminary geostatistical analysis indicates that a deterministic spatial structure may be affecting the experimental variogram for the 10 km^2 study region near DVFC. This structure may be related to local altitudinal enhancement of convection or upwind/downwind effects of local topography. Future analysis needs to identify and then de-trend these effects prior to further variography. The lack of a 'sill', large 'nugget' or 'hole' effects within the variogram for a rain forest region the size of the 0.5–10 km^2 experimental catchment does, however, indicate that the spatial patterns of above-canopy rainfall can be modelled geostatistically. This means that the spatial structure of above-canopy rainfall can be separated from the more complex sub-canopy patterns to be examined within the future analysis of Bidin (in preparation).

(6) Comparison of the proportion of the gross rainfall lost as wet-canopy evaporation for patches of undisturbed lowland dipterocarp forest near DVFC with those from recent studies undertaken within Southeast Asia indicates that differences may relate to changes in annual rainfall totals. As the data-set for this comparison is small, such a conclusion is tentative and requires further investigation. A rate of 8 percent wet- canopy evaporation for the undisturbed lowland dipterocarp forest near DVFC is low, but is not inconsistent with some published studies (e.g., UNESCO, 1978) and with rates observed by ongoing studies elsewhere in the tropics (L. A. Bruijnzeel, pers. comm.).

(7) Within the Ulu Segama forest, the moderately impacted and creeper-covered, highly damaged patches of forest have reduced rates net-precipitation and elevated levels of wet-canopy-evaporation in comparsion to the undisturbed patches. It should be noted that such changes relate to a period some 9 years post logging and during a severe ENSO-drought. The reduction in net-precipitation needs further investigation, but may relate to (a) elevated wet-canopy-evaporation from creeper-covered surfaces, (b) rougher forest surfaces, or (c) the lower canopies of disturbed forest patches receiving less rainfall due to local sheltering effects. The large differences in net-precipitation between the patches of (i) undisturbed and *Macaranga* spp. forest, and (ii) the moderately impacted cum creeper-covered forest patches, will mean that the hectare-scale spatial variability in sub-canopy rainfall will have increased in comparison to large regions of undisturbed forest. This will mean that spatial patterns of infiltation-excess overland flow and litter moisture status will be more heterogeneous on natural slopes within selectively-logged forest and, therefore, more difficult to predict. Analysis to identify which of these explanations is the most robust, is ongoing (Bidin in preparation).

Rates of transpiration might be increased or decreased by selective logging and would, therefore, either magnify or offset changes in the water-yield of a catchment resulting from changes in wet-canopy-evaporation. The 10 km^2 region near DVFC where net-precipitation has been studied, contains the 0.44 km^2 'Baru Experimental Catchment' (Chappell et al. 1998, 1999a). Bidin (in preparation) is currently undertaking the water-balance for this catchment, to assess the net result of separate changes in wet-canopy-evaporation and biological transpiration. Estimation of the catchment-average net precipitation will be undertaken by integrating the average rates per canopy category, multiplying these values by the proportion of the catchment covered by that category and then summing the products. Areal photographs, taken from 30 to 600 m above the canopy, will be used aid in the estimation of the proportion of the Baru Experimental Catchment covered by each canopy category. Despite this, uncertainties in the catchment-average net precipitation will result from uncertainties in the classification of all patches of canopy into one of the 6 discrete categories defined (Bellehumeur et al. 1997).

Acknowledgements

The authors would like to thank Rahman Merami (UMS-supported assistant) for help in the collection of the gross and net precipitation data in the DVFC locality, Johnny Larenus and Mohd. Jamal Hanapi (DVFC Hydrology Project) for the collection of the DVFC meteorological station and 'km63' data, and the Malaysian Meteorological Service for the rainfall data from the Lahad Datu and Tawau meteorological stations. Ismail Abd. bin Samat and Shaidih Abd. bin Samat (Royal Society Survey Team) are thanked for their detailed electronic survey. The comments of Dr J. F. Robin McIlveen (Lancaster University, UK), Prof Wayne T Swank (Coweeta Hydrologic Laboratory, USA), Dr L. A. (Sampurno) Bruijnzeel (Free University, Netherlands) and reviewers of the manuscript are much appreciated. M.D. Sherlock is thanked for the basemap for Figure 2. The support of Prof. Dr Noh Dalimin (Universiti Malaysia Sabah), Dr. Waidi Sinun of the Forestry Division (Innoprise Corporation) of Yayasan Sabah, and the Danum Valley Management Committee are gratefully acknowledged. Financial support for this work has been provided by Universiti Malaysia Sabah (UMS), UK Natural Environment Research Council (grant GR3/9439) and the South East Asia Rain forest Research Programme of the Royal Society of London (publication number A/306).

References

Asdak, C. Jarvis, P. G., van Gardingen, P. & Fraser, A. 1998. Rainfall interception loss in unlogged and logged forest areas of Central Kalimantan, Indonesia. J. Hydrol. 206: 237–244.

Barry, R. G. & Chorley, R. J. 1982. Atmosphere, weather and climate. Fourth Edition. Methuen, London.

Bellehumeur, C., Legendre, P. & Marcotte, D. 1997. Variance and spatial scales in tropical rain forest: Changing the size of sampling units. Plant Ecol. 130: 89–98.

Bidin, K. in preparation. Spatial and temporal structure of gross and net precipitation within a selectively-logged forest catchment. Unpublished Ph.D. thesis, University of Lancaster, Lancaster.

Borga, M. & Vizzaccaro, A. 1997. On the interpolation of hydrologic variables: formal equivalence of multiquadratic surface fitting and kriging. J. Hydrol. 195: 160–171.

Brooks, S. M. & Spencer, T. 1995. Vegetation modification of rainfall characteristics: implications for rainfall erosivity following logging in Sabah, Malaysia. J. Trop. For. Sci. 7: 435–446.

Bruijnzeel, L. A., Waterloo, M. J., Proctor, J., Kuiters, A. T. & Kotterink, B. 1993. Hydrological observations in montane rain forests on Gunung Silam, Sabah, Malaysia, with special reference to the *'Massenerhebung'* effect. J. Ecol. 81: 145–167.

Calder, I. R. 2001 Canopy processes: implications for transpiration, interception and splash induced erosion, ultimately for forest management and water resources. Plant Ecol. 153: 203–214 (this volume).

Chappell, N. A., Franks, S. W. & Larenus, J. 1998. Multi-scale permeability estimation for a tropical catchment. Hydrol. Process. 12: 1507–1523.

Chappell, N. A., McKenna, P., Bidin, K, Douglas, I, & Walsh, R. P. D. 1999a. Parsimonious modelling of water and suspended-sediment flux from nested-catchments affected by selective tropical forestry. Philos. Trans. R. Soc. Lond. Ser. B 354: 1831–1846.

Chappell, N. A., Ternan, J. L. & Bidin, K. 1999b. Correlation of physicochemical properties and sub-erosional landforms with aggregate stability variations in a tropical Ultisol disturbed by forestry operations. Soil Tillage Res. 50: 55–71.

Conway, S. 1982. Logging practices. Miller Freeman Publications, San Francisco.

Dingman, S. L. 1994. Physical hydrology. Prentice Hall, Englewood Cliffs.

Discenza, A. R., Whittaker, J., & Chappell, N. A., in preparation. Stochastic modelling of tropical rainfall occurrence processes at daily and hourly time-scales.

Douglas, I., Bidin, K., Balamurgam, G., Chappell, N. A., Walsh, R. P. D., Greer, T. & Sinun, W. 1992. The role of extreme events in the impacts of selective tropical forestry on erosion during harvesting and recovery phases at Danum Valley, Sabah. Philos. Trans. R. Soc. Lond. Ser. B 354: 1749–1761.

Douglas, I., Spencer, T., Greer, T., Bidin, K., Sinun, W. & Wong, W. M., 1992. The impact of selective commercial logging on stream hydrology, chemistry and sediment loads in the Ulu Segama Rain Forest, Sabah. . Philos. Trans. R. Soc. Lond. Ser. B 335: 397–406.

Eschenbach, C., Glauner, R., Klein, M. and Kappen, I. 1998. Photosynthesis rates of selected tree species in lowland dipterocarp rain forest of Sabah, Malaysia. Trees-Struct. Funct. 12: 356–365.

Goodrich, D. C., Faurès, J-M., Woolhiser, D. A., Lane, L. J. & Sorooshian, S. 1995. Measurement and analysis of small-scale convective storm rainfall variability. J. Hydrol. 173: 283–308.

Henderson-Sellers, A., Dickinson, R. E., Durbridge, T. B., Kennedy, P. J., McGuffe, K. & Pitman, A. J. 1993. Tropical deforestation:

modelling local to regional-scale climatic change. J. Geophys. Res. 98 (D4): 7289–7315.

Herwitz, S. R. 1985. Interception storage capacities of tropical rain forest trees. J. Hydrol. 77: 237–252.

Houze, R. A., Geotis, S. G., Marks, Jr., F. D. & West A. K. 1981. Winter monsoon convection in the vicinity of north Borneo. Part 1: structure and time variation of the clouds and precipitation. Mon. Weather Rev. 109: 1595–1614.

Johnson R. H. & Priegnitz, D. L. 1981. Winter monsoon convection in the vicinity of North Borneo. Part II. Effects on large-scale fields. Mon. Weather Rev. 109: 1619–1632.

Journel, A. G. & Huijbregts, C. J. 1978. Mining geostatistics. Academic Press, New York.

Kasran, B. 1989. Rainfall interception in dipterocarp forest of Peninsular Malaysia. Pp. 1–15. In: Regional seminar on tropical forest hydrology. Forest Research Institute of Malaysia, Kuala Lumpur.

Kripalani, R. H. & Kulkarni, A. 1997. Rainfall variability over South-east Asia – connections with Indian monsoon and ENSO extremes: new perspectives. Int. J. Clim. 17: 1155–1168.

Kripalani, R. H. & Kulkarni, A. 1998. The relationship between some large-scale atmospheric parameters and rainfall over Southeast Asia: a comparison with features in India. Theor. Appl. Clim. 59: 1–11.

Lebel, T. and Le Barbé, L. 1997. Rainfall monitoring during HAPEX-Sahel. 2. Point and areal estimation at the event and seasonal scales. J. Hydrol. 188: 97–122.

Lloyd, C. R. 1990. The temporal distribution of Amazonian rainfall and its implications for forest interception. Quart. J. Roy. Meteorol. Soc. 116: 1487–1494.

Lloyd, C. R. Gash, J. H. C., Shuttleworth, W. J. & Marques Filho, A de O. 1988. The measurement and modelling of rainfall interception by Amazonian rain forest. J. Hydrol. 43: 277–294.

Lyons, W. F. & Bonell, M. 1992. Daily mesoscale rainfall in the tropical wet/dry climate of the Townsville area, north-eastern Queensland during the 1988/89 wet season: synoptic scale airflow considerations. Int. J. Clim. 12: 655–684.

Madden, R. A. & Julian, P. R. 1972. Description of global scale circulation cells in the tropics with a 40-50 day period. J. Atmos. Sci. 29: 1109–1123.

Martin-Smith, K. 1999. Biodiversity patterns of tropical freshwater fish following selective timber extraction: a case study from Sabah, Malaysia. Ital. J. Zool. 65: 363–368.

Moliŏvá, H. & Hubert, P. 1994. Canopy influence on rainfall fields' microscale structure in tropical forests. J. Appl. Meteorol. 33: 1464–1467.

Newbery, D. McC., Campbell, E. J. F., Lee, Y. F., Ridsdale, C. E. & Still, M. J. 1992. Primary lowland dipterocarp forest at Danum Valley, Sabah, Malaysia: structure, relative abundance and family composition. Philos. Trans. R. Soc. Lond. Ser. B. 335: 341–356.

Newell, R. E. & Gould-Stewart, S. 1981. A stratospheric fountain ? J. Atmos. Sci. 12: 2789–2796.

Noguchi, S., Nik, A. R., Sammori, T., Tani, M., & Tsuboyama, Y. 1996. Rainfall characteristics of tropical rain forest and temperate forest: comparison between Bukit Tarek, Peninsular Malaysia and Hitachi Ohta in Japan. J. Trop. For. Sci. 9: 206–220.

Pannatier, Y. 1996. Variowin: software for spatial data analysis in 2D. Springer-Verlag, New York.

Pinard, M. A. & F. E. Putz. 1996. Retaining forest biomass by reducing logging damage. Biotropica 28: 278–295.

Ramage, C. S. 1964. Diurnal variation of summer rainfall of Malaya. J. Trop. Geogr. 19: 62–68.

Riehl, H. 1954. Tropical meteorology. McGraw-Hill, New York.

Sherlock, M. D., 1997. Plot-scale hydrometric and tracer characterisation of soil water flow in two tropical rain forest catchments in Southeast Asia. Unpublished PhD thesis, Lancaster University.

Shuttleworth, W. J. 1988. Evaporation from Amazonian rain forest. Proc. R. Soc. Ser. B 233: 321–346.

Sinun, W., Meng, W. W., Douglas, I. & Spencer, T. 1992. Throughfall, stemflow, overland flow and throughflow in the Ulu Segama rain forest, Sabah, Malaysia. Philos. Trans. R. Soc. Lond. Ser. B 233: 321–346.

Talip, M. A., 1996. Using a Geographical Information System (GIS) to analyse land use changes in eastern Sabah, Malaysia. Unpublished M.Sc. thesis, University of Manchester, UK.

Taylor, J. R. 1982. An introduction to error analysis. University Science Books, Mill Valley.

Thompson, B. W. 1951. An essay on the general circulation of the atmosphere over South-East Asia and the West Pacific. Quart. J. Roy. Meteorol. Soc. 77: 569–597.

UNESCO. 1978. Tropical forest ecosystems (Chapter 12). UNESCO, Paris.

Walsh, R. P. D. 1996. Drought frequency changes in Sabah and adjacent parts of northern Borneo since the late nineteenth century and possible implications for tropical rain forest dynamics. J. Trop. Ecol. 12: 385–407.

Watts, I. E. M. 1955. Equatorial weather: with particular reference to Southeast Asia. University of London Press, London.

Weickmann, K. M., Lussky, G. R. & Kutzbach, J. E. 1985. Intraseasonal (30–60 day) fluctuations of outgoing longwave radiation and 250 mb streamfuction during northern winter. Mon. Weather Rev. 111: 1838–1858.

Whitmore, T. C. 1998. An introduction to tropical rain forests. Second Edition. Oxford University Press, Oxford.

Wolter, K. & Timlin, M. S. 1998. Measuring the strength of ENSO - how does 1997/98 rank ? Weather 53: 315–324.

Wood, C. A. 1948. Report on the weather of the Borneo-Celebes region, 1946. Quart. J. Roy. Meteorol. Soc. 74: 144–160.

Young, P. C. 1998. Data-based mechanistic modelling of environmental, ecological, economic and engineering systems. Environ. Modell. Softw. 13: 105-122.

Young, P. C., Pedregal, D. J., & Tych, W. 1999. Dynamic harmonic regression. J. Forecasting 18: 369–394.

Yusop, Z. 1996. Nutrient cycling in secondary rain forest catchments of Peninsular Malaysia. Unpublished Ph.D. thesis, University of Manchester, Manchester.

Plant Ecology **153:** 231–239, 2001.

Micrometeorological conditions and canopy energy exchanges of a neotropical rain forest (Surumoni-Crane Project, Venezuela)

Jörg Szarzynski & Dieter Anhuf
Department of Physical Geography, University of Mannheim, L 9, 1-2, 68131 Mannheim, Germany

Key words: Canopy conductance, Evapotranspiration, Microclimate, Penman-Monteith equation, Transpiration, Turbulent exchanges

Abstract

Vertical profiles of air temperature, humidity, wind speed and photosynthetically active radiation were examined systematically within and above a primary Amazonian rain forest in Southern Venezuela. During daylight hours the observed gradients of temperature and humidity suggest that turbulent mixing between vegetation and the atmosphere is reasonably efficient in the top two-thirds of the forest, whereas the understorey remains partially isolated. At night, however, the canopy exhibits a significant decoupling from the overlying atmospheric layer due to substantial radiative cooling, causing a stable density stratification above. It is fairly clear that these variations in microclimate produce a spectrum of different living conditions for the flora and fauna.

In addition, the collected data provide important information concerning the turbulent exchanges of heat and water vapour. Actual evapotranspiration loss from the forest was estimated using a single-layer version of the Penman-Monteith equation including a submodel of canopy conductance. The values computed for hourly and daily periods were found to agree well with the simultaneously recorded xylem sap flow of several tree species. Thus they prove that this application is suitable to describe the environmental impact of micrometeorological and physiological factors on the complex process of evapotranspiration.

Introduction

The significant role of tropical rain forests in regulating the regional surface energy and water balance and in maintaining the rich diversity of animal and plant species is increasingly endangered by human interventions such as deforestation and the conversion into agricultural land or pasture (Mooney et al. 1980; Wilson 1992). Assessments of the large-scale impacts of these changes in vegetation are still limited by the restricted current knowledge of many critical processes involved (Shukla et al. 1990; but see Chappell et al. 2001). In part, this deficiency is related to the fact that the major interface between the vegetation and the atmosphere, the uppermost forest canopy, remained unexplored for a long time because of the difficulties of access. However, the need for tropical forest research is nowhere greater than for the upper crown area, where natural photosynthetic rates are highest (Holbrook & Lund 1995) and canopy-mediated exchanges of energy and mass largely influence the water, carbon and nutrient cycles, as well as the chemical composition of the atmosphere (Graedel & Crutzen 1994).

To investigate the structure and function of an undisturbed tropical rain forest with special reference to its upper canopy level, a collaborative research project of the Austrian Academy of Science and the Venezuelan government was established in the centre of the Venezuelan state Amazonas. Since November 1995 permanent access into the canopy is provided by a mobile crane system running on rails 120 m long. The crane stands 42 m high with a horizontal jib 40 m long. Thus, the research tool facilitates *in situ* experimentation covering an area of approximately 1.5 ha. This paper presents first results of the micrometeorological programme. Main objectives of the long-term data acquisition are three-fold, namely: (i) to provide a comprehensive data set of regional climate and forest microclimate; (ii) to determine the component fluxes of the forest energy balance including the amount of actual evapotranspiration loss; (iii) to evaluate the applicability of different measurement techniques and parameterization models.

Methods

Site description and background climatology

The crane plot is located close to the confluence of the Río Surumoni (3°10′ N; 65°40′ W; 105 m a.s.l.) and the upper Río Orinoco (Figure 1). Geologically the region is part of the ancient Guayana Shield. The physiography of the landscape is dominated by gently undulating peneplains partly interrupted by the occurrence of sheer cliffed sandstone table mountains (Huber 1995). One of these so-called 'tepuis', the large and extensive Duida-Marahuaka massif, is located some 15 km to the north of the crane plot and reaches 2358 m in elevation at the southwestern rim of the Cerro Duida and 2800 m at the highest point of the northern Cerro Marahuaka. The exposed relief has a controlling influence on local climatic conditions since it favours the occurrence of orographic rain.

The study area is covered by a primary ombrophilous lowland rain forest growing on a mosaic of deeply weathered ferralsols and arenosols (FAO-UNESCO 1988). The soils are highly acidic (pH 3.5–5), poor in nutrients and generally characterized by a low cation exchange capacity. The 1.5 ha rectangular crane plot contains 1061 marked trees (DBH > 10 cm), which might be associated with about 100 tree species (Morawetz et al. 1998). An interesting feature of the forest composition is the great abundance of *Goupia glabra* Aubl. (Celastraceae), a typical pioneer tree that is usually frequent in secondary forests. At the Surumoni 30% of the crown coverage is formed by 80 individuals of this species (Nieder et al. 2000). Other common tree species are *Dialium guianense* (Aubl.) Sandw. (Caesalpiniaceae), *Ocotea sp.* (Lauraceae), *Qualea sp.* (Vochysiaceae) and *Oenocarpus baccaba* Mart. (Arecaceae), one of the 18 palm species representing about 10% of all marked tree individuals within the plot.

The forest extends to an average height of 25 m with a stratification in shrub layer, understorey and canopy; emergent trees do not occur. The topography of the canopy is characterised by depressions of occasionally 5–6 m in the vertical vegetation profile and a considerable number of gaps occupying 10% of the plot area.

The average rainfall for this region is rather seasonal. The rainy season occurs from April-November under the influence of moist southeastern trade winds from the Amazon basin. Drier and warmer weather conditions during December-March result from northeastern trade winds. Average annual precipitation is around 2700 mm (La Esmeralda; period: 1970–1995; Dirección de Hidrología y Meteorología, Caracas). In 1996, however, in common with much of northern and central South America (WMO 1997), total precipitation in the study area was well above average (3700 mm) with devastating local flooding in June and July. Around one third of the Surumoni research plot near to the river bank was inundated. The average annual temperature is about 26 °C with a daily range of 5 to 10 °C. Specific humidity usually remains between 15 to 25 g kg^{-1}, with slight seasonal variations.

Instrumentation

Two micrometeorological configurations were employed for simultaneous measurements within and above the forest. The above-canopy system comprised profile readings of air temperature, vapour pressure and wind velocity recorded at 28.5, 31.5, 35 and 41 m height, as well as the determination of wind direction, air pressure and precipitation. The instruments were mounted on metal tubes 6 m long, which were directly fixed to the crane tower. Dry and wet bulb temperatures were measured with shielded and fully ventilated platinum-resistance psychrometers (Vector Instruments, UK) and wind speed with sensitive cup anemometers (Vector Instruments, UK). Data concerning radiative fluxes were provided by two pyranometers (Kipp & Zonen, the Netherlands) measuring the incoming and the outgoing short-wave radiation, a hemispherical pyrradiometer (Schenk, Austria) recording net all-wavelength radiation and a quantumsensor (Sky Instruments, UK) for photosynthetically active radiation (PAR) in the waveband 400–700 nm. Sensors are connected to a battery-powered 21X datalogger (Campbell Scientific, UK) computing 10-min averages from 20-s sampling intervals.

Within the forest, an innovative rope and pulley system was rigged to a 28 m high *Albizia* tree (Mimosaceae) to support the recordings of shielded temperature and relative humidity probes (Vaisala, UK) which were fixed at three heights appropriate to the forest stratification (5 – 12.5 – 21 m). Additionally

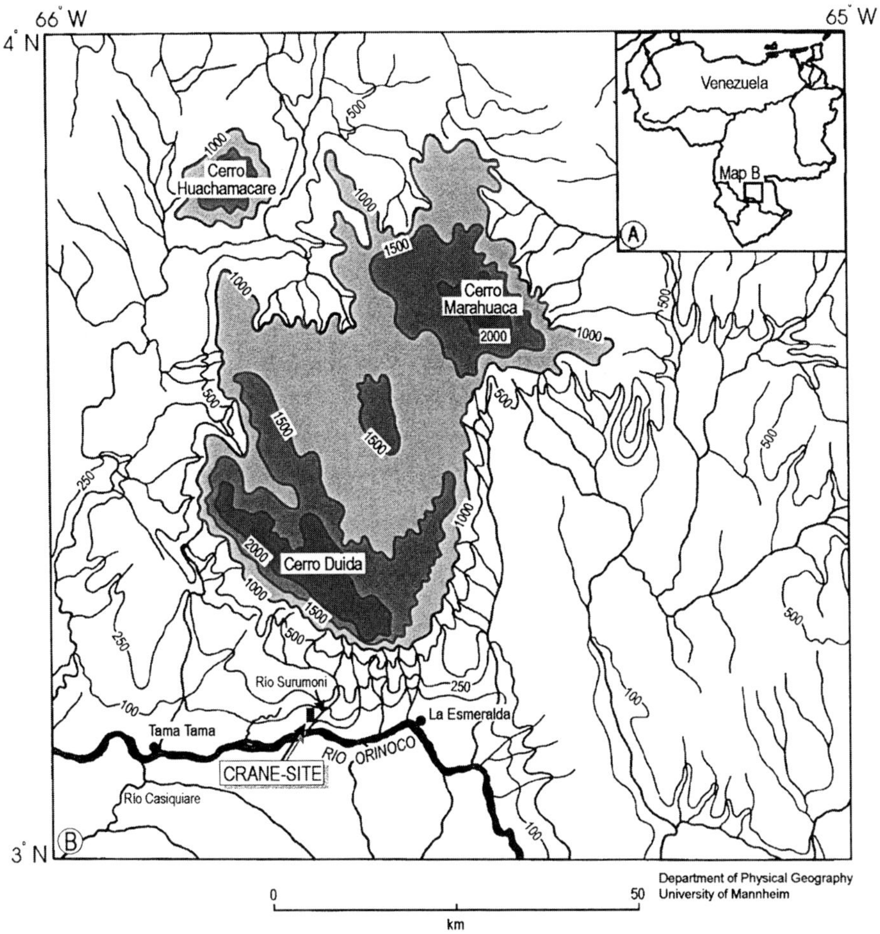

Figure 1. Map showing the study-site of the Surumoni-Project in Southern Venezuela

wind speed and PAR were measured in the middle layer (12.5 m) and in the upper canopy (21 m). Soil temperature was registered at five depths and several heat flux plates and soil moisture sensors were buried below the surface at 8 mm and 200 mm, respectively. Data were stored on a CR10 datalogger (Campbell Scientific, UK) operating with the same intervals as the above-canopy system.

Results and interpretation

Micrometeorological conditions

Of particular importance are the complex biophysical mechanisms that drive exchange processes between the vegetation and the atmosphere (Shuttleworth 1988), as well as the effects on flora and fauna (Endler 1993; Wright & Colley 1994; Théry 2001). In order to gain a first insight into the micrometeorological conditions of the Surumoni rain forest, a sequence of two consecutive but contrasting days is presented in Figure 2. A completely fine day (20 June 1997) was followed by a cloudy day in which a short convective rainstorm occurred (shown by dashed line in Figure 2 and subsequent figures).

The first day (20 June 1997) represents dry canopy conditions. During daylight hours the temperature cycle is mainly influenced by the absorption of incident solar radiation in the upper part of the canopy, leading to exponential decay of light and radiative energy with distance into the forest. The stronger heating of biomass and interstitial air in the crown region produces a stable density stratification of warmer lighter air above, and cooler denser air in the vegetation below (Figure 2a). In this situation the temperature inversion separates air in the understorey from that in the upper part of the forest and the atmosphere above. It is

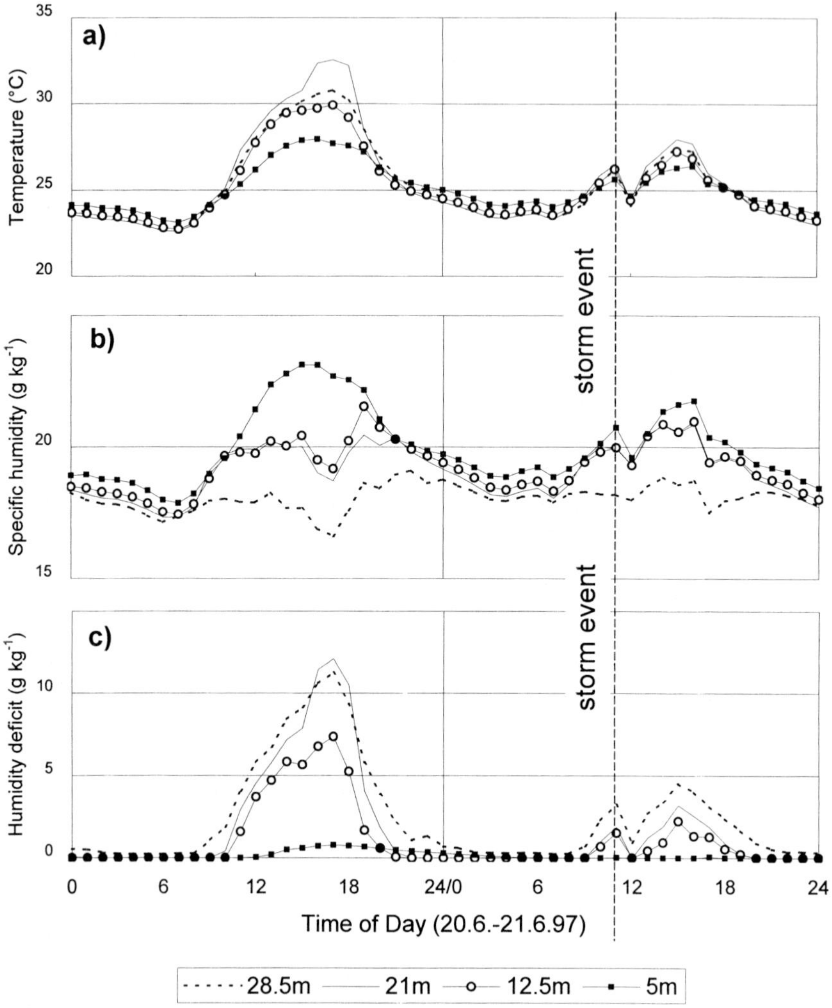

Figure 2. Daily cycles of (a) air temperature, (b) specific humidity, and (c) specific humidity saturation deficit measured at four heights within and just above the forest canopy, for the two days 20 and 21 June 1997. The dashed line indicates the short rainstorm (see text).

therefore clear that the cycle of specific humidity near the base of the forest remains unbiased of convective transport processes that occur at higher levels (Figure 2b). In addition, the diurnal amplitude of specific humidity saturation deficit in the understorey is only about 20% of that at the top of the canopy (Figure 2c). On the other hand, micrometeorological conditions in the top two-thirds of the forest follow those of the overlying atmospheric layer. This coupling during daylight hours is generally provided by turbulent mixing associated with downward wind gusts penetrating the canopy and the influence of convective structures. The latter occur when the canopy surface is heated and buoyant thermals start to rise, acting as principal agents for the transport and mixing of aerial scalars (Oke 1987).

At night, the temperature stratification of the Surumoni forest is reversed, clearly recognizable by the crossing of the temperature graphs after sunset (Figure 2a). Lower air temperatures in the canopy result from radiative cooling at the top of the tree crowns, while the loss of long-wave radiation from the vegetation lower down and therefore the decrease of temperature in the understorey are reduced by the presence of the canopy. Variations in humidity are small throughout the forest and the air is nearly saturated. Due to increasing atmospheric stability above the forest vertical air motion is progressively suppressed during the course of the night. In other words, air within the canopy exhibits a nocturnal decoupling from the atmosphere, unless substantial onset of cloudiness interrupts the radiative cooling or occasional increases in the wind speed again introduce turbulent mixing as observed above the Reserva Ducke rain forest in Brazil (Fitzjarrald & Moore 1990).

The convective rainfall event between 11:00 and 12:00 (local time) on the following day (21 June 1997) with 3.8 mm precipitation causes a rapid decline of temperature, specific humidity and specific humidity deficit at all levels (see storm event illustrated by dashed line in Figures 2 and 3). Afterwards the profiles re-establish their previous diurnal trend within two or three hours. With regard to the described stability-producing and stability-destroying mechanisms, the temperature response is of particular interest since the cooling that occurs when the canopy surface is wetted can also induce a reversal in the normal daytime temperature gradient above the forest by the advection of energy (Shuttleworth et al. 1985). Figure 3 illustrates the temperature difference measured over the height range 28.5 to 41 m. Following meteorological practice, a decrease of temperature with height corresponds to a negative gradient and an outward flux of sensible heat. This situation usually occurs during daylight hours when air in the canopy is warmer than the atmosphere above. A positive gradient consequently indicates reversed conditions and a downward flux of sensible heat appropriate to the situation at night. When the rain starts at about 11:00 the temperature gradient slightly inverts, yet no stable layers are produced above the forest, which suggests a low intensity of local energy advection (Figure 3).

Canopy energy exchanges

Assuming horizontal homogeneity, the energy balance of the forest can be written as

$$R_N = H + \lambda E + G + J + \mu Fc, \tag{1}$$

where all flux components are expressed in units of watts per square metre. In this equation R_N is the net radiation, H and E are the vertical flux densities of sensible and latent heat, λ is the latent heat of vaporization of water (J kg^{-1}), G is the soil heat flux, J is the equivalent energy flux into storage , and μFc represents the energy fixed by photosynthesis, with μ being the photochemical energy requirement (J kg^{-1}) of CO_2 fixation (F_c). As indicated earlier, R_N and G were measured directly at the crane plot. The energy equivalent of CO_2 fixation only amounts to a few percent of R_N and has therefore been neglected. The storage term J was calculated from hourly changes in the sensible and latent heat contents of the air within the forest (Denmead & Bradley 1985).

Total evapotranspiration loss (λE) from the canopy was estimated using the Penman-Monteith equation

$$\lambda E = \frac{\Delta'(R_N - G - J) + \rho c_p D g_a}{\Delta' + \gamma(1 + g_a/g_c)}, \tag{2}$$

where Δ' denotes the rate of change of saturated water vapour pressure with temperature (Pa K^{-1}), ρ is the density of dry air (kg m^{-3}), c_p is the specific heat of dry air at constant pressure (J kg^{-1} K^{-1}), D is the vapour pressure deficit (Pa), γ is the psychrometric constant (Pa K^{-1}), g_a is the aerodynamic boundary-layer conductance (m s^{-1}), and g_c is the canopy surface conductance (m s^{-1}). A detailed derivation of this formula is given by Jarvis (1981). Equation (2) is used to calculate two evaporation rates. Firstly, under dry canopy conditions the estimated rate primarily refers to the amount of transpiration with g_c finite. Secondly, in case of a totally wet canopy for which $g_c = 0$, the complete canopy saturation suppresses all transpiration whereas the interception loss; i.e., the evaporation of free water from the wet canopy surface, determines the amount of latent heat. In both situations evaporation from the soil is rather low because of the dense vegetation cover.

The aerodynamic conductance g_a is given by the expression

$$g_a = \frac{u_*^2}{u}, \tag{3}$$

where u is the above-canopy wind speed (m s^{-1}) and u_* is the friction velocity (m s^{-1}) which can be determined from wind profile measurements at several heights above the forest (Thom 1975). Some authors found that g_a is also adequately described as a simple function of above-canopy wind speed u with the empirically derived $g_a = 0.1\ u$ (see, for example, Jarvis et al. 1976; Kelliher et al. 1992; Raupach 1989). The more complex canopy surface conductance g_c represents the integrated behaviour of individual stomatal conductances in the sense of Monteith's 'big-leaf' approximation (Monteith 1965). A convenient modelling framework of g_c was derived from long-term experimental data gathered at the Reserva Ducke forest, Brazil, specifically calibrated for Amazonian rain forest (Dolman et al. 1991; Wright et al. 1996). For the use in this study the application was transfered to the climatic conditions of the Surumoni rain forest. The basis of the model is an expression which takes the parametric form

$$g_c = g_0 g_V g_R g_D g_T g_M. \tag{4}$$

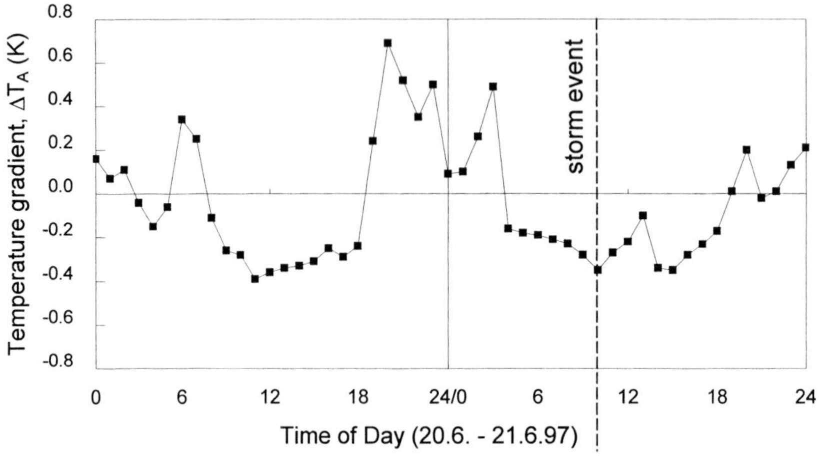

Figure 3. Gradient of air temperature ($\Delta T_{41\ m-28.5\ m}$ in K) measured over the height range 28.5 to 41 m for the two days 20 and 21 June 1997. The dashed line indicates the short rainstorm (see text).

Apart from a typical value of whole-canopy surface conductance, g_0, the model incorporates the response of g_c to environmental variables such as canopy cover (leaf area index, LAI), g_V, incident solar radiation, g_R, specific humidity saturation deficit, g_D, air temperature, g_T, and soil moisture status, g_M. A detailed mathematical description of this parameterization is published by Shuttleworth (1989). It is used to describe the canopy surface conductance during daylight hours and under dry canopy conditions. At night gc is set to the low value of 0.1 mm s^{-1} in order to simulate the physiologic response of plants to declining light.

Results of the above mentioned energy budgeting procedures including the estimation of total evapotranspiration loss via the Penman–Monteith approach are shown in Figure 4. The component fluxes of the energy balance on the two days considered demonstrate the partitioning of intercepted net radiation (R_N) into the outgoing fluxes of latent (λE) and sensible (H) heat (Figure 4a). Also illustrated is the sum of soil heat flux and the forest storage term (G). During the fine day (20 June) a great portion of available energy sustains evapotranspiration from the forest. In this case, λE corresponds to transpiration from a totally dry canopy. On the second day, after the rain at first λE decreases, but then interception loss is responsible for the increase of latent heat. The low-energy fluxes at night typify the suppression of turbulent exchanges between the forest and the atmosphere.

It is an interesting question how the calculated evapotranspiration is related to the actual physiological response of the vegetation, primarily expressed by the stomatal control of transpiring leaves. Thus, in Figure 4b the integrated daily water equivalent of λE (mm h^{-1}) is compared with the daily sap flux density (kg dm^{-2} h^{-1}) of three investigated tree species (*Goupia glabra, Dialium guianense, Ocotea sp.*) Additionally the cycle of vapour pressure saturation deficit (kPa) is shown. Sap flux density was measured using the 'constant heating method' (Granier 1985; Motzer 1998), which was successfully applied within a tropical rain forest in French Guiana (Granier et al. 1996). A quantitative comparison would require the determination of total sap flow; i.e., the product of sap flux density and sapwood cross-section. This procedure is still in process and results will be published in following papers. Nevertheless, the sap flux density can be used as an indicator of plant response to changing environmental conditions (Köstner et al. 1992; Larcher 1994).

An important implication of Figure 4b is the fact that the calculated daily course of evapotranspiration obviously corresponds to the physiological behaviour of the observed trees. This becomes evident, for example, in the afternoon of the first day when the evaporation rate remains at a constant level similar to the rate of sap flux density. Under dry canopy conditions and a sufficient soil water availability the process of transpiration is mainly driven by the net radiation, the vapour pressure deficit of the air, the efficiency of turbulent mixing, as well as morphological and physiological adaptations of the vegetation. According to the findings of Motzer (1998), at the Río Surumoni such plant-atmosphere feedback mechanisms are recognizable, for instance, by stomatal closure during periods of high solar insulation; i.e., the *midday*

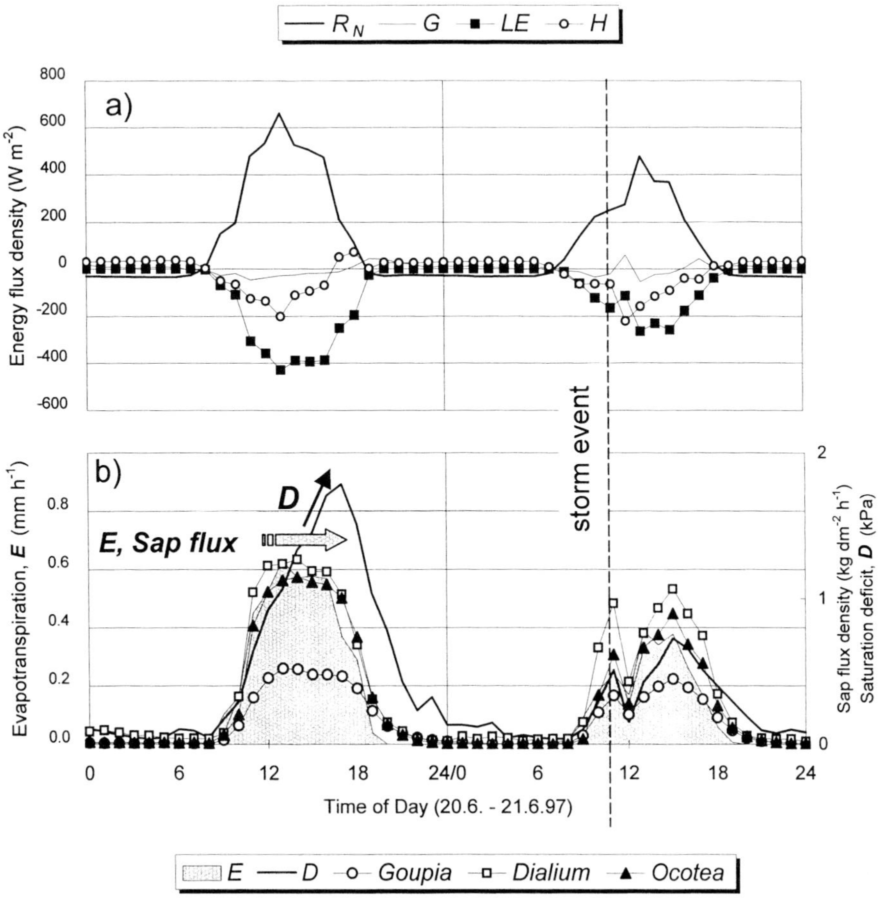

Figure 4. (a) Component fluxes of the energy balance where R_N is the net radiation, G is the sum of soil heat flux G and the forest storage term J, and LE and H are the turbulent fluxes of latent and sensible heat. (b) Total evapotranspiration loss E according to the Penman-Monteith method (grey underlayed area), vapour pressure saturation deficit D (solid line), and sap flux densities of the tree species *Goupia glabra* (open circle), *Dialium guianense* (open square), and *Ocotea sp.* (solid triangle) are presented. The dashed line indicates the short rainstorm (see text).

depression, in order to avoid high transpiration loss (cf., Roy & Salager 1992; Lösch & Tenhunen 1981; Lösch & Kazda 1992). Also short-termed oscillations of transpiration flux have been observed; e.g., induced by the temporary onset of cloudiness that interrupts radiative heating of the tree crowns. In other words, the strong radiative absorption within the canopy layer of the forest leads to high values of air temperature and vapour pressure deficit through which the evaporative demand consequently increases often exhausting the limited reservoir of available water (cf., Lösch & Schulze 1992).

During the short rainstorm on the second day the graphs of all illustrated parameters decline (Figure 4b). Although interception loss may occur immediately after the canopy surface has been wetted, in this particular situation neither the low vapour pressure deficit nor the amount of radiative energy are sufficient to support enhanced evaporation. The increase of those variables after the rain shower is a prerequisite for the reinforcement of evaporation. Finally in Figure 5 the modelled values of hourly evapotranspiration loss for June 1997 are plotted against measured sap flux density. The relation is statistically significant ($r^2 = 0.914$, $P < 0.001$), although the results are subject to a slightly higher scatter at low rates of sap flux density.

Conclusion

Micrometeorological observations within and just above an Amazonian rain forest in Southern Venezuela reveal the diurnal variability of microclimate, as well as physical mechanisms which have controlling influ-

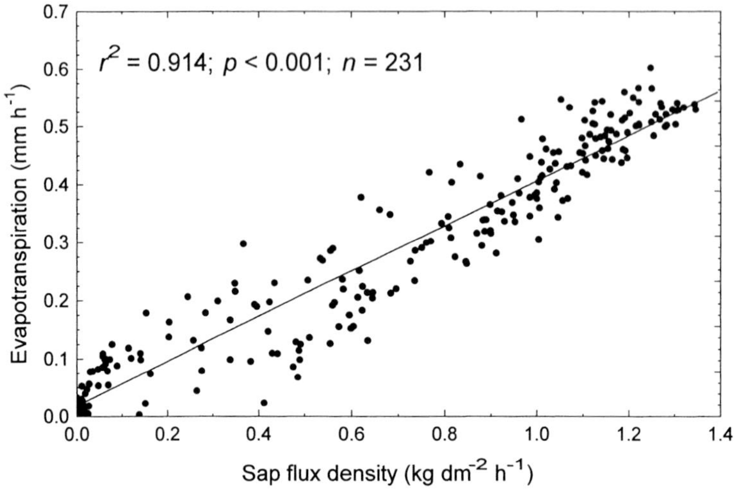

Figure 5. Relationship between calculated evapotranspiration loss using the Penman-Monteith equation and measured sap flux density for June 1997.

ence on the exchanges of mass and energy between the plant community and the atmosphere. In this context, the upper canopy forms a major interface with the atmosphere, where substantial energetic and kinetic transformations take place. Understanding these interactions is essential for determining the exact role of tropical rain forest in the context of global climatology and for estimating the environmental impact of human disturbances in the tropics. At first sight, the correspondence between calculated actual evapotranspiration and measured sap flux density suggests a reliable transferability of the energy partition model derived in the Amazon basin of Brazil. Further data analysis within this framework is necessary in order to calculate monthly and yearly amounts of total evapotranspiration loss from the Surumoni forest and to compare the results of different empirical energy balance methods. However, in combination with preliminary results from sap flow measurements and current hydrologic investigations there is increasing evidence that actual evapotranspiration at Río Surumoni consists of a higher portion of interception loss, whereas the amount of transpiration is comparatively lower (Anhuf et al. 1999). In light of data reported from study sites in the central Amazon Basin, such as the Reserva Ducke near Manaus or the experimental site near Ji-Parana (Rondônia, Brazil), this partitioning is apparently different (cf., Gash et al. 1996). Yet it seems to be plausible, since regional climatic features in the Venezuelan study area are generally prone to a higher atmospheric evaporative demand and the vegetation at Río Surumoni is characterized by several morphological and physiological adaptations with respect to the environmental conditions of this location.

Acknowledgements

This study was financially supported by the Deutsche Forschungsgemeinschaft (AN 214/2-1 and 2-2), Germany, and conducted in frame of a collaborative research project founded by the Austrian Academy of Science (AAS) and the Ministerio del Ambiente y de los Recursos Naturales Renovables (MARNR) in Venezuela. W. Morawetz (Leipzig) initiated this project and H. Winkler and members from the Konrad Lorenz Institut (Wien) provide the current administration. All of them are gratefully recognized. The authors are pleased to acknowledge the generous support of the Venezuelan ambassador E. Becker-Becker and his staff from the embassy in Bonn. Essential help and constructive discussions were gained from G. Gravenhorst, H. Kreilein and A. Ibrom (Institute for Bioclimatology, Göttingen). Substantial help was also provided by O. Huber (Caracas), the Dirección de Hidrología y Meteorología (Caracas), and the Department of Geography and the Botanical Institute in Bonn. In particular we would like to thank S. Engwald (Bonn) for his invaluable practical help during the field work in Venezuela, J. Nieder (Bonn) for reviewing the manuscript, as well as T. Motzer, R. Rollenbeck and B. Schröder (all Mannheim) for providing important data.

References

Anhuf, D., Motzer, T., Rollenbeck, R., Schröder, B. & Szarzynski, J. 1999. Water budget of the Surumoni crane-site (Venezuela). Selbyana 20 (1): 179–185.

Chappell, N. A., Bidin, K. & Tych, W. 2001. Modelling rainfall and canopy controls on net-precipitation beneath selectively-logged tropical forests. Plant Ecol. 153: 215–229 (this volume).

Denmead, O. T. & Bradley, E. F. 1985. Flux-gradient relationships in a forest canopy. Pp. 421–442. In: Hutchinson, B. H. & Hicks, B. B. (eds), The Forest-Atmosphere Interaction. Reidel Publishing Company, Dordrecht.

Dolman, A. J., Gash, J. H. C., Roberts, J. & Shuttleworth, W. J. 1991. Stomatal and surface conductance of tropical rain forest. Agric. For. Meteorol. 54: 303–318.

Endler, J. A. 1993. The colour of light in forests and its implications. Ecol. Mono. 63: 1–27.

FAO-UNESCO. 1988. Soil map of the world. Revised legend. Food and Agriculture Organization, Rome.

Fitzjarrald, D. R. & Moore, K. E. 1990. Mechanisms of nocturnal exchange between the rain forest and the atmosphere. J. Geophys. Res. 95 (D10): 16839–16850.

Gash, J. H. C., Nobre, C. A., Roberts, J. M. & Victoria, R. L. 1996. Amazonian deforestation and climate. Wiley & Sons, Chichester, New York.

Graedel, T. E. & Crutzen, P. J. 1994. Chemie der Atmosphäre. Spektrum Akademischer Verlag, Heidelberg, Berlin.

Granier, A. 1985. Une nouvelle méthode pour la mesure du flux de sève brute dans le tronc des arbres. Ann. Sci. For. 42: 193–200.

Granier, A., Huc, R. & Barigah, S. T. 1996. Transpiration of natural rain forest and its dependence on climatic factors. Agric. For. Meteorol. 78: 19–29.

Holbrook, N. M. & Lund, C. P. 1995. Photosynthesis in forest canopies. Pp. 411–430. In: Lowman, M. D. & Nadkarni, N. M. (eds), Forest canopies. Academic Press, San Diego, California.

Huber, O. 1995. Geographical and physical features. Pp. 1–61. In: Steyermark, J. A., Berry, P. E. & Holst, B. K. (eds), Flora of the Venezuelan Guayana. Vol. 1, Timber Press, Portland, Oregon.

Jarvis, P. G. 1981. Stomatal conductance, gaseous exchange and transpiration. Pp. 175–204. In: Grace, J., Ford, E. D. & Jarvis, P. G. (eds), Plants and their atmospheric environment. Blackwell Science, Oxford.

Jarvis, P. G., James, G. B. & Landsberg, J. J. 1976. Coniferous Forest. Pp. 171–240. In: Monteith, J. L. (ed.), Vegetation and the atmosphere, Vol. 2. Academic Press, London.

Kelliher, F. M., Köstner, B. M. M., Hollinger, D. Y., Byers, J. N., Hunt, J. E., McSeveny, T. M., Meserth, R., Weir, P. L. & Schulze, E.-D. 1992. Evaporation, xylem sap flow and tree transpiration in a New Zealand broad-leaved forest. Agric. For. Meteorol. 62: 53–73.

Köstner, B. M. M., Schulze, E.-D., Kelliher, F. M., Hollinger, D. Y., Byers, J. N., Hunt, J. E., McSeveny, T. M., Meserth, R. & Weir, P. L. 1992. Transpiration and canopy conductance in a pristine broad-leaved forest of *Nothofagus*: An analysis of xylem sap flow and eddy correlation measurements. Oecologia 91: 350–359.

Larcher, W. 1994. Ökophysiologie der Pflanzen. Ulmer, Stuttgart.

Lösch, R. & Tenhunen, J. D. 1981. Stomatal responses to humidity - phenomenon and mechanism. Pp. 137–161. In: Jarvis, P. G. & Mansfield, T. A. (eds), Stomatal physiology. Cambridge University Press, Cambridge.

Lösch, R. & Kazda, M. 1992. Water turnover of leaves in the rain forest canopy, the understorey and in a regrowth site of the Reserve de Campo, S. Cameroon. Pp. 206–216. In: Hallé, F. & Pascal, O. (eds), Biologie d'une canopée de Forêt Équatorial-II. Rapport de Mission: Radeau de Cimes Octubre–Novembre 1991, Reserve de Campo, Cameroun. Montpellier.

Lösch, R. & Schulze, E.-D. 1992. Internal coordination of plant responses to drought and evaporational demand. Pp. 185–204. In Schulze, E.-D. & Caldwell, M. M. (eds), Ecophysiology of photosynthesis. Berlin.

Mooney, H. A., Bjorkman, O., Hall, A. E., Medina, E. & Tomlinson, P. B. 1980. The study of the physiological ecology of tropical plants: current status and needs. BioScience 30: 22–26.

Monteith, J. L. 1965. Evaporation and environment. Symp. Soc. exp. Biol. 19: 206–234.

Monteith, J. L. (ed.) 1975, Vegetation and the atmosphere, Vol. 1. Academic Press, London.

Morawetz, W., Liebig, A. & Wesenberg, J. 1998. Ein Vegetationsprofil im Untersuchungsgelände des Surumoni-Projektes bei La Esmeralda/Oberer Orinoko. In: Dalitz, H., Haverkamp, M., Homeier, J. and Breckle, S.-W. (eds), 11. Jahrestagung der Deutschen Gesellschaft für Tropenökologie, Bielefelder Ökologische Beiträge, Bielefeld.

Motzer, T. 1998. Untersuchungen zum Transpirationsverhalten ausgewählter Baumarten eines neotropischen Tieflandregenwaldes (Projekt Surumoni, Oberer Orinoco, Venezuela). Unpublished MSc thesis, Department of Physical Geography, University of Mannheim, Germany.

Nieder, J., Engwald, S., Klawun, M. & Barthlott, W. 2000. Spatial distribution of vascular epiphytes in a lowland Amazonian rain forest (Surumoni Crane Plot) in Southern Venezuela. Biotropica 32 (3) (in press).

Oke, T.R. 1987. Boundary layer climates. Methuen, London.

Raupach, M. R. 1989. Stand overstorey processes. Phil. Trans. R. Soc. Lond. B 324: 175–190.

Roy, J. & Salager, J.-L. 1992. Midday depression of net CO_2 exchange of an emergent rain forest tree in French Guiana. J. Trop. Ecol. 8: 499–504.

Shukla, J., Nobre, C. & Sellers, P. 1990. Amazon deforestation and climatic change. Science 247: 1322–1325.

Shuttleworth, W. J. 1988. Evaporation from Amazonian rain forest. Proc. R. Soc. Lond. B 233: 321–346.

Shuttleworth, W. J. 1989. Micrometeorology of temperate and tropical forest. Phil. Trans. R. Soc. Lond. B324: 207–228.

Shuttleworth, W. J., Gash, J. H. C., Lloyd, C. R., Moore, C. J., Roberts, J., Marques-Filho, A. O., Fisch, G., Filho, V. P. S., Ribeiro, M. N. G., Molion, L. C. B., Sá, L. D. A. Nobre, J. C. A., Cabral, O. M. R., Patel, S. R. & Carvalho, J. M. 1985. Daily variations of temperature and humidity within and above Amazonian forest. Weather 40: 102–108.

Théry, M. 2001. Forest light and its influence on habitat selection. Plant Ecol. 153: 251–261 (this volume).

Thom, A. S. 1975. Momentum, mass and heat exchange of plant communities. Pp. 57–109. In: Wilson, E. O. (ed.), 1992. The diversity of life. Harvard University Press, Cambridge, Massachusetts.

WMO 1997. WMO statement on the status of the global climate in 1996. WMO-No. 858, World Meteorological Organization, Geneva.

Wright, S. J. & Colley, M. 1994. Accessing the canopy: Assessment of biological diversity and microclimate of the tropical forest canopy. United Nations Environment Programme, Nairobi.

Wright, I. R., Gash, J. H. C., Rocha, H. R. da & Roberts, J. M. 1996. Modelling surface conductance for Amazonian pasture and forest. Pp. 437–458. In: Gash, J. H. C., Nobre, C. A., Roberts, J. M. & Victoria, R. L. (eds), Amazonian deforestation and climate. Wiley & Sons, Chichester, New York.

Plant Ecology **153:** 241–250, 2001.

The influence of epiphyte cover on branch temperature in a tropical tree

Martin Freiberg
Abteilung Systematische Botanik und Ökologie, University of Ulm, Albert-Einstein-Allee 11, D-89081 Ulm, Germany (e-mail: martin.freiberg@biologie.uni-ulm.de)

Key words: Canopy, Epiphytes, French Guiana, Microclimate, Tropical rain forest

Abstract

The air temperature around three branches loaded with epiphytes in a tropical lowland rain forest in French Guiana was measured from mid-June to mid-July 1997. These data represent the first direct measurements of temperature close to epiphyte-loaded branches in a tropical lowland forest. The epiphytic biomass of one of the branches was removed after two weeks of measurements. These differences in stripped versus unstripped branch temperatures were positive during day time. A maximum difference of 4.8 °C was measured on a single sunny day; the maximum of four mid-day hours averaged over all days was 2.3 °C. Maxima were determined at locations close to the branch, representing high accumulations of humus, while nearly no differences were detected 75 cm away from the branch. At night, the temperature was cooler after epiphyte biomass removal; differences were up to −0.5 °C. Close to the branch, evaporation of stored water in humus cools the environment while latent heat transformation on epiphyte surfaces raises temperature. Shading under a branch prevents heating. Epiphytes reduce air circulation, keeping unheated air on branch undersides. At night, auto-heating of humus may occur through increased rates of respiration.

Introduction

Epiphytes are abundant in the canopy trees of many wet tropical rain forests. The epiphyte community includes living vascular and non-vascular epiphytes, as well as dead organic matter such as litter and humus, and can accumulate up to several tons per hectare (Pócs 1980; Nadkarni 1984; Hofstede et al. 1993; Coxon & Nadkarni 1995). This material stores rain water and retains nutrients and thus plays an important role in the ecology of tropical rain forests. Studies have focused on the chemical and hydrological properties of epiphytic biomass, especially in montane forests (Pócs 1980; Vance & Nadkarni 1990; Veneklaas et al. 1990; Nadkarni & Matelson 1991, 1992; Ingram & Nadkarni 1993; Hofstede et al. 1993).

The documentation of microclimatical gradients in tropical forests are based on data collected from towers, so that the sensors are located at great distances from branches and the epiphytic biomass itself (Haddow & Corbet 1961; Cachan 1963; Allen et al. 1972; Molión 1987; Parker 1995). Results of these studies show an increase in temperature and a decrease in humidity from the forest floor towards the canopy. However, the epiphytic biomass may influence the microclimatic conditions within the canopy (Freiberg 1997), so microclimatic measurements taken directly at the growth sites of the epiphytes are needed to understand epiphyte ecology. In epiphyte-rich canopies, distances of even a few decimeters reveal significant microclimatic gradients. Different infraspecific phenotype expression of epiphytes located within single trees reflect these gradients (Freiberg 1996). In contrast to the temperature measurements made with the help of towers, there was only a difference of 1 °C between crown humus and forest floor soil in a montane forest (Bohlman et al. 1995). The term 'ramosphere' was introduced to define the spatial region in the canopy which is influenced microclimatically by the epiphytic biomass, in contrast to the 'aerosphere', which is the spatial region between the ramospheres of canopy branches and which is microclimatically not influenced by epiphytic biomass (Freiberg 1996).

These studies led to the hypothesis that ramospheres of major canopy branches have distinct microclimatical gradients, which are influenced by living and dead epiphytic biomass (Freiberg 1996). In this study, I tested this hypothesis by measuring the temperature around epiphyte-loaded branches, and the shift in temperature following removal of the epiphytic biomass in the canopy of a tropical lowland forest. A new measuring device was designed to gather the required data. The results were used to define the spatial border between ramosphere and aerosphere.

Material and methods

Study site

The investigation was carried out near the Aratai Station in the Reserve Naturelle Des Nouragues, French Guiana, South America (45 m, 3°59′24″ N, 52°35′27″ W). The mean air temperature measured in the forest 1 m above the forest floor between June 1997 and December 1998 was 24.9 °C. The minimum temperature recorded in this period was 20.8 °C, the maximum was 33.1 °C. The average annual precipitation is ca. 3000 mm. The dryest months are September and October, but long term observations show that these months may be humid as well (P. Gauchet, unpubl. data).

The phorophyte

A single *Virola michelii* Heckel (Myristicaceae) tree was chosen to measure the microclimate around canopy-level branches. The tree was located 120 m from the Arataye river on a little hill; the base of the trunk was always at least 10 m above the water level of the river. The tree was 52 m high and its first major branches occurred at 33 m. The maximum crown diameter, measured 37 m above the ground, was 20 m. This tree was chosen because mode of branching was perpendicular to the trunk, so that the inner portions of the branches were growing nearly horizontally, and branches had similar architectural features. Moreover, the epiphytic biomass had similar structure and dimensions. The tree was climbed using modified single rope techniques (Perry 1978; Perry & Williams 1981).

Analysis of epiphytic biomass

After recording the microclimatical data for a period of 2 weeks in June 1997, the epiphytic biomass of one branch was removed 1 m towards the trunk and 1 m towards the periphery from the sensor setup. This biomass was later separated into five fractions: litter, humus, bryophytes, roots, and shoots and leaves of each vascular epiphyte species. The only lichens that occurred were crustose species that could not be separated from the bark. The dry weight of the fractions was determined after 3 days of air- and oven-drying (ca. 65 °C) to a constant weight.

The experimental setup

The microclimate around the branches was measured with a central data logger (ModulAT 386, 25 Mhz, Jumptec, Deggendorf, Germany) that was located on a platform within the tree, and 12 acrylic glass tubes containing the sensors for air temperature. Temperature was measured by SMT160-30 IC sensors, LJK Technology, England. Three silicium solar panels were installed above the canopy to provide power for the entire system. The sensor signals were transported via cable to the data logger and were stored at 5-min intervals.

A complete measuring set consisted of four acrylic glass tubes, which were installed perpendicular on both sides, above, and underneath the branches (Figure 1). The sensors had been fixed within the tubes so that they were exposed to the environment at distances from the branches of 1 cm, 12 cm, 25 cm, 50 cm, 75 cm, and 100 cm. Sinter-bronce filters (Krebsöge, Radevormwald, Germany) were used for radiation shielding. Discs of PVC (10 cm diameter) were mounted horizontally 3 cm above the sensitive area of the sensors for rain protection.

Three branches were equipped with this setup. They were named the experimental branch (E), the undisturbed branch (U), and the reference branch (R). Branch E was the one from which the epiphytes load was removed (see above). Branch E is illustrated in Figure 1. The positions of all three measuring sets within the canopy were 36–37 m above the forest floor, 3–3.5 m away from the trunk, approximately 7 m away from the horizontal edge of the crown, and 15 m below the upper periphery. Only the upper 10 m of this tree emerged above the surrounding canopy. The direction of the experimental branch was 340 °, the undisturbed branch 232 ° and the reference branch 274 ° on the compass, and the branch diameters ranged between 21 cm and 28 cm.

Figure 1. Photograph inside the canopy of the *Virola michelii* tree showing the experimental branch E after removing the epiphytic biomass 1 m on each side of the sensor tubes. The length of one acrylic glass tube is 1 m.

Analysis of microclimate

The temperature data of branches U and E sampled two weeks before (13.06.-29.06.1997; $T_{U,b}$ and $T_{E,b}$) and two weeks after (02.07.-14.07.1997; $T_{U,a}$ and $T_{E,a}$) the removal of the epiphytic biomass of branch E were considered as means of hours in the analysis presented here. The average daily rainfall was 97 ml before and 95 ml after the removal. The two days that were used to remove the epiphytic biomass and the transition times between night and day were excluded from the statistical analysis. The total day time (08:00 to 16:59), the central day time before the afternoon rain (10:00 to 14:59), and night time (20:00 to 4:59) were analyzed separately.

To compare the temperature before and after the removal, the datasets had to be independent from an influence of changes in the ambient temperature during the four weeks of measuring. Therefore, instead of comparing absolute temperatures, the temperature differences between each sensor of both branches U and E and a reference temperature were calculated ($\Delta T_{E-R,b} = T_{E,b} - T_{R,b}$; $\Delta T_{E-R,a} = T_{E,a} - T_{R,a}$; $\Delta T_{U-R,b} = T_{U,b} - T_{R,b}$; $\Delta T_{U-R,a} = T_{U,a} - T_{R,a}$). As reference, the average temperatures of all sensors of the branch R measured at exactly the same time were used. This approach assumed that a change in the ambient temperature would affect the temperature of the reference and the experimental branch in the same way. For both branches, the statistical difference between the data sets before ($\Delta T_{E-R,b}$; $\Delta T_{U-R,b}$) and after ($\Delta T_{E-R,a}$; $\Delta T_{U-R,a}$) the epiphyte removal was tested with a t-test. In a further step, all data of both branches before the removal were subtracted from data after the removal ($\Delta(\Delta T_{E-R})_{a-b} = \Delta T_{E-R,a} - \Delta T_{E-R,b}$; $\Delta(\Delta T_{U-R})_{a-b} = \Delta T_{U-R,a} - \Delta T_{U-R,b}$). In the case of branch E, temperature shifts due to the epiphyte biomass should be detected, while the data of branch U should serve as control.

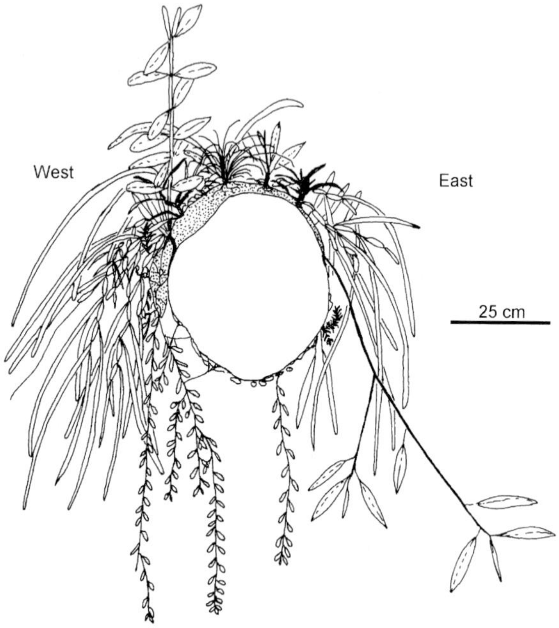

Figure 2. Schematic cross-section of the experimental branch E before removing the epiphytic biomass. A depth of about 20 cm of the central section of the branch is shown. The accumulated humus layer is represented by the dotted area on the upper surface of the branch.

Table 1. Biomass composition of removed epiphytic biomass 1 m on each side of the sensor setup of branch E (Figure 1).

Fraction	Dry weight [kg m^{-2}]	Percent of total dry weight (%)
Living epiphytes	**1175**	**67**
Bryophytes	31	2
Vascular epiphytes	1144	65
Green parts	697	39
Roots	447	26
Dead organic matter	**570**	**33**
Litter	193	11
Humus	377	22
Total	**1745**	**100**

Results

Accumulation of epiphytic biomass

The surface area of the 2 m section of branch E, was 1.75 m^2. The total biomass was 1.75 kg m^{-2} branch surface. *Maxillaria alba* (Hook.) Lindl. (Orchidaceae) accounted for 41% of the green parts of living vascular biomass, *Sphyrospermum buxifolium* Poepp. & Endl. (Ericaceae) for 9%, *Clusia grandiflora* Splitg. (Clusiaceae) for 7% and *Codonanthe crassifolia* (Focke) Morton (Gesneriaceae) for 6%. Each of the other vascular epiphyte species accounted for less than 5% (Table 1).

Substrate and epiphytes were not equally distributed around the branch (Figure 2). The diameter was irregularly ellipsoid in shape, the horizontal diameter being shorter than the vertical diameter. The larger part of the eastern hemisphere and the lower side of the branch was covered only with crustose lichens. Some epiphytes, especially *Sphyrospermum buxifolium* and *Maxillaria alba*, hung about 60 cm down the branches (Figure 2). The density of epiphytic growth was highest on the western side. On the underside of the branch, only liana and hemiepiphyte roots grew.

Temperature in the canopy

The temperature averaged over the four weeks of all sensors of branch U was 28.7 °C during mid-day, and 23.2 °C during the night. The minimum temperature measured on branch E before removing the epiphytic biomass was 20.2 °C 100 cm off the western side of the branch; after the removal it was 20.0 °C 50 cm underneath the branch. The maximum temperature measured on branch E before removing the epiphytic biomass was 31.0°C 75 cm off the eastern side of the branch; after the removal it was 33.8°C 10 cm above the branch.

During day time, the temperature after the removal of the epiphytic biomass ($\Delta T_{E-R,a}$) was generally higher than before ($\Delta T_{E-R,b}$), leading to a positive value of $\Delta(\Delta T_{E-R})_{a-b}$. The interval between 13:00 and 13:59 and the sensor closest to the branch revealed the largest difference (Figure 3). The maximum difference reached up to 3.6 °C averaging all days (Figure 3a), and up to 4.8 °C comparing a single sunny day before with a sunny day after the removal (Figure 3b). At greater distances from the branch the differences decreased rather quickly, and 12 cm off the branch they were just 1.2 °C on all days and 1.5 °C on a sunny day (Figure 3a and 3b). More than 75 cm off the branch the differences were nearly negligible (Figure 3). On rainy days, the differences at 1 cm and 12 cm off the branch closely resembled each other (Figure 3c), and were significantly smaller than on sunny days. The decrease at 16:00 on the sunny day (Figure 3b) was due to a brief episode of rain.

During night time, the temperature close to the branch after the removal of the epiphytic biomass ($\Delta T_{E-R,a}$) was generally lower than before ($\Delta T_{E-R,b}$), leading to a negative value of $\Delta(\Delta T_{E-R})_{a-b}$. At 75 cm off the branch, there was nearly no difference between before and after the removal.

During the mid-day hours all sensors close to branch E up to 50 cm off the branch on its lower and western side showed temperature differences between before and after the removal of the epiphytic biomass were significant (t-test, $p < 0.05$; Figure 4). Performing this calculation for the total day time, the difference was significant ($p < 0.001$) for distances up to 12 cm off the branch for the western and lower side, while it was only significant ($p < 0.001$) for the first centimeters on the eastern side. In general, the significance levels of the average of single hours (Figure 3) exceeded those of the five mid-day hours (Figures 4 and 5), which exceeded those of the total day time. On the upper side, the difference was significant ($p < 0.05$) for distances up to 75 cm off the branch (Figure 5), but, in contrast to the western and lower sites, it was positive only at the first and negative at the following sensors, indicating that the temperature above the branch with epiphytes was higher than without epiphytes.

At night time, differences between before and after the epiphyte removal were significant ($p < 0.001$) up to 12 cm off the branch (Figure 5). However, the difference was negative (−0.2 °C to −0.5 °C), showing that the temperature decreased after the removal of the epiphytic biomass. Moreover, the absolute value of the difference $\Delta(\Delta T_{E-R})_{a-b}$ at night was smaller in the daytime. There was no a significant change in temperature on branch U after the removal date of epiphytic biomass at day time (Figure 4) nor at night time (Figure 5).

Discussion

The epiphytic biomass on branch E (1.75 kg m^{-2}), was similar to values reported from major branches of tropical lowland canopy trees (1.3–1.9 kg m^{-2}), in Amazonian Ecuador (M. Freiberg & E. Freiberg, unpubl. data). Higher values (up to 13 kg m^{-2}), have been measured in tropical montane forests (Nadkarni 1984; Hofstede et al. 1993; M. Freiberg & E. Freiberg, unpubl. data).

The influence of epiphytes and accumulated humus on the microclimatic environment of branches in the canopy of tropical forests has been discussed, but only a few preliminary studies have been carried out (Johansson 1974; Bohlmann et al. 1995; Freiberg 1997). Positive correlations between humus accumulations and relative humidity as well as negative correlations between humus accumulations and temperature were expected. The data presented here were the first direct measurements of temperature close to epiphyte-loaded branches in a tropical lowland forest. Because the temperature of the ramosphere was significantly different

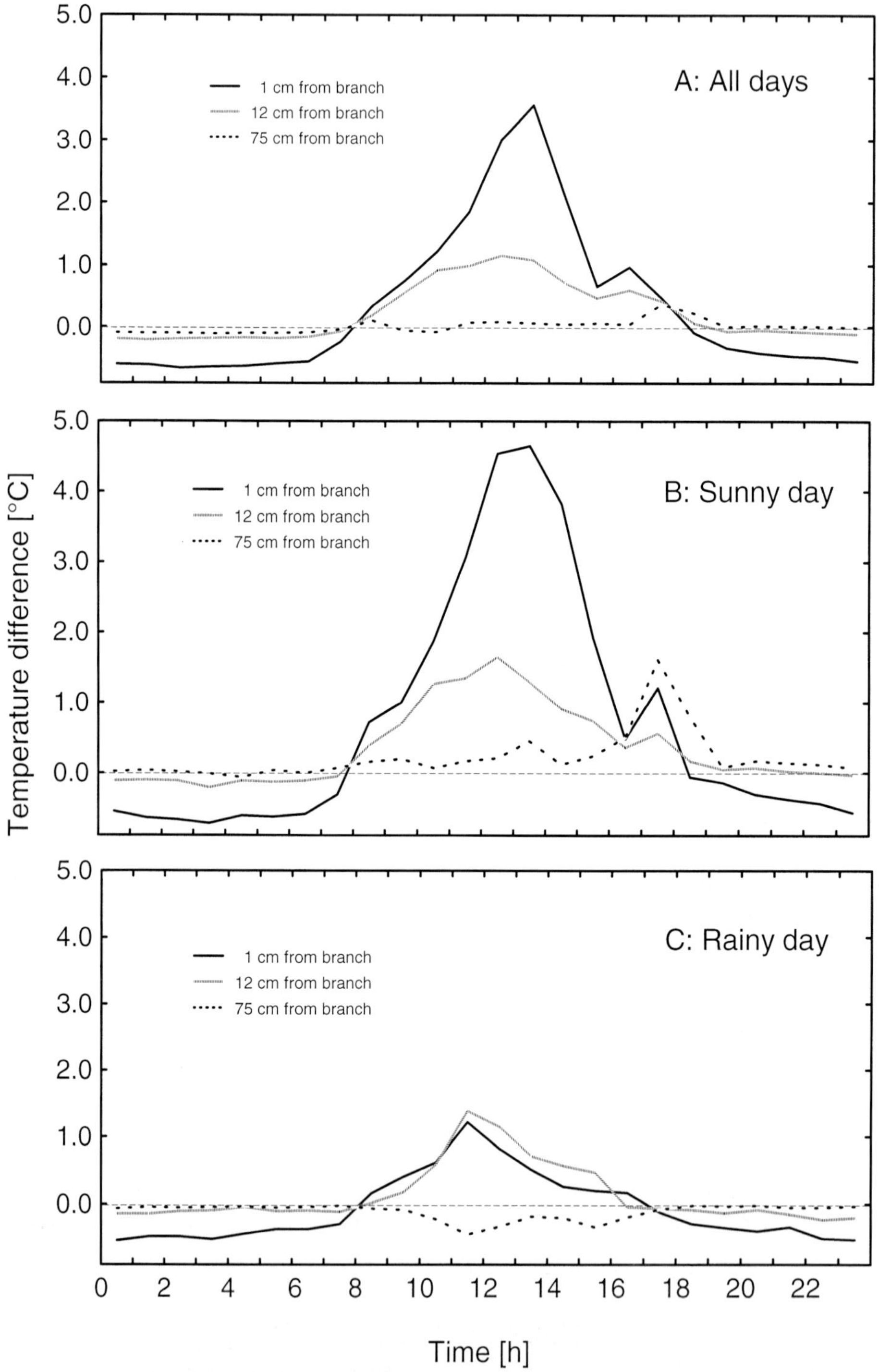

Figure 3. Hourly temperature difference $\Delta(\Delta T_{E-R})_{a-b}$ between before and after removing the epiphytic biomass of three western sensors of the experimental branch E. (a) Average of all hours, (b) average for a single sunny day, and (c) average for a single rainy day. The decrease in difference at 15:00 recognizable in (a) and (b) was due to a short thunderstorm. The dashed line represents zero temperature difference.

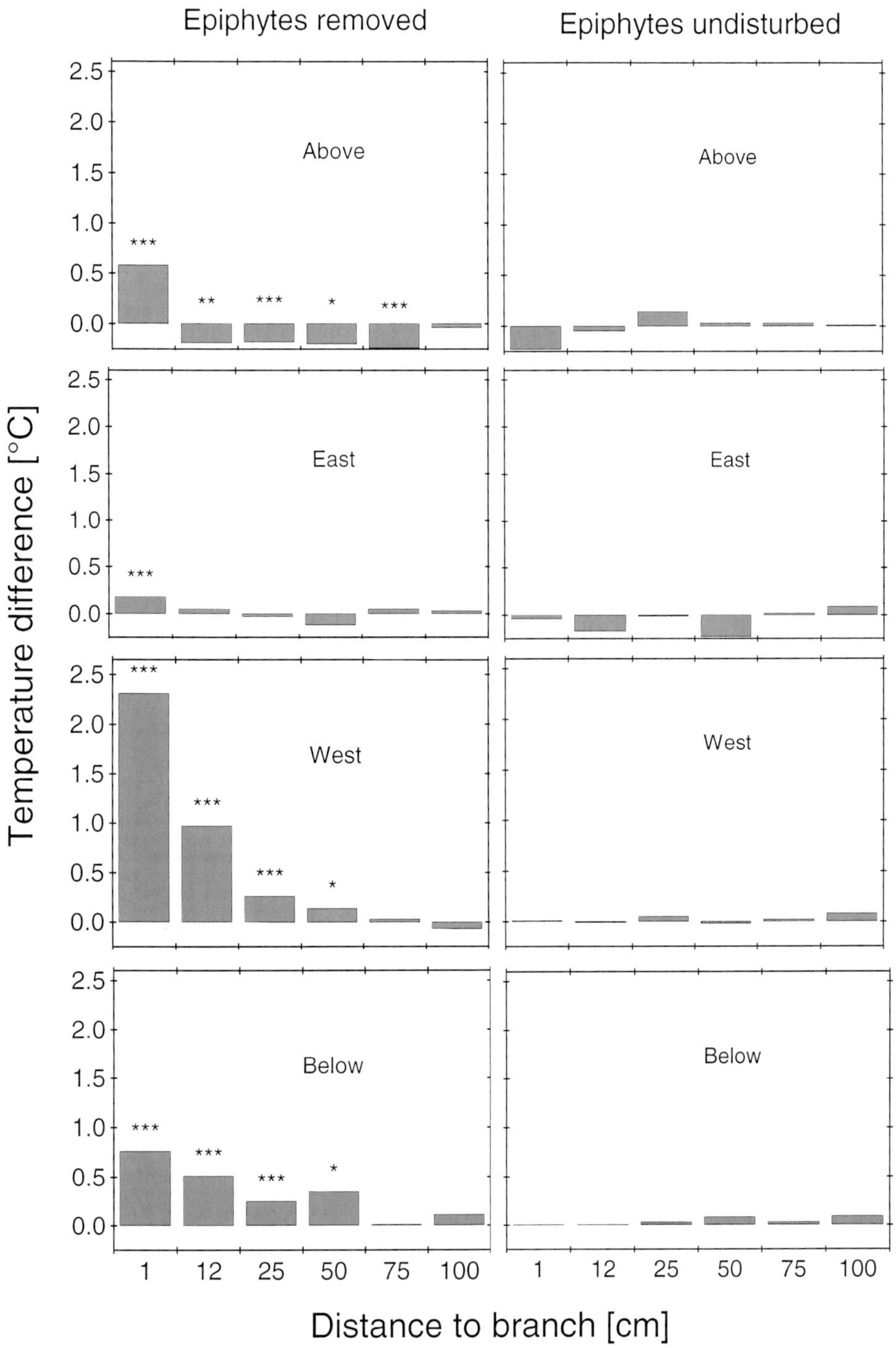

Figure 4. Temperature difference $\Delta(\Delta T_{E-R})_{a-b}$ before and after the removal of the epiphytic biomass for the experimental branch E (left column) and the undisturbed branch U (right column). Data calculated for mid-day hours (10:00 to 14:59). Asterisks represent level of significance of the datasets of a particular sensor between before and after the removal date (t-test; $* = p < 0.05, ** = p < 0.01, *** = p < 0.001$).

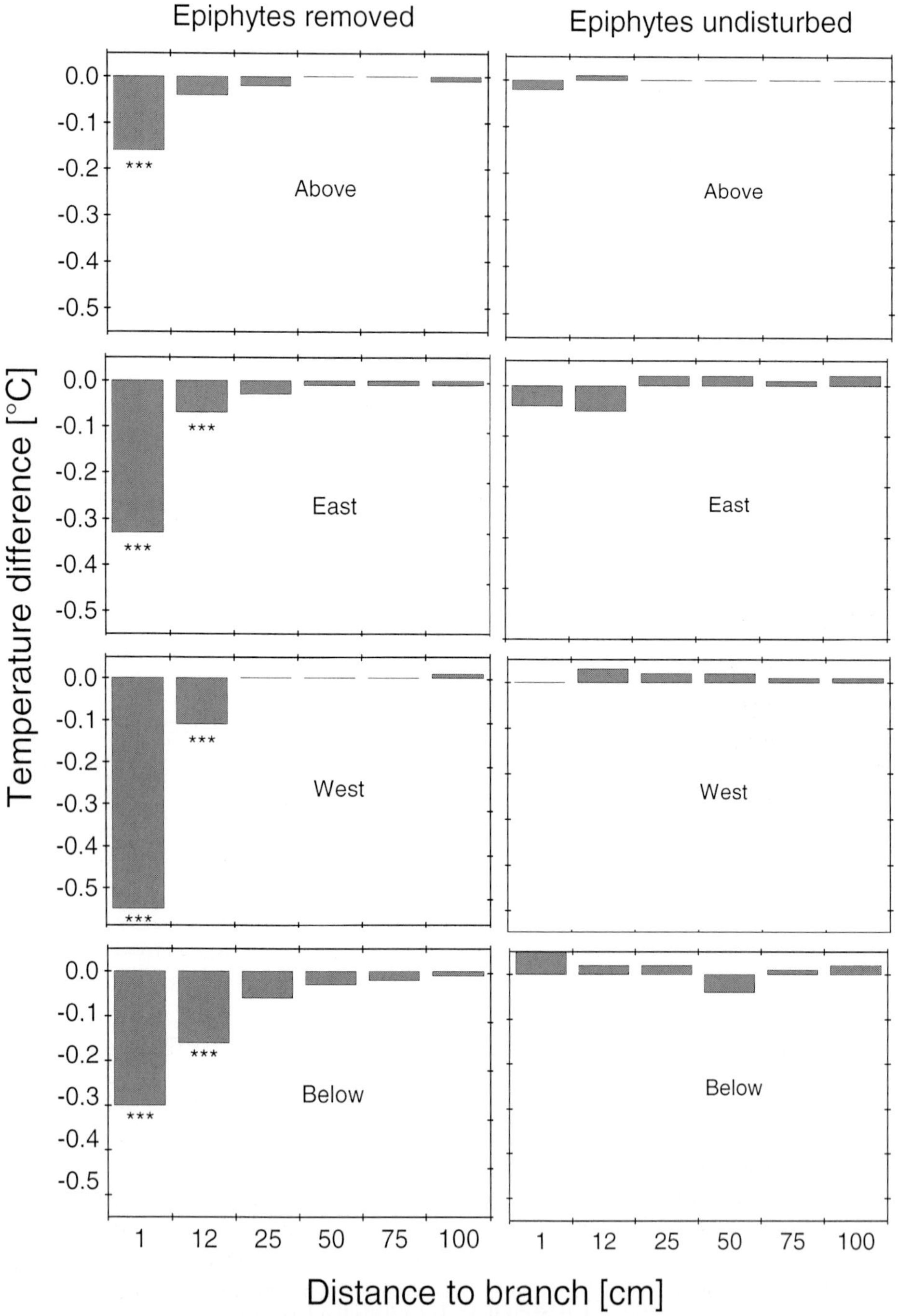

Figure 5. Temperature difference $\Delta(\Delta T_{E-R})_{a-b}$ before and after the removal of epiphytic biomass for the experimental branch E (left column) and the undisturbed branch U (right column). Data calculated for night hours (20:00 to 4:59). Asterisks represent level of significance of the datasets of a particular sensor between before and after the removal date (t-test; $*** = p < 0.001$).

after the removal of the epiphytes, but did not occur on the branch where the epiphytes were kept undisturbed, the removal of the accumulated epiphytic biomass was the main reason for this temperature shift.

The increase in temperature due to epiphyte removal was most significant during the mid-day hours and on the western side of the branch. The epiphytic biomass was also much denser on the western than on the eastern side, probably caused by more rain water being stored on the western side. Consequently, the cooling effect through evaporation of water stored in the dead organic matter was more pronounced on this side. Additionally, the eastern side dried out faster because in the morning hours it was exposed to full sun, while the western side was still shaded by the epiphytic vegetation. In the afternoon, the western side would be fully exposed, but because clouds frequently developed and regular rain occurred, heat development was smaller and penetration into the epiphytic biomass was less deep. Presumably, the western side retained more water and stayed cooler.

Underneath the branch close to the bark, a temperature difference $\Delta(\Delta T_{E-R})_{a-b}$ of 0.8 °C was measured. Fifty cm underneath the branch this difference was still only 0.4 °C. The steepness in temperature difference from the branch towards the aerosphere here was smaller than that measured on the western side of the branch, but the spatial extension of significant difference was the same. Because humus accumulation on the underside of the branch was negligible, humus could not be responsible for the cooler temperature on the underside. Therefore, the cooler temperature was assumed to be mainly due to shading by the branch and the epiphytic vegetation. The epiphytes hanging down the flanks of the branch, especially *Maxillaria alba* and *Sphyrospermum buxifolium*, reduced air circulation and kept a cooler temperature environment, due to the shade of the branch above. The epiphytes somewhat separated the cooler air underneath the branch from the warmer air of the aerosphere.

Unexpectedly, there was a sign inversion in temperature difference a few centimeters off the upper branch (Figure 4). The temperature differences on the upper side of the branch are thus explained by two contradictory effects, which compensate each other a few centimeters above the branch at the point of sign inversion. The first centimeters are influenced by the cooling effect through evaporation of water stored in the accumulated humus. Further above, the shading influence of the epiphytic vegetation became more important. The surface of the epiphytes transformed the incoming radiation to latent heat. After removing the epiphytes on the branch E, no further latent heat was produced and thus air temperature was comparatively lower than before.

At night, the temperature was warmer before the removal of the epiphytes than afterwards. Only temperature differences measured with the sensors closest to the branch were significant (Figure 5) while no significant differences were detected for branch U (Figure 5). Therefore, I conclude that at night temperature differences were due to the removed epiphytic biomass. Latent heat production could not be the reason for the nocturnal temperature difference. Moreover, a possible cooling of the bark via sapflow, an effect which occurred diurnally through transport of cold ground water into the canopy, could be neglected at night as well. Therefore the epiphytic material itself was a source of heat, which may originate from intensive respiration activity of decomposers in the accumulated humus. This would be supported by the fact that the largest differences were detected at highest humus accumulations at the western side of the branch. However, there was a considerable difference of −0.3 °C at the first centimeters and −0.15 °C at 12 cm underneath the branch, where no such biomass accumulated. Therefore, the hypothesis that this location receives its heat indirectly from the flanks of the branch while the hanging epiphytes function like an insulator preventing heat exchange with the aerosphere at night must be tested.

The extension of the ramosphere was defined as the area which was influenced by epiphyte biomass on the microclimate (Freiberg 1996). The thicker the accumulation of humus and the taller the epiphytes grow, the larger is the extension of the ramosphere. This study shows that the border limits were not circular nor regular distributed around the branch, as has been proposed earlier (Freiberg 1997). While accumulated humus and its water storage capacity were the most important factors, the border of the ramosphere was also influenced by shade provided by epiphytes and the branch itself. The transformation of radiation into latent heat is more important on the upper side of the branch than the cooling effect of water stored in accumulated humus. The cooling and heating effects described above influence each other and create a particular microclimate around a canopy branch. Further measurements need to be done of other branches, as well as additional microclimate data gathered on substrate and air humidity, and also wind speed, before the findings presented in this paper can generalized.

Acknowledgements

G. Gottsberger, University of Ulm, supported all phases of this research. I. and L. Schmidt, Mainz, helped to construct the measuring tubes. U. Rascher, University of Darmstadt, Germany, helped to install the measuring system in the canopy. P. Gauchet, Cayenne, French Guiana, gave permission to work at the station and helped with logistics. E. Freiberg, University of Ulm, Germany, patiently supported my work and reviewed and discussed the manuscript. The financial support of Mrs. Cherbon, the Anschubförderung of the University of Ulm, Germany, and the DFG (grant no. FR 1150/3-1) made this project possible. I gratefully thank all persons and institutions named above and two anonymous reviewers.

References

Allen, L., Lemon, E. & Müller, L. 1972. Environment of a Costa Rican forest. Ecology 53: 102–111.

Bohlman, S. A., Matelson, T. J. & Nadkarni, N. M. 1995. Moisture and temperature patterns of canopy humus and forest floor soil of a montane cloud forest, Costa Rica. Biotropica 27: 13–19.

Cachan, P. 1963. Signification écologique des variations microclimatiques verticales dans la forêt sempervirante de basse Cote D'Ivoire. Ann. Fac. Sci. Dakar 8: 89–155.

Coxson, D. S. & Nadkarni, N. M. 1995. Ecological roles of epiphytes in nutrient cycles of forest ecosystems. Pp. 495–543. In: Lowman, M. D. & Nadkarni, N. M. (eds), Forest canopies. Academic Press, San Diego.

Freiberg, M. 1996. Phenotype expression of epiphytic Gesneriaceae under different microclimatic conditions in Costa Rica. Ecotropica 2: 49–57.

Freiberg, M. 1997. Spatial and temporal pattern of temperature and humidity of a tropical premontane rain forest tree in Costa Rica. Selbyana 18: 77–84.

Haddow, A. J. & Corbet, P. S. 1961. Observations of certain environmental factors at different levels. Trans. Roy. Entomol. Soc. London 113: 257–269.

Hofstede, R. G., Wolf, J. H. D. & Benzing, D. H. 1993. Epiphytic biomass and nutrient status of a Colombian upper montane rain forest. Selbyana 14: 37–45.

Ingram, S. W. & Nadkarni, N. M. 1993. Composition and distribution of epiphytic organic matter in a neotropical cloud forest, Costa Rica. Biotropica 25: 370–383.

Johansson, D. R. 1974. Ecology of vascular epiphytes in West African rain forest. Acta Phytogeogr. Sueci. 59: 1–136.

Molión, L. C. B. 1987. Micrometeorology of an Amazonian Rain Forest. Pp. 255–272. In: Dickinson, R.E. (ed.), The Geophysiology of Amazonia. Vegetation and climate interactions. John Wiley & Sons, New York.

Nadkarni, N. M. 1984. Epiphyte biomass and nutrient capital of a neotropical elfin forest. Biotropica 16: 249–256.

Nadkarni, N. M. & Matelson, T. 1991. Fine litter dynamics within the tree canopy of a tropical cloud forest. Ecology 72: 2071–2082.

Nadkarni, N. M. & Matelson, T. 1992. Biomass and nutrient dynamics of epiphytic litterfall in a neotropical montane forest, Costa Rica. Biotropica 24: 24–30.

Parker, G. G. 1995. Structure and microclimate of forest canopies. Pp. 73–106. In: Lowman, M.D. & Nadkarni, N.M. (eds), Forest canopies. Academic Press, San Diego.

Perry, D. R. 1978. A method of access into the crowns of emergent and canopy trees. Biotropica 10: 155–157.

Perry, D. R., & Williams, J. 1981. The tropical rain forest canopy: a method providing total access. Biotropica 13: 283–285.

Pócs, T. 1980. The epiphytic biomass and its effect on the water balance of two rain forest types in the Uluguru mountains (Tanzania, East Africa). Acta Bot. Acad. Sci. Hungaricae 26: 143–167.

Vance, E. D. & Nadkarni, N. M. 1990. Microbial biomass and activity in canopy organic matter and the forest floor of a tropical cloud forest. Soil Biol. Biochem. 22: 677–684.

Veneklaas, E. J., Zagt, R. J., Van Leerdam, A., van Ek, R., Broekhoven, A. J. & van Genderen, M. 1990. Hydrological properties of the epiphyte mass of a montane tropical rain forest, Colombia. Vegetatio 89: 183–192.

Plant Ecology **153:** 251–261, 2001.

Forest light and its influence on habitat selection

Marc Théry
UMR 8571 CNRS, Muséum National d'Histoire Naturelle, Laboratoire d'Ecologie Générale, 4 avenue du petit château, 91800 Brunoy, France (e-mail: thery@mnhn.fr)

Key words: Niche differentiation, Species divergence, Tropical forest, Visual communication

Abstract

Light filtered through the forest canopy is the most variable physical factor in tropical forests, both in space and time. Vegetation geometry, sun angle, and weather generate five light environments, which greatly differ in intensity and spectrum. Forest light spectra can directly affect photosynthesis, plant morphogenesis, visual communication, and the effectiveness of plant-animal interactions. For animals, the apparent simplicity of five light environments is complicated by different types of contrast with the optical background which greatly modify the conspicuousness of visual signals. The purpose of this paper is to describe peculiarities of light in tropical forest, and to review the effects of light intensity and especially quality on plants and animals. Ecophysiological adaptations of plants to cope with contrasting light environments operate at daily, seasonal and life time-scales. Ambient light quality acts as a signal for both animals and plants, and consequences on plant growth, colour display, and signal design are examined. An analysis of the range of spectral parameters along a deforestation gradient is presented, testing if sites with more variation in light could support more species which are light-environment specialists. It is suggested that light quality measurement may be used to estimate the structural impact of forest exploitation, and that gives us the information necessary for a functional explanation of anthropogenic effects on tropical forest diversity.

Introduction

There has already been considerable research and recent syntheses on the subject of light in tropical forests, and on its consequences for animals and plants (e.g., Hailman 1977, 1979; Endler 1993; Vasquez-Yanes & Orozco-Segovia 1993; Chazdon et al. 1996; Fleishman et al. 1997; Bongers & Sterck 1998; Bradbury & Vehrencamp 1998 and references therein). Light measurements are relatively scarce in forest canopies (e.g., Hailman 1977, Endler 1993; Fleishman et al. 1997), mostly due to the difficult access and to the high spatial and temporal variation of light, but light filtered through the forest canopy has been measured in various ways. In plants, knowledge of light effects has considerably progressed with the identification of the wavelength-sensitive photoreceptors, a task which is still under way (reviews in Smith 1995; Fankhauser & Chory 1997; Tanada 1997; Ahmad 1999). Similarly, questions about crypsis and conspicuousness of colour signals have been only appropriately addressed in tropical forests following the development of spectroradiometry (e.g., Endler 1990, 1991, 1993; Fleishman et al. 1993, 1997; Endler & Théry 1996; Craig et al. 1996; Regan et al. 1996, 1998) and a better understanding of visual systems (e.g., Osorio & Vorobyev 1996; Fleishman et al. 1997; Regan et al. 1996, 1998).

In this paper, the characteristics of light and the variation of light environments in tropical forest are summarized. Results show that variation of both light intensity and quality is perceived by wavelength-sensitive photoreceptors of plants, and that the repair of plant damage by high UV radiation also depends on the quality and timing of light in the environment. Adaptation of plants to light damage is examined, suggesting important consequences for niche differentiation. The effects of light environments on crypsis and conspicuousness of colour signals are reviewed, showing the potential for habitat selection and sensitivity to forest disturbance. Because forest light spectra can directly affect plant and animal growth and reproduction, light measurement may provide a functional explanation of anthropogenic effects on biodiversity. To test this hypothesis, a recent study showing that spectral diversity predicts bird and plant species diversity along a human disturbance gradient is reported.

Light availability in tropical forests

Two types of ambient light-flux measurements can be taken, each differing in the acceptance angle of the sensor: 'irradiance' (i.e., the total amount of light incident on a surface measured at 180° and including diffuse light) and 'radiance' (i.e., the energy flux emitted from a specific area at a small acceptance angle – see Endler 1990; Bradbury & Vehrencamp 1998). Total irradiance (i.e., total light intensity) is obtained by integrating the irradiance spectra between 400 nm and 700 nm, yielding photosynthetically active radiation (PAR). The radiance of a colour patch in a particular light environment is the product of the irradiance (ambient light) spectrum and the intrinsic reflectance spectrum of the measured patch.

Total irradiance received by tropical forest canopy is more intense than at higher latitudes, due to higher solar declination closer to the equator (Mulkey et al. 1996). Light intensity frequently reaches instantaneous values of 1800–2200 μmol m^{-2} s^{-1} at midday, with highest measured values of 2600 μmol m^{-2} s^{-1} (Wright & Colley 1994; Chazdon et al. 1996). However, cloudy weather can reduce PAR by 75% or more (Lee & Downum 1991). Although annual variation in photoperiod and sun angle yield maximal sun radiance at the summer solstice, highest irradiance is usually received by the canopy during the dry season because of reduced cloud cover (Chazdon & Fetcher 1984; Rich et al. 1993; Mulkey et al. 1996). As radiance of clouds is larger than that of blue sky (Endler 1993), light intensity in forest shade increases when the sun is obscured by a cloud (Anderson 1964; Evans 1966; Endler 1993). With cloudy weather, ambient light also contains relatively more photons at photosynthetically active wavelengths than forest shade with clear sky (Endler 1993).

Light intensity is greatly reduced when passing through the highest canopy layer, and then gradually decreases further in the understorey (Yoda 1974; Koop & Sterck 1994). Light extinction can be as high as 94% over the first 5 m (Johnson & Atwood 1970). Mean light intensity levels of 0.1–1.9% of full sunlight have been reported at ground level, where sunflecks contribute to the highest mean value (review in Bazzaz & Pickett 1980). This reduction of light quantity coincides with drastic changes of direct sunlight spectrum induced by selective transmission and reflection by leaves. The ultraviolet (UV) part of the spectrum is almost completely absorbed, as well as most of the visible wavelengths with the exception of those between 520 and 620 nm (review in Endler 1993), and a larger proportion of the infra-red (IR) is reflected and transmitted by leaves (e.g., more than 30% reflectance and 40% transmittance between 700 and 1100 nm – Gates 1965). Despite the high spatial and temporal heterogeneity of light in forests, forest geometry, weather, and sun angle define a limited number of spectral light environments (Endler 1993; Figure 1). At dawn and dusk, and during transitions with daylight, when the sun is less than about 10° above the horizon, ambient light is enriched in the purple wavelength; the deficiency in middle wavelengths (570–630 nm) is caused by light absorbence by atmospheric ozone, particularly around 604 nm (Rozenberg 1966; McFarland & Munz 1976; Endler 1993). Light in large gaps and open areas is whitish. Light in small gaps is yellow-orange (through hole that subtend a solid angle of less than 2° to the vertical, e.g., from a 30 m canopy this would represent a 1 m diameter sun patch; Endler 1993). The spectrum of 'forest' shade is greenish to yellow-green, because most of the ambient light is from sunlight reflected by, and transmitted through, leaves between 520 and 620 nm, with little or no ambient light from the sky, and none at all from the sun. In 'woodland' shade, where forest canopy is more discontinuous and a larger proportion of blue sky is visible from the forest floor than in forest shade, the light spectrum is bluish to bluish-grey. The spectrum of small gaps, woodland shade, and forest shade converge to white with cloudy weather (Endler 1993; Figure 1). Spatial and temporal variations of these five heterogeneous light environments define the availability of light to forest plants and animals. However, the relative simplicity of five light environments is much more complex for animals: crypsis and conspicuousness will also be affected by transmission and reception of light signals (review by Bradbury & Vehrencamp 1998). Some of these effects for animals, like background radiance and contrast, countershading, or pattern contrast, are presented later in the paper.

Light is also polarized in tropical rain forest (Shashar et al. 1998). Among light environments, woodland shade and open areas are dominated by

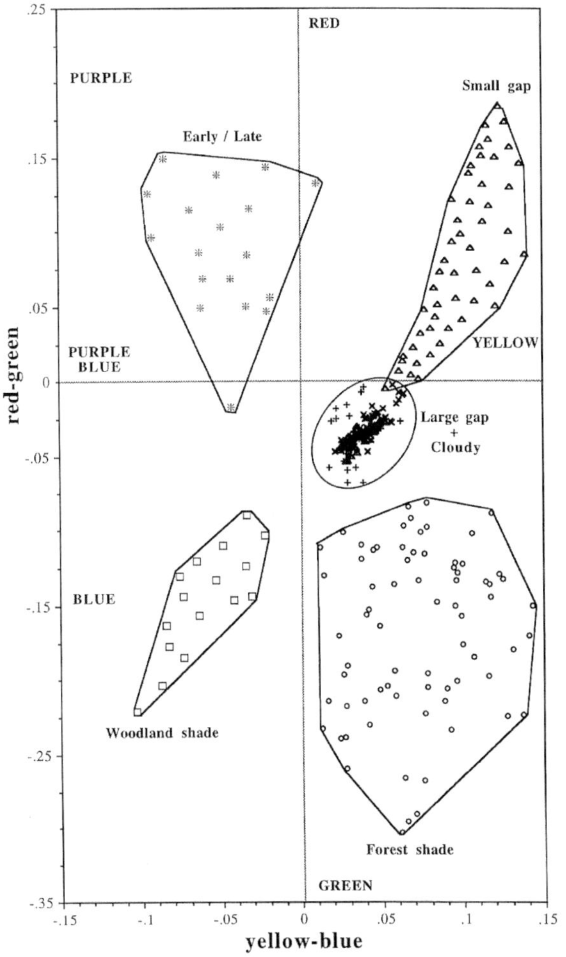

Figure 1. Colour variation of ambient light in primary tropical rain forest of French Guiana. Each point represents a single irradiance spectrum from which hue angle and chroma have been extracted using the segment classification method of Endler (1990) between 400 nm and 700 nm. The ordinate represents relative brightness of red and green spectral segments, and the abscissa is the difference between yellow and blue segments. The distance from the origin is chroma (colour saturation) and the angle clockwise from the top of the ordinate is hue (aspect of colour identified by words like 'red', 'green', and 'blue'). The direction of vectors approximately corresponding to Munsell (1975), hues are indicated by colour names. Light was measured between 11:00 and 13:00 about 50 cm above the forest floor at the Nouragues CNRS research station (4°05′ N-52°42′ W), French Guiana. A PS-1000 Ocean Optics portable spectroradiometer was used with an irradiance detector head (cosine-corrected, 180° acceptance angle) connected to a 200 μm diameter fused silica fibre optic. Light environments were selected before measurement, and outlines and names of light environments are indicated on the graph. No cloud was visible in clear sky conditions, and blue sky was totally absent in cloudy conditions. Small gaps measured 30–90 cm in diameter; the transition between small gaps and large gaps is shown by decreasing chroma of small gaps. The gradients between small gaps and forest shade, woodland shade and forest shade, early/late and large gaps, and clear sky and cloudy conditions are not shown here.

celestial polarization, whereas small gaps and forest shade are depolarized. Nevertheless, polarized light penetrating forest understorey is still measurable in dense forest shade. Cloud coverage has little or no effect on light polarization in small gaps, forest shade or woodland shade, but depolarizes the light field in open areas and in large gaps. When light arrives through a small gap and not directly from the sun (woodland shade with small gap, see 'canopy hole shade' in Shashar et al. 1998), light is polarized at the orientation of the light source, and is not depolarized by cloud coverage.

Differences in the spectrum of forest shade have been related to leaf thickness, with thinner leaves leading to greater 'chroma' (colour saturation – Endler 1993). Because canopy leaves are thicker than understorey leaves (e.g., Oberbauer & Strain 1986; Lee et al. 1990; Strauss-Debenedetti & Berlyn 1994), forests with denser canopies may show lower chroma in forest shade. Sun and shade plants appear to have similar optical properties, but shade plants are more efficient than sun plants in absorbing light for photosynthesis (e.g., Lee & Graham 1986; Lee et al. 1990). To my knowledge, only one study has considered the influence of position within trees on the optical properties of leaves (Poorter et al. 1995). This study showed that the efficiency of light absorption increased as light level decreased, the efficiency of light absorption per unit leaf chlorophyll remained relatively constant, but the optical properties of leaves showed no significant difference with light environment.

Differences between light environments are much greater than variation in leaf colour. Consequently, the same light habitats are found in very different forests, although the relative frequency of these habitats will depend upon the forest and canopy geometry (Endler 1993, 1997). Forests with more opened canopies have more woodland shade, and those with denser and more continuous canopies have more forest shade. Within the same forest, both the frequency and size of gaps increase with height off the ground (Hubbell & Foster 1986; Koike & Syahbuddin 1993; Van der Meer et al. 1994). Therefore, canopy light habitats mostly consist of a mosaic of large gaps and woodland shade (Endler 1993). Woodland shade, which is only found in the canopy and in large treefall gaps in wet forests with dense canopy cover, can reach lower levels in drier forests with thin or defoliated crowns. The understorey of dense tropical forest is, in contrast, a mosaic of forest shade and small gaps.

Because of vegetation cover, the blue part of the spectrum, present in the canopy, disappears as distance from the canopy increases. In absolute intensity, the UV and IR wavelengths are also strongly reduced in the understorey compared to the canopy (e.g., Gates 1965). However, the intensity of near-IR wavelengths (far-red around 730 nm) is increased relatively to red wavelengths (around 660 nm) strongly reducing the red/far-red ratio, to which plants are sensitive, compared to open areas (Holmes & Smith 1977a, b; Lee 1987; Schmitt & Wulff 1993). Leaves absorb the most UV (e.g., Lee 1987), except in open or disturbed canopies where more diffused UV-rich light can penetrate (Brown et al. 1994). Endler (1993) also predicts more UV in woodland shade than in forest shade. As diffuse light contributes more to understorey light with cloudy weather than with clear sky, there may be relatively more UV in the understorey when clouds obstruct the sun.

Wavelength-sensitive perception of light by plants

As plant progress from seed to the adult stage, they have to respond to a wide range of light intensities and qualities over a variety of time scales, from seconds to daily cycles, to annual cycles and yearly variations (reviews in Bazzaz & Pickett 1980; Roach & Wulff 1987; Vasquez-Yanes & Orozco-Segovia 1993; Long et al. 1994; Chazdon et al. 1996; Bongers & Sterck 1998). Light variations are perceived by three photoreceptor systems responding in the UV (UV-B photoreceptors), blue/near-UV (cryptochrome photoreceptors), and red/far-red (phytochrome photoreceptors) spectral ranges (reviews in Fankhauser & Chory 1997; Ahmad 1999). Blue light responses include inhibition of production of chloroplasts and pigments, inhibition of the rate of hypocotyl growth, phototropism, stomatal opening, induction of gene expression, and morphogenesis (reviews in Endler 1993; Kendrick & Kronenberg 1994; Fankhauser & Chory 1997; Horwitz & Berrocal 1997; Ahmad 1999). A novel flavoprotein, NPH1, has been proposed as a photoreceptor for phototropism (Christie et al. 1998). The relative intensity of red to far-red light has been shown to affect seed germination, hypocotil elongation, tillering rate, growth and elongation of stems, expansion rates of leaves, flowering, fruiting, and photosynthetic rates (reviews in Endler 1993; Kendrick & Kronenberg 1994; Smith 1995; Fankhauser & Chory 1997; Ahmad 1999). Even the high irradiance responses, which mostly require prolonged exposition to high light intensity between 710 and 720 nm (e.g., inhibition of hypocotyl elongation, leaf movement, flowering), are controlled by wavelength-sensitive photoreceptors in the red/far-red and blue spectral regions (Tanada 1997). Nevertheless, there appears to be complex interactions between and within these types of photopigments (Smith 1995; Fankhauser & Chory 1997; Horwitz & Berrocal 1997; Tanada 1997; Ahmad 1999). Phytochrome interacts with cryptochrome photoreceptors to potentiate blue-light signalling, and impacts on phototropism downstream of the NPH1 mediated signalling pathway (Ahmad 1999). Cryptochrome functions in photomorphogenesis and participates in phototropism through interaction with the NPH1 pathway (Ahmad 1999). Complete studies of these phenomena have been conducted in *Arabidopsis*, but little is known about plants in the wild, and even less about tropical canopy plants. However, in all cases studied, wavelength-sensitive photoreceptors perceive the intensity and colour of ambient light, allowing plants to respond to environmental cues.

Damage caused to plants by high irradiance in the canopy

The avoidance of high light-intensity stress is obtained by chloroplast movements, leaf movements (which include greater leaf angles at the top of the canopy), increased albedo, and photoinhibition (reviews in Demmig-Adams & Adams 1992; Long et al. 1994). Because of direct sunlight, canopy plants have to deal with high UV radiation in the canopy, which damages DNA (review in Britt 1996). Photoreactivation of DNA, obtained during subsequent exposures to UV-A and blue light, is the main line of defence plants have against UV-induced damage (Britt 1996). It reflects the action of photolyase enzymes, which are mediated by phytochrome pigments through successive expositions to red and UV lights. Therefore, the repair of plant damage by high UV radiation also depends on the quality and timing, as well as quantity, of light in their environment (Britt 1996).

Adaptations of plants to constraints of canopy light and consequences for habitat selection

Light is both a fundamental requirement and a constraint for canopy organisms. Many tree species with access to deep soil water produce leaves and flowers during the dry season, benefiting from increased irradiance (Wright & Van Schaik 1994) or because it constitutes a relatively enemy-free temporal space. Similarly, several tree species benefit from this seasonal opportunity to produce leaves and improve carbon gain (review in Mulkey et al. 1996). Light is often in excess in the canopy, well above the light-saturation point of canopy species (Fetcher et al. 1987; Doley et al. 1988). However, light intensity can sometimes be limited by cloud cover, which frequently reduces irradiance on exposed leaves below photosynthetic light saturation (Oberbauer & Strain 1986; Pearcy 1987; Zotz & Winter 1993; Hogan et al. 1995), and which limits tree growth in wet forests during years of highest cloud cover (Schulz 1960; Alvim 1964).

Both the morphology and physiology of leaves show adaptations to canopy light. Leaves exposed in the canopy are thicker, more vertical, and have higher specific leaf mass, lower chlorophyll density and higher a/b ratio than understorey leaves (Oberbauer & Strain 1986; Givnish 1987; Pearcy 1987; Bongers & Popma 1988; Fetcher et al. 1994). Shade leaves being, in contrast, relatively enriched in chlorophyll b, their low chlorophyll a/b ratio shows increased chlorophyll and light harvesting per chloroplast (review in Chazdon et al. 1996). Canopy leaves have higher light-saturated photosynthetic rates than conspecific leaves positioned lower in the canopy, and pioneer species have higher light-saturation points than climax trees or shade plants (review in Chazdon et al. 1996). Light quality varies with light environments and position within the forest, so it is possible that morphogenetic processes may differ among species living in different light environments (Endler 1993). For example, critical levels of light intensity or quality could be reached in the canopy for shade-tolerant or shade-demanding species, but not for light-demanding species. Because of canopy light characteristics, canopy species should be better adapted to high irradiance, UV, and IR radiation. This is potentially a strong feature of habitat selection and niche differentiation.

Among tropical forest trees, shade-tolerant and light-demanding species are well recognized (review in Denslow 1987). Specific differentiation of trees in light habitats has been proposed (Denslow 1980, 1987) in contrast to the null hypothesis of random occupation of light habitats (Hubbell & Foster 1986). However, adult trees of both types occur in a wide variety of light intensities, and most species are not associated with a particular light habitat (e.g., Lieberman et al. 1989, 1995; Welden et al. 1991). In a recent review of gap-size niche differentiation by tropical rain forest trees, Brown & Jennings (1998) showed that there are little evidence to support species differentiation according to gap size or light variation, and reviewed the numerous problems involved in detecting niche differentiation, notably inappropriate measurement and use of various gap sizes and light conditions in experiments. Similarly, Bongers & Sterck (1998) showed that there are too few data presently available on the light levels of trees larger than small saplings to form conclusion on the effective differentiation of tree species in light habitats compared with random occupation.

Pittendrigh (1948), who distinguished 'exposure', 'sun' and 'shade-tolerant' species, has described the vertical partitioning of light in epiphytic bromeliads. Each species group is characterized by morphological and physiological adaptations to light and water levels (Pittendrigh 1948; Benzing 1995). Species of the 'exposure' and 'sun' group are characterized by crassulacean acid metabolism (a photosynthetic pathway adapted to drought and high irradiance). Shade-tolerant species, in contrast, often have lower light-compensation and light-saturation points typical of C_3 photosynthesis (Benzing 1995). Because light availability decreases with cloud cover, the shade group is not represented at the wettest sites, and the exposure group prevails at sites with lower precipitation (Griffiths & Smith 1983). The dense growth of epiphytes in cloud forest can also be explained by variations of light intensity and spectrum compared to other tropical forests (Endler 1993). Physiological adaptation of not only epiphytes (Benzing 1995), but also of hemiepiphytes (Williams-Linera & Lawton 1995) and non-vascular epiphytes (Rhoades 1995), potentially contribute to species space-partitioning in relation to light availability.

Crypsis and conspicuousness of colour signals, and consequences for habitat selection

The transmission efficiency of a visual signal can be enhanced by four types of contrast with the optical background: brightness, colour, pattern and movement contrasts (Bradbury & Vehrencamp 1998). Each of these strategies for conspicuousness has an opposed contrast-reducing strategy for crypticity. In all cases, brightness, colour components of the ambient light irradiance, and background radiance spectra, greatly affect the conspicuousness of a colour signal (Hailman 1979; Lythgoe 1979; Endler 1986, 1990, 1993; Endler & Théry 1996; Fleishman et al. 1997). These visual contrasts can be used in communication between animals (predation and crypsis; recognition of species, sex, and age; sexual selection), and in communication between animals and plants (herbivory, pollination, seed dispersal).

Brightness contrast is most efficient when backgrounds are extremely light or extremely dark, and may be influenced by the direction of illuminating light (Cott 1940; Bradbury & Vehrencamp 1998). For example, in *Phylloscopus* warblers from habitats of different brightness, numerous white patches increase brightness contrast in species from dark habitats, whereas species from bright habitats have no white patch and rely more on acoustic communication (Marchetti 1993). For animals illuminated from above, such as those in forest canopy, countershading can be used for concealment by lightening the ventral side, and reverse countershading can be used for conspicuousness by darkening the ventral side (Cott 1940; Hailman 1979). This greatly depends on the position of the viewer, either potential mate or predator (Endler 1990; Bradbury & Vehrencamp 1998). Colour contrast can be obtained by difference in either hue or chroma (colour saturation) between the visual signal and its background (Endler 1990). Therefore, the most conspicuous colours are those which are rich in ambient light wavelengths but poor in the background reflectance spectra (Endler 1990). If the ambient light is strongly coloured, as it often occurs in forests, animals should use a colour signal that matches the ambient light peak and surround themselves with complementary colours to maximize brightness contrast. In addition, animals can use complementary colours whose cut-offs are centred on the region of greatest ambient light intensity (Endler 1992). Forest canopy vegetation provides a strongly patterned background that can be exploited for either contrast or camouflage (e.g., disruptive colouration, Cott 1940). The visual background moves in a slow and sinusoidal motion when the wind blows through the vegetation, which animals can utilize to hide from prey (e.g., vine snakes; Fleishman 1985). Contrast against this moving background can also be used to detect prey or to increase conspicuousness of a display (e.g., *Anolis* lizards, Fleishman 1992).

Canopy light consists of a mosaic of large gaps and woodland shade which greatly modifies brightness and colour contrasts between signals and backgrounds. In open areas, and above and within the canopy, ambient light is whitish and rich in UV, and the background mostly consists of bright blue and UV sky. Therefore, conspicuous signals in such large gaps should be black for brightness contrast (Hailman 1979). No easy prediction can be made for colour contrast in white light, although a slight disadvantage to blue and possibly green colouring, and also to utilization of a blue background, may result from the reduction of light irradiance below 450 nm (Endler 1993). In canopy shade (woodland shade), ambient light is relatively rich in blue and UV, and background colour mostly consists of green leaves reflecting little UV (Fleishman et al. 1997). Consequently, colour signals in canopy shade should be blue, blue-green or UV to maximise brightness contrast, with some red or orange for increased contrast with the green background (Endler 1993). Because canopy light irradiance is richer in UV than forest shade, and because background radiance is green and poor in UV in both habitats, there should be more UV signals for communication in the canopy compared to heavily shaded environments (as shown by Fleishman et al. (1993) in *Anolis* lizards of Puerto Rico). Because woodland shade is most often found in the canopy, and forest shade in the understorey, there should be a vertical stratification of colours, with blue and blue-green hues used more often in the canopy and red or orange closer to the floor (Endler 1993). There is support for this hypothesis among 233 rain forest bird species of French Guiana, with brighter and bluer reflectance spectra in birds living within the canopy, and darker and redder colours of understorey birds (D. Gomez pers. com.). Colour saturation (chroma) can also potentially affect conspicuousness in different light habitats: patches on the body which are highly chromatic will be less sensitive to changes in ambient light than those with low chroma (which will change dramatically in both hue and chroma in different light environments) (Endler 1993, 1997). Consequently, species that need to be cryptic in different light habitats should have low chroma patches that change with the background as ambient light fluctuates. Species that need to be conspicuous could choose low chroma if they display in specific light environments, or high chroma that retain their contrast and hue independent of light environment (Endler 1997). If colour signals are brighter and more chromatic in the canopy, they will also be relatively insensitive to spectral variation of light habitats. However, too few species have been studied to confirm or invalidate this hypothesis on chromatic variation (Endler 1997).

Different flower, fruit or animal colours are more conspicuous in some tropical forest light environments and backgrounds, and less in others (Endler 1993; Craig 1996; Endler & Théry 1996). For example, it is noteworthy that blue fruits are most often displayed in open areas of high light intensity (Cooper et al. 1986). This may be explained by increased conspicuousness of blue signals in woodland shade compared to other light habitats (Endler 1993, 1997), or by the fact that it takes a 100 fold increase in light intensity to stimulate the blue cones in primate retinas that it does to stimulate the cones sensitive to green or red light (Zrenner & Gouras 1981), bearing in mind that primates may be important dispersers of the seeds from such fruit. However, care should taken in generalizing because of important specific differences in colour vision systems, particularly in the UV (Neumeyer 1991; Jacobs 1992; Bennett et al. 1994; Goldsmith 1994; Tovée 1995). Evidence does not fully support the hypothesis that fruit colour increases conspicuousness in particular light environments, and several other hypotheses could explain the evolution of fruit colour (review in Willson & Whelan 1990). Interpretation of the UV-reflectant signals should also be done with great care, since for example UV-reflectant waxy bloom of fruit does not appear to increase detectability by UV-sensitive frugivores (Willson & Whelan 1989), but may in fact have a photoprotection role. Studies conducted on the relationship between colour vision systems and specific plant colour signals concluded that visual systems optimize discrimination of green leaves by dichromatic mammals (Lythgoe & Partridge 1989), the discrimination of flowers by hymenopterans (Chittka & Menzel 1992), and the detection of fruit by primates (Osorio & Vorobyev 1996; Regan et al. 1996, 1998). In terrestrial vertebrates, visual systems also optimize detection of objects against a background dominated by green radiance (Fleishman et al. 1997). As a consequence of distribution patterns of polarized light in forest, various functions of polarized vision (e.g., navigation) must be limited to open areas or to the canopy, whereas other functions can be used even under canopy cover (e.g., determination of surface orientation; object detection – Shashar et al. 1998).

Few studies have investigated the effects and use of light environments in tropical forest animals. Endler (1987, 1991) showed that guppies (*Poecilia reticulata*), small fishes that live in tropical rain forest streams, mostly display during twilight, and that predation takes place in all other light environments. Guppies have highest colour and brightness contrast when they display in purple light at dawn and dusk, and lowest contrasts in the other light environments, especially forest shade. This increases conspicuousness of males to females during courtship, and decreases visibility to predators at other times and places. This variation of conspicuousness with light environments is further increased by the differences in colour vision between guppies and their predators. In anoline lizards of Puerto Rico, Fleishman et al. (1993) showed that UV patterns used for display are only present in canopy species benefiting from high UV levels in ambient light, these signals contrasting with the green vegetation that reflects little UV. Both this study of lizards and that of Marchetti (1993) on birds support the idea that light availability influences habitat selection and species divergence. The influence of light on bird courtship display was studied in lek-breeding manakins and in the cock-of- the-rock (Théry 1987, 1990a, b; Théry & Vehrencamp 1995; Endler & Théry 1996). These birds maximise their conspicuousness to mates, and minimise their conspicuousness to predators, by selecting light habitats in times and places that generate these differences in conspicuousness (Endler & Théry 1996). Each species optimizes its visibility by different combinations of display postures, colours and light habitats (Théry 1990a; Endler & Théry 1996). Among the bird species studied by Endler & Théry (1996), males of any one species use either brightness or colour contrast, but not both. In five sympatric species of lekking manakins, the disposition of colour patches on upper- or underparts can be explained by display height and light incidence in relation to vegetation geometry (Théry 1990a). In the white-throated manakin, light properties of the display site and attendance of the male determine the attractiveness of male display to females (Théry & Vehrencamp 1995). As in guppies, colour and/or brightness contrast is highest at times and places of courtship display to females, and colours are used for mate choice. Another example of the use of light habitats is given by the variation of web colour in *Nephila clavipes*, a large orb-spinning spider widely distributed throughout the New World tropics and subtropics (Craig et al. 1996). Yellow is very attractive to herbivorous and pollinating insects, and this spider attracts and intercepts more insects by constructing webs with yellow pigments (which appear golden in high light intensity). The web colour is adapted to differences of both light intensity and quality, and spiders can adjust pigment production to produce golden webs in a variety of ambient light spectra. In contrast to guppies and lekking birds, which select particular light habitats for specific colour display, this spider adapts its colour signal to a wide range of light environments. In that sense, it is potentially less sensitive to changes of light environment (Endler 1997), which may explain its very wide distribution in forested, edge, and non-forested habitats. Light-environment specialists will, in contrast, show greater sensitivity to modifications of their preferred light environment.

Alteration of light by deforestation, and consequences for biodiversity

Because logging alters the fraction of the sky illuminating objects within the forest (Endler 1997), it also alters the relative abundance of light environments. As disturbance and sky fraction increase, the frequency of woodland shade and large gaps increases at the expense of forest shade and small gaps. As the number and diversity of light environments decrease with the intensity of disturbance, the measurements of light spectra may therefore provide an estimate of the structural impact of forest exploitation (Endler 1997). Because forest light spectra can directly affect animal and plant reproduction and growth, its measurement may also provide a functional explanation of anthropogenic effects on tropical forest diversity. To test this hypothesis, a recent study measured ambient light spectra as well as carrying out a census of manakin birds (Pipridae) and melastome plants (Melastomataceae) (both which are known to have precise light-environment requirements). Irradiance was measured as in Figure 1, and its variation sampled in circles 20 m in diameter. A total of 34 circles were sampled at three sites in primary and disturbed rain forests with similar climate, dominant plant families and bird communities. The gradient of forest disturbance ranged from totally undisturbed forest to different levels of selective logging, an old clear-cut and an active quarry. The ranges of spectral parameters (hue, chroma, total brightness and red/far-red ratio) were measured and analysed in the expectation that sites with more variation in light could support more species which are light-environment specialists.

Positive correlations were found between hue range and manakin diversity, and between the range of red/far-red ratio and melastome diversity (Théry & Endler unpublished). There was no significant relationship between species numbers and means of hue, chroma, total brightness or red/far-red ratio, and with the ranges of chroma or brightness. There was no significant correlation between manakins and melastomes species numbers. The positive correlations are both an indicator of general habitat loss through anthropogenic disturbance and a direct measure of habitat requirements. However, changes in light environment also correlate with important changes in temperature, humidity and physical structure. It will take controlled experiments to show that change of light habitat itself is the causal factor. Nevertheless, the restriction of light habitats through alteration of forest shade leads directly to the expansion of few high-light demanding species, and to a loss in biodiversity through the disappearance of numerous forest shade or small gap specialists. Melastome fruit constitutes a large proportion of the diet of manakins, which in turn are major dispersers of melastome seeds (review in Krijger et al. 1997), and so these results may not only reflect habitat specialization in manakins and melastomes, but also a direct ecological linkage between the two. Changes in forest light quality may affect each group directly (e.g., melastome germination and development, crypsis and conspicuousness of manakins) or indirectly (e.g., interruption of breeding due to fruit scarcity, inefficient detection of fruit in altered visual conditions). Synergistic effects induced by pollination or seed-dispersal mutualisms are expected to multiply the consequences of light disturbance, and finally to lead to elimination from logged forest (Théry & Endler unpubl. data). This study of effects of human disturbance on forest light and species diversity shows that it is crucial to investigate the use, distribution and importance of light environments in conservation projects.

Acknowledgements

This research has been funded by the Centre National de la Recherche Scientifique and the Muséum National d'Histoire Naturelle (ECOTROP URA 1183 and UMR 8571), and by the French Ministry of Environment (grant SOFT 94.082 to M. Théry & C. Erard). I would like to thank J. A. Endler for collaboration and radiance calibration of the spectrometer, P. Charles-Dominique and Silvolab-Guyane for logistical support in French Guiana, the European Science Foundation and the Tropical Canopy Research Programme chairman K. E. Linsenmair for the organization of stimulating meetings in Vienna (Austria), San Feliu de Guixols

(Spain), Venezuela and Oxford (UK). I am grateful to A. J. Davis, L. J. Fleishman, I. Cuthill, and an anonymous reviewer for their comments which greatly improved this manuscript.

References

Ahmad, M. 1999. Seeing the world in red and blue: insight into plant vision and photoreceptors. Curr. Opin. Plant Biol. 2: 230–235.

Alvim, P. de T. 1964. Tree growth periodicity in tropical climates. Academic Press, San Diego.

Anderson, A. C. 1964. Light relations of terrestrial plant communities and their measurement. Biol. Rev. 39: 425–486.

Bazzaz, F. A. & Pickett, S. A. T. 1980. Physiological ecology of a tropical succession: a comparative review. Ann. Rev. Ecol. Syst. 11: 287–310.

Bennett, A. T. D., Cuthill, I. C. & Norris, K. J. 1994. Sexual selection and the mismeasure of color. Am. Nat. 144: 848–860.

Benzing, D. H. 1995. Vascular epiphytes. Pp. 225–254. In: Lowman, M. D. & Nadkarni, N. M. (eds), Forest canopies. Academic Press, San Diego.

Bongers, F. & Popma, J. 1988. Is exposure-related variation in leaf characteristics of tropical rain forest species adaptive? Pp. 191–200. In: Werger, M. J. A., van der Aart, P. J. M., During, H. J. & Verhoeven, J. T. A. (eds), Plant form and vegetation structure. SPB Academic Press, The Hague.

Bongers, F. & Sterck, F. J. 1998. Architecture and development of rain forest trees: responses to light variation. Pp. 125–162. In: Newberry, D. M., Prins, H. H. T. & Brown, N. D. (eds), Dynamics of tropical communities. Blackwell Science, Oxford.

Bradbury, J. W. & Vehrencamp, S. L. 1998. Principles of animal communication. Sinauer Associates, Sunderland.

Britt, A. B. 1996. DNA damage and repair in plants. Ann. Rev. Plant. Physiol. Plant. Mol. Biol. 47: 75–100.

Brown, M. J., Parker, G. G. & Posner, N. 1994. A survey of ultraviolet-B radiation in forests. J. Ecol. 82: 843–854.

Brown, N. D. & Jennings, S. 1998. Gap-size niche differentiation by tropical rain forest trees: a testable hypothesis or a broken down bandwagon? Pp. 79–94. In: Newberry, D. M., Prins, H. H. T. & Brown, N. D. (eds), Dynamics of tropical communities. Blackwell Science, Oxford.

Chazdon, R. L. & Fetcher, N. 1984. Photosynthetic light environments in a lowland tropical rain forest in Costa Rica. J. Ecol. 72: 553–564.

Chazdon, R. L., Pearcy, R. W., Lee, D. W. & Fetcher, N. 1996. Photosynthetic responses of tropical forest plants to contrasting light environments. Pp. 5–55. In: Mulkey, S., Chazdon, R. L. & Smith, A. P. (eds), Tropical plants ecophysiology. Chapman & Hall, New York.

Chittka, L. & Menzel, R. 1992. The evolutionary adaptation of flower colours and the insect pollinators' colour vision. J. Comp. Physiol. A 171: 171–181.

Christie, J. M., Reymond, P., Powell, G. K., Bernasconi, P., Raibekas A. A., Liscum, E. & Briggs, W. R. 1998. *Arabidopsis* NPH1: a flavoprotein with the properties of a photoreceptor of phototropism. Science 282: 1698–1701.

Cooper, H. M., Charles-Dominique, P. & Viénot, F. 1986. Signification de la coloration des fruits en fonction de la vision des vertébrés consommateurs. Mem. Mus. Nat. Hist. Nat. 132: 131–143.

Cott, H. B. 1940. Adaptive coloration in animals. Methuen, London.

Craig, C. L., Weber, R. S. & Bernard, G. D. 1996. Evolution of predator-prey systems: spider foraging plasticity in response to the visual ecology of prey. Am. Nat. 147: 205–229.

Demmig-Adams, B. & Adams, W. W. III. 1992. Photoprotection and other responses of plants to high light stress. Ann. Rev. Plant Physiol. Plant Mol. Biol. 43: 599–626.

Denslow, J. S. 1980. Gap partitioning among tropical rain forest trees. Biotropica 12: 47–55.

Denslow, J. S. 1987. Tropical rain forest gaps and tree species diversity. Ann. Rev. Ecol. Syst. 18: 431–451.

Doley, D., Unwin, G. L. & Yates, D. J. 1988. Spatial and temporal distribution of photosynthesis and transpiration by single leaves in a rain forest tree. Austr. J. Plant Physiol. 15: 317–326.

Endler, J. A. 1986. Defense against predation. Pp. 109–134. In: Feder, M. E. & Lauder, G. V. (eds), Predator-prey relationships: perspectives and approaches from the study of lower vertebrates. University of Chicago Press, Chicago.

Endler, J. A. 1987. Predation, light intensity, and courtship behaviour in *Poecilia reticulata*. Anim. Behav. 35: 1376–1385.

Endler, J. A. 1990. On the measurement and classification of colour in studies of animal colour patterns. Biol. J. Linn. Soc. London 41: 315–352.

Endler, J. A. 1991. Variation in the appearance of guppy color patterns to guppies and their predators under different visual conditions. Vis. Res. 31: 587–608.

Endler, J. A. 1992. Signals, signal conditions, and the direction of evolution. Am. Nat. 139: S125–S153.

Endler, J. A. 1993. The color of light in forests and its implications. Ecol. Monogr. 63: 1–27.

Endler, J. A. 1997. Light, behavior and conservation of forest-dwelling organisms. Pp. 330–356. In: Clemmons, J. R. & Buchholz, R. (eds), Behavioral approaches to conservation in the wild. Cambridge University Press, Cambridge.

Endler, J. A., & Théry, M. 1996. Interacting effects of lek placement, display behavior, ambient light, and color patterns in three Neotropical forest-dwelling birds. Am. Nat. 148: 421–452.

Evans, G. C. 1966. Model and measurement in the study of woodland light climates. Pp. 53-76. In: Bainbridge, R., Evans, G. C. & Rackham, O. (eds), Light as an ecological factor. Blackwell, Oxford,

Fankhauser, C. & Chory, J. 1997. Light control of plant development. Ann. Rev. Cell Dev. Biol. 13: 203–229.

Fetcher, N., Oberbauer, S. F. & Chazdon, R. L. 1994. Physiological ecology of plants. Pp. 128–141. In: McDade, L., Bawa, K. S., Hespenheide, H. & Hartshorn, G. S. (eds), La Selva: ecology and natural history of a Neotropical rain forest. University of Chicago Press, Chicago.

Fetcher, N., Rojas, G & Strain, B. R. 1987. Efectos del régimen de luz sobre la fotosintesis y el crecimiento en plantulas de arboles de un bosque lluvioso tropical de Costa Rica. Rev. Biol. Trop. 35 (suppl.): 97–110.

Fleishman, L. J. 1985. Cryptic movement in the vine snake *Oxybelis aeneus*. Copeia 1985: 242–245.

Fleishman, L. J. 1992. The influence of the sensory system and the environment on motion patterns in the visual displays of anoline lizards and other vertebrates. Am. Nat. 139: S36–S61.

Fleishman, L. J., Bowman, M., Saunders, D., Miller, W. E., Rury, M. J. & Loew, E. R. 1997. The visual ecology of Puerto Rican anoline lizards: habitat light and spectral sensitivity. J. Comp. Physiol. A 181: 446–460.

Fleishman, L. J., Loew, E. R. & Leal, M. 1993. Ultraviolet vision in lizards. Nature 365: 397.

Gates, D. M. 1965. Energy, plants, and ecology. Ecology 46: 1–13.

Givnish, T. J. 1987. Leaf and canopy adaptations in tropical forests. Pp. 51–84. In: Medina, E., Mooney, H. A. & Vasquez-Yanes, C. (eds), Physiological ecology of plants in the wet tropics. Tasks for Vegetation Science 12, Junk, The Hague.

Goldsmith, T. H. 1994. Ultraviolet receptors and color vision: evolutionary implications and a dissonance of paradigms. Vision Res. 34: 1479–1487.

Griffiths, H. & Smith, J. A. C. 1983. Photosynthetic pathways in the Bromeliaceae of Trinidad: relations between life forms, habitat preference and the occurrence of CAM. Oecologia 60: 176–184.

Hailman, J. P. 1977. Optical signals: animal communication and light. Indiana University Press, Bloomington.

Hailman, J. P. 1979. Environmental light and conspicuous colors. Pp. 289–354. In: Burtt, E. H. J. (ed.), The behavioral significance of color. Garland STPM Press, New York.

Holmes, M. G. & Smith, H. 1977a. The function of phytochrome in the natural environment. I. Characterization of daylight for studies in photomorphogenesis and photoperiodism. Photochem. Photobiol. 25: 533–538.

Holmes, M. G. & Smith, H. 1977b. The function of phytochrome in the natural environment. II. The influence of vegetation canopies on the spectral energy distribution of natural daylight. Photochem. Photobiol. 25: 539–545.

Hogan, K. P., Smith, A. P. & Samaniego, M. 1995. Gas exchange in six tropical semi-deciduous forest canopy tree species during the wet and dry seasons. Biotropica 27: 324–333.

Horwitz, B. A. & Berrocal, G. M. 1997. A spectroscopic view of some recent advances in the study of blue light photoreception. Bot. Acta 110: 360–368.

Hubbell, S. P. & Foster, R. B. 1986. Biology, chance and history and the structure of tropical rain forest tree communities. Pp. 314–329. In: Diamond, J. & Case, T. J. (eds), Community ecology. Harper & Row, New York.

Jacobs, G. H. 1992. Ultraviolet vision in vertebrates. Amer. Zool. 32: 544–554.

Johnson, P. L. & Atwood, D. M. 1970. Aerial sensing and photographic study of the El Verde rain forest. Pp. B63–B78. In: Odum, H. T. & Pigeon, R. F. (eds), A tropical rain forest; a study of irradiation and ecology at El Verde, Puerto Rico. Division of Technical Information, U.S. Atomic Energy Commission, Oak Ridge.

Kendrick, R. E. & Kronenberg, G. H. M. 1994. Photomorphogenesis in plants, 2 nd edn. Kluwer Academic Publishers, Dordrecht.

Koike, F. & Syahbuddin. 1993. Canopy structure of a tropical rain forest and the nature of an unstratified upper layer. Funct. Ecol. 7: 230–235.

Koop, H. & Sterck, F. J. 1994. Light penetration through structurally complex forest canopies: an example of a lowland tropical rain forest. Forest Ecol. Manag. 69: 111–122.

Krijger, C. L., Opdam, M., Théry, M. & Bongers, F. 1997. Courtship behaviour of manakins and seed bank composition in a French Guianan rain forest. J. Trop. Ecol. 13: 631–636.

Lee, D. W. 1987. The spectral distribution of radiation in two Neotropical rain forests. Biotropica 19: 161–166.

Lee, D. W., Bone, R. A., Tarsis, S. L. & Storch, D. 1990. Correlates of leaf optical properties in tropical forest sun and extreme-shade plants. Amer. J. Bot. 77: 370–380.

Lee, D. W. & Downum, K. R. 1991. The spectral distribution of biologically active solar radiation at Miami, Florida, USA. Int. J. Biomet. 35: 48–54.

Lee, D. W. & Graham, R. 1986. Leaf optical properties of rain forest sun and extreme shade plants. Amer. J. Bot. 73: 1100–1108.

Lieberman, M., Lieberman, D. & Peralta, R. 1989. Forests are not just Swiss cheese: canopy stereogeometry of non-gaps in tropical forests. Ecology 70: 550–552.

Lieberman, M., Lieberman, D., Peralta, R. & Hartshorn, G. S. 1995. Canopy closure and the distribution of tropical forest tree species at La Selva, Costa Rica. J. Trop. Ecol. 11: 161–178.

Long, S. P., Humphries, S. & Falkowski, P. G. 1994. Photoinhibition of photosynthesis in nature. Ann. Rev. Plant Physiol. Plant Mol. Biol. 45: 633–662.

Lythgoe, J. N. 1979. The ecology of vision. Clarendon Press, Oxford.

Lythgoe, J. N. & Partridge, J. C. 1989. Visual pigments and the acquisition of visual information. J. Exp. Biol. 146: 1–20.

Marchetti, K. 1993. Dark habitats and bright birds illustrate the role of the environment in species divergence. Nature 362: 149–152.

McFarland, W. N. & Munz, F. W. 1976. The visible spectrum during twilight and its implications to vision. Pp. 249–270. In: Evans, G. C., Bainbridge, R., Rackham, O. (eds), Light as an ecological factor, II. Blackwell Scientific, Oxford.

Mulkey, S. S., Kitajama, K. & Wright, S. J. 1996. Plant physiological ecology of tropical forest canopies. Trends Ecol. Evol. 11: 408–412.

Munsell, A. H. 1975. Munsell book of color. Munsell Color Co., Baltimore.

Neumeyer, C. 1991. Evolution of colour vision. Pp. 284–305. In: Cronley-Dillon, J. R. & Gregory, R. L. (eds), Vision and visual dysfunction: evolution of the eye and visual systems. Macmillan Press, London.

Oberbauer, S. F. & Strain, B. R. 1986. Effects of canopy position and irradiance on the leaf physiology and morphology of *Pentaclethra macroloba* (Mimosaceae). Amer. J. Bot. 73: 409–416.

Osorio, D. & Vorobyev, M. 1996. Colour vision as an adaptation to frugivory in primates. Proc. R. Soc. London B 263: 593–599.

Pearcy, R. W. 1987. Photosynthetic gas exchange responses of Australian tropical forest trees in canopy, gap and understory microenvironments. Funct. Ecol. 1: 169–178.

Pittendrigh, C. S. 1948. The bromeliad-*Anopheles*-malaria complex in Trinidad. I. The bromeliad flora. Evolution 2: 58–89.

Poorter, L., Oberbauer, S. F. & Clark, D. B. 1995. Leaf optical properties along a vertical gradient in a tropical rain forest canopy in Costa Rica. Amer. J. Bot. 82: 1257–1263.

Regan, B. C., Viénot, F, Charles-Dominique, P., Pefferkorn, S., Simmen, B., Julliot, C. & Mollon, J. D. 1996. The colour signals that fruits present to primates. Invest. Ophtalmol. Vis. Sci. 37: S 648.

Regan, B. C., Julliot, C., Simmen, B., Viénot, F., Charles-Dominique, P. & Mollon, J. D. 1998. Frugivory and colour vision in *Allouata seniculus*, a trichromatic platyrrhine monkey. Vision Res. 38: 3321–3327.

Rhoades, F. M. 1995. Nonvascular epiphytes in forest canopies: worldwide distribution, abundance, and ecological roles. Pp. 353–408. In: Lowman, M. D. & Nadkarni, N. M. (eds), Forest canopies. Academic Press, San Diego.

Rich, P. M., Clark, D. B., Clark, D. A. & Oberbauer, S. F. 1993. Long-term study of solar radiation regimes in a tropical wet forest using quantum sensors and hemispherical photography. Agric. Forest Meteo. 65: 107–127.

Rozenberg, G. V. 1966. Twilight, a study in atmospheric physics. Plenum Press, New York.

Schmitt, J. & Wulff, R. 1993. Light spectral quality, phytochrome and plant competition. Trends Ecol. Evol. 8: 47–51.

Schulz, J. P. 1960. Ecological studies on rain forest in northern Suriname. Noord-Hollandsche Uitgevers Maatschappij, Amsterdam.

Shashar, N., Cronin, T. W., Wolff, L. B. & Condon, M. A. 1998. The polarization of light in a tropical rain forest. Biotropica 30: 275–285.

Smith, H. 1995. Physiological and ecological function within the phytochrome family. Ann. Rev. Plant Physiol. Mol. Biol. 46: 289–315.

Strauss-Debenedetti, S. & Berlyn, G. P. 1994. Leaf anatomical responses to light in five tropical Moraceae of different successional status. Amer. J. Bot. 81: 1582–1591.

Tanada, T. 1997. The photoreceptors of high irradiance in plants. Physiol. Plant. 10: 451–454.

Théry, M. 1987. Influence des caractéristiques lumineuses sur la localisation des sites traditionnels, parade et baignade des manakins (Passeriformes, Pipridae). C. R. Acad. Sci. Paris sér. III 304: 19–24.

Théry, M. 1990a. Influence de la lumière sur le choix de l'habitat et le comportement sexuel des Pipridae (Aves: Passeriformes) en Guyane Française. Rev. Ecol. (Terre Vie) 45: 215–236.

Théry, M. 1990b. Display repertoire and social organization of the white-fronted and white-throated manakins. Wilson Bull. 102: 123–130.

Théry, M. & Vehrencamp, S. L. 1995. Light patterns as cues for mate choice in the lekking white-throated manakin *Corapipo gutturalis*. Auk 112: 133–145.

Tovée, M. J. 1995. Ultra-violet photoreceptors in the animal kingdom: their distribution and function. Trends Ecol. Evol. 10: 455-460.

van der Meer, P. J., Bongers, F., Chatrou, L. & Riéra, B. 1994. Defining canopy gaps in a tropical rain forest: effects on gap size and turnover time. Acta Oecol. 15: 701–714.

Vasquez-Yanes, C. & Orozco-Segovia, A. 1993. Patterns of seed longevity and germination in the tropical rain forest. Ann. Rev. Ecol. Syst. 24: 69–87.

Welden, C. W., Hewett, S. W., Hubbell, S. P. & Foster, R. B. 1991. Sapling survival, growth and seedling establishment: relationship to canopy height in a Neotropical forest. Ecology 72: 85–96.

Williams-Linera, G. & Lawton, R. O. 1995. The ecology of hemiepiphytes in forest canopies. Pp. 255–283. In: Lowman, M. D. & Nadkarni, N. M. (eds), Forest canopies. Academic Press, San Diego.

Willson, M. F. & Whelan, C. J. 1989. Ultraviolet reflectance of fruits of vertebrate-dispersed plants. Oikos 55: 341–348.

Willson, M. F. & Whelan, C. J. 1990. The evolution of fruit color in fleshy-fruited plants. Am. Nat. 136: 790–809.

Wright, S. J. & Colley, M. 1994. Accessing the canopy: assessment of biological diversity and microclimate of the tropical forest canopy: Phase I, UNEP.

Wright, S. J. & van Schaik, C. P. 1994. Light and the phenology of tropical trees. Am. Nat. 143: 192–199.

Yoda, K. 1974. Three-dimensional distribution of light intensity in a tropical rain forest of West Malaysia. Jap. J. Ecol. 24: 247–254.

Zotz, G. & Winter, K. 1993. Short-term photosynthesis measurements predict leaf carbon balance in tropical rain forest canopy plants. Planta 191: 409–412.

Zrenner, E. & Gouras, P. 1981. Characteristics of the blue sensitive cone mechanism in primate retinal ganglion cells. Vision Res. 21: 1605–1609.

Plant Ecology **153:** 263–277, 2001.

Methods to assess tropical rain forest canopy structure: an overview

Frans Bongers
Department of Environmental Sciences, Silviculture and Forest Ecology Group, Wageningen University, P.O. Box 342, 6700 AH Wageningen, The Netherlands (e-mail: frans.bongers@btbo.bosb.wau.nl)

Key words: Canopy structure, Forest texture, Methodology, Overview, Scale, Structure spectra, Tropical rain forest, Vertical structure

Abstract

Forest canopy structure (*sensu latu*) is the combination of forest texture (the qualitative and quantitative composition of the vegetation as to different morphological elements), and forest structure (*sensu strictu*, the spatial arrangement of these elements). Scale is an aspect of major importance. At a regional scale forest types can be distinguished, like broadleaf or coniferous forest. At local scale, distribution and size and shape of tree crowns, and the spatial distribution of leaves and branches within tree crowns determine to a large extent the canopy structure. Which components and sub-components are used, and also the scale at which their spatial arrangements are studied, is of great importance for the possible outcome of the analysis of canopy structure. This is specially the case when canopy structure is needed as a correlate to ecological questions, e.g., on habitat specificity of animals, or epiphytes. Methods available for describing and analysing canopy structure are discussed. At large scale levels remote sensing data are used to describe differences in structure. High-resolution radar images are used to describe canopy structure in detail and over large areas. Repeated measurements over time can be used for monitoring purposes. Ways to measure the three dimensional structure of (components within) individual trees in detail are being developed, and are coupled to physiological models. Currently, use of such methods is only feasible for small plants. Forest tomography (where the vegetation occupation and empty spaces are determined in horizontal and vertical slices of the forest) is proposed as a way to describe vertical and horizontal structure. Vegetation cover and occupation is analysed above grid points in a forest. As an example the vertical structure of a Cameroonian forest is described at several levels of detail. The research question asked should govern completely the choice of the parameters and the methods used for the description of forest canopy structure.

Introduction

The forest canopy is conceived as a unique subsystem of the forest in which a major part of many forest processes take place, like photosynthesis, flowering and fruiting. The interaction between the forest and the atmosphere occurs at the canopy level, and the canopy is also a special habitat for many forest dwellers. The general term forest canopy is often used but hardly ever defined. That is not necessarily a problem, but for comparative reasons definitions and de-limitations of forest canopy are needed. These are scarce, however. The upper boundary, also called outer canopy (Parker 1995) or canopy surface (Birnbaum 1997), is relatively easy to distinguish as it is the outer contact of the forest with the environment above. This can range from high (40 m or more) above the ground to ground level (in a canopy gap). The lower boundary is far more difficult because there is no clear lower limit. In Figure 1 several ways of delimiting the canopy are given. The lower canopy limit can be assigned to the lowest part of the crowns that touch the canopy surface or to the lowest part of all crowns. The canopy can be seen as the total of the crowns that touch the canopy surface, but also as the total of all crowns in a forest. Another possibility is to delimit the canopy to the volume that is occupied by the upper and lower crowns. An extreme viewpoint is that the canopy consists of the whole forest volume above the ground.

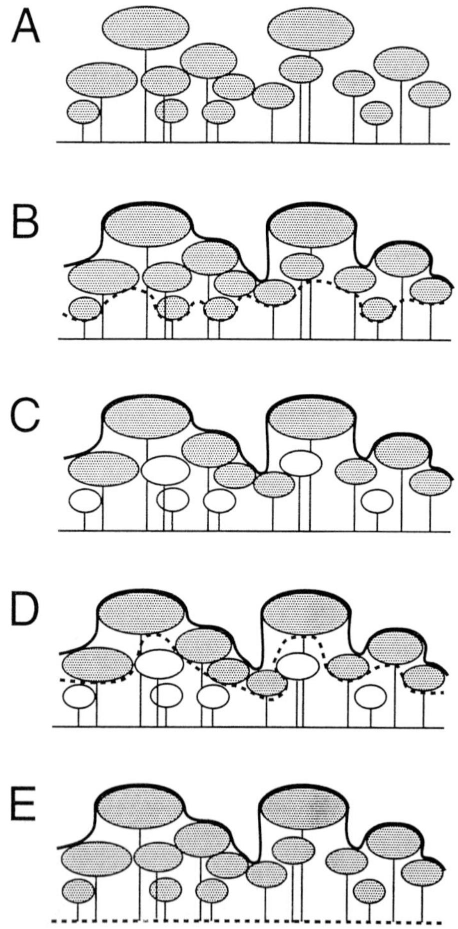

Figure 1. Five ways of defining the forest canopy. The forest canopy is (A) the collection of all crowns, (B) the whole volume between and including the upper and lower crowns, (C) the collection of crowns touching the canopy surface, (D) the whole volume between and including the canopy surface crowns, and (E) the whole above-ground forest volume.

In this paper the canopy is conceived as the total above ground part of the forest. In line with this, forest canopy structure can be apprehended as the physical structure, the spatial arrangement of all the above-ground elements in a forest community (cf., Campbell & Norman 1989). It, however, can also be comprehended as the functional structure, the structure of the relations between the elements in the forest canopy, including plants and animals (Reagan & Waide 1996). In this paper the focus is on the physical structure.

A complete description of the structure (*sensu latu*) of a forest canopy would require the specification of the position, size and orientation of each element of surface in the canopy. Barkman (1979) distinguishes between texture and structure (*sensu strictu*) of vegetation. Texture is the qualitative and quantitative composition of the vegetation as to different morphological elements (in the widest possible sense) regardless of their arrangement. Structure is the spatial (horizontal and vertical) arrangement of these elements. This distinction between texture and structure is not widespread. In comparing forests and forest types, textural comparisons alone distinguish well: overall spectra of, e.g., growth forms, leaf sizes, plant families, architectural tree models, pollination types and ecological groups, show quantitatively the differences between the elements in the forests concerned. Spectra give, at a certain level of abstraction, the building elements of the forest. For many study goals, that is sufficient detail. The measurement of structure involves the location of all these elements in three-dimensional (3-D) space and thus involves a lot of extra work. This, however, is important for studies on abiotic environmental patterns in the forest (for instance light availability), spatial location of plants and animals that are related to particular habitats in forest trees, and movement patterns of animals.

Forest canopy texture and structure are complex because of the large number of different elements, at several levels of organisation. The complexity is ever increasing as we proceed from individual organs to plants to plant communities. Each higher level contains the elements of the lower levels and thus at each organisation level canopy structure can be described in terms of the elements of the next lower level, and in terms of elements of the organisational level in question.

Apart from canopy structure and canopy texture, the terms canopy architecture and canopy physiognomy are also used. Canopy architecture is the above-ground forest structure in terms of the architecture of individual tree crowns. Examples are Hallé & Oldeman (1970), Edelin (1984), Barthélémy et al. (1991), Bouchon et al. (1997), Sterck (1997), Ackerly (1996), Aiba & Kohyama (1997) and many others. Sometimes it is also used as a more general term (Oldeman 1990; Hallé 1995; Van der Sanden 1997) for the whole forest canopy. Canopy physiognomy refers to the form and shape of individual crowns or to the sizes of the leaves (e.g., Richards 1996), and as such is part of forest texture.

The spatial description of canopy structure generally consists of a representation of vertical and/or horizontal structure. Vertical structure is the variation of canopy characteristics along the vertical axis and can be projected on one axis (for instance averages for a defined surface) (e.g., Brokaw & Lent 1998; Popma et al. 1988; Parker 1995), and on two axes (for instance the vertical variation projected on the ground surface). In many studies, the 3-D variation is projected on a plane, built up from a vertical and one horizontal axis. This can be called a profile diagram. These diagrams can vary from very abstract cubes in 2-D space, with or without vegetation, to more or less accurate drawings of trees, lianas, palms and other forest elements in their natural setting. The classic profile drawings (Richards 1996) are an example of the latter. Recently, Vester (1997) used this approach to study and show the development of tree architecture in the context of secondary forest succession in Colombia.

The horizontal structure of the forest canopy is a projection of a structural feature (e.g., canopy surface height, leaf area index) on a horizontal plane. This is a mapping of structural features of the forest at a certain level of detail. Examples include maps of forest structural types (for instance forest formation maps) and more detailed maps of forest vegetation types (e.g., Duivenvoorden & Lips 1993). Within a forest type several structural patches can be distinguished like successional patches based on canopy height (Whitmore 1984), on tree architecture combined with height (Torquebiau 1986), or on tree density combined with height (Riéra et al. 1998). More detailed mapping includes mapping of individual trees (Hubbell & Foster 1986a) and individual crowns (see Oldeman 1990 for many examples of crown projection maps).

This paper aims at an overview of the different ways the forest canopy structure is approached. First the importance of clear aims and questions will be emphasised, and then the importance of scale therein. Then forest canopy structure at different levels of organisation will be described, from levels at large scale (whole forest structural types) to levels at small scale (e.g., leaves). Finally, some conclusions will be drawn and promising directions for the study of forest canopy structure indicated.

Canopy structure: what do you want to know?

The variability in the ways in which forest canopy structure can be conceived, measured and described, is extremely large. A major point of focus is the level of integration at which the canopy structure is considered. There is a large range of organisation levels (Table 1), from large scale (forests in a landscape mosaic) to small scale levels (forest canopy structure from the point of view of leaves occupying 3-D space in the forest). Because the different organisation levels need different methods and techniques for their study it is of prime importance to define exactly the goals and questions of a particular research.

Several questions have to be asked:

- What do you want to know about structure (*sensu latu*)? Structure as a subject to describe, or structure as a correlate for other aspects, e.g., as a basis for epiphyte studies?
- At what spatial scale do you want to work (e.g., at the local scale of a tree, or at a regional scale of a forested landscape)?
- What do you define as components making up the structure (e.g., leaves, flowers, individuals, forest patches)?
- Is the spatial 3-D location of the elements important?
- Is canopy structure conceived as the occupied or as the empty space in the forest canopy (e.g., the empty space is available for colonisation)
- At which levels of organisation are the components determined (e.g., leaves at the level of individuals, species, life forms, forest patch, forest)?
- Do you want to follow structure over time (canopy dynamics)?
- Do you depart from physical structure of from functional structure?
- Do you conceive canopy structure as habitat for e.g., smaller plants or animals?

Forest canopies vary on spatial scales (from centimeters to kilometers) as well as on temporal scales (from seconds to decades). To describe the canopy structure adequately we need information on this variation (average, deviations, extremes), but details may be interesting at one level but not at another one. For instance, no study was found in which a vertical structure diagram is shown with the (horizontal) variability quantified per height class. For that horizontal stratification is needed.

The accuracy with which data on canopy structure are collected determine for a large part the outcome and the level of interpretation.

Spatial distribution of forest, forest types and forest patches

A forest is a patch in a landscape level mosaic, and in general it consists of a mosaic of patches of different forest types, and also a mosaic of patches within each

Table 1. Levels of integration for the analysis of forest canopy structure, and stratification possibilities.

Level of integration	Stratification into components
Forest in a landscape mosaic	• forest – non forest, at various spatial scales
Forest types	• different (types of) forest communities
Within-forest patches	• successional development phases
	• local environmental differences in e.g., soil conditions
Individuals	• species (genera, families, life forms, architectural models, guilds)
	• size (height, diameter, developmental phase)
	• resource availability (light, soil, water)
Plant parts (nested within individuals)	• crown (ramification levels, reiteration complexes, age classes, leaves, branches, flowers, fruits)
	• stem (position and type of buttresses, bark, form)
	• roots
Plant organs, aboveground (without taking individuals into consideration)	• leaves
	• metamers
	• growth units
	• branches
	• stem
	• buttresses
	• flowers
	• fruits
	• seeds
	• branching points

forest type (e.g., successional phases). These mosaics form one aspect of the structure of the forest canopy at different scale levels. To identify these patches several techniques are available.

At large scale, remote sensing techniques can be used to identify forest patches. At a small scale, several other techniques are available (see below). In this paragraph the focus will be on the use of remote sensing techniques for determining forest patches. We can distinguish between airborne (aerial photographs from plane, helicopter, ultra-light aircraft, canopy raft, air-balloon; different types of photographs: optical and radar) and space-born techniques (from satellites, optical and radar). The resolution of the system determines the spatial level of distinction. Airborne techniques generally have a higher resolution than space-born ones (1-10 m and 0.01 - 5 km respectively). Space-born techniques have the advantage of frequent temporal sampling, but their coarse spatial resolution is disadvantageous, except for large scale approaches. Most remote sensing techniques measure within the optical window of electromagnetic radiation, where the influence of atmospheric conditions is high. Radar systems on the other hand measure within the microwave window, and are relatively independent from atmospheric conditions. This is a strong advantage in moist tropical regions.

Forest vs. non forest

At the landscape mosaic level the differences between forest and non-forest (thus including large scale deforestations) can be determined using optical remote sensing; e.g., Landsat or SPOT (FAO 1993). The same results can be achieved with radar (Sicco Smit 1988; Thomson & Dams 1990; Hoekman et al. 1996). Small scale deforestation (e.g., from shifting cultivation or selective logging) can only be detected with large spatial resolutions. Using radar, 0.25 ha shifting cultivation plots were detected (Thomson & Dams 1990; Jorritsma 1993), and also plots with three levels of selective logging disturbance (Thomson & Dams 1990). The resolution of the pictures is therefore crucial here.

Forest types

Forest types generally cannot be identified with space-born systems (except in extreme cases of mono-specific forest plantations or naturally occurring mono-dominant forests, Nezry et al. 1993). Airborne systems, in contrast, are able to show such differences. Types that were strictly related to physiographic differences have been distinguished for some time using radar (Sicco Smit 1988). The introduction of high spatial resolution radar systems currently enables discrimination of forest types based on differences in canopy architecture. Van der Sanden (1997), for instance, showed clear differences between six forest types in Guyana, analising the textural patterns of high resolution radar images. The textural patterns are congruent with canopy elements such as individual tree crowns, tree crown clusters or canopy openings, but some 'noise' is also present, sometimes blurring the real pattern (e.g., different sides of a crown can result in different grey patterns). For a lowland tropical forest area in Colombia Hoekman & Quinones (1998a,b) showed that several forest types could be well defined using radar.

For more complex forest structures sometimes visual inspection of images yields better results than pixel-by-pixel digital classification (Tuomisto et al. 1994). An important reason is the mostly gradual transitions from one type to another, often leading to high proportions of mis-classified pixels (Van der Sanden 1997). Large scale visual inspection is tedious however, but at local spatial scales that is not necessarily a problem. By using extensive ground data and a combination of radar frequencies Hoekman & Quinones (1998b) drastically reduced the number of misinterpretations, using an automated system.

Small scale regeneration and disturbance patches

Forest regeneration types are difficult to assess using space-born optical systems: they are too similar to either agricultural field or primary forest (Van der Sanden 1997). Using high resolution radar sometimes regeneration types can be distinguished, based on differences in canopy architecture. Small scale disturbances like natural or logging canopy gaps are also difficult to detect, in contrast with large scale disturbances and logging roads (Van der Sanden 1997). However, ongoing studies in Indonesia show that using high resolution equipment, and short wavelengths, very small logging roads can be detected, as well as (emergent) tree crowns (D. H. Hoekman pers. comm.). These new developments, with very high resolution equipment, are very promising as they approach the level of detail normally only found with airborne photographs. The radar's independence of atmospheric conditions is a great advantage. Such radar images can be used to monitor logging activities, including illegal selective logging of large individuals (Hoekman & Varekamp 1998).

Spatial distribution of the forest canopy surface

The roughness of the upper surface of a forest is an important structural parameter, indicating the length of the forest to macro-environment interface. It can be analysed using various techniques. Aerial photographs can be used to detect relative heights of the upper forest surface using stereoscopic measurements. This is quite laborious, however, and new techniques (e.g., image digitising; Birnbaum 1999) are being developed (Trichon et al. 1998; Trichon 2001). Also the recent high resolution radar images (see above) can be developed in such a way that relative canopy surface heights can be derived (Varekamp & Hoekman 1998, 1999).

Another method to construct the surface of the forest canopy is the combination of measurements of the height of the vegetation (either the height of the upper canopy trees in a certain area, or the height of the highest crowns or leaves above points in a grid) with soil topographical data. Measuring poles are often used for low-stature forest (up to 15 m), and inclination measurers for larger trees. Range finders are frequently used for grid measurements (Brockelman 1998). A new device has been developed by Birnbaum (1997), who used a photo-camera with a 400 mm tele-lens. For detailed distance readings, a potentiometre was connected to a special ring attached to the tele-objective. He was able to measure very accurately the upper-height of the forest. Due to optical diffraction this was also possible in relatively dense vegetation, as long as a part of the upper height could be detected. Currently laser distance measurers are in use as well but in dense forests this is very difficult because it needs to be unobstructed over the whole length from measurer to object.

These methods can be used at different scales. A well known study is that of Hubbell & Foster (1986a–c), who determined upper canopy heights over a 5 × 5 m grid in a 50 ha plot in Panama. Hubbell and Foster used 5 m height classes and re-measured the grid points after 5 years. They showed the canopy surface for the whole 50 ha and were able to determine sizes and depths of canopy gaps (at their 5×5 m resolution). By re-measuring the grid points they were able to describe changes in the upper canopy systematically and accurately. Brockelman (1998) measured forest canopy height in three forests in Thailand over 10 × 10 m grid points in 1 ha. He showed frequency distributions of canopy surface heights and found that the three forests were very different, not only in height but also in height variation. He suggested the use of standard deviation and coefficients of variation as a measure of the distinctiveness of the upper forest canopy.

Essentially the same method was used to analyse the upper canopy structure in and around large and small openings in the canopy of French Guianan rain forest (Van der Meer 1995, 1997). Leaf occupation was measured in imaginary cylinders above 1 m points along two perpendicular transects across the canopy gaps, using different height ranges, from small (0.25 m) ones near the ground to large (5 m) ones high above. Van der Meer showed that the small gaps closed mainly from the sides and that the large gaps closed mainly from below.

Spatial distribution of individuals

One way of looking at canopy structure is to take individual plants (trees, shrubs, palms, lianas) and to measure some morphological aspects of them. By combining this information with the co-ordinates and size of the individuals and their crowns, a 2-D or 3-D picture of the forest can be constructed, as has been done frequently (Oldeman 1974; Koop 1989; Richards 1996). With respect to the crowns several levels of detail can be distinguished. Crown forms can be based on general measures (depth, width, volume; Philip 1994) or on more detailed characteristics of the crown form and crown structure (architectural models *sensu* Hallé & Oldeman 1970; mono- and multilayered crowns *sensu* Horn 1971; crown ideotypes *sensu* Bruenig 1976). For this, more detailed information is needed on crown structure of the species in question, and may be also on the development phase of the tree. A further level of detail is combining this individual information with general information on species level (leaf information, flower information, fruit information, branching type, allometry). Lowman (1995) used a profile diagram to represent 'hot spots' in the canopy with respect to the level of herbivory.

Generalisations of spatial distribution of individuals for a certain area can be given in the form of frequency distributions for classes of values for structural characteristics, for instance classes of individuals with a certain leaf size, tree diameter, tree height, crown size, crown form, growth form, architectural model (Bongers et al. 1988; Vester & Cleef 1998). In vertical structure diagrams, results are presented in height classes (Holdridge et al. 1971; Kira 1978; Popma et al. 1988; Parker 1995). Although the vertical dimension is shown clearly, the horizontal variation is not taken into account, unless subplots are used. In cases with quantitative values for subplots results can be shown as averages and their variation.

With respect to the construction of vertical structure diagrams, several important questions need to be answered: How large should the height classes be? How much area is to be lumped into the diagram? Should the structure diagram be horizontally stratified? And, if so, how should stratification be done (per forest type, per developmental forest patch)? What should be put into the diagram? Biomass, volumes, number of individuals, species, families, life forms, space occupied (by, e.g., leaf area), specific habitat availability? For the Los Tuxtlas rain forest in Mexico (Popma et al. 1988) several vertical structure diagrams were drawn up, based on individuals of trees, shrubs and palms: number of individuals, basal area, crown cover, percentage of individuals of deciduous species, percentages of individuals with compound leaves, leaf size classes, importance of some families, species number, and various diversity measures. Three of them are shown in Figure 2. In a comparative way maps (in fact horizontal structure diagrams) are constructed in which information on spatial distribution of individuals per plot is lumped into classes (Riéra et al. 1998; Torquebiau 1986). Areas can be compared with respect to frequency distributions of the classes.

Individual tree crowns can be detected on remotely sensed images when they are large compared to the spatial resolution of the image, and are part of the upper canopy, preferably of emergents trees. Recently, radar systems with high spatial resolution (1 to 3 m) became available for civil applications (D. H. Hoekman pers. comm.) and can be used for such purposes

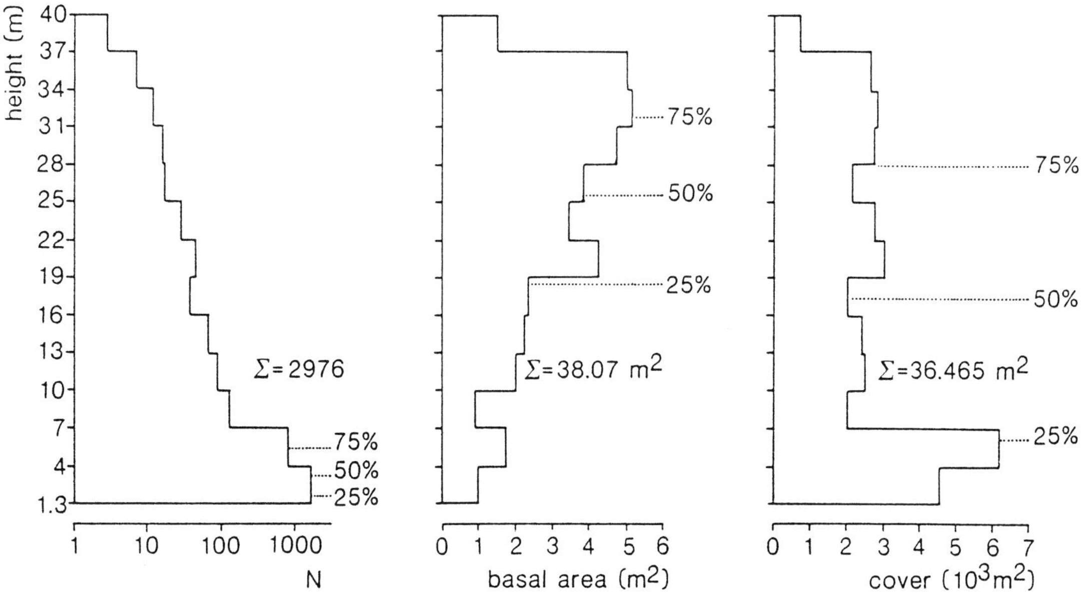

Figure 2. Vertical distribution of various structural features (according to tree height) in one hectare of tropical rain forest, Los Tuxtlas, Mexico. (A) tree density, (B) basal area, (C) foliage cover. Only trees >= 1 cm diameter at breast height. Cumulative percentages for each parameter are shown, e.g., 75% of all individuals were <= 6 m tall, 50% of total basal area was contributed by trees > 26 m tall. There was only one distinct layer of foliage: the one contributed by understorey palms (after Popma et al. 1988).

(e.g., Hoekman & Varekamp 1998). Large scale aerial photographs can be (and are) used in this respect. Currently such photographs are not only made from airplanes but also from low flying helicopters (Hallikaninen et al. 1990), ultra-lights (Vooren & Offermans 1985), and hot-air balloons (Laumonier et al. 1992; Trichon et al. 1998). A nice study is a long term assessment of individual tree crown dynamics of the Australian tropical tree *Toona ciliata* (Herwitz et al. 1998). Using stereoscopic air photos from 1976 and 1994 they were able to show the mortality and individual changes in crown size of this species. This method can thus be used to detect detailed changes in canopy structure at the individual tree crown level. A general problem, however, is the exact location of the individual trees, specially after such a long period. With the use of the recent very sophisticated types of Global Positioning Systems, however, this should not be a large problem anymore.

Distribution of organs within individual crowns

In addition to the location of trees and crowns, the locations of tree parts can also be described: i.e., the three dimensional architecture of a tree. As a tree consists of a trunk, branches of different order, leaves, flowers and fruits all these may need to be located in 3-D space. To locate and describe the 3-D structure of individual crowns we can distinguish between a topological approach (branching patterns and connection between plant units, Hallé et al. 1978) and a geometrical approach (the spatial location, orientation, size and shape of the vegetation elements, Ross 1981).

In the geometrical approach, the location of the structures in question are explicitly measured. Ashton (1978) determined the 3-D distribution of leaf area and leaf mass in crowns of several trees in Malaysia, using a clipping technique. The work on branching patterns is purely topological. (e.g., Fisher 1986). Chiariello et al. (1987) used a combination of topology and geometry: they measured the orientation and inclination of axes, their diameters, and the positions of leaves non-destructively for crowns of two species of *Piper* shrubs in Mexico. In a same way Ackerly & Bazzaz (1995) measured small individuals of four neotropical pioneer species in their study on the relation between crown architecture and light interception. It is obvious that this method is difficult to apply in larger trees. Because several ways are currently available to enter tree crowns (Moffett & Lowman 1995; Sutton 2001) such methods can be used for (parts of) large tree crowns as well.

Tree topology can be described at different scale levels, from single scale (for instance in terms of only metamers, only extention units or only gross branches, Room et al. 1994, 1996; Chiariello et al. 1987; Bell 1991) to multiscale (taking into account several scales at the same time, Godin et al. 1997). Combining topology with angles and orientations of segments can lead to 3-D representations of trees. Another way to do this is combining topology with three dimensional digitising of plant entities: the exact location of every entity can be measured (Sinoquet et al. 1997; Room et al. 1996). Various devises are currently available. Sinoquet et al. (1997) for instance use a digitiser that is based on magnetic fields generated by a transmitter and measured by a receiver that is put at the to-be-measured location. This device can be used for small trees but not for tree entities high above the ground. Also maximum distance between transmitter and receiver is only 4 m.

The AMAPmod system is able to describe exactly the 3-D structure of trees (Godin et al. 1997; de Reffye & Houllier 1997), and can be used to generate 3-D pictures as well as 3-D analyses. An example is the 3-D architecture of the Walnut tree (Sinoquet et al. 1997). Two individual free standing and symmetric trees were analysed completely. Data collection is tedious (between 2 and 4 weeks per tree), but the resulting 3-D image was shown to be very accurate. Analyses gave information on the spatial position (with respect to soil and tree trunk) of branches of different sizes, leaf area density (amount of leaf area in a cubic meter) and fruit density (number of fruits in a cubic meter) (summarized in 2-D in Figure 3).

Such representation could be the basis for, e.g., locating herbivory, density of herbivorous insects, light extinction in a crown, photosynthesis, transpiration etc. The dynamic aspect is also very promising, where measurements are done again after some time, not only for describing tree architectural dynamics but specially in relation to external effects like changes in light, or damage by wind or herbivores. For an overview and theoretical basis of this modelling and simulation work on tree architecture see Bouchon et al. (1997).

The AMAPmod system is currently too tedious for tropical rain forest canopy analysis, but it will have more promise in the near future as more and more devises become available for easy access to the canopy (Moffett & Lowman 1995). For specific canopy studies the system could be used to locate small scale 3-D canopy structure. Recently, Pearcy and Yang (1996, 1998) developed the YPLANT method for very accurate 3-D descriptions of the canopy of small plants aimed at estimating light interception. YPLANT requires input of the angles and azimuths of the petiole, leaf surface and internode extending from this node, and the azimuth of the midrib, for each node of the plant.

Well-known descriptions of the spatial structure of individual trees are the studies on the spatial distribution of epiphytes in tree crowns, following Johansson (1974), in which the tree is divided into five general parts, the buttress part, the free bole, branching part of the bole and the thickest branches, the second level branches, and the finer outer branches (Ter Steege & Cornelissen 1989; Benzing 1995a, 1995b). These classes are, however, too broad for detailed studies on habitat specificity of epiphytic vegetation (Wolf 1993).

Canopy structure seen as the 3-D distribution of plant organs

At the organ level we have to know where, e.g., leaves, fruits, flowers, branches, buds are located in 3-D space. This can be determined using 3-D point quadrat samples (points or volumes in space in which the presence or abundance of elements are determined, e.g., along vertical lines in a grid system, cf. MacArthur & Horn 1969) or via a stratification by individuals, also known as the dispersed-individual-plant (DIP) method (Ross 1981, see above).

Leaf area has received considerable attention, specially since the classical paper by MacArthur & Horn (1969) in which they used point quadrat samples to get profiles of foliage density, and also calculated leaf area index (the average number of leaf layers in a forest). The distribution of leaf area index in a forest is often used to indicate vertical forest structure, mainly in modelling studies like those in which light extinction through the canopy is modelled. These can be measured directly by estimating the amount of leaves, or indirectly by the amount of light at certain points in the forest, compared to points above the forest. As these aspects and methods have received considerable attention in earlier reviews (see Campbell & Norman 1989; Norman & Campbell 1989) they will not be discussed here.

In 3-D point quadrat samples, probes are used to determine the number and location of intersections of the probe with canopy elements. The probe can be a physical (metal aluminum, plastic) rod that can be

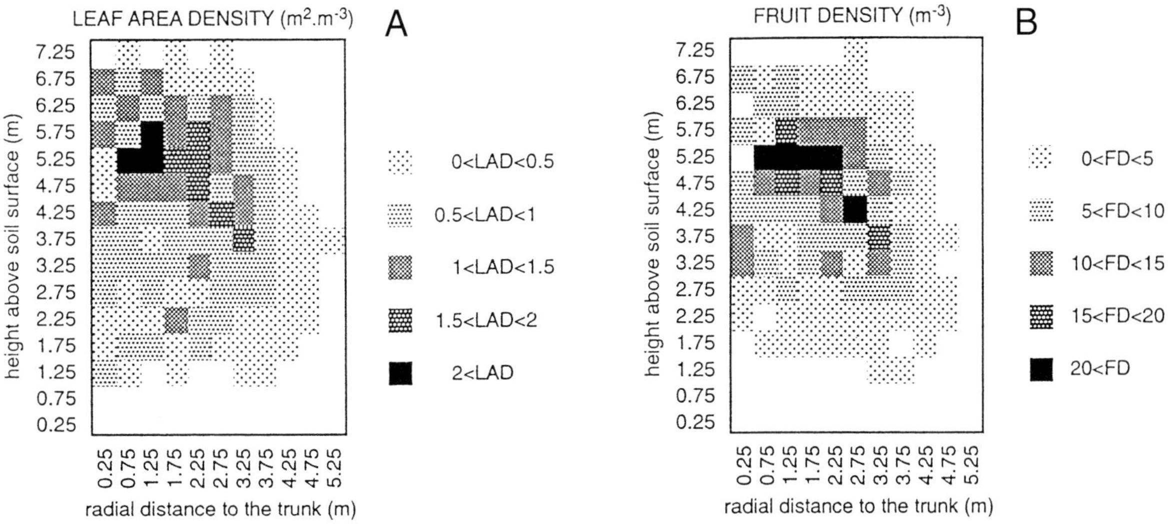

Figure 3. Spatial distribution of the foliage density (A) and of fruit density (B) around the main stem of Walnut tree. LAD is leaf area density, FD is fruit density (after Sinoquet et al. 1997).

pushed up into the canopy at fixed points, or a non-physical one (laser or optical instruments). Sterck et al. (2001) for instance used this method to determine the number of leaf layers in crowns of trees up to 15 m in height of eight Malaysian rain forest species. A laser probe can be used to measure distances between the instrument and the canopy element to be measured, but it is limited to only the first intersections. Recently optical instruments like telescopic-lenses have been used. Birnbaum (1997, 2001) determined various locations of tree crowns and sub-crowns within the high forest canopy. The use of this method is easy and quick. Breugel & Wanders (1996) used a physical rod and optical range finders to locate leaf mass above points in a grid system with the aim to determine the canopy structure in a rain forest in southern Cameroon. Van der Meer (1997) used the same method to determine changes in leaf occupation in a 3-D grid system in small and large canopy gaps. He quantified differences in canopy closure in relation to gap size.

In cases where leaves are the habitat for algae or small animals, such as mites (Walter & O'Dowd 1995), 3-D data on plant organ distribution can be compared with 3-D animal distribution data.

It may be interesting to know where in 3-D space vegetation is present, for all parts of the vegetation (leaves, branches, stems), and where it is not present (and thus where room for extension, i.e. growing space or colonisation space, is available). The method of looking at vegetation occupation above a grid point is one way of studying this, with either predefined height classes (Van der Meer 1997) or without predefined height classes (Breugel & Wanders 1996). In fact this is a kind of forest tomography (Figure 4), in which horizontal and vertical slices of the forest are taken and vegetation occupation as well as empty spaces are determined. For example, in Figure 4, two horizontal and two vertical slices are given for a forest in southern Cameroon. The vertical slices indicate canopy as well as understorey gaps. Only sporadically does a canopy gap extend through all levels in the forest. The horizontal slices show data for two levels in the forest, one at 5–10 m and one at 30–35 m. It is striking here that the overall cover at these levels is the same (48–49%). However, it is important to note that a non-empty cell does not mean that the whole cell is filled with vegetation. Lower actual occupation results are given when taking that difference into account (occupation 26% for the 30–35 m slice and only 13% for the 5–10 m slice). The 5–10 m class thus has a less dense vegetation. Using a combination of horizontal and vertical slices also the number of canopy gaps, gap sizes, and depth into the forest canopy of canopy gaps can be determined (cf. Hubbell & Foster 1986b).

In the same way the number, the location, and the size of understorey gaps can be determined. Although far less studied than overstorey gaps, the gaps in the understorey are also ecologically very important (Hallé et al. 1978; Riéra 1982; Connell et al. 1997; Birnbaum 1997).

In Figure 5, for example, some theoretical possibilities are given of vertical canopy patterns. The veg-

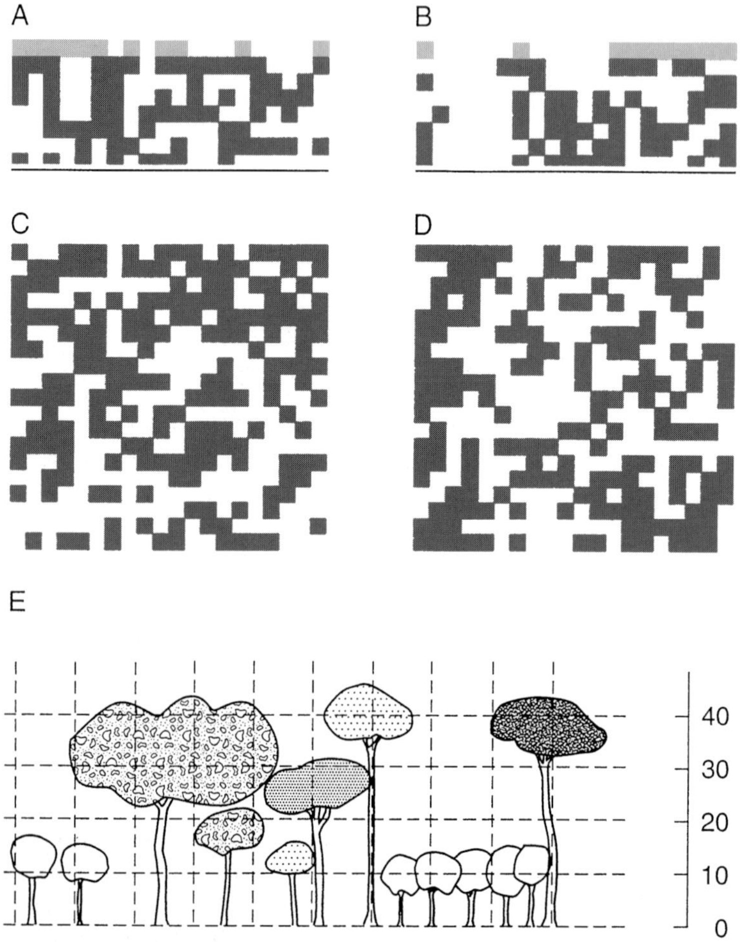

Figure 4. Forest tomography. A forest plot can be sliced both horizontally and vertically (E). These slices give detailed information on vegetation occupation and open space in the forest. Presentations can be stratified, depending on the type of information needed. Two vertical and 2 horizontal slices of a 1 ha plot of tropical rain forest in south Cameroon are presented as examples. The grid size is 5 × 5 m. (A) and (B) show 100 m long vertical slices. (C) gives a horizontal slice at 5–10 m, total cover of that slice is 49%, occupation 13% (see text). (D) gives a horizontal slice at 30–35 m high in the forest, total cover is 48%, occupation is 26%. (Breugel, Wanders, Parren & Bongers unpubl. data).

etation occupation above grid points is put in height classes. In the case of one height class, two possibilities exist for occupation above a certain grid point (i.e., vegetation present and no vegetation present), for two height classes there are four possibilities, for three classes there are 2^3=8 possibilities, for four classes there are 2^4=16 possibilities, and so forth. For eight height classes of 5 m in a 40 m high forest this gives 256 possible combinations. Combining several of such combinations will give interesting information on the variation in the vertical forest structure. The data on the frequency of each pattern in 1 ha of forest in south Cameroon is given (Breugel, Wanders, Parren & Bongers unpubl. data). The availability of canopy gaps extending through all vegetation levels (*sensu* Brokaw 1982) was very low (3.3%), but the number of less deep overstorey gaps (*sensu* Hubbell & Foster 1986b) and understorey gaps was very high (using all class systems, Figure 5). Figure 5 clearly shows the level of detail in vertical structure that can be achieved using such a grid system, and also that with more detail the complexity increases dramatically.

For three height classes, our data are compared (Table 2) with those of three other forests, one in temperate North America (Parker 1995) and two in tropical Australia (Connell et al. 1997). Canopy gaps extending through all vertical classes were very scarce in all forests (around 3%). No gaps at all occurred in 30–40% of the points. Overall the temperate forest is slightly more open then the tropical ones, but variation

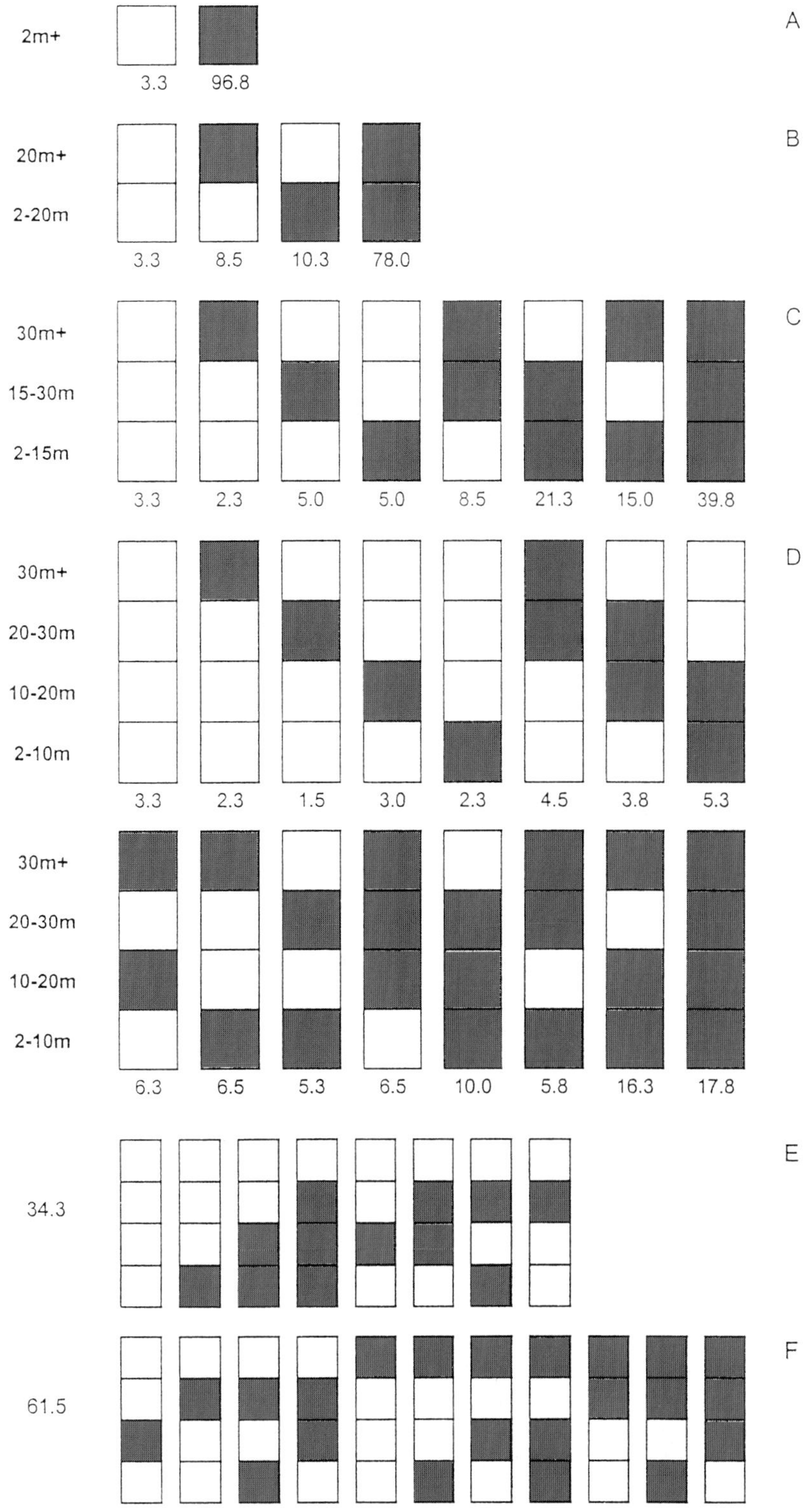

Figure 5 Vegetation occupation along a vertical forest axis. Vertical vegetation patterns are determined by presence (shaded) or absence (empty) of vegetation in height classes. A different number of vertical classes are used (1 to 4, A to D). The total number of different vertical patterns is 2^n where n is the number of height classes distinguished. This clearly indicates that the distinction of a high number of height classes give an extremely high number of different vertical patterns. The data presented are from a 1 ha forest plot in south Cameroon and represent the percentage of grid points (5 × 5 m) with a certain vertical profile. Combinations of vertical profiles give values for overstorey (E) and for understorey (F) gaps. (Breugel, Wanders, Parren & Bongers unpubl. data).

Table 2. Patterns of vertical vegetation structure with 3 height classes. Four different gap types are recognised and total percentage of each type is given for forests in different areas. Note that one vegetation pattern (empty-closed-empty) is present in both overstorey and understorey gap types.

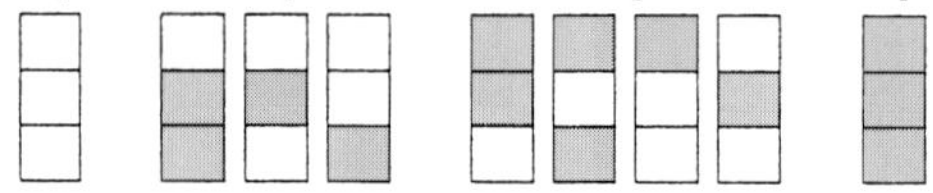

Gap types:	All level Gaps	Overstorey Gaps	Understorey Gaps	No Gaps
Appalachian TMD[1]	2.9	26	57.2	27.9
Queensland STRF[2]	2	14	48†	36
Queensland TRF[2]	3	25	38†	33
SW Cameroon TRF[3]	3.3	10.5	45	44
SW Cameroon TRF[3]	2.3	24	27	51

TMD= Temperate mixed deciduous forest; STRF= subtropical rain forest; TRF=tropical rain forest.
Limits of height classes: 2–9.9 m, 10–19.9 m, 20 m and above.
†Only first three patterns are counted as understorey gaps.
Data from [1] Parker (1995), [2] Connell et al. (1997), [3] Breugel, Wanders, Parren & Bongers (unpubl. data).

is large. Understorey gaps occurred in almost half of all points, and are thus extremely abundant. Too few sites were studied to draw any general conclusions. The abundance of understorey gaps indicate that these deserve more detailed study, not only of occurrence and size, but also on implications with respect to, for instance, light levels.

Connection between empty cells can lead to understorey open spaces: these are free to be invaded by neighbouring tree crowns; are open for flying animals; and are in areas where no light interception and light extinction occurs. These type of analyses could also be performed for specific types of organs (e.g., flowers), which would give plant-animal studies a spatial dimension.

Conclusions

The description of forest canopy structure is complicated. Firstly, a large variety of canopy parameters, at various levels of integration, can be used as a basis for such a description. Secondly, in many forests the canopy (at least a large part of the canopy) is beyond our reach, despite the many new techniques that are available to reach higher levels in the forest. This makes accurate descriptions of 3-D structure difficult and taking a lot of time, energy and money.

It is hard to make generalisations for specific aspects of forest canopy structure. Very few studies have used standardised procedures for describing and analysing forest structure. Partly this is due to the differences in questions and in focus of the studies, but this is also due to the lack of protocols. Because tropical forest canopy research is a relatively young field such protocols do not abound (but see Théry et al. 1998).

What are interesting developments with respect to the analysis of forest canopy structure? Firstly, recent developments in high spatial resolution remote sensing techniques will lead to small scale characterisation of canopy structure, including small forest gaps and individual crown characteristics. These techniques may be incorporated into detailed monitoring systems of changes in forest structure resulting from, for example, illegal logging practices. Secondly, plant topological research that leads to detailed characterisations of the form, structure and architecture of trees (and palms, lianas, shrubs) can be used. These characterisations can be used in studies on habitat structure for canopy dwelling plants and animals. Thirdly, the description of occupied and empty space in the forest can be used at several levels of scale, from landscape to individual crowns. In the near future these recent developments will lead to a much more precise analysis of forest canopy structure.

Acknowledgements

This paper was initially presented at the European Science Foundation Tropical Forest Canopies Symposium at Oxford, UK, December 1998. I would like to thank the symposium organisers K.E. Linsenmair, A. Davis and M. Speight for the invitation to present this paper. M. van Breugel and P. Kostense have prepared and drawn part of the figures. The paper greatly benefited from the helpful comments of M. van Breugel, H. Bartelink and D. Hoekman on an earlier version of the manuscript, and from two anonymous reviewers.

References

Ackerly, D. D. 1996. Canopy structure and dynamics: integration of growth processes in tropical pioneer trees. Pp. 619–658. In: Mulkey, S. S., Chazdon, R. L. & Smith, A. P. (eds), Tropical forest plant ecophysiology. Chapman & Hall, London.

Ackerly, D. D. & Bazzaz, F. A. 1995. Seedling crown orientation and interception of diffuse radiation in tropical forest gaps. Ecology 76: 1134–1146.

Aiba, S. & Kohyama, T. 1997. Crown architecture and life-history traits of 14 tree species in a warm-temperate rain forest: significance of spatial heterogeneity. J. Ecol. 85: 611–624.

Ashton, P. S. 1978. Crown characteristics of tropical trees. Pp. 591–615. In: Tomlinson, P. B. & Zimmerman, M. H. (eds), Tropical trees as living systems. Cambridge University Press, Cambridge.

Barkman, J. J. 1979 The investigation of vegetation texture and structure. Pp. 123–160. In: Werger, M. J. A. (ed.), The study of vegetation. Dr. W Junk Publishers, The Hague.

Barthélémy, D., Edelin, C. & Hallé, F. 1991. Canopy architecture. Pp. 1–20. In: Raghavendra, A. S. (ed.), Physiology of trees. John Wiley and Sons Inc, New York.

Bell, A. D. 1991. Plant form. An illustrated guide to flowering plant morphology. Oxford University Press, Oxford.

Benzing, D. H. 1995a.Vascular epiphytes. Pp. 225–254. In: Lowman, M. D. & Nadkarni, N. M. (eds), Forest canopies. Academic Press, San Diego.

Benzing, D. H. 1995b. The physical mosaic and plant variety in forest canopies. Selbyana 16 (2): 159–168.

Birnbaum, Ph. 1997. Modalités d'occupation de l'espace par les arbres en foret guyanaise. Thése de doctorat de L'Université Paris VI, Paris.

Birnbaum, Ph. (in press). Protocols to study the canopy surface of a tropical rain forest. In : H. Balslev (ed.), Methods for studying tropical rain forest structure. Aarhus University Press, Aarhus.

Bouchon, J., de Reffye, Ph. & Barthélémy, D. (eds). 1997. Modélisation et simulation de l'architecture des végétaux. INRA editions, Paris.

Bongers, F., Popma, J., Meave del Castillo, J. & Carabias, J. 1988. Structure and floristic composition of the lowland rain forest of Los Tuxtlas, Mexico. Vegetatio 74: 55–80.

Bouchon, J., de Reffeye, Ph. & Barthelemy, D. (eds). 1997. Modelisation et simulation de l'architecture des vegetaux. Science Update, INRA Editions, France.

Breugel, M. van & Wanders, T. 1996. Ecological aspects of an African tropical rain forest: gaps, composition, structure and light. MSc. thesis, Wageningen Agriculture University, Wageningen.

Brockelman, W. Y. 1998. Study of tropical forest canopy height and cover using a point-intercept method. Pp. 521–53. In: Dalmeier, F. & Comiskey, J.A. (eds), Forest biodiversity research, monitoring and modeling: conceptual background and old world case studies. Man and Biosphere Series Volume 20, Paris.

Brokaw, N. V. L. 1982.The definition of treefall gap and its effect on measures of forest dynamics. Biotropica 14: 158–160.

Brokaw, N. V. L. & Lent, R. A. 1998. Vertical structure. In: Hunter, M. L. Jr. (ed.), Maintaining biodiversity in forest ecosystems. Cambridge University Press, Cambridge.

Bruenig, E. F. 1976. Tree forms in relation to environmental conditions: an ecological viewpoint. Pp. 139–156. In: Cannell, M. G. R. & Last, F. T. (eds), Tree physiology and yield improvement. Academic Press, London.

Campbell, G. S. & Norman, J. N. 1989. The description and measurement of plant canopy structure. Pp. 1–20. In: Russel, G., Marshall, B. & Jarvis, P. G. (eds), Plant canopies: their growth, form and function. Society for Experimental Biology, Seminar Series 31. Cambridge University Press, Cambridge.

Chiariello, N. R., Field, C. B. & Mooney, H. A. 1987. Midday wilting in a tropical pioneer tree. Funct. Ecol. 1: 3–11.

Connell, J. H., Lowman, M. D. & Noble, I. R. 1997. Subcanopy gaps in temperate and tropical forests. Austr. J. Ecol. 22 (2): 163–168.

Duivenvoorden, J. F. & Lips, J. M. 1993. Ecología del paisaje del Medio Caquetá. Estudios en la Amazonia Colombiana. Bogotá.

Edelin, C. 1984. L'architecture monopodiale: l'exemple de quelques arbres d'Asie tropicale. Thése d'Etat, Université Montpellier II, Montpellier.

FAO 1993. Forest resources assessment 1990. Tropical countries. Forestry Papers no. 112. Rome.

Fisher, J. B. 1986. Branching patterns and angles in trees. Pp. 493–523. In: Givnish, T. J. (ed.), On the economy of plant form and function. Cambridge University Press, Cambridge.

Godin, Ch., Costes, E. & Caraglio, Y. 1997. Exploring plant topological structure with the AMAPmod software: an outline. Silva Fennica 31 (3): 357–368.

Hallé, F. 1995. Canopy architecture in tropical trees: a pictorial approach. Pp. 27–44. In: Lowman, M. D. & Nadkarni, N. M. (eds), Forest canopies. Academic Press, San Diego.

Hallé, F. & Oldeman, R. A. A. 1970. Essay sur l'architecture et la dynamique de croissance des arbres tropicaux. Masson, Paris.

Hallé, F., Oldeman, R. A. A. & Tomlinson, P. B. 1978. Tropical trees and forests: an architectural analysis. Springer-Verlag, Berlin.

Hallikaninen, M. T., Tares, T., Hyppa, E., Somersalo, E. Ahola, P., Toikka, M., Pulliainen, J. & Sieber, A. J. 1990. Helicopter-borne measurements of radar backscatter from forests. Int. J. Rem. Sens. 11 (7): 1179–1191.

Herwitz, S., Slye, R. E. & Turton, S. M. 1998. Redefining the ecological niche of a tropical rain forest canopy tree species using airborne imagery: long-term crown dynamics of *Toona ciliata*. J. Trop. Ecol. 14 (5): 683–703.

Hoekman, D. H., van der Sanden, J. J. & Bijker, W. 1996. Radar remote sensing of tropical rain forests: The AIRSAR-93 campaign in Guyana and Colombia. Delft (BCRS, Netherlands Remote Sensing Board), BCRS report no. 96-02.

Hoekman, D. H. & Quinones, M. J. 1998a. Forest type classification by airborne SAR in the Colombian Amazon. Pp. 225–231. In: Proceedings of the Second International Workshop on Retrieval of Bio- and Geophysical Parameters from SAR data for Land Applications, 21–23 October 1998, ESA-ESTEC, Noordwijk, The Netherlands. Vol. SP-441.

Hoekman, D. H. & Quinones, M. J. 1998b. Land cover type and biomass classification using AirSAR data for evaluating of monitoring scenarios in the Colombian Amazon. Pp. 215–223. In: Proceedings of the Second International Workshop on Retrieval of Bio- and Geophysical Parameters from SAR data for Land Applications, 21–23 October 1998, ESA-ESTEC, Noordwijk, The Netherlands, Vol. SP-441.

Hoekman, D. H. & Varekamp, C. 1998. High resolution single-pass interferometric radar observation of tropical rain forest trees. Pp. 233–239 In: Proceedings of the Second International Workshop on Retrieval of Bio- and Geophysical Parameters from SAR data for Land Applications, 21-23 October 1998, ESA-ESTEC, Noordwijk, The Netherlands, Vol. SP-441.

Holdridge, L. R., Grenke, W. C., Hatheway, W. H., Liang, T. & Tosi, J. A. Jr. 1971. Forest environments in tropical life zones: a pilot study. Pergamon Press, Oxford.

Horn, H. S. 1971. The adaptive geometry of trees. Monographs in population biology 3. Princeton. University Press, Princeton.

Hubbell, S. P.& Foster, R. B. 1986a. Biology, chance, and history and the structure of tropical rain forest tree communities. Pp. 314–329. In: Diamond, J. & Chase, J. T. (eds), Community ecology. Harper & Row, New York.

Hubbell, S. P. & Foster, R. B. 1986b. Canopy gaps and the dynamics of a neotropical forest. Pp. 77–98. In: Crawley, M. J. (ed.), Plant ecology. Blackwell Scientific Publications, Oxford.

Hubbell, S. P. & Foster, R. B. 1986c. The spatial context of regeneration in a neotropical rain forest. Pp. 395–412. In: Crawley, M., Edwards, P. J. & Gray, A. (eds), Colonization, succession, and stability. Blackwell Scientific Publications, Oxford.

Johansson, D. 1974. Ecology of vascular epiphytes in West African rain forest. Acta Phytogeographica Suecica 59: 1–129.

Jorritsma, M. 1993. SAREX-92 Araracuara: radar observation of tropical forests. MSc. Thesis Wageningen Agricultural University, Wageningen.

Kira, T. 1978. Community architecture and organic matter dynamics in tropical lowland rain forests of Southeast Asia with special reference to pasoh Forest, West Malaysia. Pp. 561–590. In: Tomlinson, P. B. & Zimmerman, M. H. (eds), Tropical trees as living systems. Cambridge University Press, Cambridge.

Koop, H.1989. Forest dynamics, SILVISTAR: a comprehensive monitoring system. Springer-Verlag, Berlin.

Laumonier, Y., Trichon, V., Gastellu-Etchegorry, J-P. & Birnbaum, Ph. 1992. Hétérogénéite géométrique et radiométrique de la canopée: essai de cartographie. Pp. 271–288. In: Hallé, F. & Pascal, O. (eds), Biologie d'une canopée de forêt équatoriale II. Fondation ELF, Paris.

Lowman, M. D. 1995. Herbivory as a canopy process in rain forest trees. Pp. 431–455. In: Lowman, M. D. & Nadkarni, N. M. (eds), Forest canopies. Academic Press, San Diego.

MacArthur, R. H. & Horn, H. S. 1969. Foliage profiles by vertical measurements. Ecology 50: 802–804.

Moffett, M. W. & Lowman, M. D. 1995. Canopy access techniques. Pp. 3–26. In: Lowman, M. D. & Nadkarni, N. M. (eds), Forest canopies. Academic Press, San Diego.

Nezry, E., Mougin, E., Lopez, A., Gastellu-Etcherrgorry, J. P. & Laumonier, Y. 1993. Tropical vegetation mapping with combined visible and SAR spaceborn data. Int. J. Rem. Sens. 14 (11): 2165–2184.

Norman, J. M. & Campbell, G. S. 1989. Canopy structure. Pp. 301–325. In: Pearcy, R. W., Ehleringer, J., Mooney, H. A. & Rundel, P. W. (eds), Plant physiological ecology. Field methods and instrumentation. Chapman and Hall, London.

Oldeman, R. A. A. 1974. L'architecture de la forêt guyanaise. Memoires ORSTOM 73.

Oldeman, R. A. A. 1990. Forests: elements of silvology. Springer-Verlag, Berlin.

Parker, G. G. 1995. Structure and microclimate of forest canopies. Pp. 73–106. In: Lowman, M. D. & Nadkarni, N. M. (eds), Forest canopies. Academic Press, San Diego.

Pearcy, R. W. & Yang, W. 1996. A three-dimensional crown architecture model for assessment of light capture and carbon gain by understorey plants. Oecologia 108: 1–12.

Pearcy, R. W. & Yang, W. 1998. The functional morphology of light capture and carbon gain in the Redwood forest understorey plant *Adenocaulon bicolor* Hook. Funct. Ecol. 12: 543–552.

Philip, M. S. 1994. Measuring trees and forests. 2nd edition. Wallingford CAB International.

Popma, J., Bongers, F. & Meave del Castillo, J. 1988. Patterns in the vertical structure of the tropical lowland rain forest of Los Tuxtlas, Mexico. Vegetatio 74: 81–91.

Reagan, D. P. & Waide, R. B. 1996. The food web of a tropical rain forest. University of Chicago Press, Chicago.

Richards, P. W. 1996. The tropical rain forest, an ecological study, Second edition. Cambridge University Press, Cambridge.

Riéra, B. 1982. Observations sur les chablis, Piste St. Elie, Guyane. Bulletin de liaison du groupe de travail sur l'écosysteme forestier guyanais 6: 165–183, ORSTOM, Cayenne.

Riéra, B., Pelissier, R. & Houllier, F. 1998. Caractérisation d'une mosaique forestiére et de sa dynamique en forêt tropicale humide sempervirente. Biotropica 30 (2): 251–260.

Room, P., Mailette, L. & Hanan, J. 1994. Modular and metamer dynamics and virtual plants. Adv. Ecol. Res. 25: 105–157.

Room, P., Hanan, J. & Prusinkiewicz, P. 1996. Virtual plants: new perspectives for ecologists, pathologists and agricultural scientists. Trends Plant Sci.: 33–38.

Ross, J. 1981.The radiation regime and architecture of plant stands. Dr. W. Junk Publishers, The Hague.

Reffye, Ph. de & Houllier, F. 1997. Modelling plant growth and architecture: some recent advances and applications to agronomy and forestry. Current Sci. 73 (11): 984–992.

Sicco Smit, G. 1988. A practical application of radar imagery for tropical rain forest vegetation mapping. Pp. 249–263. In: Küchler, A. W. & Sonneveld, I.E. (eds), Vegetation mapping. Kluwer Academic Publishers, Dordrecht.

Sinoquet, H., Rivet, P. & Godin, Ch. 1997. Assessment of the three dimensional architecture of Walnut trees using digitising. Silva Fennica 31 (3): 265–273.

Sterck, F. J. 1997. Trees and light. Tree development and morphology in relation to light availability in a tropical rain forest in French Guiana. PhD thesis Wageningen Agricultural University, Wageningen.

Sterck, F. J., Bongers, F. & Newbery, D. M. 2001. Tree architecture in a Bornean rain forest: intraspecific and interspecific patterns. Plant Ecol. 153: 279–292 (this volume).

Sutton, S. L. 2001. Alice grows up: canopy science in transition from Wonderland to Reality. Plant Ecol. 153: 13–21 (this volume).

Ter Steege, H. & Cornelissen, J. H. C. 1989. Distribution and ecology of vascular epiphytes in lowland rain forest of Guyana. Biotropica 21: 331–119.

Théry, M., Stevens, A-D., Hoppe, J. R., Charles-Dominique, P. & Schuchman, K-L. 1998. Angiosperm pollination and seed dispersal, a review. Ecotropica 4: 69–91.

Thomson, M. D. & Dams, R. V. 1990. Forest and land cover mapping from SAR: a summary of recent tropical studies. Pp. 509–548. In: Proceedings of the 23rd International Symposium on Remote Sensing of Environment. Bangkok 18–26 April 1990.

Torquebiau, E. F. 1986. Mosaic patterns in Dipterocarp rain forest in Indonesia, and their implications for practical forestry. J. Trop. Ecol. 2: 301–325.

Trichon, V., Guillemyn, D. & Fromard, F. 1998. Cartographie d'une canopée de forêt tropicale par photographies aériennes numériques et analogiques. Pp. 26–36. In: Hallé, F. (ed.), Biologie d'une canopée de forêt équatoriale III. Pro-Natura International & Opération Canopée.

Trichon, V. 2001. Large-scale aerial photographs for the study of crown architecture and phenology of rain forest trees. Plant Ecol. 153: 301–312 (this volume).

Tuomisto, H., Linna, A. & Kalliola, R. 1994.Use of digitally processed satellite images in studies of tropical rain forest vegetation. Int. J. Rem. Sens. 15 (8): 1595–1610.

Van der Meer, P. J. 1995. Canopy dynamics of a tropical rain forest in French Guiana. PhD thesis Wageningen Agricultural University, Wageningen.

Van der Meer, P. J. 1997. Vegetation development in canopy gaps in a tropical rain forest in French Guiana. Selbyana 18 (1): 38–50.

Van der Sanden, J. J. 1997. Radar remote sensing to support tropical forest management. PhD thesis Wageningen Agricul tural University. Tropenbos Guyana Series 5, Tropenbos Guyana Programme, Georgetown.

Varekamp, C. & Hoekman, D. H. 1998. An inversion algorithm for automatic retrieval of tree crown characteristics from high-resolution interferometric SAR data. Pp. 241–244. In: Proceedings of the Second International Workshop on Retrieval of Bio- and Geophysical Parameters from SAR data for Land Applications, 21–23 October 1998, ESA-ESTEC, Noordwijk, The Netherlands, Vol. SP-441.

Varekamp, C. & Hoekman, D. H. 1999. Interferometric phase difference segmentation usinf Fourier parameterised deformable models. In: Fifth annual conference of the advanced school for computing and imaging, 15–17 June, 1999. Heijen, The Netherlands.

Vester, H. F. M. 1997. The trees and the forest: The role of tree architecture in canopy development: a case study in secondary forests (Araracuara, Colombia), PhD. thesis, University of Amsterdam, Amsterdam.

Vester, H. F. M. & Cleef, A. M. 1998. Tree architecture and secondary tropical rain forest development. A case study in Araracuara, Colombian Amazonia. Flora 193: 75–97.

Vooren, A. P. & Offermans, D. M. J. 1985. An ultra-light aircraft for low cost, large scale stereoscopic aireal photographs. Biotropica 17: 84–88.

Walter, D. E. & O'Dowd, D.J. 1995. Beneath biodiversity: factors influencing the diversity and abundance of canopy mites. Selbyana 16 (1): 12–20.

Whitmore, T. C. 1984. Tropical rain forests of the Far East. 2nd edition. Clarendon Press, Oxford.

Wolf, J. H. D. 1993. Ecology of epiphytes and epiphytic communities in montane rain forests, Colombia. PhD. thesis, University of Amsterdam, Amsterdam.

Plant Ecology **153:** 279–292, 2001.

Tree architecture in a Bornean lowland rain forest: intraspecific and interspecific patterns

Frank J. Sterck[1,2], Frans Bongers[1] & David M. Newbery[2]
[1]*Silviculture and Forest Ecology Group, Department of Environmental Sciences, Wageningen University, P.O. Box 342, 6700 AH Wageningen, the Netherlands (E-mail: f.j.sterck@bio.uu.nl; frans.bongers@btbo.bosb.wau.nl);* [2]*Geobotanical Institute, Bern University, Altenbergrain 21, 3013 Bern, Switzerland (E-mail: david.newbery@ips.unibe.ch)*

Key words: Branch architecture, Crown shape, Dipterocarpaceae, Euphorbiaceae, Leaf display, Life-history

Abstract

Intraspecific and interspecific architectural patterns were studied for eight tree species of a Bornean rain forest. Trees 5–19 m tall in two 4-ha permanent sample plots in primary forest were selected, and three light descriptors and seven architectural traits for each tree were measured. Two general predictions were made: (1) Slow growing individuals (or short ones) encounter lower light, and have flatter crowns, fewer leaf layers, and thinner stems, than do fast growing individuals (or tall ones). (2) Species with higher shade-tolerance receive less light and have flatter crowns, fewer leaf layers, and thinner stems, than do species with lower shade-tolerance. Shade-tolerance is assumed to decrease with maximum growth rate, mortality rate, and adult stature of a species.

Two light descriptors (crown position index and available space, but not canopy height) indicated higher light conditions for trees with more rapid growth, and for trees of greater height. Light levels were higher for species with high maximum growth rates and with greater adult stature.

Most intraspecific architectural patterns conformed to the predictions: total leaf area and the number of leaf layers increased with increasing height and higher growth rates, and crown length/tree height ratio and stem slenderness respectively increased and decreased with growth rate. Yet, crown width/tree height ratio and relative crown length did not change with tree height, nor did they with previous growth. Slow growing (and short) trees may not have the reserves to invest in further horizontal crown growth, and to avoid leaf self-shading sufficiently within their relatively narrow crowns.

Predictions on interspecific architectural patterns were not supported by the data. Species that were expected to be more shade-tolerant (lower maximum growth, lower mortality, and shorter stature) had deeper crowns, greater leaf areas, and more leaf layers, than did less shade-tolerant species. These patterns may be explained by lower loss rates of branches and leaves of the more shade-tolerant species. These species avoid leaf self-shading by distributing their leaves at the crown periphery. The role of lateral light appears to be more important than hitherto realized. The crown width/height ratio and height/dbh ratio were negatively correlated, both intraspecifically and interspecifically. It is suggested that trees co-ordinate their crown and stem growth so that they maintain their stability at small safety margins in the forest understorey.

Introduction

Trees of tropical rain forest vary greatly in their architectural characters. These characters depend upon genetic constraints, previous growth, and tree size (Hallé et al. 1978; Bongers & Sterck 1998). Light plays a major determining role in the architectural patterns of tropical rain forest trees because (i) it is the main resource limiting growth in the forest understorey (Chazdon & Fetcher 1984; Fetcher et al. 1994), (ii) species have differing light requirements (Veneklaas & Poorter 1998) and their individuals encounter different light environments during their lifetimes (Clark & Clark 1992; Sterck et al. 1999), and (iii) architectural form determines the way light is intercepted and determines tree stability mechanically, both of which finally affect tree performance (Chazdon 1986; Pearcy & Yang 1996; Valladares 1999). Several hypotheses have been put forward to explain the relationships between intraspecific and interspecific architectural patterns and light environment, but rarely have they been tested explicitly (Horn 1971; Oldeman & Van Dijk 1989; Valladares 1999).

We believe that there is no *a priori* reason to distinguish between hypotheses of intraspecific architectural patterns, and those of interspecific patterns. If we pool patterns within and between species, and compare all trees in low-light environments with those in high-light environments, the following conflicting views emerge from the literature: (1a) Trees in low-light environments (further referred to as 'low-light trees') invest proportionally more of their acquired resources in horizontal crown growth than in vertical crown growth, while trees of high-light environments ('high-light trees') do the opposite (Kohyama & Hotta 1990; Aiba & Kohyama 1997). Low-light trees spread their leaves over a larger horizontal plane and thus reduce leaf self-shading (Horn 1971; Ashton 1978), and in this way they increase their chances of encountering high-light patches (Shukla & Ramakrishnan 1986; O'Brien et al. 1995). (1b) Other studies, however, have suggested that low-light trees grow more slowly than high-light ones, and cause increased self-shading by packing their leaves more densely in their smaller crowns (Shukla & Ramakrishnan 1986; Küppers 1994). (2a) Low-light trees have longer leaf longevity and higher leaf allocation (King 1991; Reich et al. 1992), and greater leaf area, than do high-light trees. (2b) Alternatively, low-light trees face a reduction in leaf area because of their slow growth, and of their slow rate of leaf production in particular (King 1994; Sterck 1999). (3) Low-light trees have thinner stems than high-light trees because the former produce generally denser wood, which gives a mechanical stability to the stem equivalent to the thicker, less-dense wood of the latter, at a given tree height and stem diameter (Putz et al. 1983; Sterck & Bongers 1998). In this last respect there is general agreement. We now extend and reformulate these views to predict, first, intraspecific architectural patterns in relation to growth and size, and, second, interspecific patterns in relation to shade-tolerance, and light requirements of the species.

The effects of growth and size on intraspecific architectural patterns may actually be determined by the carbon budget. In the case of growth effects, trees with higher carbon budgets have more carbon available for growth and can allocate this carbon in various ways. They generally produce relatively thicker stems (Holbrook & Putz 1989), have higher (net) production rates for leaves, shoots, roots, and branches, and they are thus able to produce deeper crowns with more leaves (Sterck 1999). In the tropical forest understorey, particularly, carbon budgets and growth rates of trees generally reflect the availability of light (Fetcher et al. 1994). The role of the carbon budget in the effects of size is still not well understood (Cannell & Dewar 1994). On the one hand, trees will generally receive more light as they become taller (Bongers & Sterck 1998, although the opposite pattern may occur, temporarily, over short time scales – Clark et al. 1993; Sterck et al. 1999) and, accordingly, they increase their maximal photosynthetic rates (Kitajima 1996). As a consequence, trees may achieve higher net carbon budgets and change their architecture with height in a similar way as they would when responding to a range of light levels at a given size. On the other hand, taller trees have to maintain an increasing woody body and root structure, i.e., greater respiration costs to be maintained by a relatively small leaf area (Givnish 1986). This will reduce their net carbon budget and could result in changes in architecture with tree height that correspond to those that occur across decreasing light levels. In the literature there is still little evidence for either of these two possible effects of size on tree architecture and leaf display (Sterck 1997).

Interspecific architectural patterns are also related to the degree of shade-tolerance characteristic to the species. Tree species with low shade-tolerance generally grow fast and only survive at high levels of light (Ackerly 1996). In comparison, species with high shade-tolerance grow more slowly, and persist for longer time periods at low light levels. Tropical forest tree species appear to show a general trade-off between the growth rate when resources do not limit growth (maximum rate of growth) and the ability to survive in shade (Veneklaas & Poorter 1998). This is also reflected in the positive interspecific correlation between growth rate and mortality rate in forest saplings (Kobe 1996). Species with high growth and mortality rates are thus expected to exhibit architectural characteristics that are selected under high light, while species with lower rates exhibit those that are selected under low light environments (Fig-

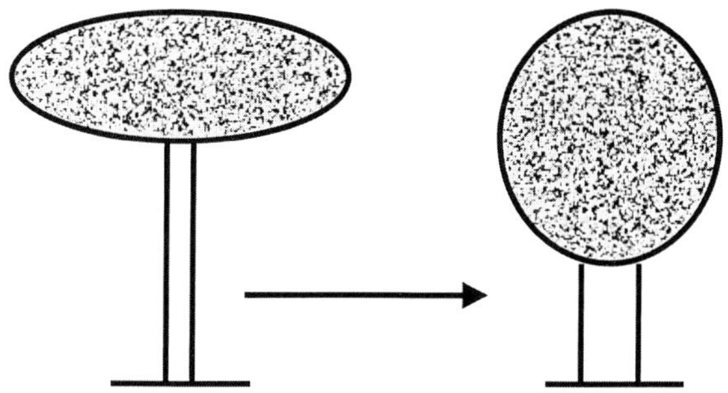

Low carbon budget ('low-light')			High carbon budget ('high-light')
Within species			
Growth rate	low	→	high
Tree height	short	→	tall
Among species			
Maximum growth rate	low	→	high
Mortality rate	low	→	high
Adult height (stature)	short	→	tall

Figure 1. Predictions for intraspecific and interspecific architectural patterns. The arrows indicate the predicted directions of change with improved carbon budget (or increased light availability).

ure 1). Patterns may, furthermore, be correlated with adult stature. Successful individuals of species with tall adult stature spend a considerable part of their lives in the highly lit overstorey and may therefore show high-light features there. In contrast, individuals of most species of short adult stature spend their lives in the shaded understorey and would thus benefit from low-light features.

The hypothesized patterns remain unsolved because empirical studies have not distinguished between the effects of growth (as the outcome of light affecting carbon budget) and tree height (Coleman et al. 1994). Most studies have been confined to seedlings or very small saplings, and usually considered only a few species (Veneklaas & Poorter 1998). Our study concerns the architectural patterns in the large sapling to small tree size range, which is typical of understorey conditions, and compares these patterns among eight tree species of a Bornean rain forest. Referring to Figure 1, the following questions were addressed: (1) Do individuals within species with slow growth (or of short stature) encounter low-light environments, and do they have flatter crowns, fewer leaf layers, and thinner stems, than individuals with faster growth (or of taller stature) encountering high-light environments? (2) Do species with low maximum growth rates (and short adult stature) encounter low-light environments, and do they have also flatter crowns, fewer leaf layers, and thinner stems, than species with higher maximum growth rates (and taller adult stature) encountering high-light environments?

Methods

Site and species

Tree architectural characters were studied for eight co-existing species of a Bornean lowland rain forest near to the Danum Valley Field Centre, Sabah, Malaysia (4°58′ N, 117°48′ E, Marsh & Greer 1992). This site is characterized by humid warm climate with temperatures between 18 (morning) and 35 °C (afternoon), and with average annual rainfall of 2669 mm in 1985–1998 (Walsh & Newbery 1999). The forest was classified as a *Parashorea malaanonan* type following Fox (1972). Two 4-ha plots were used for this study. Both plots were inventoried in 1985–1986, and again in 1995–1996, and ca. 18.000 trees $\geq$ 10 cm girth at breast height (dbh $\geq$ 3.1 cm) were tagged, measured, sampled, and identified to species. The plots have been described for their structure, composition (Newbery et al. 1992), associated topography and soils (Newbery et al. 1996), and for their dynamics 1986-1996 (Newbery et al. 1999). The canopy dominants in the two

Table 1. Life-history traits for the eight study species using data of populations of each species from two 4-ha plots of a tropical rain forest, Danum Valley Field Centre, Sabah, Malaysia (Newbery et al. 1999). Maximum sizes were derived for the populations, and growth rates (dbh increments between 1986 and 1996) for all trees with a dbh between 3.1 and 19.0 cm. These dbh limits correspond with the height limit of 5–19 m (see method section). *n* is given for trees ≤ 19 m.

Species	n	Stem[a]	Stature		Growth		Mortality[c]
			Dbh (cm)	Height (m)	Median ($mm\ yr^{-1}$)	95 Perc[b] ($mm\ yr^{-1}$)	($\%\ yr^{-1}$)
Euphorbiaceae							
Baccaurea stipulata J. J. Sm.	205	P	37	18	0.41	1.37	1.07
Mallotus wrayi King ex. Hk. f.	1865	P	17	18	0.39	1.28	1.56
Mallotus penangensis Muell. Arg.	173	O	23	21	0.57	1.59	1.12
Aporosa falcifera Hook. f.	189	O	31	22	0.97	3.20	1.31
Dipterocarpaceae							
Parashorea malaanonan (Blco). Merr.	87	O	56	34	0.79	3.35	2.46
Shorea fallax Meij.	249	O	101	49	0.64	4.26	2.46
Shorea johorensis Foxw.	71	O	127	55	4.40	10.01	5.74
Shorea parvifolia Dyer	79	O	121	49	6.66	14.09	5.11

[a]P= plagiotropic, constructed by axes that cannot be distinguished from branches in the crown, O = orthotropic, clearly distinct from branches in the crown.
[b]95 percentile values are presented, thus correcting for outliers.
[c]Annualized mortality rate (Sheil et al. 1995, Newbery et al. 1999).

plots reach 60 m in height (Table 1), and the canopy surface is open and rough although ground level gaps are very scarce (Newbery et al. 1992).

The eight species studied belong to two families, and differ in adult stature, maximum growth rates, population mortality rate, and type of branch architecture (Table 1). The short adult stature species belong to the Euphorbiaceae, and the tall ones to the Dipterocarpaceae. These two plant families dominate, and are strongly representative of, the understorey and overstorey of the Danum Valley forest respectively (Newbery et al. 1992, 1996).

Selection of individuals and measurements

In October 1996, 5–19 m tall individuals were selected for each species (n = 31–51). Trees were selected from populations in both 4-ha plots, using a table of random numbers. Trees were excluded if they had an abrupt (>50%) decrease in stem diameter with height. Such a diameter change was most probably an indicator of recent heavy damage: to have retained such trees would have introduced a further complicating factor. Accordingly, 7–17% of individuals per species was excluded.

Tree height and the height of the lowest crown leaf were measured with a telescopic measuring pole (for trees ≤ 16 m) or range finder (for trees >16 m). Crown radii were measured in eight directions (N, NE, E, etc.), using a clinometer (mean difference = 10 cm, 95% of the measurements with a precision of ±20 cm on the basis of independent 80 repeated measurements). The following traits were calculated: average crown width/tree height ratio, crown length/tree height ratio, relative crown length (crown length/average crown width ratio), and stem slenderness (height/dbh ratio).

The number of leaf layers was determined at five fixed locations, four of them halfway between the stem centre and crown edge at the four cardinal directions, and one above the stem. The telescoping measuring pole was pushed up through the foliage and the number of 'leaf hits' was counted. For trees taller than the pole (16–19 m) one leaf layer was added to compensate for the leaf layer(s) in the top of the crown that could not be reached. Internal crown cover was assessed visually by the vertical projection of the crown. Five classes were used: 1, 0 – 20% covered by leaves; 2, > 20 - 40%; 3, > 40 - 60%; 4, > 60 – 80%; and 5, > 80 – 100%.

For a sub-sample of trees (n = 10 per species) the number of leaves was counted and a sample of leaves (n = 30 per tree) was collected. Selected trees varied widely in their number of leaf layers and internal crown cover. Individual leaf areas were measured with a Delta Image Analysis System (Eijkelkamp 1991), and total leaf area per tree crown was calculated as the

Table 2. ANOVA-table representing the procedure to test for significant contributions to the model for species (dummy variables) and interaction terms (species × tree height, species × Previous annual growth). See Method section for explanation.

Model		SS	df	F	R^2	F change	p change
All variables included	Regression	1.26	23	7.6	0.41		
	Residuals	1.79	246				
	Total	3.05	269				
Remove height interactions	Regression	1.23	16	10.7	0.40	0.64	0.723
Remove growth interactions	Regression	1.23	16	10.7	0.40	0.60	0.754
Remove species	Regression	0.56	2	29.9	0.18	12.82	0.0001

Trait = constant + height + growth + species_{1-7} + $(\text{species} * \text{height})_{1-7}$ + $(\text{species} * \text{growth})_{1-7}$.
Significance of the whole model: $p < 0.0001$.
Data are shown for the trait crown width – height ratio.

Table 3. Correlations between crown position and dbh growth (partial tau coefficients), and between tree height and three light descriptors (Spearman's rank correlations) for eight tree species of a Bornean lowland rain forest, Danum Valley, Sabah. Only trees 5–19 m tall are included. Dbh growth data for 1986–1996 (Newbery et al. 1999). Partial tau coefficients (τ) correct for associated changes in tree height, but significance levels cannot be calculated. See text for explanation of light descriptors.

	n	Crown position and Dbh growth	Tree height and		
		τ	Crown position	Canopy height	Space available
Baccaurea stipulata	45	0.06	0.49***	−0.23	0.18
Mallotus wrayi	44	0.15	0.36**	−0.18	0.10
Mallotus penangensis	36	0.34	0.63***	−0.05	0.43**
Aporosa falcifera	41	0.22	0.66***	0.02	0.53***
Parashorea malaanonan	25	0.27	0.69***	−0.18	0.29*
Shorea fallax	39	0.35	0.79***	0.12	0.53***
Shorea johorensis	21	0.30	0.69***	0.04	0.68***
Shorea parvifolia	19	0.48	0.66***	−0.04	0.30*

*, $P \leq 0.05$; **, $P \leq 0.01$; ***, $P \leq 0.001$.

product of the number of leaves and mean leaf area. This measured total leaf area was then regressed on an estimate, i.e., the product of the crown area, the number of leaf layers, and the internal crown cover. Using natural logarithms of both measured total leaf area and the mentioned product a linear relationship was fitted ($F = 206$, df = 72, $r^2 = 0.77$, $p < 0.0001$). This model, ln(total leaf area) = 0.70 ln(product), was then used to predict total leaf area for all individuals.

Three descriptors of the light environment were determined for each tree. Firstly, the crown position was estimated by two independent observers. Seven different classes of crown position were distinguished: 1, no direct light; 1.5, low lateral light; 2.0, medium lateral light; 2.5, high lateral light; 3, some overhead light; 4, full overhead light; 5, crown completely exposed (after Clark & Clark 1992). The classifications of both observers were averaged to give one value per tree. Secondly, the amount of open space above each crown was determined, using the telescopic measuring pole and/or the range finder. For this measure trees were assigned to one of three classes : 1, < 1 m free space above at least half of the crown area; 2, 1–5 m free space above at least half of the crown area; 3, > 5 m free space above at least half of the crown area. Finally, the height of the forest canopy above each individual tree was determined, using the range finder. Because the upper canopy was often barely visible with the range finder, trees were assigned to five approximate canopy height classes: 1, < 10 m; 2, 10 - < 20 m; 3, 20 - < 30 m; 4, 30 - < 40 m; 5, ≥ 40 m.

Analysis

Interspecific and intraspecific patterns in the light environment were tested. Since the three light descriptors were recorded on ordinal scales, non-parametric tests were used. The descriptors were thus compared among species using the Kruskall–Wallis test, and then the Dunnet test for pair-wise species' comparisons. Within species, each of the light descriptors was correlated with tree height, using Spearman's rank correlation (r_s) coefficient. Crown position was correlated with previous growth (dbh increment in 1986–1996), using Kendall's partial correlation (τ) which removed the correlation between previous growth and tree height.

The dependency of architectural traits on species, previous growth and tree height was explored using a separate multiple regression model for each trait. Each model included a number of independent variables; tree height, previous growth, species (codes as 'dummy variables'), and interaction terms for species × tree height and species × previous growth. Previous growth and tree height were independent, because trees were selected on the basis of their height at the end of the experiment in 1996. Prior to the analysis, the architectural traits were $\log_{10}$-transformed to normalize their variances.

Each regression model was simplified as far as possible in the following way (Table 2). The contributions of species' interaction terms to the regression were tested for their significance, and than the species effects. The full model first included tree height, previous growth, the species as seven 'dummy variables', and the interaction terms (species × height and species × previous growth). For all architectural traits a similar result was found: species had a significant effect ($p < 0.01$), but the two interactions terms did not ($P > 0.05$). The model could thus be simplified in all cases by removing the interaction terms, to allow the effects of height and previous growth to be tested irrespective of species. Analysis of covariance (ANCOVA) was then used to test for the predicted intraspecific and interspecific architectural patterns. The intraspecific patterns were based on the regression coefficients of the architectural trait on tree height or on previous growth, taking the other as a covariate. The interspecific patterns were based on differences in the intercepts of the ANCOVA model.

A further test was performed to test for interspecific architectural patterns. Species-specific maximum growth rates, mortality rates, and measures of adult statures (Table 1) were correlated with species-specific architectural traits (Pearson correlation, r). The intercepts in the ANCOVA analysis were taken as the specific architectural trait values. Interspecific differences in the intercept represent 'average' architectural trait differences between species, given that the effects of tree height and previous growth were similar for all species. Partial parametric correlation coefficients (r_p) were calculated in order to remove the phylogenetic effect of plant family.

Results

Light environment

The intraspecific architectural patterns were similar among species: crown position was positively correlated with the growth rate between 1986 and 1996 (Table 3). The correlations were weak, possibly because crown positions were determined after the period of growth. Crown position and space available above individuals increased with tree height for all species (Table 3). The correlations were weakest (crown position) or not significant (available space) for *B. stipulata* and *M. wrayi*. These weak correlations resulted from the low crown position values (≤ 2) across all heights for these two species. Forest canopy height did not change with tree height.

There were few interspecific differences in crown position and available space (Table 4). The *Shorea* spp. tended to have the highest crown positions and the largest space available above their crowns, in particular *S. johorensis* and *S. parvifolia*. There were no significant differences in crown position or available overhead space among the other species. The species also did not differ significantly in the forest canopy height over the individuals sampled.

Architecture: intraspecific variation

The effects of tree height and previous growth on architectural traits were significant, and did not statistically differ among species (Table 5). With increasing height, the crown width/tree height ratio decreased while the crown length/tree height ratio did not change. This pattern indicated an increase in relative crown length with height. The stem slenderness did not change significantly over the investigated height range. Traits of leafiness (total leaf area, internal crown cover, number of leaf layers) generally increased with tree height. The effects of previous

Table 4. Medians of three light descriptors compared between eight species of a Bornean lowland rain forest, Danum Valley, Sabah. Sample sizes are shown in Table 3. See text for explanation of light descriptors (crown position, canopy height, and space available).

Species	Crown position	Canopy height	Space available
Kruskall–Wallis, $P=$	0.003	0.33 n.s.	0.02
Baccaurea stipulata	2 a	5	2 a
Mallotus wrayi	2 ab	4	2 a
Mallotus penangensis	2 ab	4	2 ab
Aporosa falcifera	2 ab	5	2 ab
Parashorea malaanonan	1.75 a	4	2 ab
Shorea fallax	2 ab	3	3 ab
Shorea johorensis	2.5 b	4	3 ab
Shorea parvifolia	2.25 b	4	3 b

Species are different for a trait when the medians do not share any of the same letters (Dunnet T3 test).

growth on crown and stem dimensions differed from the effects of tree height. Trees with faster previous growth generally had larger crowns and less slender stems. Relative crown length was unaffected by previous growth (Table 5). The effect of previous growth on traits of leafiness was similar to the effect of tree height: leafiness increased with higher previous growth rates.

Architecture: interspecific variation

Tree species differed significantly for most architectural traits (Table 5). There were some general differences between the dipterocarps and the euphorbs: dipterocarps had shallower crowns and smaller total leaf areas than euphorbs. *S. parvifolia*, the fastest growing species, had the shallowest crown, the lowest internal crown cover, the lowest number of leaf layers, and the smallest total leaf area of all species. *S. parvifolia* was followed in these respects by *S. johorensis*, the second fastest growing species. *B. stipulata* and *M. wrayi* had the highest crown width/height ratio, two of the three highest crown length/height ratios and total leaf areas, and they tended to have the smallest stem slenderness ($p = 0.10$). Finally, *M. penangensis* differed strongly from all of the other species: it had the deepest and (relatively) narrowest crowns, with the highest number of leaf layers and greatest internal crown cover.

Several architectural traits were significantly correlated with the mortality rate (crown length/height ratio, number of leaf layers) and the maximum growth rate (crown length/height ratio, stem slenderness, total leaf area, internal crown cover) of the species (Figure 2), but not with the two measures of adult stature (height and dbh, data not shown). Except for stem slenderness all relations with mortality rates and with maximum growth rates were negative (Figure 2). All patterns were significant when species were pooled, and most of them also held when the effect of plant family was partialled out.

Discussion

Intraspecific patterns for light descriptors

Light descriptors indicate that current light levels were higher for trees with the higher growth rates during the past 10 yr, as well as for trees of taller stature (Table 6). The consideration of growth as the outcome of light affecting carbon budget is supported by the correlation between light level and previous growth (Table 3). The correlation is weak, however, possibly due to light changes between 1986 and 1996 (see Rich et al. 1993; Sterck et al. 1999).

From the modest to strong correlation between our light descriptors and tree height we infer that trees may gradually receive more light as they get taller. At the same time trees face increased carbon sinks (e.g., maintenance of increasing woody parts and roots) as they get taller. The net effect for changes in the carbon balance with tree height therefore remains unclear.

Intraspecific patterns for architectural traits

Trees with slow previous growth were predicted to have flatter crowns, fewer leaf layers, and thinner stems than did trees with faster previous growth. These predictions are only partly supported by the data (Table 6). Slow growers indeed had less leaf area and fewer leaf layers than did fast growers, but they did not have shallower crowns. These patterns are similar to differences between shade trees and trees grown in the open: shade trees reduced their leafiness considerably because their long leaf life-span did not compensate for their slower leaf production rates (Sterck 1999). The result was that the shade trees had lower numbers of leaf layers, just as our slow growing trees at Danum. There seems little need for reducing self-shading by investments in further horizontal crown (i.e., branch) extension. In addition, the low carbon

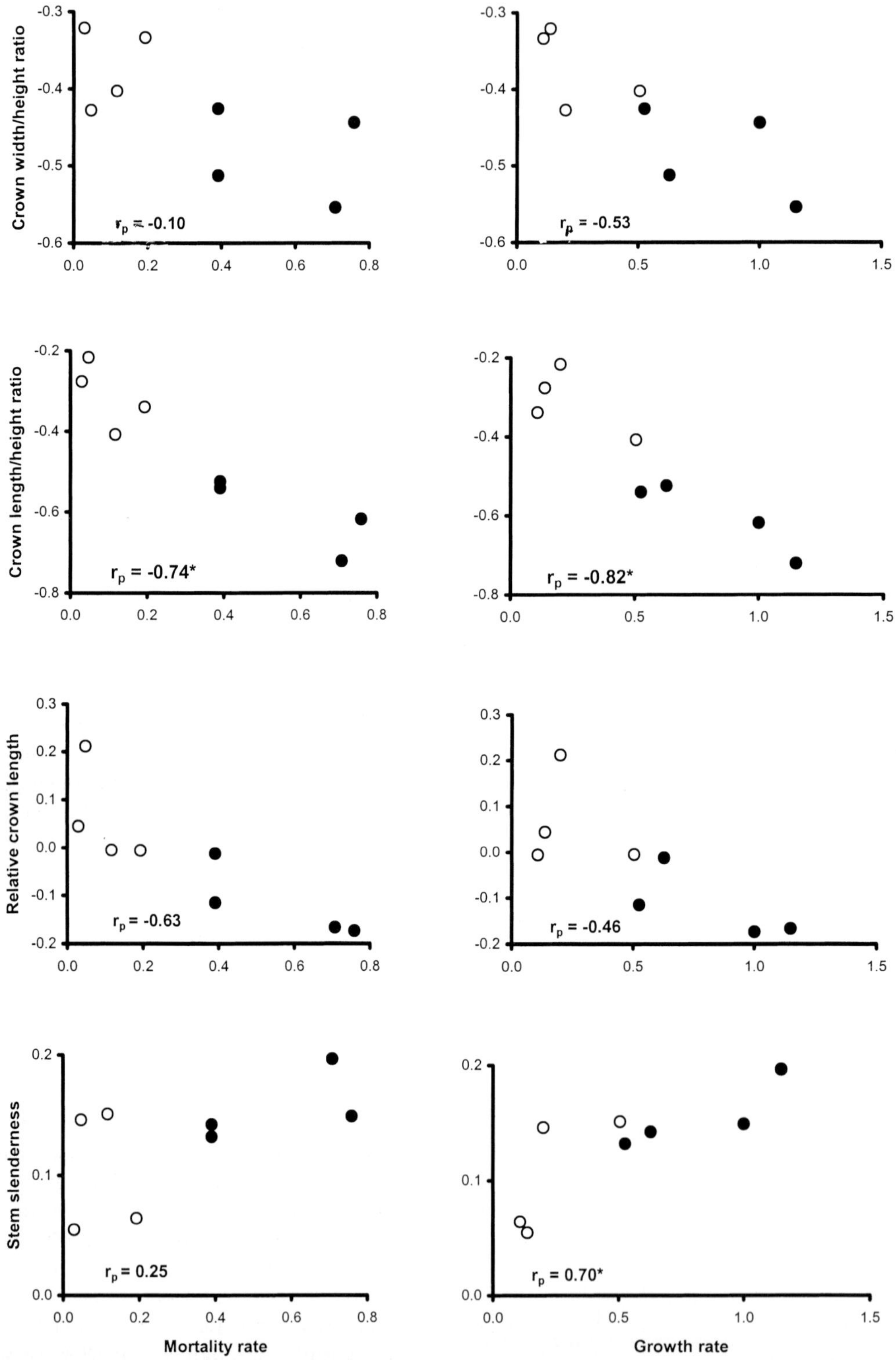

Figure 2. Scatter plots for interspecific patterns between architectural traits and life-history traits for eight tree species of Bornean rain forest at Danum Valley, Sabah. Architectural trait values are the ($\log_{10}$-transformed) intercept values of each species in the ANCOVA analysis (see Table 5 for the back transformed intercept values). These values may be negative because of the $\log_{10}$-transformation. The life-history traits were mortality rates and the maximum growth rates, calculated as the 95% percentile values of the population rates (see Table 1), and they were also $\log_{10}$-transformed. Partial correlation coefficients (r_p) were calculated between architectural traits and life-history traits, to correct for the effect of plant family (df = 5, *, $p \leq 0.05$, (*), $p \leq 0.10$). Euphorbs are represented as open circles and dipterocarps as closed circles.

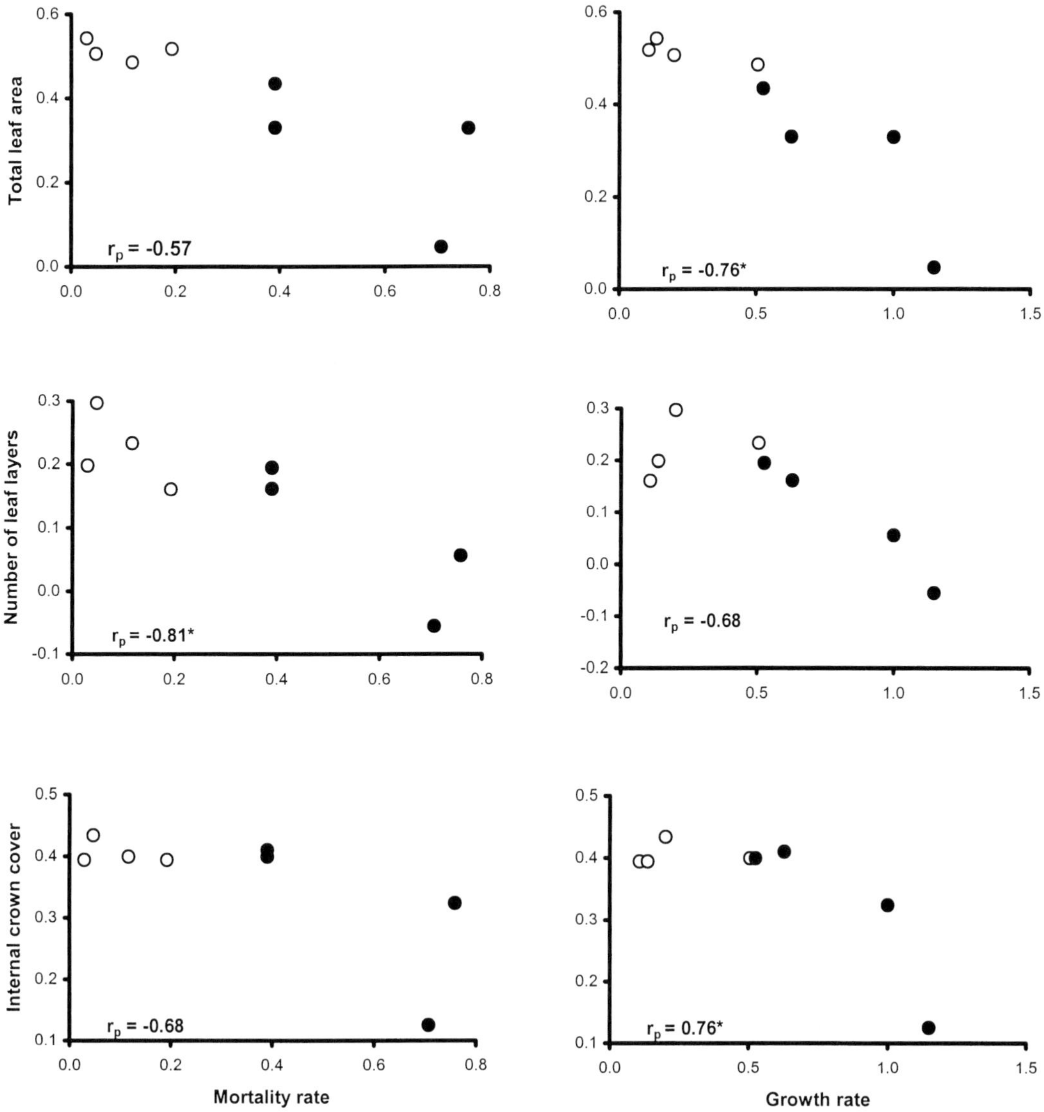

Figure 2. Continued.

budgets of shade trees may limit their crown extension, in both horizontal and vertical direction.

The slow growers had thinner stems (higher stem slenderness) at a given tree height than fast growers (Table 5). Slow growers may have had thin stems due to relatively large investments in crown development and leaf area, and may thus have enhanced their probability of survival in the shade under a limited carbon budget (e.g., King 1994; Veneklaas & Poorter 1998). Additionally, their thin stems may reflect a mechanical (thigmomorphogenic) response to lower load and bending stresses owing to their relatively small crowns (e.g., Mattheck & Kubler 1997). These two alternative explanations, based on the data here, remain hypothetical.

Trees were predicted to have deeper crowns, more leaf layers, and relatively thicker stems (decreasing stem slenderness) as they got taller. Taller trees had relatively deeper crowns because, whilst the crown length/tree height ratio did not change with tree height, the crown width/tree height ratio decreased. This is consistent with relative crown length patterns found in a number of Neotropical rain forest trees (King 1996; Sterck & Bongers 1998). Taller trees also had larger total leaf areas and more leaf layers, but did not have lower stem slenderness. From other studies it appears

Table 5. Architecture of 5–19 m tall trees compared among eight species of a Bornean rain forest, Danum Valley, Sabah. Variables were $\log_{10}$-transformed. ANCOVA was used with species as the main factor. Because the effects of height and annual growth on the independent variables were not significantly different between species, differences between species are indicated by the intercept values (back-transformed intercept values are presented, and values with different letters are significantly different, $p < 0.05$). The effects of previous annual growth and tree height were the same for all species and are given. Sample sizes are shown in Table 3.

Species	Crown width/ height (m/m)	Crown length/ height (m/m)	Relative crown length (m/m)	Stem slenderness (m/cm)	Total leaf area (m^2)	Internal crown cover (class values)	Number of leaf layers
Baccaurea stipulata	0.48 c	0.53 de	1.11 bc	1.13 a	3.49 c	2.48 b	1.58 bc
Mallotus wrayi	0.46 c	0.46 de	0.99 b	1.16 a	3.29 bc	2.48 b	1.45 bc
Mallotus penangensis	0.37 b	0.61 e	1.63 c	1.40 ab	3.21 bc	2.72 b	1.98 c
Aporosa falcifera	0.40 b	0.39 cd	0.99 b	1.42 ab	3.06 bc	2.51 b	1.71 bc
Parashorea malaanonan	0.37 b	0.29 bc	0.77 ab	1.35 a	2.72 bc	2.51 b	1.57 bc
Shorea fallax	0.31 ab	0.30 bc	0.97 b	1.39 ab	2.14 b	2.52 b	1.45 bc
Shorea johorensis	0.36 b	0.24 b	0.67 a	1.41 ab	2.13 b	2.11 b	1.14 b
Shorea parvifolia	0.28 a	0.19 a	0.68 a	1.57 b	1.11 a	1.33 a	0.88 a
Height (m) $\times 10^{-2}$	−1.59 ***	ns	1.10 **	ns	4.43 ***	1.05 ***	2.75 ***
Annual growth (cm yr^{-1}) $\times 10^{-2}$	8.85 ***	8.76 ***	ns	−3.61 **	16.16 ***	7.60 ***	6.48 **

that species differ for their stem slenderness patterns with tree height (Rich et al. 1986; Alvarez-Buylla & Martinez-Ramos 1992; Claussen & Maycock 1995; Sterck & Bongers 1998). A reduction in the ratio at maturity seemed to be typical for shade-tolerant tree species (e.g., Kira 1978; Rich et al. 1986), but was not found for our ('shade tolerant') understorey species.

In our study we were able to separate the effects of tree height from the effects of previous growth on architectural traits. However, the underlying causes of changes in architectural traits with tree height remain complicated. The effects of tree height may be driven by the correlated carbon budget, (light) resource levels, and/or bio-mechanics.

Interspecific patterns for light descriptors

The results indicated interspecific differences in light habitat. The *Shorea* species received more light and had more available free space for upward growth. This was most strongly the case for *S. johorensis* and *S. parvifolia*, the species that had the highest maximum growth rates, and the highest mortality rates. From this we suggest that these species experienced higher mortality due to light suppression than did the other species, while they could outgrow the other species in high-light environment and with more space availability. This is consistent with the idea of a trade-off between high maximum growth rates and high mortality rates due to light limitation in less shade-tolerant species (Kobe 1996; Veneklaas & Poorter 1998).

The other interspecific pattern amongst the light descriptors is related to height: while most species had more light and space available as they grew taller, *B. stipulata* and *M. wrayi* did not. These latter two species differed in a number of characters from the other species: they had the lowest maximum growth rates, achieved their maximal (adult) height, and became reproductive in the studied height range (F.J. Sterck pers. obs.). They may be more frequently overgrown by neighbouring trees than the other species that still expand vertically in this height range. The two species also differed from the other species in their stem development. The stem did not differ from the branches (all plagiotropic), while those of the other species did (orthotropic stem and plagiotropic branches). It remains unclear how these growth patterns relate to the observed pattern in light descriptors. In general, these results support our prediction that species with lower maximum growth rates, lower mortality rates, and smaller adult stature, experience lower light levels.

Interspecific patterns for architectural traits

We predicted that species with lower maximum growth rates, lower mortality rates, and shorter stature would have flatter crowns, fewer leaf layers, and thinner stems. Some of the observed patterns were op-

Table 6. Predictions for environmental and architectural patterns and the factual evidence for these patterns. Symbols indicate an increase (+), an decrease (−), or no change (0) of an environmental or architectural variable. Predictions refer to intraspecific and interspecific variation.

	Predictions	Evidence			
		Intraspecific		Interspecific	
		Height	Growth	Stature	Growth[c]
Environment[a]					
Crown position	+	+	+	+	+
Space available	+	+	nt	+	+
Canopy height	−	0	nt	0	0
Allometry[b]					
Crown width/height	−	−	+	−	0
Crown length/height	+	0	+	−	−
Relative crown length	+	+	0	−	−
Stem slenderness	−	0	−	0	0
Leaf display[b]					
Total leaf area	+	+	+	−	−
Internal crown cover	+	+	+	0	−
Number of leaf layers	+	+	+	0	−

nt = not tested.
[a]See Tables 3 and 4.
[b]See Table 5, Figure 2.
[c]Maximum growth, the same pattern applies to mortality rate m (Table 1).

posite to these predictions. Dipterocarps had shallower crowns, smaller total leaf areas, and fewer leaf layers than euphorbs (Table 5). This was particularly true for two dipterocarp species, *S. parvifolia* and *S. johorensis*. In general, these patterns seemed to be unrelated to the adult stature of the species, while they were related to the maximum growth rates and mortality rates of the species. A similar broad conclusion was drawn by Aiba & Kohyama (1997) for trees in of a Japanese warm-temperate rain forest. In more detail, relative crown length, total leaf area, the number of leaf layers, and internal crown cover in our study decreased with the increasing mortality rate and increasing maximum growth rate of the species, although not always significantly (Figure 2). Across species stem slenderness increased with maximum growth rate, indicating that fast growing species, especially dipterocarps, put on more height per unit dbh than slow growing species (euphorbs). The patterns were strongest when species were pooled, but also held when the effect of plant family was removed.

How to explain these patterns? In general, species that are less shade-tolerant (and require more light) produce leaves, branches, and stem length at a higher rate than more shade-tolerant species (Shukla & Ramakrishnan 1986; King 1994). If this is applied to our species, the less shade-tolerant species should have high rates of leaf and branch shedding at the base of the crown, to produce shallow crowns with few leaf layers and little leaf area. In terms of the carbon balance physiology, there seems no direct advantage in such a produce-and-discard strategy for the high light environment in which they usually live. Moreover, these trees would suffer from light suppression, as a reduced carbon budget would not allow them to produce enough new leaves and branches to compensate for the leaf and branch loss (King 1994). These arguments accord with the high mortality rates of less shade-tolerant species, and also with their reduced crown sizes under slower growth.

Interestingly, an alternative ecological explanation has been proposed by Campbell & Newbery (1993). They studied the same species and mentioned that shallow crowns, owing to rapid branch shedding rates, allow dipterocarps to reduce liana colonization from the ground (see also Putz 1984). They found, for a given stem dbh, that dipterocarps had lower liana loads than euphorbs. Moreover, liana colonization was lowest for the two faster growing species (with the shallowest crowns), *S. johorensis* and *S. parvifolia*.

Our results are consistent with the idea that these dipterocarps, and in particular the fast growing ones, may reduce liana colonization through fast growth rates and rapid branch (and leaf) turnover rates.

Euphorbs had deeper crowns, more leaf layers, and lived at lower levels of light availability than did the dipterocarps. *M. penangensis* showed the strongest pattern in this respect: it had the deepest and narrowest crown with the highest number of leaf layers, since it kept its leaf-supporting plagiotropic branches attached for long stretches of the stem. It formed a crown with a more or less hollow cylinder of foliage, with the leaves distributed over the crown periphery only (F. J. Sterck, unpubl. data). While trees of *M. penangensis* might experience high leaf self-shading when light comes mainly from above, they increase their light interception and minimize leaf self-shading when light comes more from the side. Analogously, deep crowns are typically found at higher latitudes where light comes more from the side than from directly above (Kuuluvainen 1992). Strikingly, diurnal light fluctuations showed relative high peaks in light halfway through the morning and the afternoon in a Neotropical forest (Oberbauer et al. 1988). Unfortunately, there are no such data for the Danum forest, or for any other Asian forest. In response to lateral light, tropical trees seem to adjust leaf distributions and leaf angles to maximize light interception (Ackerly & Bazzaz 1995; Pearcy & Yang 1996). Finally, the development pattern of *M. penangensis* seems to be associated with low branch (and leaf) loss rates, which permits this species to keep its crown in shape at relatively low growth rates in the understorey. In general, the same arguments might hold for the crown shape and leaf distribution of the other euphorb species, but more observations are needed. That the forest at Danum is still aggrading and is quite open and rough at mid-height plus the plots being largely located on ridge slopes suggests that lateral light might be of more importance compared to overhead light than previously realized. Tree architecture at Danum point to this notion.

B. stipulata and *M. wrayi* had the highest crown width/height ratios. As mentioned earlier, these two species differ from the other species in a number of properties: their low maximum growth rates, their short maximal (adult and reproductive) height, and their stem development. Possibly, the development of a wide crown at an early stage of their life helps these trees to shade shorter neighbours, and to establish their adult crown in a stratum that is usually very crowded with competing tree crowns (Lieberman & Lieberman 1994). In other forests, species of short adult stature also had wider crowns than species of tall adult stature (Aiba & Kohyama 1997; Sterck & Bongers 1998). At the same time, *B. stipulata* and *M. wrayi* had thicker stems for a given height than the other species. Apparently, the species with the wider crowns have thicker stems (Figure 3), which may reflect a thigmomorphogenic response to compensate for increased load and bending stresses (Mattheck & Kubler 1997). The generally small safety margins of trees in the forest understorey (Claussen & Maycock 1995; Sterck & Bongers 1998) suggest that responses are crucial to resist increased loading and bending stresses (imposed by growing crowns). This interspecific pattern is also reflected within species, and seems a general feature of trees world-wide (Mattheck 1991).

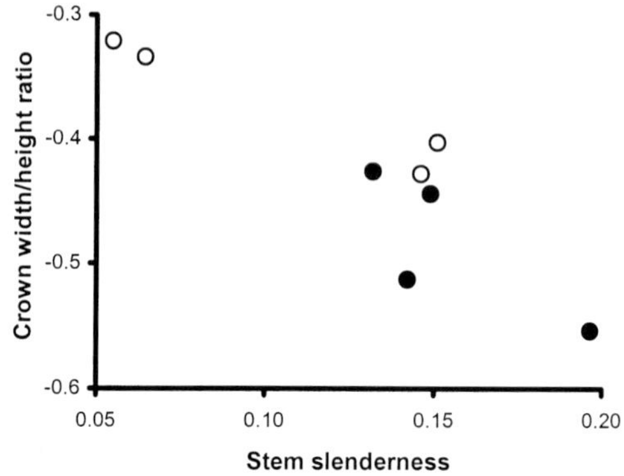

Figure 3. A species comparison for crown and stem dimensions for eight tree species of Bornean rain forest at Danum Valley, Sabah. The partial correlation coefficient r_p corrects for the effect of plant family ($r_p = 0.86$, df $= 5$, $p \leq 0.01$, two-tailed). Architectural trait values are the ($\log_{10}$-transformed) intercept values of each species in the ANCOVA analysis. These values may be negative because of the $\log_{10}$-transformation. Euphorbs are represented as open and dipterocarps as closed circles.

Acknowledgements

We thank Reinier Hille Ris Lambers, Donald Kennedy, and Hubert Petol for giving field support; Han Olff, Lourens Poorter, and Peter Stoll for their helpful comments on earlier drafts of this manuscript; the Economic Planning Unit Office of the Prime Minister's office for permission to do research in Sabah, and the staff at Danum Valley. This paper is publication no. A/312 of the Royal Society's South-East Asia Rain Forest Research Programme. This study was supported by the CEC/DG12 grant, no. TS3-CT94-0328.

References

Ackerly, D. D. 1996. Canopy structure and dynamics: integration of growth processes in tropical pioneer trees. Pp. 619–658. In: Mulkey, S. S. Chazdon, R. L. & Smith, A. P. (eds), Tropical forest plant ecophysiology. Chapman & Hall, London.

Ackerly, D. D. & Bazzaz, F. A. 1995. Seedling crown orientation and interception of diffuse radiation in tropical forest gaps. Ecology 76: 1134–1146.

Aiba, S. & Kohyama, T. 1997. Crown architecture and life-history traits of 14 tree species in a warm-temperate rain forest: significance of spatial heterogeneity. J. Ecol. 85: 611–624.

Alvarez-Buylla, E. R. & Martinez-Ramos, M. 1992. Demography and allometry of *Cecropia obtusifolia*, a Neotropical pioneer tree - an evaluation of the climax - pioneer paradigm for tropical rain forests. J. Ecol. 80: 275–290.

Ashton, P. S. 1978. Crown characteristics of tropical trees. Pp. 591–615. In: Tomlinson, P. B. & Zimmermann, M. H. (eds), Tropical trees as living systems. Cambridge University Press, Cambridge.

Bongers, F. & Sterck, F. J. 1998. The architecture of tropical rain forest trees: responses to light. Pp. 125–162. In: Newbery, D. M. Prins, H. H. T. & Brown, N. (eds), Dynamics of tropical communities. Blackwell Science, Oxford.

Campbell, E. J. F. & Newbery, D. M. 1993. Ecological relationships between lianas and trees in lowland rain forest in Sabah, East Malaysia. J. Trop. Ecol. 9: 469–490.

Cannell, M. G. R. & Dewar, R. C. 1994. Carbon allocation in trees: a review of concepts for modelling. Adv. Ecol. Res. 25: 59–104.

Chazdon, R. L. 1986. The costs of leaf support in understorey palms: economy versus safety. Am. Nat. 82: 9–30.

Chazdon, R. L & Fetcher, N. 1984. Light environments of tropical forests. Pp. 27–36. In: Medina, E. Mooney, H. A. & Vasquez-Yanes, C. (eds), Physiological ecology of plants of the wet tropics. W. Junk, The Hague.

Clark, D. A. & Clark, D. B. 1992. Life-history diversity of canopy and emergent trees in a Neotropical rain forest. Ecol. Monogr. 62: 315–344.

Clark, D. B., Clark, D. A. & Rich, P. M. 1993. Comparative analysis of microhabitat utilization by saplings of nine tree species in Neotropical rain forest. Biotropica 25: 397–407.

Claussen, J. W. & Maycock, C. R. 1995. Stem allometry in a North Queensland tropical rain forest. Biotropica 4: 421–426.

Coleman, J. S., McConnaughay, K. D. M. & Ackerly, D. D. 1994. Interpreting phenotypic variation in plants. Trends Ecol. Evol. 9: 187–191.

Eijkelkamp, B. V. 1991. Data Image Analysis System: a manual. Eijkelkamp B. V. Enschede.

Fetcher, N., Oberbauer, S. F. & Chazdon, R. L. 1994. Physiological ecology of plants. Pp. 128–141. In: McDade, L. A., Bawa, K. S., Hespenheide, H. A. & Hartshorn, G. S. (eds), La Selva: ecology and natural history of a Neotropical rain forest. University of Chicago Press, Chicago.

Fox, J. E. D. 1972. The natural vegetation of Sabah and natural regeneration of dipterocarp forests. Unpublished Ph.D. Thesis, University of Wales.

Givnish, T. J. (ed.). 1986. On the economy of plant form and function. Cambridge University Press, Cambridge.

Hallé, F. Oldeman, R. A. A. & Tomlinson, P. B. 1978. Tropical trees and forests: an architectural analysis. Springer-Verlag, Berlin.

Holbrook, N. M. & Putz, F. E. 1989. Influence of neighbors on tree form: effects of lateral shading and prevention of sway on the allometry of *Liquidambar styraciflua* (sweet gum). Am. J. Bot. 76: 1740–1749.

Horn, H. S. 1971. The adaptive geometry of trees. Princeton University Press, Princeton.

King, D. A. 1991. Correlations between biomass allocation, relative growth rate and light environment in tropical forest saplings. Funct. Ecol. 5: 485–492.

King, D. A. 1994. Influence of light level on the growth and morphology of saplings in a Panamanian forest. Am. J. Bot. 81: 948–957.

King, D. A. 1996. Allometry and life history of tropical trees. J. Trop. Ecol. 12: 25–44.

Kira, T. 1978. Community architecture and organic matter dynamics in tropical lowland rain forests of Southeast Asia with special reference to Pasoh Forest, West Malaysia. Pp. 561–590. In: Tomlinson, P. B. & Zimmermann, M. H. (eds), Tropical trees as living systems. Cambridge University Press, Cambridge.

Kitajima, K. 1996. Ecophysiology of tropical tree seedlings. Pp. 559–596. In: Mulkey, S. S., Chazdon, R. L. & Smith, A. P. (eds), Tropical forest plant ecophysiology. Chapman & Hall, London.

Kobe, R. K. 1996. Intraspecific variation in sapling mortality and growth predicts geographic variation in forest composition. Ecol. Monogr. 66: 181–201.

Kohyama, T. & Hotta, M. 1990. Significance of allometry in tropical saplings. Funct. Ecol. 4: 515–521.

Küppers, M. 1994. Canopy gaps: competitive light interception and economic space filling – a matter of whole-plant allocation. Pp. 111–144. In: Caldwell, M. M. & Pearcy, R. W. (eds), Exploitation of environmental heterogeneity by plants: ecophysiological processes above and below ground. Academic Press, San Diego.

Kuuluvainen, T. 1992. Tree architectures adapted to efficient light utilization: is there a basis for latitudinal gradients? Oikos 65: 275–284.

Lieberman, M. & Lieberman, D. 1994. Patterns of density and dispersion of forest trees. Pp. 106–119. In: McDade, L. A., Bawa, K. S., Hespenheide, H. A. & Hartshorn, G. S. (eds), La Selva: ecology and natural history of a Neotropical rain forest. University of Chicago Press, Chicago.

Marsh, C. W. & Greer, A. G. 1992. Forest land-use in Sabah, Malaysia: an introduction to the Danum Valley. Phil. Trans. Roy. Soc. London B 335: 331–339.

Mattheck, C. I. 1991. Trees: the mechanical design. Springer-Verlag, Berlin.

Mattheck, C. I. & Kubler, H. 1997. Wood – the internal optimization of trees. Springer-Verlag, Berlin.

Newbery, D. M., Campbell, F., Lee, Y. F., Ridsdale, C. E. & Still, M. J. 1992. Primary lowland dipterocarp forest at Danum Valley, Sabah, Malaysia: structure, relative abundance and family composition. Phil. Trans. R. Soc. London B 335: 341–356.

Newbery, D. M., Campbell, E. J. F., Proctor, J. & Still, M. J. 1996. Primary lowland dipterocarp forest at Danum Valley, Sabah, Malaysia. Species composition and patterns in the understorey. Vegetatio 122: 193–220.

Newbery, D. M., Kennedy, D. N., Petol, G. H., Madani, L. & Ridsdale, C. E. 1999. Primary forest dynamics in lowland dipterocarp forest at Danum Valley, Sabah, Malaysia, and the role of the understorey. Phil. Trans. R. Soc. London B 354: 1763–1782.

Oberbauer, S. F., Clark, D. B., Clark, D. A. & Quesada, M. 1988. Crown light environments of saplings of two species of rain forest emergent trees. Oecologia 75: 207–212.

O'Brien, S. T., Hubbell, S. P., Spiro, P., Condit, R. & Foster, R. B. 1995. Diameter, height, crown, and age size relationships in eight Neotropical tree species. Ecology 76: 1927–1939.

Oldeman, R. A. A. & van Dijk, J. 1991. Diagnosis of the temperament of rain forest trees. Pp. 21–89. In: Gómez-Pompa, A., Whitmore, T. C. & Hadley, M. (eds), Rain forest regeneration and management. UNESCO, Paris.

Pearcy, R. W. & Yang, W. 1996. A three-dimensional crown architecture model for assessment of light capture and carbon gain by understory plants. Oecologia 108: 1–12.

Putz, F. E. 1984. How trees avoid and shed lianas. Biotropica 16: 19–23.

Putz, F. E., Coley, P. D., Lu, K., Montalvo, A. & Aiello, A. 1983. Uprooting and snapping of trees: structural determinants and ecological consequences. Can. J. For. Res. 13: 1011–1020.

Reich, P. B., Walters, M. B. & Ellsworth, D. S. 1992. Leaf life span in relation to leaf plant and stand characteristics among diverse ecosystems. Ecol. Monogr. 62: 365–392.

Rich, P. M., Clark, D. B., Clark, D. A. & Oberbauer, S. F. 1993. Long-term study of solar radiation regimes in a tropical wet forest using quantum sensors and hemispherical photography. Agric For. Meteorol. 65: 107–127.

Rich, P. M., Helenurm, K., Kearns, D., Morse, S. R., Palmer, M. R. & Short, S. 1986. Height and stem diameter relationships for dicotyledonous trees and arborescent palms of Costa Rican tropical wet forest. Bull. Torr. Bot. Club 133: 241–246.

Sheil, D., Burslem, D. F. R. P. & Alder, D. 1995. The interpretation and misinterpretation of mortality rate measures. J. Ecol. 83: 331–333.

Shukla, R. P. & Ramakrishnan, P. S. 1986. Architecture and growth strategies of tropical trees in relation to successional status. J. Ecol. 74: 33–46.

Sterck, F. J. 1997. Trees and light: tree development and morphology in relation to light availability in a tropical rain forest in French Guiana. PhD thesis, Wageningen Agricultural University, Wageningen.

Sterck, F. J. 1999. Crown development in tropical rain forest trees in gaps and understorey. Plant Ecol. 143: 89–98.

Sterck, F. J. & Bongers, F. 1998. Ontogenetic changes in size, allometry, and mechanical design of tropical rain forest trees. Am. J. Bot. 85: 266–272.

Sterck, F. J., Clark, D. B., Clark, D. A. & Bongers, F. 1999. Light fluctuations, crown traits, and response delays for tree saplings in a Costa Rican lowland rain forest. J. Trop. Ecol. 15: 83–95.

Valladares, F. 1999. Architecture, ecology, and the evolution of plant crowns. Pp. 122–177. In: Pugnaire, F. & Valladares, F. (eds), The handbook of functional plant ecology. Marcel Dekker, New York.

Veneklaas, E. J. & Poorter, L. 1998. Growth and carbon partitioning of tropical tree seedlings growing in contrasting light environments. Pp. 337–336. In: Lambers, H. Poorter, H. & Van Vuuren, M. I. (eds), Inherent variation in plant growth: physiological mechanisms and ecological consequences. Backhuys Publishers, Leiden.

Walsh, R. P. D. & Newbery, D. M. 1999. The ecoclimatology of Danum, Sabah, in the context of the world's rain forest regions, with particular reference to dry periods and their impact. Phil. Trans. R. Soc. London B 354: 1869–1883.

Plant Ecology **153:** 293–300, 2001.

Canopy surface topography in a French Guiana forest and the folded forest theory

P. Birnbaum
756, Av. Prof. E. Jeanbrau, 34090 Montpellier, France (e-mail : ph-birnbaum@magic.fr)

Key words: Canopy model, Eco-unit, Forest dynamic, Fractal, Tree height, Tropical forest

Abstract

The canopy surface is an undulating surface that follows the irregular contours of the upper tree crowns and defines the inner and the outer limits of the forest volume. In French Guiana, the height of the canopy surface was surveyed in both a primary and a 20-years old clear-felled secondary forest plot. The topographic surface was displayed in a three-dimensional mesh, where X and Y are horizontal co-ordinates, and Z is the canopy height measured from the ground with an optical telemeter. The statistical dispersion of Z-data, and the spatial tree height variations, are interpreted at different levels of ecosystem organisation, from forest type (primary or secondary forest) to the trees themselves, following the folded forest model theory (Oldeman 1992, 1994). The vertical growth of trees creates a convex pattern in the relief of canopy surface, whereas gaps make 'concavities' which delimit impact of perturbation on the forest structure. These events are either the result of the dynamic of single trees (emergent and decayed trees), or arise from the dynamic of a group of trees working together (group of emergent trees or complex gaps). At the plot scale, the elementary events, convexities and concavities, are gathered on similar topo-sequences, and form canopy units either higher or lower than the average canopy height. This study suggests that the topography of the canopy surface is defined by a complex nested system from trees, to groups of trees, to canopy units, within a delimited floristic and physical environment.

Introduction

The canopy surface is the visible part of the forest seen from an aerial photograph. It is an undulating surface that follows the irregular contours of the upper tree crowns, and defines, as a membrane, the inside and the outside of the forest ecosystem. This confined environment is composed of leaves, branches, flowers or fruits of different canopy trees, and also lianas and epiphytes species. The composition and the structure of the canopy surface as a whole evolve in time and space along with the biology of each organism and the ecology of communities.

Because trees constitute the main structure of the canopy surface, the height variation at the scale of individual tree (the 'tree level') has been frequently used to analysis forest dynamics. Whitmore (1978) defines three canopy phases according to the relative height of trees; small trees belong to the 'gap phase'; those of medium height are in the 'building phase' while taller trees delimit the 'adult phase' (Aubréville 1938; Richards 1952; Hallé et al. 1978). Richards (1952) proposed an absolute value of 30 m height to delimit trees of the upper strata, while Brokaw (1982) fixed a lower limit of 2-m height to characterise gaps in the canopy of tropical lowland forests. Torquebiau (1986) adopts the same absolute boundary in a Sumatran rain forest, and distinguishes three supplementary canopy phases (i.e., 'Aggrading', 'Steady state' and 'Degrading') which relate to the relative height of trees and the architectural parameters of the crown. The limits of canopy phases remain ambiguous because:

- the forest growth cycle is a continuous process that is subdivided into convenient arbitrary phases (Whitmore 1978);
- the height of trees is a function of the local environmental conditions, notably the soil properties,

the drainage, and the water supply (Campbell et al. 1986; Kahn & de Granville 1992; Sabatier et al. 1997);
- canopy height is dependent of the forest type in the vegetation succession process. In a confined floristic and physical environment, the average height of primary forest canopy trees is generally higher than that in secondary forest (Richards 1952; Rollet 1974);
- the height of the canopy surface is dependent of the nature of the tree species. Canopy tree species in Asian tropical rain forest are commonly higher than African canopy trees, which are in turn higher than those in Neotropic rain forest (Richards 1952; Rollet 1974).

In order to define and understand forest structure, canopy topography can be arranged into a hierarchical system. Conceptually, the canopy surface of vegetation can be seen as a folded green carpet. At the landscape level, folds reflect the shift in the succession process from fields, savannahs, secondary and primary forests. Each fold can be folded in turn into 'eco-units' (Oldeman 1983), which can be folded further by individual trees. At the tree level, folds are created by crowns, crowns are then folded by sub-crowns and so on. Ultimately folds are created by chloroplasts within the leaf structure. By growing folds at each level in this way, the vegetation carpet tends to capture more light, thereby increasing the photosynthetic surface. This hierarchical nesting constitute the basic of the folded forest theory expressed by Oldeman (1992, 1994).

The purpose of this paper is to describe and interpret the relief of a French Guiana canopy surface in the context of the folded forest theory. Canopy height was measured from the ground independently of the tree crown delimitation and analysed as a continuous topographic surface evolving from a secondary to a primary forest plot. Height variations were studied as a nested system, from the tree to the landscape level.

Study area

This study was carried out in French Guyana, at the field station 'Piste de Saint Elie', created in 1977 for experimental studies of forest ecology, regeneration and soil erosion (Sarrailh 1980). This area is located at 05°17′ N and 53°03′ W, and is covered by a tropical rain forest. The rain season is from January to July, with generally a short interruption in March. The annual precipitation is from 2.5 to 4 m. The most abundant trees belong, in decreasing order to the Lecythidaceae, Caesalpiniaceae and Chrysobalanaceae (Puig & Lescure 1981). The floristic diversity for trees with DBH $>$ 10 cm ranges from 140 to 190 species ha^{-1} for 19 plots of 1 ha each (Prévost & Sabatier 1996). Within these plots, a small strip of 1 ha (20 $\times$ 500 m) was delimited to mark the primary forest plot of the present study. The plot is oriented East-West, and runs through several topographic features, from a plateau to three successive topo-sequences, including slopes and ridges (cf., Figure 2a). In addition, a 0.3 ha plot (20 $\times$ 150 m) located close to the primary forest plot was demarcated in a secondary forest which has been clear-felled 20 years ago for regeneration studies.

Methods

The canopy surface relief was investigated on both the 1 ha plot of primary forest (20 $\times$ 500 m) and the 0.3 ha (20 $\times$ 150 m) secondary forest plot every two meters, in both X and Y horizontal co-ordinates. At each of these positions, the canopy height (Z) was measured as the vertical distance from the ground to the upper height of the forest. The vertical Z measurement was taken using an optical telemeter designed around a tele-lens. In this system, focusing is done through a reflex camera, and is translated into a distance measurement by using a potentiometer fixed to the lens with a gear system (Birnbaum 1997). Before use, the device is calibrated using a set of distance measurements made on the horizontal plane. The focusing error of the optical telemeter fits an exponential equation relative to the measured distance, and produce less than 2% of random deviation for distance measurements up to 40 m, and a maximum error of 4% at 60 m. The telemeter was oriented vertically and placed on a tripod at each X, Y co-ordinate. The total set of points (11 $\times$ 251 $=$ 2761 on the 1 ha plot, and 11 $\times$ 76 $=$ 836 on the 0.3 ha of secondary forest plot) was used to analyse both the statistical and the spatial distribution of canopy height on plots.

Results

At the landscape level, the height distribution of the canopy surface in the two forest types is very different. In the primary forest plot, the surface undulates in the range class 0–55 m, with an average height of

30.2 m (Figure 1). The minimum value corresponds to the absence of trees, and the maximum defines the height of the tallest tree in the plot (i.e., *Ruizterania albiflora* (Warm.) Marc.-Berti., Vochysiaceae). In the secondary forest, the average height is only 17.2 m, and tallest trees never exceed 30 m in height. Sixty per cent of the total set of points are in the average range 15–20 m whereas the maximum height frequency in primary forest is about 25% in the average class 30–35 m. The ratio between the canopy surface area and the ground surface area (Figure 1) reflects the global distribution of height frequencies in forest plots. The closer the ratio is to 1, the more the canopy surface is horizontal and the canopy relief non-existent. In the secondary forest, the canopy/ground area ratio is 1.2, while the primary canopy surface is more than twice the ground area. The canopy surface of the primary forest plot is therefore characterised by a higher average height, a larger range of height variations, and a more folded relief than in the secondary forest plot.

In the primary forest, the canopy surface undulates on both sides of the average height. However, the non normal distribution (Kolmogorov–Smirnov one sample test, $p < 0.002$) indicates the absence of symmetry around the average height. In this case, the absence of symmetry can be attributed to the relative importance of tallest trees, which were measured several times in the sample if they had a broad crown. Canopy height measurements above the average height (30 m) account for 53.75% of the total set of points. They are dispersed in such a way that the topography of the canopy surface forms two major units : higher canopy units (trees higher than the average height) covering the first part of the plot, and lower canopy units (trees lower than the average height) mainly grouped on the second half of the plot (Figure 2b). These variations are not directly linked to the topo-sequence. The plateau is completely covered with high canopy trees. On the first slope, these trees constitute a network in which small patches of lower canopy units are enclosed. In contrast, the ridge and second slope are almost entirely covered with canopy trees low in height. On the last topo-sequence, the canopy surface is a blend of both canopy units, mainly lower on the down slope and higher on the ridge.

At the level of canopy units, the height variations of canopy trees create new undulations in the ranges 0–30 and 30–55 m (Figure 2c). Differences in the vertical growth of trees create ‘emerged events’ or convexities at the canopy surface in high canopy units. In some cases, these convexities represent emergent domes resulting from an isolated emergent tree (i.e., at one end of the primary plot, *Eperua grandiflora* (Aubl.) Benth., Caesalpiniaceae, area = 95 m^2, $H_{\max}$ = 47 m and at the other end the *Ruizterania* previously cited). In other cases, convexities are composed of several juxtaposed crowns, as illustrated in the Y range 40–80 m by a complex of five different tree species (*Eperua falcata* Aubl., *Swartzia polyphylla* DC., *Dicorynia guianensis* Amsh., *Vouacapoua americana* Aubl. (Caesalpiniaceae) and *Goupia glabra* Aubl., Celastraceae, area = 560 m^2, $H_{\max}$ = 52.4 m). In such groups of emergent trees, crowns are juxtaposed and fit together in such a way that the inter-crowns space is minimal. In lower canopy units, perturbations have the opposite effect on the canopy relief. Canopy phases lower than 2 m height cover 1% of the forest plot (99.45 m^2 of the total planar area) and delimit two gaps in the plot. Located at Y = 90 m, the first gap is an isolated event in a forest part where the canopy is mainly higher than the average height (cf., Figure 2b). The second gap is located on the first ridge (Y = 290 m) and is included in the large lower canopy unit covering the entire topo-sequence, from the ridge to the down-slope. The Brokaw’s limit represents only a very restricted part of the canopy surface perturbations arising from gaps. Perturbations create sharpened concavities in the canopy surface, where 2-m is an arbitrary limit for the epicentre (Oldeman 1978; Figure 2c). However, gap disturbances also extend from the epicentre to the boundaries of the concavity and reflect the impact of the perturbations on the forest at the margin of the gap. Other concavities are formed by remnant perturbations arising from older gaps, and are marked in the field by ground marks from broken and fallen trunks (Birnbaum 1997). Small and circular concavities arise from the fall of branches or the death of a single tree. These events are predominant in the first half of the primary forest plot, making some vertical holes (50–100 m^2) in a canopy surface mainly higher than the average height. Larger concavities (200–400 m^2) arise from disturbances following the death of several juxtaposed trees. They are notably more frequent in the second half of the primary plot where they form complex entities composed of several gaps of different ages.

Sharpened events (convexities and concavities) occur in the extreme classes of the height measurement distribution (cf., Figures 1b and 2c). Concavities arise mainly in the range 0–25 m (24.7%) whereas convexities are mostly concentrated in height classes above 35 m (28%). The main area of the canopy surface be-

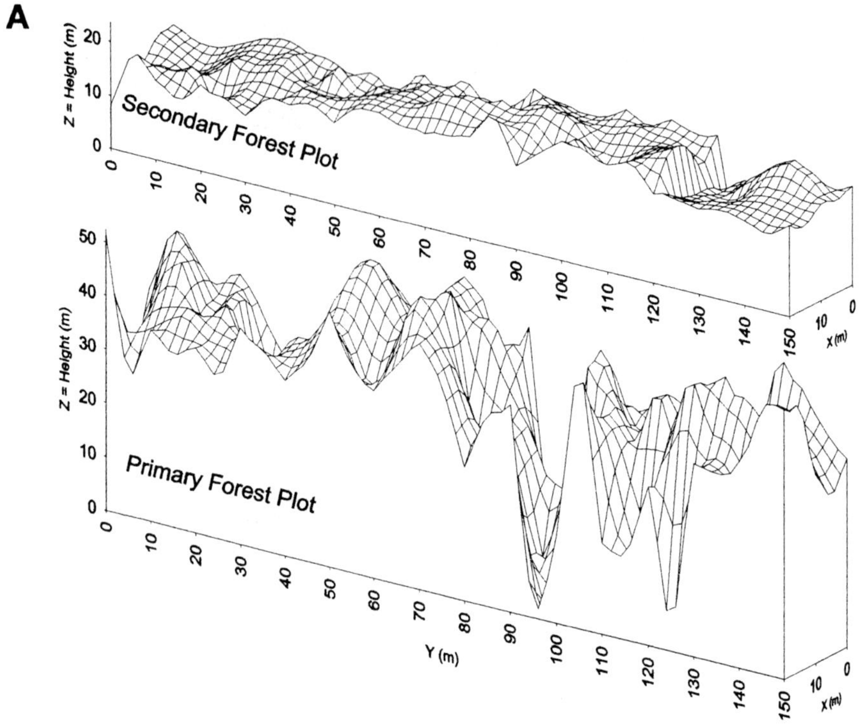

B

	Secondary	Primary
Average Height (m)	17.2	30.2
Maximum Height (m)	23.7	53.3
Ground area (m²)	3000	10000
Canopy area (m²)	3716.7	21042.9
Ratio Canopy/Ground area	1.2	2.1

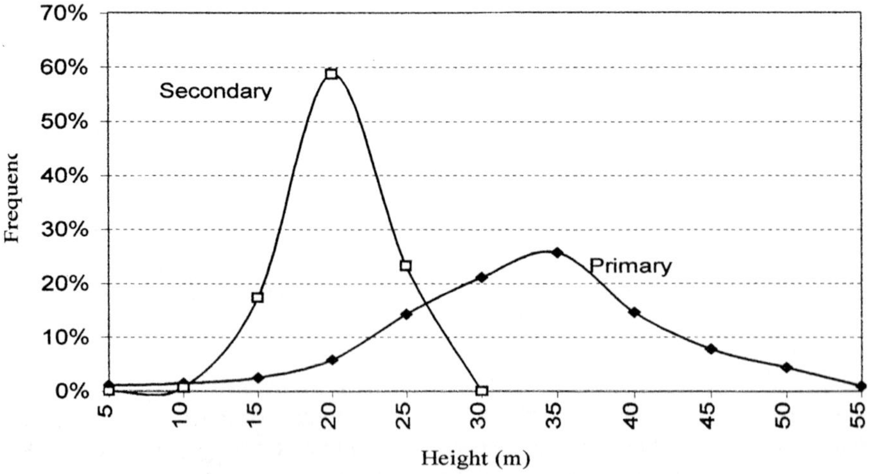

Figure 1. Canopy surface relief at the landscape level in a French Guiana rain forest. (A) Representation of the canopy topography of 20 years old clear-felled secondary forest plot (0.3 ha), and part of a primary forest plot (1 ha). Note that X, Y and Z have the same geometric scale in the two samples. (B) Distribution and statistical properties of canopy height measurements (n=2761 points in primary forest plot, n=836 points in secondary forest plot).

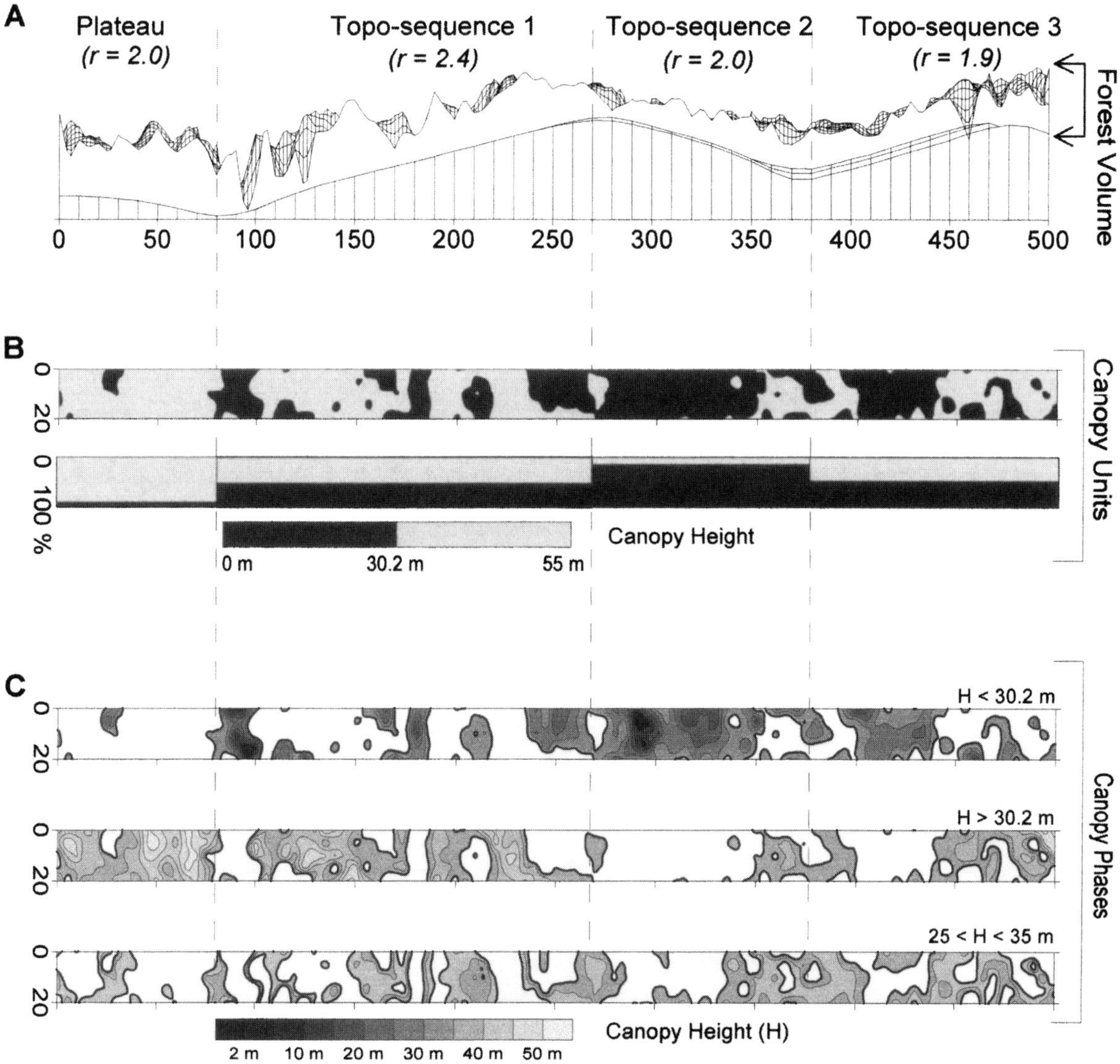

Figure 2. Canopy surface relief at the tree-scale in the primary forest plot. (A) The plot runs from a plateau to three successive topographic features (topo-sequences) including slopes and ridges. The canopy surface and the ground delimit the forest volume (302 943 m^3) while r, the ratio of their respective areas, represents the fold intensity (see text). (B) Dispersion and proportion of canopy units. High canopy units delimit patches of canopy trees higher than the average height (30.2 m). They are essentially dispersed in the first part of the plot, and dominant on the plateau. Lower canopy units are the result of different perturbation events (e.g., treefalls), and predominate in the second topo-sequence. On the other topo-sequences (1 and 3), the canopy surface is a blend of both canopy units. (C) Canopy phases enclosed within the canopy units. Perturbations form concavities in the canopy surface topography of the low canopy units. In contrast, the vertical growth of adjacent trees creates groups of emergent trees that form convexities above the average canopy height. These two opposing events are enclosed in a horizontal network centred on the average plot height. Contour maps are produced by interpolation to the nearest neighbour.

longs to the intermediate range 25–35 m (47.3%) and forms a more or less horizontal network, illustrated in the Y-plot range 350–450 m (Figure 2c), in which sharpened events are enclosed. This medium class doesn't necessary constitute an intermediate stage in forest dynamics. Some of the trees will grow isolated or within a group to progressively delimit convexities, whereas others have already stopped their vertical growth and will eventually be overlapped by emergent trees or will die and contribute to a new gap.

Conclusions and discussion

The forest type (i.e., primary or secondary) defines the average height and the ratio of undulations. The average height of 30 m is a value frequently found in other mature forests of French Guiana (Benoist 1924; de Granville 1994) and more generally in Amazonian 'terra firme' mature forests (Davis & Richards 1933; Campbell et al. 1986; Duivenvoorden 1996). This average height relates to the nature of the tree species that compose the canopy surface as well as the environmental conditions (Oldeman 1983; Duivenvoorden 1996). In a mature forest, the high diversity of trees species than can contribute to the canopy surface increases the height variations, and consequently the ratio of undulations. The increase of the canopy surface area is a common feature of mature forests. Parker et al. (1992) obtain a comparable canopy/ground area ratio to data presented here of 2.5 where canopy surface area was measured from a crane in a Panamanian mature rain forest. In contrast, the tree species diversity of secondary forests is low (Brown & Lugo 1990) and the relief as a whole tends towards a horizontal plane. The shift between an undulated to a horizontal canopy surface reflects a transformation in the canopy floristic composition; this transformation is illustrated by the invaded forests of Tahiti island, where an introduced tree species, *Miconia calvescens* DC. (Melastomataceae), has changed the canopy surface from a diversified and undulating one to a horizontal and monospecific surface (Birnbaum 1995).

The ground topography is a physical factor that both directly and indirectly influences the height of trees (Torquebiau 1981), forest dynamics (Guillaumet 1987), and the floristic diversity (Basnet 1992; Sabatier et al. 1997). In the present study, we cannot conclude that trees growing on ridges are smaller than trees of the plateau or slope because the forest dynamic is different in each topotype. However, the canopy surface of the plateau is a homogenous canopy unit, higher than the average height, and composed of emergent trees, working in the same dynamic. In a similar way, the canopy surface of the second topo-sequence (from the ridge to the foot-slope) is actually lower than the average height, and homogeneously composed of concavities created by several perturbations of different ages. Dynamic events are associated on similar topotype to form alternatively a higher and a lower canopy unit. The blend of the both canopy units observed on the first and third topo-sequences illustrate a transition in this dynamic process.

The forest dynamics directly or indirectly determine canopy relief. Depending on local height variations, the relief of the canopy surface can be dissociated in three major canopy phases: (1) concavities that represent the impact of present and past perturbations. Area covered by the gap's epicentre (i.e., 1% of the primary forest plot) corresponds to other gaps quantification in the French Guianan forest (1.1% at Piste of St Elie; Riera 1994; 1.13% of a 12 ha plot in Nouragues field station; Van der Meer 1995). Gap closure occurs with the vertical growth of trees which fill the epicentre and the periphery of the concavity; (2) the convexities that delimit organised and stable patches of trees in the forest canopy where groups of emergent trees constitute extreme phases of the forest dynamics where the substitution process (i.e., the replacement of dead trees without perturbation) can be fully developed (Pascal 1995); (3) the average height of the canopy, which represents the 'intrinsic properties' of the forest type: trees within this canopy phase have either reached their maximum height development or follow an intermediate dynamic system following previous events. The canopy surface relief in primary forest follows a combination of these canopy phases (1-3) resulting from the dynamics of a single tree (isolated emergent; decaying tree) or a group of juxtaposed trees (group of emergent trees; large gap of several fallen trees).

The relief of the canopy surface is defined by a complex integrated system from the tree itself to the groups of trees, and to canopy units. Considering groups of trees as the equivalent of Oldeman's 'eco-unit' (i.e., a surface of forest on which, at one moment in time, vegetation has begun to develop, Oldeman 1983), we can class the canopy unit as a group of several eco-units linked to a topo-sequence, and working within the same dynamic system. However, we cannot certify that trees enclosed in such groups have begun their development at the same moment in time. Furthermore, the distinction of group is only based on tree heights, and there do not necessary reflect the development, architecture, species composition and ecophysiological functioning of the whole forest. For these reasons, we cannot conclude that groups of trees and perturbations are equivalent to Oldeman's eco-units.

Height variations and undulations are nested per integration level from the intra-individual to the inter-individual scale. At each level, dynamic events are symmetrically arranged around the average height, stretching the surface upwards during vertical growth, and depressing it down to the ground level during degrading phases. At the intra-individual scale, the growth and the death of young shoots at the extremities of canopy tree crowns (Blanc 1992) extensively modify the relief of the canopy surface. At a different scale, the height variations of the canopy surface at the landscape level (e.g., mature forest, secondary forest, logged forest, swamp forest, etc.) reflect the overall dynamic of the vegetation. At each level of the biological organisation (sub-crown, crown, forest landscape, etc.), the canopy surface undulates in space following height variation, and in time following the dynamics of components within a delimited physical environment. The scale of observation (i.e., the inter-distant point in the present study, the pixel size in remote sensing studies) reflect one level of these nested system (in the present study, the scale of observation relate to crowns in secondary forest plot and crowns/sub-crowns in primary forest plot).

By growing folds, the forest ecosystem increases the exchange surface area with the atmosphere. The increase in folds (i.e., as shown in primary forest) indicates that the photosynthetic blanket tends to capture more light. Interactions occur from the outside to the inside of the forest ecosystem through the absorption of resources (photons, H_2O, CO_2) and the rejection of transformed energy (O_2) from the photosynthetic carpet to the atmosphere. In the folded forest model the forest ecosystem can be considered as a delimited volume where the canopy surface is a folded membrane regulating chemical reactions and transfers of information between the inside and the outside. Thus, the folded forest model is a promising theory that allows the primary productivity of vegetation to be interpreted at different levels of ecosystem organisation, particularly by using modern technologies such as remote sensing.

Acknowledgements

The present study was conducted at the Centre ORSTOM at Cayenne (French Guiana) during my thesis work in 1997. I am very grateful to C. Collin, the Director at this time, for his help in making my stay as efficient and comfortable as possible. The present work would have not been possible without the invaluable support and helpful discussions with D. Sabatier, M-F. Prévost and J.F. Molino during and after the field work. I thank Professor Oldeman and Dr Davis who have commented encouragingly on the present paper. E.D.F. and the Foundation Ushuaïa in Paris provided financial support.

References

Aubréville, A. 1938. La forêt coloniale: les forêts d'Afrique occidentale française. Ann. Acad. Sci. Col. 9: 1–245.

Basnet, K. 1992. Effect of topography on the pattern of trees in Tabonuco (*Dacryodes excelsa*) dominated rain forest of Puerto Rico. Biotropica 24: 31–42.

Benoist, R. 1924. La végétation de la Guyane française. Bull. Soc. Bot. France 71: 1169–1177.

Birnbaum, P. 1995. Histoire de l'invasion d'une plante introduite, *Miconia calvescens*, dans une île polynésienne, Tahiti. J. d'Agr. Trad. Bot. Appl. 26: 283–295.

Birnbaum, P. 1997. Modalités d'occupation de l'espace par les arbres en forêt guyanaise. Unpublished PhD thesis, Université Pierre et Marie Curie (Paris VI), Paris.

Blanc, P. 1992. Comment poussent les couronnes d'arbres dans la canopée ? Pp. 155–172. In: Hallé, F. & O. Pascal (eds), Biologie d'une canopée de forêt équatoriale. II. Fondation ELF.

Brokaw, N.V.L. 1982. The definition of treefall gap and its effect on measures of forest dynamics. Biotropica 14: 158–160.

Brown, S. & Lugo, A. E. 1990. Tropical secondary forests. J. Trop. Ecol. 6: 1–32.

Campbell, D. G., Daly, D. C., Prance, G. T. & Maciel, U. N. 1986. Quantitative ecological inventory of Terra firme and Varzea tropical forest on the Rio Xingu, Brazilian amazon. Brittonia 38: 369–393.

Davis, T. A. W. & Richards, P. W. 1933. The vegetation of the Moraballi creek, British Guiana: an ecological study of a limited area of tropical rain forest. Part I. J. Ecol. 21: 350–384.

Duivenvoorden, J. F. 1996. Patterns of tree species richness in rain forests of the middle Caqueta area, Colombia, NW Amazonia. Biotropica 28: 142–158.

Granville de, J.-J. 1994. Les formations végétales primaires de la zone intérieure de Guyane. In: Raymond, H. L. (ed.), Forêt guyanaise. Gestion de l'écosystème forestier et aménagement de l'espace régional. Actes du II ème Congrès Régional de l'Environnement, Cayenne. Nature guyanaise, Sepanguy/CCEE. 21–40.

Guillaumet, J.-L. 1987. Some structural and floristic aspects of the forest. Experientia, 43: 241–251.

Hallé, F., Oldeman, R. A. A. & Tomlinson, P. B. 1978. Tropical trees and forest, an architectural analysis. Springer-Verlag, Berlin.

Kahn, F. & Granville de, J.-J. 1992. Palms in forest ecosystems of Amazonia. Springer-Verlag, Ecological studies, Berlin, p. 95.

Oldeman, R. A. A. 1978. Architecture and energy exchange of dicotyledonous trees in the forest. Pp. 535–560. In: Tomlinson, P. B. & Zimmermann, M. H. (eds), Tropical trees as living systems. Proceedings of the 4th Cabot symposium, Petershaw. Cambridge University Press, Cambridge.

Oldeman, R. A. A. 1983. Tropical rain forest, architecture, silvigenesis and diversity. Pp. 139–150. In: Sutton, S. L., Whitmore, T. C. & Chadwick, A. C. (eds), Tropical rain forest: ecology and management. Blackwell Science, Oxford.

Oldeman, R. A. A. 1992. Architectural models, fractals and agroforestry design. Agric. Ecosys. Env. 41: 179–188.

Oldeman, R. A. A. 1994. Sur les ecosystèmes forestiers: quatre principes sylvologiques. L'Italia Forestale e Montana 49: 1–16.

Parker, G. G., Smith, A. P. & Hogan, K. P. 1992. Access to the upper forest canopy with a large tower crane. Sampling the treetops in three dimension. Bioscience 42: 664–670.

Pascal, J.-P. 1995. Quelques exemples de problèmes posés à l'analyse et au modélisateur par la complexité de la forêt tropicale humide. Revue d'Ecologie (La Terre et la Vie) 50: 237–249.

Prévost, M. F. & Sabatier, D. 1996. Variations spatiales de la richesse et de la diversité du peuplement arboré en forêt guyanaise. Pp. 263–275. In: Guillaumet, J.-L., Belin, M. & Puig, H. (eds), Phytogéographie tropicale. Réalités et perspectives. Colloque de Phytogéographie tropicale, réalités et perspectives, Paris.

Puig, H. & Lescure, J.-P. 1981. Etude de la variabilité floristique dans la région de la piste de St Elie. Bull. ECEREX 3: 25–29.

Richards, P. W. 1952. The tropical rain forest: an ecological study. Cambridge University Press, Cambridge.

Riera, B. 1994. Dynamique de la forêt guyanaise. Pp. 105–110. In: Raymond, H. L. (ed.), Forêt guyanaise. Gestion de l'écosystème forestier et aménagement de l'espace régional. Actes du II ème Congrès Régional de l'Environnement, Cayenne. Nature guyanaise, Sepanguy/CCEE.

Rollet, B. 1974. L'architecture des forêts denses humides sempervirentes de plaine. C.T.F.T., Nogent sur Marne.

Sabatier, D., Grimaldi, M., Prévost, M.-F., Guillaume, J., Godron, M., Dosso, M. & Curmi, P. 1997. The influence of soil cover organisation on the floristic and structural heterogeneity of a Guianan rain forest. Plant Ecol. 131: 81–108.

Sarrailh, J. M. 1980. L'écosystème forestier guyanais. Etude écologique de son évolution sous l'effet des transformations en vue de sa mise en valeur. Bois For. Trop. 189: 31–36.

Torquebiau, E. F. 1981. Analyse architecturale de la forêt de Los Tuxtlas (Veracruz), Mexique. Unpublished PhD thesis, Université Montpellier II, Montpellier.

Torquebiau, E. F. 1986. Mosaic patterns in Dipterocarp rain forest in Indonesia, and their implications for practical forestry. J. Trop. Ecol. 2: 301–325.

Van der Meer, P. J. 1995. Canopy dynamics of tropical rain forest in French Guiana. Unpublished PhD thesis, Wageningen Agricultrural University, Wageningen.

Whitmore, T. C. 1978. Gaps in the forest canopy Pp. 639–655. In: Tomlinson, P. B. & Zimmermann, M. H. (eds), Tropical trees as living systems. Proceedings of the 4th Cabot symposium, Petershaw. Cambridge University Press, Cambridge.

Plant Ecology **153:** 301–312, 2001.

Crown typology and the identification of rain forest trees on large-scale aerial photographs

Valérie Trichon
Laboratoire d'Ecologie Terrestre, UMR 5552 CNRS / UPS, Centre de Télédétection, B.P. 4 072, F-31 029 Toulouse cedex 4, France

Key words: Architecture, Canopy, French Guiana, Paracou, Phenology, Texture

Abstract

This paper assesses the potential of large-scale aerial photographs for the identification of rain forest trees. Colour slides at 1: 3000 scale were acquired in French Guiana, above the canopy of a 25 ha study site where trees had already been identified. Firstly, the stereoscopic analysis of photographic prints served to establish a typology of the crowns, based on seven main classes of criteria: crown size, crown status, crown contour, crown architecture, foliage cover, foliage texture and colour, completed where possible by information on phenology. The terminology chosen was based on those proposed in previous studies. Secondly, a smaller area of 5 ha was delimited in the field, on which 15 tree categories (vernacular names) represented by 5 crowns or more on the photographs were selected. For each category, a standard crown was described using terminology previously defined. Twelve tree categories, including first and second class commercial timbers, displayed specific characteristics allowing them to be identified on aerial photographs. Further analyses will be undertaken in the future to measure the success of this identification method. This method may be applied for the recognition of particular species of interest: commercial, rare, endemic or key-stone species. Phenological data, as well as information on crown architectural development, can also be retrieved from aerial photographs, bringing new prospects for a better knowledge of crowns biology and their functional role in the forest ecosystem.

Introduction

The study of the biology and the ecology of rain forest trees is not an easy task. In such forests, which are characterized by a high species richness, most of the species are sparsely distributed, and have differing flowering periods. Furthermore, the crowns of the taller trees can be completely obscured from the ground. This hampers the collecting of fertile twig samples that are necessary for species identification, and also the study of spatial distribution, phenological behaviour, reproductive biology or crown development. For these reasons, having access to the rain forest canopy is a dream for many botanists, ecologists and foresters. An aerial view of the rain forest offers a valuable complement to the usual ground measurements, allowing observations to be made over large areas.

Up to now, large-scale aerial photographs (at 1:5000 and above) were mainly used in tropical rain forest for the recognition and study of only a few number of selected species. Clément & Guellec (1974) successfully used colour slides in Gabon to identify the Okoumé (*Aucoumea klaineana*). Aerial photographs were also used by Vooren & Offermans (1985) for the study of the aging process in *Piptadeniastrum africanum*, South-West Ivory Coast, and by Herwitz et al. (1998), who studied the long-term crown dynamics of *Toona ciliata* in the Queensland forest of Australia. Difficulties were encountered by Sayn-Wittgenstein et al. (1978) when they tried to identify a large number of forest timber species in Surinam. The authors underlined the necessity to improve the standard descriptions of the crowns used for the identifications. Following this pioneer work, a comparable investigation was conducted in the Queensland forests in Australia (Myers & Benson 1981; Myers 1982), and among the 111 selected tree species, 24 could be identified from aerial photographs with more than 75% success. Despite these encouraging results, recent similar studies seldom appear in the literature.

The present paper reports the first results from a study carried out in French Guiana. Firstly, by comparison with the studies of Sayn-Wittgenstein et al. (1978) and Myers (1982), the work focuses on the description criteria observed by stereoscopy used for recognizing the species. Changes and additions to previous terminologies used to describe species are examined. Secondly, a description of the standard crown of the most frequently recorded species on the photographs is given, using previously defined criteria. A validation of the description method will be undertaken in the next phase of study. The usefulness of large-scale aerial photographs for retrieving phenological data and information on tree architectural development in the canopy are examined, followed by a discussion on the functional implications of crown characteristics in the rain forest ecosystem.

Material and methods

Study site

The study plot belongs to the experimental site of Paracou (5°15′ N; 52°56′ W), French Guiana, established in 1984 by the CIRAD-Forêt[1]. Trees above 10 cm dbh are labeled on 25 ha of primary forest, and their location, diameter and vernacular name are recorded in a G.I.S.

Field work

The photographs were acquired at the end of the dry season, in October 1996, during the 'Opération Guyane 96' scientific mission (Trichon et al. 1998). The flights were done using a hot-air airship on three successive days, between 6.30 and 7.30 am due to the sensitivity of this airship to atmospheric disturbances. Altitude of the flights was about 130 m above the ground (about 100 m above the mean level of the canopy). The photographs were taken along parallel transects, allowing stereoscopic views between two successive photos. We used a FM2 Nikon® camera with a 35 mm lens, installed on the skirt of the nacelle, and loaded with colour slides (Fuji Sensia, 400 ASA). The approximate scale of the slides is 1:3000. About 15 ha of the study plot was covered by the photographs.

Before the aerial survey began, silver mylar balloons filled with helium, attached to a nylon yarn, were directed through the trees until they emerged from the canopy. Located at every intersect of 1 ha subplots, the silver spots served as spatial landmarks, both for the navigation during the survey and for further mapping of the canopy (by producing photomosaics).

After the aerial survey, photomosaics were used in the field: (1) to recognize from the ground crowns seen on the photographs (mainly by referring to the shape, size and location of the crowns), and (2) to note down the identification number of the corresponding tree. This field work was conducted in March 1998.

Photograph analyses

The photographs were analyzed by stereoscopy. We chose to work with photographic prints (scale 1: 600), which show more detail of crown architecture.

We established a list of crown description criteria to express the diversity of crowns in the canopy, and to guide the identification of tree species. The criteria were defined by comparison with those previously used by Myers (1981), which had been based on those proposed by Sayn-Wittgenstein et al. (1978). As far as possible, we kept the same terminology as used in these previous studies. However, our typology (i.e., the way criteria are organized in classes and sub-classes) is somewhat different from previous ones, which will be discussed later in the paper.

We used these criteria to propose a standard description of the species that have an occurrence of at least five clearly visible crowns on the photographs. For each species, we recorded the characteristics of each crown following the defined criteria. Standard crown descriptions then focuses on the marked characteristics which are stable for a given species. The botanical names used in this paper follow those given by Boggan et al. (1997).

[1] Forest department of the International Research Center for Agronomy and the Development. CIRAD-Fôret, BP 5035, 34032 Montpellier cedex 1, France

Results and discussion

Crown typology

The aerial photographs show the extreme diversity of crown shapes in the canopy (Figures 1–4). Seven classes of criteria are defined. These classes are: crown size, crown status, crown contour, crown architecture, foliage cover, foliage texture, and colour. Each class includes a various number of sub-classes, in which the different proposed criteria are exclusive (i.e., the choice of one criterion excludes the others). For example, a crown contour cannot be *round* and *elongated* (same sub-class), but it can be *round* and *regular* (two different sub-classes). One further class is added, relating to phenology, where the phenological characteristics of the crown at the date of the photograpic survey are recorded. The terminology used, and the definition of the criteria, are shown in Table 1.

Below, we compare this typology of tree crowns to those proposed by Myers (1982) and Sayn-Wittgenstein et al. (1978).

– The criteria of 'crown size' and 'crown status' correspond to the class 'crown size and dominance' of Myers (1982) and Sayn-Wittgenstein et al. (1978). However, we adapt the size limit used in the Queensland forest to classify *small, medium* and *large* crowns. The upper limit for small crowns is increased from 5 to 10 m, and the lower limit for large crowns from 10 to 15 m. These limits better fit to this French Guianan rain forest canopy.

– The criteria of 'crown architecture' were referred to as 'crown shape and structure' by Myers (1982). Myers also included a conical shape in the same class, mainly used to describe *Agathis* species in the Queensland forest. This is not useful in our case as this shape do not appear on our study site. However, it is an example of using a specific term to describe a canopy which may not fit into the standard groupings. In his class 'crown shape and structure', Myers identified some types that we do not include in our study. For example, his definition of a *compact* crown included other criteria (e.g., a crown which is not segmented, with no large openings, etc.) and was more the definition of a whole crown-type than a criterion in itself. We prefer to limit the list to simple criteria that could be subsequently combined together to define a crown type. Also, Myers defined the *open* crown as 'a crown including major gaps, or areas with very little foliage, within its general perimeter'. We describe the crown openness more precisely under the class 'foliage cover', as it will be discussed later.

– The textural criteria are the most difficult to express without some form of quantification. Here, we also adopt terminology that differs in some way from that used by Myers (1982). We look at the 'foliage texture' rather than the crown texture, the latter being partly linked to architectural criteria. For example, a *billowy* or a *tufted* crown will have a *rough* texture at the crown level (as defined by Sayn-Wittgenstein et al. 1978), but may also have, for example, either a *smoky* texture (no textural design) or a *loosely granular* texture (presence of a textural design) at the foliage level, as shown in Table 1. A next step for this work will be to define textural indices that will help to quantify these different textures.

– The 'colour' on the photographs is very sensitive to variation in photographing conditions (atmospheric composition, environmental light, film used) and development processes. However, in contrast to black and white images, colour has been used successfully in association with other criteria to help recognize tree species (Sayn-Wittgenstein et al. 1978). Myers (1982) used two different systems of colour description, the ISCC-NBS system and the Munsell colour system. However, there is still a problem when comparing the colour of crowns when using different photographic surveys. As a first step, we propose rough categories of tones to be used together: *green, yellow, blue, grey, red*, and two criteria of intensity: *light* and *dark*. The accuracy of the colour description using this method and its success in comparing photos taken using different methods or film-type will be assessed in subsequent work.

– The criteria of 'crown contour' has not been defined clearly in previous published studies. Only the notions of *round* and *irregular* contours were included in the Myers' definition of two criteria of crown shape, the *rounded* crown and the *open* crown respectively. Contour can be dissociated from the architecture of the crown, and may help to describe the crown shape, and so we put the contour criteria as a single class, subdivided in several sub-classes (Table 1).

– The criteria of 'foliage cover' has also not been used in previous studies, although the idea was included in Myers' classes 'crown shape and structure' and 'visibility of branches', and in the class 'crown density' of Sayn-Wittgenstein et al. In our typology, foliage characteristics are distinguished from foliage spatial distribution in the crown (i.e., branches can be visible when foliage is *opaceous* and *discontinu-*

Table 1. Crown typology and definition of each criteria.

Major classes of criteria	Sub-classes and criteria	Definition
Crown size	small	$L < 10$ m
	medium	$10 \text{ m} \leq L \leq 15$ m when L is the maximum width of the crown
	large	$L > 15$ m
Crown status	Dominant	above the mean canopy level, receiving side light
	Co-dominant	part of the mean canopy level, crowded on the sides, receiving little side-light
	Supressed	under the mean canopy level, lower than its neighbours, receiving no side-light
Crown countour	Round	
	Elliptic	
	Elongated	
	or	
	Regular	
	Lobate	
	Irregular	contour has deep invaginations
Crown architecture	Flat	more-or-less horizontal surface
	Rounded	definite curvature of the crown surface, highest point in the middle
	Billowy	surface rising and rolling in prominent waves reminiscent of a cumulus cloud
	Tufted	individual branches (branching habit > 45°) extending upward and outward
	Star-shaped	branches or big componed leaves disposed in verticilles or in a spirale
	Layered	several separated segments
	or	
	Segmented	with individual crownlets separated by more or less prominent fissures (a phenomenon called 'crown shyness' (Ng 1977))
	Not segmented	not like above
	or	
	Multiple	two or more large division within the crown, each component resembling an individual crown
	Single	not like above, crown with a well-defined center, even in the case of discontinuous foliage
	or	
	Shallow	the foliage is concentrated to the top of the tree
	Deep	the foliage is vertically distributed, from the top to approximately more than 1/3 of the total tree height
Foliage cover	Opaceous	nothing is visible through the foliage
	Light	branches or any other lower elements are distinguishable through the foliage
	or	
	Continuous	the foliage is regularly distributed within the crown
	Discontinuous	the foliage is concentrated in some parts of the crown, branches are visible in other parts
Foliage texture		No textural design is discernable at this scale of observation:
	Smooth	opaceous and dense foliage with small compacted leaves
	Smoky	opaceous or light foliage with leaves not compacted
		Textural design is discernable:
	Grainy	repetition of small particles giving a mottled appearance to the crown. These particles can be individual leaves
	Granular	repetition of larger particles. These particles can be clusters of leaves or crownlets
	Loosely granular	the granular textural design is loosely repeated
	Dotted	the foliage is light and the leaves are separated, appearing like individual spot
Colour	Green, Yellow, Blue, Grey, Red,	rough colour tones (can be combined together)
	or	
	Light	
	Dark	
Phenology	Leaf shed	
	Flowering	
	Fruiting	
	Flushing	
	or	
	Synchronism	of the phenological phenomenon at the species and/or at the crown levels
	Disynchronism	

ous, or when foliage is *light* and *continuous*, which correspond to very different crown characteristics).

The crown typology described here cannot be considered as definitive. It can be expected to be modified and improved as the interpretation methods evolve following further field studies.

Species description

During seven days of field work (March 1998) an area of 5 ha of forest was surveyed and 279 crowns were recognized on photographs from the ground. This represents 78 different tree categories (a category corresponds to a vernacular or commercial name) and a larger but undefined number of species (a tree category may include more than one species). Due to the floristic richness of the forest, 34 tree categories were recorded only once, and 63 were represented by less than five individuals. Only trees recorded five times or more were used in analysis (15 categories). Three of these categories showed a great variety of crown forms and so were not included in our analysis (these categories recover under their vernacular name more than one species each, and include the black Maho (*Eschweilera* spp.), the Chawari (*Cariocar* spp.), and the Pepe boïti (*Podoluma benai* or *Prieurella prieurii*)). The description of standard crowns for the 12 categories under analysis are given in Table 2. The category Symphonia recovers two species: *Symphonia globulifera* (usually restricted to hydromorphic soils) and another species, more likely to be found on well-drained soils, which is called *Symphonia* sp.1 (Loup 1994). 'Lebi koko' is a vernacular name that covers many different species, but the description given in Table 2 refers to *Licania alba* or liciana robusta. 'Maho cigare' may refer to different *Couratari* species; the standard description corresponds to *Couratari multiflora*. Carapa is for Carapaguianensis or Carapa procera. Kimboto is for Pradosia cochlearia or Pradosia ptychandra. Carapa is for Carapa guianensis or Carapa procera. Kimboto is for Pradosia cochlearia or Pradosia ptychandra. The other tree categories refer to a single species each.

The crown typology defined in Table 1 is based on thirteen sub-classes of criteria, belonging to seven major classes (plus the class 'phenology', when phenological aspects of the crowns are observed). Theoretically, every crown can be defined by thirteen criteria taken in each sub-class. In practice however, certain criteria are much more relevant (more obvious or marked) than others in particular crowns. Accordingly, the thirteen possible criteria are usually reduced to a smaller number when characterizing a given crown (Table 2). When certain criteria vary among the trees of a same category, it is recorded so as to aid the future identification of each species on aerial photographs.

This list of species recognizable on aerial photographs from this study includes first class commercial timbers, such as the Angélique (*Dicorynia guianensis*) and the Wacapou (*Vouacapoua americana*), as well as second class timbers: Carapa (*Carapa guianensis*), Goupi (*Goupia glabra*), Gonfolo (*Qualea rosea, Ruizterania albiflora*), Grignon (*Ocotea rubra*), Manil marécage (*Symphonia globulifera*) (Détienne et al. 1990). Some other commercial species that were not studied here due to lack of replicates, will probably be surveyed in more extensive studies. For example, the single specimens of Satiné rubané (*Brosimum paraense*, Moraceae) and Bagasse trees (*Bagassa* spp., Moraceae) observed on our images (a precious wood and a first class timber tree) display respectively marked crown characteristics. *Brosimum paraense* displays a tufted architecture with the ultimate twigs drooping, and the crown has an invaginated contour and an opaceous and discontinuous foliage with a grainy texture. The Bagasse tree displays a crown with a round contour, and a typical punctuated texture with a light and discontinuous foliage. The results from these commercial timber species should convince the foresters of the value of this approach for assessing the commercial value of a forest site.

It is possible that this work, mainly focused here on commercial timbers, may be extended to the recognition of other particular species of interest (i.e., ecological indicators, endemic or rare species, or key-stone species playing a major role in ecosystem functioning (Bond 1993)). For example, the palm *Euterpe oleracea* constitutes widely distributed and easily recognizable homogeneous populations, indicating the presence of swampy areas within the forest site (Figure 3).

The next step of this identification work will be to validate each description by (1) interpreting other photographs outside the 5 ha training area, and (2) measuring the success of the identification results. The efficiency of the method for species identification will also depend on the possibility of recognizing the species at lower scales (i.e., from photographs covering a more extensive area of forest) and thus at lower cost. The influence of scale for species identification will be undertaken in future studies.

Table 2. Standard crown description of the more abundant tree species.

Tree category (with number of useful records) Latin name Family	Crown size & status	Crown contour & architecture	Foliage cover & texture	Colour and phenology	Key features and conclusion
Gonfolo rose (18) *Qualea rosea* Vochysiaceae	large co-dominant to emergent	single crown with round & regular contour, or: multiple crown with elliptic & irregular projection crown rounded, segmented	opaceous foliage continuous foliage smooth texture	dark green pale-coloured flowers, abundant, scattered or absent, depending on the tree. Some case of intra-crown disynchronism of the flowering	Despite the variability from the single to the multiple crowns, the basic structure of the Gonfolo rose is stable: rounded and regular opaceous crownlets, giving a 'cauliflower' shape. Among the easiest species to identify from above.
Diaguidia (12) *Sclerolobium melinonii* Caesalpiniaceae	medium co-dominant	round & regular projection single crown, deep and layered, rounded crownlets at different height level in the canopy	opaceous foliage continuous foliage grainy texture	light yellowish green some crowns leafless at the end of the twigs	All the observed Diaguidia shared stable crown characteristics (layered crown with a grainy texture), allowing *a priori* the identification of this species from above.
Kimboto (10) *Pradosia cochlearia* Pradosia ptychandra Sapotaceae	large emergent	regular projection, elliptic (smaller crowns) to round (larger crowns) single crown, layered with two obvious layers: upper layer rounded and centered on the trunk, lower layer extended outward the center (this is less pronounced in older trees)	opaceous to light foliage discontinuous foliage, concentrated at the end of the branches, white portion of branches visible smoky texture	colour very variable depending on the trees, from grey-green to orange, through yellow	Probably among the easiest crowns to recognize in the canopy, thanks to the big size, the emergent position, and the specific colour.
Wacapou (10) *Vouacapoua americana* Caesalpiniaceae	medium co-dominant to suppressed	round & regular projection deep single rounded crown, more or less markedly segmented in a few big crownlets	opaceous & continuous foliage granular texture due to clusters of leaves within crownlets	typically a bluish-green colour	Despite its co-dominant to suppressed position, the Wacapou is likely to be easily identified from above, thanks to stable crown characteristics (rounded shape, granular texture) and a typical colour.
Symphonia (8) *Symphonia globulifera* *Symphonia* sp.1 Clusiaceae	small suppressed to emergent	round and irregular projection star-shaped and single crown	more or less opaceous and discontinuous foliage, with leaves concentrated at the end of lateral twigs smoky texture	bluish green (suppressed trees) to light green (taller trees), apparent branches and twigs are pale Intra-species disynchronism of the flowering	The Manil marécage is easy to identify thanks to its particular star-shaped crown.
Lebi koko (7) Licania robusta *Licania alba* Chrysobalanaceae	small suppressed	round & regular projection rounded and single crown	light & discontinuous foliage grainy texture	mixed colour: green and orange (due to the tomentum on the lower face of the leaves, or to the presence of flowers ?)	Easy to identify when visible, thanks to the particular mottled colour.
Angélique (7) *Dicorynia guianensis* Caesalpiniaceae	large co-dominant to emergent	elliptic projection, regular to lobate deep single crown, slightly layered, rounded, one tree segmented	more or less opaceous grainy texture except for one tree (granular texture due to the segmented structure)	rather dark green, more or less greyish defoliation commonly observed with an intra-species disynchronism	Intra-species disynchronism of the defoliation process may hamper the identification of all the trees of this species at a given date.

Table 2. continued

Tree category (with number of useful records) Latin name Family	Crown size & status	Crown contour & architecture	Foliage cover & texture	Colour and phenology	Key features and conclusion
Maho cigare (7) *Couratari multiflora* Lecythidaceae	medium emergent	round & regular, to elliptic & lobate projection rounded and single crown	light & discontinuous foliage, branches visible in open parts dotted texture	light green, turning to yellow white branches	This species is characterized by a small crown compared to its height, and by its dotted texture.
Gonfolo gris (6) *Ruizterania albiflora* Vochysiaceae	large emergent	round to elliptic projection, regular to irregular crown rounded, segmented in small crownlets, single crown with large fissures (maybe on the way towards a multiple crown)	opaceous foliage continuous foliage granular texture due to small crownlets	light green, slightly yellowish One case of intra-crown disynchronism during flowering period	The Gonfolo gris share some characteristics with the Gonfolo rose: rounded, segmented and opaceous crowns. But texture and colour differ, allowing *a priori* differentiation of the two species.
Grignon (6) *Ocotea rubra* Lauraceae	large co-dominant to emergent	elongated and lobate projection, to irregular single and deep crown, billowy	opaceous and discontinuous foliage, irregularly clustered loosely granular texture	light green	The bright colour and the very irregular shape of the large crowns allow the Grignon to be recognized, especially if the tree is old.
Carapa (5) Carapa procera *Carapa guianensis* Meliaceae	small co-dominant to emergent, often isolated in a gap	elongated & irregular projection multiple crown, tufted, each orthotropic branch displaying a star-shape crownlet	opaceous foliage	dark green	The tufted and star-shaped branches of the Carapa crowns are easy to detect in the canopy.
Goupi (5) *Goupia glabra* Celastraceae	large co-dominant or partly isolated at the edge of a gap	elliptic, rather irregular projection single and deep crown, tufted, orthotropic branches drooping at the top	more or less opaceous and discontinuous foliage, part of branches or twigs are visible smoky texture	bluish green, turning to grey	The colour and the tufted shape of the crowns seem to be stable characteristics, and may be enough on their own to identify the Goupi.

Crown phenology

This section examines disynchronism of phenological aspects, both at species and crown levels. A phenological disynchronism appears when all trees of a same species (intra-species disynchronism), or when all parts of a single crown (intra-crown disynchronism), do not show a same phenological status at a given period.

Intra-species disynchronism

An intra-species disynchronism was observed in *Qualea rosea* (Figure 1). Some were covered with pale flowers in October 1996, whereas others were not flowering or intermediary stages were seen. This observation relates to the temporal scale used, and the apparent disynchronism may be due to the short period of the aerial survey.

Field work conducted in March 1998 brought additional results, where it was possible to survey changes in foliation from the ground. The degree of foliation in *Dicorynia guianensis* (Figure 1) presented an intra-species disynchronism in October 1996. One tree was entirely leafless on the photographs, another one had a few leaves, whereas others were covered with leaves. Eighteen months after, in March 1998, the disynchronism was noticeable from the ground. The tree that was leafless in October 1996 was partly foliated, and the tree that had a few leaves on the photographs was entirely leafless. This agrees with a study carried out in French Guiana by Loubry (1994), who concluded that each *Dicorynia guianensis* has his own periodicity of leaf shed, ranging from 10 to 35 months.

Intra-crown disynchronism

In October 1996, several *Qualea rosea* and *Ruizterania albiflora* presented flowers in only part of their crown, as it is frequently observed in these two species (Loubry 1994).

Phenological disynchronism at the crown level is understood in part to be a consequence of the architectural development of the adult tree, where most species increase their size by duplicating their elementary architecture through a process called 'reiteration' (Oldeman 1974). Each duplicated part, or 'reiterat', may be autonomous, which may be expressed through phenological behaviour (Hallé 1986). However, this phenomenon is complex (Loubry 1994), and in the Paracou forest the trees that show such phenological disynchronism may be used to further investigate the links between phenology and the reiteration process.

The difficulty of completing regular observations on rain forest trees over long periods and wide areas is in part responsible for the general confusion concerning phenological patterns at the species level. According to Newstrom et al. (1994), this situation also results from a lack of standardization in terms and methods, and from the different time scales and organization levels (population, community) at which the observations generally occur. Repetitive acquisitions of aerial photographs, linked to environmental data, would help to understand phenological cycles in the canopy, and to assess, in particular, the contribution of the endogenous versus exogenous factors.

As the timing of flowering can be an isolating mechanism in plant speciation, the spatial recognition of flowering in a population can lead to the interpretation of plant mating systems and gene flow. Repetitivity of aerial surveys may also be of great value in assessing the spatial and temporal food availability (flowering, fruiting or leaves flushing) for predators and pollinators in the canopy (key-stone species that directly relate to the trophic function of the ecosystem).

Tree architectural development in the canopy

We know little about the architectural development of canopy and emergent trees. Indeed, the 'architectural model' of a tree, a concept coined by Hallé & Oldeman (1970), is studied only in young trees. As previously noted, most of the species increase their size by means of a reiteration process. The crown shape of an adult tree will thus depend on the reiteration strategy used, which is very diverse among species (Hallé et al. 1978; Castro e Santos 1980; Edelin 1986). Moreover, even if the reiteration strategy, like the architectural model itself, is genetically fixed at the species level, environmental factors can modify the final shape of the tree (Barthélémy et al., 1997). For all these reasons, knowing the architectural model of a tree is not enough on its own to predict the shape of its crown in the canopy (Fisher & Hibbs 1982).

The architectural models of some species described in this paper (*Carapa guianensis, Symphonia globulifera, Symphonia* sp.1, *Goupia glabra*) are known from the literature. The tufted crowns of *Carapa* sp. (Figure 4) resemble clusters of stars. This corresponds to the Scarrone model that describes *Carapa guianensis* (Hallé & Oldeman 1970): big pinnate leaves inserted in a spiral on orthotropic branches,

Figure 1. Aerial photograph showing the following species: 1, 4, 7, 8: *Vouacapoua americana*; 2, 9 (flowering): *Qualea rosea*; 3, 5, 6: *Pradosia sp.*; 10: *Dicorynia guianensis*. Note the bright balloon on tree 2. Photograph D. Guillemyn. Scale 1: 600.

Figure 2. Aerial photograph showing the following species: 1, 3: *Licania sp.*; 2: *Goupia glabra*; 4: *Symphonia* sp.; 5: *Qualea rosea*. Photograph D. Guillemyn. Scale 1: 600.

themselves verticillated on the trunk. *Symphonia globulifera* and *Symphonia* sp.1, observed on the photographs, have star-shaped crowns (Figures 2 and 3) that express the Massart model (Barthélémy 1988; Loup 1994). This model is defined by a monopodial trunk with a rhythmic growth, producing regular tiers of plagiotropic branches (Hallé & Oldeman 1970). *Symphonia globulifera* is a very polymorphic species which has two extreme growth possibilities: a lack of reiteration in the canopy, or a complex crown constituted of many reiterats; whereas reiteration is rarely expressed by adult trees in *Symphonia* sp.1. (Loup 1994). *Goupia glabra* displays the Roux model (monopodial trunk with a spiraled phyllotaxy, diffuse or continuous ramification and plagiotropic branches, Hallé & Oldeman 1970). It is difficult however to recognize these characteristics in the adult crowns, a difficulty that may be due to a more or less opaceous and continuous foliage cover hiding the branches in this species, described by Loubry (1994) as 'sempervirent'. If the foliage is opaceous in most of the crowns in the canopy, the spatial arrangement of the branches and their insertion position on the trunk may be seen when crowns display a light foliage and/or a discontinuous foliage, or are leafless. In this way, aerial views of tree crowns may complement ground observations, and add to our understanding of the architectural development of tree species in the canopy.

Crown architecture and functional meaning

The terminology chosen here to describe the architecture of the crowns follows closely that used previously for the forests of Queensland (Myers 1982) and Suri-

Figure 3. Aerial photograph showing the following species: 1: *Euterpe oleracea*; 2: *Symphonia sp.* (flowering). Photograph D. Guillemyn. Scale 1: 600.

Figure 4. Aerial photograph showing the following species: 1: *Carapa sp.* (cluster of star-shaped crownlets, left of a bright balloon). Photograph D. Guillemyn. Scale 1: 600.

nam (Sayn-Wittgenstein et al. 1978). This, added to other analyses of aerial photographs acquired in Cameroon and Indonesia (unpubl. data), lead to the conclusion that the same typology can be used for the description of any rain forest canopy, as long as a comparable scale of observation is followed.

Similar types of crown architecture are found in every rain forest canopy. The more spectacular of these is probably the 'cauliflower' shape (*Qualea rosea*, Figure 1). Distinctive shapes such as these can be used to investigate the ecological meaning of crown architecture, as discussed by Hallé et al. (1978) and Oldeman & Van Dijk (1991). Oldeman & Fundter (1986) investigated the functional meaning of crown architecture by comparing the status of Vochysiaceae and Dipterocarpaceae families, in America and South-East Asia respectively. The authors suggested that these two families, sharing, among other characteristics, big emergent trees with a cauliflower shape crown, constituted the same kind of ecological guild on both continents. Through the surface area and the spatial arrangement of leaves, the architecture of the crown of an emergent tree influences its interception of incident light (within-tree shading) as well as the radiation conditions available for the rest of the ecosystem below (between-tree shading) (Terborgh 1985; Kuuluvainen & Pukkala 1987). The crowns also act as biological filters for other abiotic factors such as rain, moisture, temperature and wind. In addition, the availability of ecological niches for animals or plants living in the canopy may also depend on the architecture of the crown. For all these influences that crowns possibly exert, it may be surmised that species, or groups of species, that display a similar crown shape may, in general, insure the same function in the ecosystem.

The similarity in crown characteristics was also examined for various pioneer species by Ashton (1976). Broad leaved and flat crowns, displaying a regular granular foliage texture on aerial photographs, are observed in *Didymopanax morototoni* (Araliaceae) and *Cecropia* spp. (Cecropiaceae) in America, *Musanga cecropioides* (Moraceae) in Africa and *Macaranga* spp. (Euphorbiaceae) in Asia. Architectural convergences seen in architectural models between taxonomically or geographically remote groups (Hallé & Oldeman 1970), therefore also exist in crown shape. Such comparisons should be investigated in other taxa, crown types, or ecological guilds, such as the star-shape crowns of many Clusiaceae (e.g., *Symphonia globulifera*, America), Annonaceae (e.g., *Xylopia nitida*, America) and Myristicaceae (e.g., *Pycnanthus angolensis*, Africa, *Virola surinamensis*, America), or the flat crowns of many Fabaceae.

Conclusion

This paper has investigated the potential of large-scale aerial photographs for the identification of rain forest trees. In a 5 ha test area, 12 tree categories were described using a typology of crown characteristics. These species, mainly commercial timbers, are very likely to be identifiable from aerial photographs, thanks to stable characteristics at the species level, and/or striking characteristics.

The success of these results will be quantified in the next phase of this study, using the photographs from outside the test area. Photographs acquired at different scales on neighbouring sites will also be analyzed, aimed at (1) extending the description and identification of more species of interest, and (2) assessing the scale limit for species identification. A step towards systematization of the identification procedure will then be attempted (e.g., through the creation of a dichotomous key based on crown criteria; or through the quantification of textural design on digital images).

To conclude, aerial photographs should not be devoted to species identification only. Architectural studies focused on tree crowns should be further developed. Together with observations on tree status and phenological behaviour, such studies should improve the understanding of crown architectural development in the canopy, and may bring new information regarding the functional meaning of crowns in the rain forest ecosystem.

Acknowledgements

The author thanks Silvolab-Guyane (Kourou) and the CIRAD-Forêt (Montpellier) who sponsored this study, as well as F. Blasco, head of the Laboratoire d'Ecologie Terrestre (Toulouse), for his understanding and support. The photographic data were acquired from the hot-air airship of the Océan vert society (60, rue Truffaut, Paris); D. Guillemyn (Laboratoire d'Ecologie Terrestre) played an important role as the photographer. Sincere acknowledgments are addressed to F. Delsanti for his substantial contribution in the field as a M.Sc. student. Part of the photographic analyses was achieved during the author's stay at the Laboratoire de Biologie et Ecophysiologie (Besançon), where she found good logistic support and a warm welcome. Many thanks to the European Science Foundation for sponsoring an exciting Tropical Forest Canopies Conference (Oxford, 12–16 December 1998), and especially to A. Davis and M. Speight for their personal investment. Referees are sincerely acknowledged for their constructive comments on the paper.

References

Ashton, P. S. 1978. Crown characteristics of tropical trees. Pp. 591–615. In: Tomlinson, P. B. & Zimmerman, M. H. (eds), Tropical trees as living systems. Cambridge University Press, Cambridge.

Barthélémy, D. 1988. Architecture et sexualité chez quelques plantes tropicales: le concept de floraison automatique. Unpublished PhD Thesis, Université Montpellier II, Montpellier.

Barthélémy, D., Caraglio, Y. & Costes, E. 1997. Architecture, gradients morphogénétiques et âge physiologique chez les végétaux. Pp. 89–136. In: Bouchon, J., de Reffye, P. & Barthélémy D. (eds), Modélisation et Simulation de l'Architecture des Végétaux. Science Update. INRA Editions, Paris.

Boggan, J., Funk, V., Kelloff, C., Hoff, M., Cremers, G. & Feuillet, C. 1997. Checklist of the plants of the Guianas (Guyana, Surinam, French Guiana). 2nd ed. Smithsonian Institution, Washington DC.

Bond, W. J. 1993. Keystone species. Pp. 237–253. In: Schultze, E-D & Mooney, H. A. (eds), Biodiversity and ecosystem function. Ecological studies Vol. 99. Springer-Verlag, Berlin.

Clément, J. & Guellec, J. 1974. Utilisation des photographies aériennes au 1:5000 en couleur pour la détection de l'Okoumé dans la forêt dense du Gabon. Bois et Forêts des Tropiques 153: 3–22.

Castro e Santos (de), A. 1980. Essai de classification des arbres tropicaux selon leur capacité de réitération. Biotropica 12 (3): 187–194.

Détienne, P., Fouquet, D. & Parant, B. 1990. Les bois guyanais. Propriétés et utilisation. Bois et Forêts des Tropiques 219: 125–143.

Edelin, C., 1986. Stratégie de réitération et édification de la cime chez les conifères. Pp. 159–170. In: L'Arbre. Compte-rendu du colloque international l'Arbre, Montpellier, 9–14 Septembre 1985. Naturalia monspeliensia (Hors Sér.), Montpellier.

Fisher, J. B. & Hibbs, D. E. 1982. Plasticity of tree architecture: specific and ecological variations found in Aubreville's model. Am. J. Bot. 69: 690–702.

Hallé, F. 1986. Deux stratégies pour l'arborescence : gigantisme et répétition. Pp. 159–170. In: L'Arbre. Actes du colloque international l'Arbre, Montpellier, 9–14 Septembre 1985. Naturalia Monspeliensia (Hors Sér.), Montpellier.

Hallé, F. & Oldeman, R. A. A. 1970. Essai sur l'architecture et la dynamique de croissance des arbres tropicaux. Masson & Cie, Paris.

Hallé, F., Oldeman, R. A. A. & Tomlinson, P. B. 1978. Tropical trees and forests. Springer-Verlag, Berlin.

Herwitz, S., Slye, R. & Turton, S. 1998. Redefining the ecological niche of a tropical rain forest canopy tree species using airborne imagery: long-term crown dynamics of *Toona ciliata*. J. Trop. Ecol. 14: 683–703.

Kuuluvainen, T. & Pukkala, T. 1987. Effect of crown shape and tree distribution on the spatial distribution of shade. Agric. Forest Meteorol. 40: 215–231.

Loubry, D., 1994. Déterminisme du comportement phénologique des arbres en forêt tropicale humide de Guyane française. Unpublished PhD Thesis, Université Paris VI, Paris.

Loup, C. 1994. Essai sur le déterminisme de la variabilité architecturale des arbres Unpublished PhD Thesis, Université Montpellier II, Montpellier.

Myers, B.J. 1982. Guide to the identification of some tropical rain forest species from large-scale colour aerial photographs. Aust. For. 45 (1): 28–41.

Myers B. J., & Benson, M. L. 1981. Rain forest species on large-scale colour photos. Photogramm. Eng. Rem. S. 41: 505–513.

Newstrom, L. E., Frankie, G. V., & Baker, H. G. 1994. A new classification for plant phenology based on flowering patterns in lowland tropical rain forest trees at La Selva, Costa Rica. Biotropica 26 (2): 141–159.

Ng, F. S. P. 1977. Shyness in trees. Nat. Malaysiana 2: 34–37.

Oldeman, R. A. A. 1974. L'architecture de la forêt guyanaise. Mém. 73, O.R.S.T.O.M., Paris.

Oldeman, R. A. A. & Van Dijk, J. 1991. Diagnosis of the temperament of tropical rain forest trees. Pp. 21–65. In: Gomez-Pompa, A., Whitmore, T. C. & Hadley, M. (eds), Rain forest regeneration and management. Man and the Biosphere series, Unesco & the Parthenon Publishing Group, Paris.

Oldeman, R. A. A. & Fundter, J. M. 1986. On Diptero-Vochysian tree strategies. Pp. 191–208. In: L'Arbre. Actes du colloque international l'Arbre, Montpellier, 9–14 Septembre 1985. Naturalia Monspeliensia (Hors Sér), Montpellier.

Sayn-Wittgenstein, L., Milde (de), R. & Inglis, C. J. 1978. Identification of tropical trees on aerial photographs. Forest Management Institute, Canada Information Report, FMR-X-113.

Terborgh, J. 1985. The vertical component of plant species diversity in temperate and tropical forests. Am. Nat. 126: 760–777.

Trichon, V., Guillemyn, D. & Fromard, F. 1998. Cartographie d'une canopée de forêt tropicale par photographies aériennes numériques et analogiques. Pp. 26–36. In: Hallé, F. (ed.), Biologie d'une canopée de forêt équatoriale III. Pro-Natura International & Opération Canopée, Paris.

Vooren, A. P., & Offermans, D. M. J. 1985. An ultra-light aircraft for low-cost, large-scale stereoscopic aerial photographs. Biotropica 17 (1): 84–88.

Plant Ecology **153:** 313–317, 2001.

The management implications of canopy research

Nigel E. Stork
Cooperative Research Centre for Tropical Rainforest Ecology and Management, James Cook University, P.O. Box 6811, Cairns Qld 4870, Australia (e-mail: nigel.stork@jcu.edu.au)

Key words: Biodiversity, Canopy cranes, Forests, Last biological frontier, Tropical rain forests

Abstract

So little is known about about what goes on above our heads in rain forests and what lives there that the rain forest 'canopy' has been called the 'last biological frontier'. Until very recently, researchers had very limited access to the canopy, even in temperate forests and, not surprisingly, we are only just beginning to discover this part of the world. A new breed of scientists, canopy biologists, has evolved and have accelerated the rate of discovery. The implications for management of forests of this new voyage of discovery are enormous. At present conservationists, forest managers, environmental policy makers, and other research users have not even started to consider how to integrate the resulting new information for future planning and sustainable management and use of forests.

The Earth Summit in 1992 drew attention to several of the principal biological dilemmas of the 20th century: biodiversity and forest loss, land conversion, climate change and ozone depletion. Understanding the importance of forest canopies for these issues is essential and researchers need to consider how better to coordinate their activities and deliver their results if they are to influence decision making at all levels. One new initiative shows some promise in this regard. Recently the United Nations Environment Program has supported the International Canopy Crane Network. This network of crane sites around the world will focus on providing an improved scientific basis for the importance of the forest canopy with respect to the Conventions on Biological Diversity, Climate Change and Ozone Depletion.

Introduction

We are only just beginning to understand what has been called the 'last biological frontier', that is that canopy of forests, and in particular the canopy of tropical forests. The implications of this fact are only just beginning to be understood. Research into canopies has increased exponentially in the last few years, fuelled by the interest of scientists and because of international and national concern about a number of issues, particularly the loss of biological diversity and the causes and impacts of environmental change. Increasingly scientists and managers alike are grappling with how this new research can be used to improve the sustainable management and use of forests. As someone who plays an important role in influencing policymakers, McNeely (1998) presented his viewpoint on this issue at the Second International Canopy Conference in Florida. Here, I present my viewpoint as a practising canopy biologist and as someone who interacts with policymakers on a regular basis.

If canopy researchers are to have an impact on the way forests are managed and used then they need to consider the relevance of their work and how to make an impact. The real issue is not that canopy research is not of importance to conservation and management and sustainable use but that as biologists we have little appreciation of the issues that policy makers are addressing and we usually are not skilled in working with those that use or potentially make use of our research. As others have observed (e.g., McNeely 1998), researchers and policy-makers speak completely different languages. Some would think we are different species! Policy makers try to simplify issues and are usually interested in the short term. They are often driven by the political whims of the government and therefore have short time horizons. Scientists, in contrast, realise that the issues are never simple and are looking for long term research and benefits. Policy makers are usually very good at working with and through the media whereas scientists are usually very distrustful of the media and always feel they have been inaccurately reported. So the solution is that we need to understand much more about the needs and concerns of policy makers and other research users if we are to ensure that the canopy research we do is used and useful.

Often the users of our canopy research information are those who manage forest resources and implement policy. From my experience, many of these are trained in resource management and can understand much of the basic science. However they have limited time and resources and often have to make rapid decisions on issues. They need whatever scientific information is available to help their decision making and cannot afford to sit on the fence on issues. While they often respect researchers needs and concerns that all scientific results and conclusions need to undergo peer review and be published before being released, they are the ones who need the information immediately. This can often lead to conflict. Too often researchers carry out a piece of scientific research and then at the end look to see if it has some management implications. Involvement of managers at the concept stage of the research right through to completion increases harmony in the research partnership and better understanding of the differing needs and pressures of those involved.

McNeely (1998), as someone working for an international conservation organisation, identifies some of the key issues where he believes canopy science can play an important role: sustainable use, climate change, extinction rates, ecosystem integrity, interactions among species and disturbance. For example, he asks what changes in the forest canopy tell managers about the effects of exploitation, what can canopy studies contribute to knowledge about extinction of species, or what is the role of the forest canopy in climate change? He questions whether we should have a canopy crane in a disturbed forest - in practice we do as the Australian canopy crane is in a cyclone affected location. His suggestions on further research for all these issues are all very valuable. Here I expand on two of these areas and add a third, discovery of canopy biodiversity, which I discuss first. Finally, I discuss the importance of networking to improve the development of canopy science and its application.

Canopy biodiversity and the voyage of discovery

In 1998 the President's Council of Advisors on Science and Technology recommended that the USA increase spending on biodiversity from \$480 to \$660 million a year. Of this they recommended that \$130 million should be spent on the discovery and description of new species. For my part, I believe this voyage of discovery should focus on tropical rain forests and in particular, on the canopy of tropical rain forests. The accelerated use in the English language of the term biodiversity (Wilson & Peter 1988) in the mid1980s was in large part due to the discovery that the canopy of tropical forests held a hitherto unknown fauna and flora. In 1981 Terry Erwin used his analysis of the first large samples of insects from the rain forest canopy to suggest that we had underestimated the number of species on Earth by a factor of ten or more (Erwin 1981). He suggested that there might be 30 million species of insects in tropical forests rather than 1–2 million as previously believed and that most of these undiscovered and undescribed species were found only in the canopy (but see Stork 1997, 1999). His global estimate was very timely as attention was focussed on species extinction rates which were being calculated on the basis of extrapolation from forest loss. The idea that we might be losing millions of species rather than just a few hundred has concentrated the minds of politicians and conservationists alike.

So how rich is the canopy fauna and flora? Numerous taxonomists from the Natural History Museum in London collaborated to study the first canopy sample of insects from Borneo and revealed an immensely rich and previously unknown fauna (Stork 1991). One of these, Noyes (1984), sorted 739 morphospecies of Chalcidoidea from just 1455 specimens. Barnard et al. (1986) and Stork et al. (1997) found that even in well known places such as Richmond Park in England what previously were thought to be rare species of insects were very abundant in the canopy. The same also appears to be true for relatively well know groups such as mammals. Malcolm (1994) found that the commonest species of mammal in an area of Brazilian rain forest near Manaus was found in the canopy. Furthermore, this species was previously considered to be extremely rare.

Two studies have shed some light on just how large a contribution the canopy makes towards species richness in forests. More than 6000 species of beetles were collected in an intensive survey of the beetle fauna of the rain forest in a small area of north Sulawesi

(Hammond et al. 1997). Analysis of these beetles indicate that about one fifth of these beetle species were restricted to the canopy (i.e., canopy specialists) and almost two fifths were ground specialists. In other words, at least for beetles, the ratio of canopy to ground specialists is something like 1:2. In a study of Australian rain forests, Rodgers and Kitching (1998) examined the Collembola and mite fauna in suspended soils associated with *Asplenium* tree ferns and compared these with the soil fauna on the ground. They found that of the twenty-five species of *Collembola* in the canopy, two were not found in the ground layer. Both these studies indicate that Erwin's ideas on the richness and endemism of the canopy fauna are exaggerations. However, there are still a very large number of species around the world that are only found in the canopy and usually have never been seen before.

In practice though, canopy biologists have still some way to go to demonstrate with any scientific rigour that the canopy fauna is as rich as it has sometimes been portrayed, and that a high proportion of the biota found there is unique to this stratum. In my experience, we need more precision and less 'arm waving' for the canopy biota and its conservation to become a real issue for many managers and policy makers. Of course in one sense all forest species are canopy dependent – if we remove the canopy then very few forest species on the ground can survive!

Sustainable use of the canopy

We have still yet to establish and capture the full social and economic benefits of rain forests canopies and the impacts of their loss through deforestation. There are many examples of how indigenous peoples around the world use animals and plants collected from the rain forest. The future more realistically lies in the potential bioprospecting of useful products from the rain forest canopy. Now that technology has sufficiently advanced to allow chemical bioprospecting of even tiny insects not just large plants, then there is a huge potential for bioprospecting for useful chemicals from canopy organisms. I have no doubt that the increase in bioprospecting from the canopy will piggy back on the discovery of new canopy organisms. One novel form of bioprospecting was adopted by the recent Radeau des Cimes expedition in French Guiana which was in part sponsored by a pharmaceutical company interested in finding possible new scents and perfumes from flowers in the canopy. The canopy biologists in the dirigible were joined by two 'noses' who specialise in distinguishing fine differences between the smells of perfumes. The 'noses' guided the canopy dirigible to where they could smell new or interesting scents. I have no doubt that intensive studies of canopy organisms will reveal a huge variety of potentially useful chemicals and that the prospecting for these is only limited by access to the canopy and the technology to handle so many samples.

Tourism is one of the most rapidly growing global industries and canopy tourism is beginning to expand as more novel ways of experiencing and accessing the canopy are developed. Increasingly tourists are becoming more selective and demanding in the type of tourism experience they expect and are prepared to pay for. In Australia, for example, nature-based tourism is now a huge industry and one of the leading tourism ventures is a canopy experience, Skyrail, set in the tropical rain forests of north Queensland. Skyrail is a seven and a half kilometre cable-way which travels across lowland rain forest. Visitors ride in six seater gondolas above the canopy of the world heritage rain forest. This venture has been a huge success carrying hundreds of thousands of visitors a year. Skyrail has won many national and international awards both as a successful tourism venture but also for the way it has been carefully integrated into the natural environment. Many countries are now seeking the expertise of Skyrail in developing their own canopy cable-way plans. Others in Australia have used canopy walkways to catering for different kinds of visitor experiences. The canopy walkway in the Valley of the Giants, four hours drive south of Perth, has been constructed to allow even wheelchair access to the tops of tingle and karri trees. This venture was opened two years ago and is regarded as a financial success.

The advantages of such canopy ventures are numerous. First, they provide the general public with real and safe access to the canopy and can give a greater understanding of the conservation significance of forest canopies. The public are more interested in nature and science than we often realise and we should remember that the views of the public have a great deal of influence on politicians particularly when elections are near! Second, they provide a revenue stream for private enterprise, management of forests and for further research. Third, they provide new means of accessing the canopy for researchers. And fourth, by concentrating tourism activities on selected areas of forest they reduce visitor impacts on areas of forest that are of greater conservation value.

With careful design and planning all can benefit from such enterprises.

Other issues

The focus of attention for rain forest canopies has also shifted more to their role in the maintenance of global climates and of climate change. The work of Shuttleworth (1984) and others in rain forests on energy balances have been particularly important in this respect. However, we need much finer detailed research on the macro, meso, and micro climate of canopies if we are to give greater precision to the role of canopies in climate change. Another limiting factor has been the use of just a few sites for data collection. The establishment of a network of canopy sites is essential and below I discuss the importance of the international canopy crane network in this regard.

Much greater progress has been made on the importance of the canopy and its fauna and flora in studies on the impact of logging, particularly of old-growth forests. In Europe the importance of red data book-listed saproxylic insects as indicators of ancient forests has been recognised but this is not yet the case for tropical forests (Grove & Stork 1999). At present we also have little information on the canopy biota of plantation forests and how this differs from native forests (but see e.g., Ozanne et al. 1997; Lawton et al. 1998; Foggo et al. 2001).

In spite of the increasing scientific effort and resulting literature on edge effects and forest fragmentation (e.g., Laurance & Bierregaard 1997), the canopy has been little studied in this regard. This is perhaps surprising as the greatest effects of fragmentation are on the canopy either through loss of canopy or impacts on the canopy biota and the processes, such as pollination and herbivory, that take place there. How far edge effects extend from a forest edge at canopy level is still unclear.

Networking

The results of research from single sites (with notable exceptions) seldom have significant impact on policy makers and resource managers. Networks can allow researchers to carry out the same experiments at a series of sites or to share data across sites, thus enhancing the value of the research. Networks can allow exchange of ideas, and provide new training opportunities. Joint communication across networks can also improve communication with a range of research users. Potentially one of the most useful linkages between canopy research and our potential research users is through the International Canopy Crane Network. The United Nations Environment Program has provided financial and other support to the Smithsonian Tropical Research Institute in Panama to help establish the first canopy crane. Subsequently, UNEP have helped STRI in getting aid from the Danish Government for the establishment of a second crane. The first International Canopy Crane Conference was held in Panama in February, 1997 and UNEP played an important role in this meeting (Stork et al. 1997). Representatives from the three secretariats of the Conventions on Biological Diversity, Climate Change, and Depletion of the Ozone Layer spoke at this meeting and stressed the importance of canopy research in understanding the issues and problems associated with each of these conventions. Here we have a mechanism for ensuring that our research is delivered to those policy makers that need it most. The International Canopy Crane Network established at this meeting has now expanded to include cranes in Panama, U.S.A., Japan, Australia, Venezuela, and soon will include sites in Malaysia and Ivory Coast. The canopy crane researchers are now establishing protocols for a variety of research areas such that the results from the individual cranes will be comparable to provide more global perspective. These sites will also provide the first long term monitoring network for the canopy. The value of the International Canopy Crane Network was enhanced by the strong support from other canopy biologists for this network as a focus for future canopy biology efforts, at a joint National Science Foundation/European Science Foundation canopy biology meeting in Oxford in November 1999.

Acknowledgement

I thank the convenors of the conference for inviting me to present this paper and for commenting on earlier manuscript versions.

References

Barnard, P. C., Brooks, S. J. & Stork, N. E. 1986. The seasonality and distribution of Neuroptera, Raphidioptera and Mecoptera on oaks in Richmond Park, Surrey, as revealed by insecticide known-down sampling. J. Nat. Hist. 20: 1321–31.

Erwin, T. L. 1982. Tropical forests: their richness in coleoptera and other arthropod species. Coleopt. Bull. 36: 74–75.

Foggo, A., Ozanne, C. M., Speight, M. R. & Hambler, C. 2001. Edge effects and tropical forest canopy invertebrates Plant Ecol. 153: 347–359 (this volume).

Grove, S. J. & Stork, N. E. 1999. The conservation of saproxylic insects in tropical forests: a research agenda. J. Insect Cons. 3: 67–74.

Hammond, P. M. 1995. Practical approaches to the estimation of the extent of biodiversity in speciose groups. Pp. 119–136. In: Hawksworth, D. L. (ed.), Biodiversity measurement and estimation. Chapman & Hall, London.

Hammond, P. M., Stork, N. E. & Brendell, M. J. D. 1997. Tree-crown beetles in context: a comparison of canopy and other ecotone assemblages in a lowland tropical forest in Sulawesi. Pp. 184–223. In: Stork, N. E., Adis, J. & Didham, R. K. (eds), Canopy arthropods. Chapman & Hall, London.

Lawton, J. H., Bignell, D. E., Bolton, B., Bloemers, G. F., Eggleton, P., Hammond, P. M., Hodda, M., Holt R. D., Larsen, T. B., Mawdlsey, N. A., Stork, N. E., Srivastava, D. S. & Watt, A. D. 1988. Biodiversity inventories, indicator taxa and effects of habitat modification in tropical forest. Nature 391: 72–76.

Laurance, W. F. & Bierregaard, R. O. Jr. (eds). 1997. Tropical forest remnants: ecology, management and conservation of fragmented communities. University of Chicago Press, Chicago.

Malcolm, J. R. 1994. The small mammals of amazonian forest fragments: pattern and process. Unpublished PhD Thesis, University of Florida.

McNeely, J. A. 1998. The science of the canopy: who needs the information and how they can use it. Selbyana 19: 135–140.

Noyes, J. S. 1984. In a fog. Chalcid Forum 3: 4–5.

Ozanne, C. M. P., Hambler, C., Foggo, A., & Speight, M. R. 1997. The significance of edge effects in the management of forests for invertebrate biodiversity. In: Stork, N. E., Adis, J. & Didham, R. K. (eds), Canopy arthropods. Chapman & Hall, London.

Rodgers, D. & Kitching, R. L. 1988. Vertical stratification of rain forest collembolan (Collembola: Insecta) assemblages: description of ecological patterns and hypotheses concerning their generation. Ecography 21: 392–400.

Shuttleworth, J. W. 1984. Observations of radiation exchange above and below Amazonian forest. Quart. J. Roy. Meteorol. Soc. 110: 1163–1169.

Stork, N. E. 1991. The composition of the arthropod fauna of Bornean lowland rain forest trees. J. Trop. Ecol. 7: 161–180.

Stork, N. E. 1997. Measuring global biodiversity and its decline. Pp. 41–68. In: Wilson, E. O., Reaka-Kudla, M. L. & Wilson D. E. (eds), Biodiversity II: understanding and protecting our biological resources. Joseph Henry Press, Washington D.C.

Stork, N. E. 1999. The magnitude of biodiversity and its decline. Pp. 3–32 In: Cracraft, J. & Novacek, M. (eds), The living planet in crisis: biodiversity, science and policy. American Museum of Natural History, New York.

Stork, N. E. & Hammond, P. M. 1997. Sampling arthropods from tree-crowns by fogging with knockdown insecticides: lessons from studies of oak tree beetle assemblages in Richmond Park (UK). Pp. 3–26. In: Stork, N. E., Adis, J. & Didham, R. K. (eds), Canopy arthropods. Chapman & Hall, London.

Stork, N. E., Wright, S. J. & Mulkey, S. S. 1997. Craning for a better view: the canopy crane network. Trends Ecol. Evol. 12: 418–420.

Wilson, E. D. & Peter, F. M. 1988. Biodiversity. National Academy Press, Washington D.C.

Plant Ecology **153:** 319–333, 2001.

Neotropical bats in the canopy: diversity, community structure, and implications for conservation

Elisabeth K. V. Kalko[1,2] & Charles O. Handley, Jr.[2]
[1]*University of Ulm, Department of Experimental Ecology (Bio III), Albert-Einstein Allee 11, 89069 Ulm, Germany (e-mail: elisabeth.kalko@biologie.uni-ulm.de) and Smithsonian Tropical Research Institute, P. O. Box 2072, Balboa, Panamá;* [2]*National Museum of Natural History, Mammals, MRC-108, Washington, DC 20560, USA*

Key words: Brazil, Chiroptera, Guild, Inventory methods, Mistnetting, Vertical stratification

Abstract

We compare results of parallel ground and canopy netting of bats (Microchiroptera) in three adjacent forest sites near Belém, Brazil, to document possible differences in vertical distribution of species. We caught 1871 individuals representing 49 species of three families (Emballonuridae, Phyllostomidae, Vespertilionidae). Capture effort, totaling 1955.5 mistnet hours in several cycles over a two-year period, was similar for ground and canopy nets. The canopy rigs yielded more species ($n = 41$) than the ground nets ($n = 35$), but both samples were characterized by rank abundance curves with similar shape and with a dominance of frugivores (Phyllostomidae). Nearly half ($n = 24$) of the species were captured in numbers too small ($n < 6$) to allow firm classification, but differences in capture frequencies of some of the better-sampled species in high and low nets reveal vertical stratification. Species-specific differences in diet, foraging strategies, roost sites, and sampling bias contribute to this pattern. As a result of the differential use of space among bats, alterations of forest structure are likely to result in changes in structure and function of local bat communities, but our limited knowledge of natural history and ecology of many species limits definition of changes. We see a critical need for further research into the extent to which habitat complexity influences species richness and abundance of bats. This information is especially important in view of the need to develop and apply conservation-oriented programs to maintain biodiversity. A review of recent improvements in techniques for inventorying bats shows that a combination of methods, including mistnetting and acoustic monitoring, is mandatory for such studies.

Introduction

Spatial heterogeneity has long been recognized as an important factor promoting diversity of animals and plants, particularly in species rich tropical lowland forests. Among vertebrates, forest birds provide well documented examples of coexistence of species by means of vertical stratification (e.g., Bell 1982; Brosset 1996; Cody 1996; Erard 1987, 1990; Pearson 1971, 1975; Roth 1976). Many bird communities characteristically segregate into groups of species living predominantly at ground level, at intermediate levels, or in the forest canopy. For other vertebrate taxa – particularly bats – the extent to which forest communities are spatially structured is less obvious.

Mainly because of the difficulty of accessing the canopy, most inventories of forest bats have relied exclusively on mistnets set at ground level (e.g., Findley 1993; Fleming et al. 1972; Handley et al. 1991; Kalko et al. 1996a). The few studies that have included the canopy in their sampling routine show distinct differences in capture frequencies between forest levels for some species. For instance, flying foxes (Pteropodidae) of the Old World tropics are caught mostly in the canopy and subcanopy levels and seldom in the understory (e.g., Cosson 1995; Francis 1989, 1990, 1994; Gaskell 1984; Ingle 1993). Large body size and high wing loading (relationship of body size to wing area) adapt many of the flying foxes for sustained flight and long-distance commutes necessary to harvest patchily distributed foods, mainly fruits and flowers (e.g., Fleming 1993; Norberg & Rayner 1987), but reduce manoeuvrability and limit access to the obstacle-cluttered forest interior. In contrast, their ecological counterparts in the Neotropics, frugivorous and nectarivorous species of the family Phyllostomidae are mostly smaller in size and have shorter, broader wings, rendering them more manoeuvrable and giving them easier access to the forest interior. Furthermore, distribution and abundance of fruits in the understory (for examples, fewer fruits in the understory of Malaysian forests compared with other tropical forests) may also contribute to the observed differences in vertical distribution between communities of Old and New World fruit bats (Francis 1994).

Because comprehensive studies of bats using parallel ground and canopy sampling are so few, particularly in highly diverse Neotropical forests (e.g., Ascorra et al. 1996; Bernard 1997; Bonaccorso 1979; Brosset & Charles-Dominique 1990; Handley 1967; Simmons & Voss 1998), data are still inadequate to identify general patterns of habitat use for whole communities of bats and to address the role of forest canopies in structuring them. Given the large numbers of bats in local communities and the key roles they play in tropical ecosystems by dispersing seeds, pollinating flowers, and controlling insect populations (e.g., Findley 1993; Fleming 1993), a better understanding of factors that contribute to their diversity is crucial. This is especially important for tropical forest ecosystems increasingly threatened by habitat alteration, fragmentation, and deforestation. Because of their structural complexity, canopies are among of the most diverse parts of tropical forests, but because of their inaccessibility, they are the least understood.

Methods

Ground and canopy nets were set at three sites in large fragments of tall undisturbed forest in an ecological research reserve, Área de Pesquisas Ecológicas do Guamá (APEG) on the grounds of the Brazilian government agricultural research and experiment station, Instituto de Pesquisas e Experimentação Agropecuárias do Norte (IPEAN), on the southeastern outskirts of the city of Belém, Pará, Brazil. One site was in terra firme (upland) forest, overlooking the second site in várzea (a tidal white-water swamp) on the north shore of the Rio Guamá. The third site, about a kilometer away, across agricultural land and further from the river, was in an igapó (a non-tidal black-water swamp). At that time, the sites were bordered on one side by primary forest and on the other by agricultural lands.

Access to the swamp sites was by means of boardwalks and bridges constructed by Philip Humphrey and Thomas Lovejoy in the mid-1960s to study migratory birds. Along with the boardwalks, they erected several dozen mistnet rigs for sampling birds in the forest canopy. These ingenious rigs, invented by Humphrey and David Bridge, consisted of three or four 12 x 2 m mistnets, each suspended below the next above in tandem, from a high line strung between appropriate trees (Humphrey et al. 1968; Lovejoy 1967). The nets reached heights of 20-30 m in the canopy. For our bat study, we selected rigs that had canopy cover, were not in treefall gaps, and faced openings between canopy and subcanopy. With each 3- or 4-net canopy rig, we situated an equal number of nets at ground level as nearly beneath the canopy rig as the shrub layer and terrain would permit.

One of us (Handley) visited the IPEAN forest reserves in 1965, 1968, and 1969 to study bats, especially those of the canopy. At that time, tropical forest canopy bats had not been sampled anywhere else in the world. The brief 1965 episode, a reconnaissance, has been described by Handley (1967). Here we report on the results of the intense netting in 1968 and 1969 in the APEG várzea forest, the terra firme forest, and the Mocambo igapó forest.

The igapó forest was sampled on 23 nights in 1968 (February-March) and on ten nights in 1969 (January-March) at five capture stations, each with a 3-net canopy rig and three ground nets. The terra firme forest was netted on three nights in 1969 (February-March) at a single capture station with a 4-net canopy rig and four ground nets. The várzea forest was netted on 13 nights in 1969 (January-March) at eleven capture stations with a total of twelve 3-net canopy rigs and twelve 3-net sets of ground nets. Overall, netting nights in 1968 and 1969 totaled 49. Nets were operated continuously from sunset until about midnight, weather permitting. They were closed during heavy rains, which occurred on 28 of 49 nights for a total of 42.5 h (mean on nights with rain: 1.5 h). To minimize bias of net shyness, several netting cycles were conducted at each site and each capture station was operated only one night in each cycle, in an irregular pattern.

An observer tried to keep track of bats that entered the canopy nets as they were lowered amd while

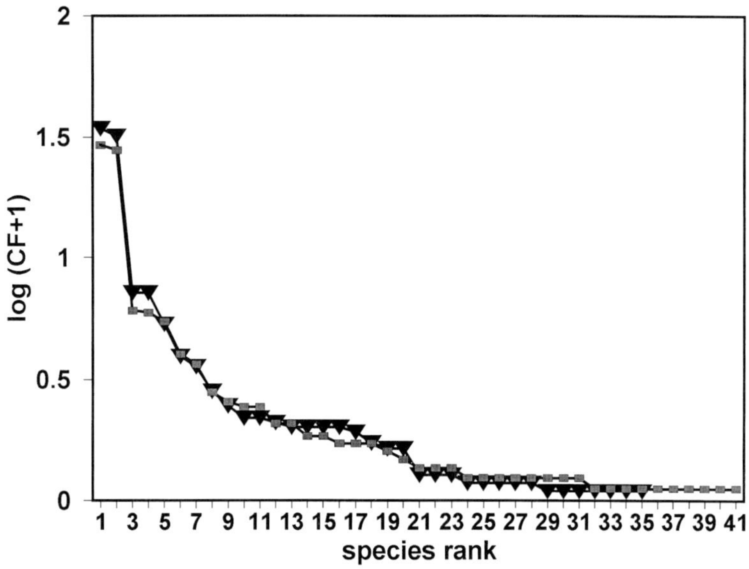

Figure 1. Rank abundance curves of all bats captured at the APEG study site near Belém, Brazil, in mistnets at ground level (triangles) and at canopy level (rectangles). CF = capture frequency in bats/100 mnh (= mistnet hours).

they were down for removal of bats captured at higher levels. Nevertheless, a few bats no doubt slipped unnoticed into the nets while they were down and gave the impression of having been captured at higher levels.

Climatic conditions in Belém in the 1960s were hot, but usually not very humid. The driest months were August to November, with a ten-year monthly rainfall average of 107–149 mm (IPEAN pers. com.). This study took place during the wettest months, January through March, when the ten-year monthly rainfall average was 623–685 mm (IPEAN pers. com.).

The data recorded for each bat were species, age, sex, reproductive condition, evidence of feeding, site, net number, net height, and time of capture. Before releasing them at the study site, 1803 bats were marked in the 1968 and 1969 sessions, most with wing membrane tattoos and a few with aluminum forearm bands.

To compare distribution of ground and canopy captures, we calculated capture frequency of each species as captures of individuals per 100 mistnet hours (1 mnh: one 12 m mistnet open for one hour) for each night, transformed to log abundances. We added 1 to keep the log abundances positive. Because the three forest sites were close to each other and were rather similar in structure (tree height, canopy closure), and because recapture data show that individual bats move between the sites, we have grouped all canopy net captures into one sample and ground net capture into another in order to obtain sample sizes large enough to show consistent patterns in the use of ground and canopy strata.

For each species, we calculated the mean and the standard deviation of capture frequency in ground and canopy mistnets for the total sample to test where the bats were more or less frequent. We assumed a Gaussian distribution of data and used a two-tailed Student's t-test to establish the confidence level (p) at which canopy and ground mistnet frequencies could be assumed to differ. If this difference was significant at $p < 0.05$ and sample size was $n \geq 6$, we assigned the bats to the category 'canopy' or 'ground' according to their means and percentage of canopy captures. If the difference was not significant, the bats were attributed to the category 'none' (no preference). In case of a sample size $n < 6$, we did not assign the bats to any category ('−').

To facilitate comparisons of ecological attributes, we assigned each bat to a guild, following Kalko et al. (1996a) and Schnitzler & Kalko (1998). Based on the original guild concept of Root (1967), each guild is composed of species that forage in similar habitats in similar ways for similar foods. Because flight, motor coordination, and sensory capabilities largely determine which foraging areas can be used efficiently by bats, we define habitat with respect to the proximity of a bat to obstacles or clutter. 'Uncluttered space' is the open area above the canopy or ground. 'Background cluttered space' is at forest edges, in larger gaps, and in openings between canopy and subcanopy. Bats in 'highly cluttered space' forage very close to or within vegetation. With regard to feeding modes, we distinguish between bats that take food from surfaces (gleaning) or catch it in the air (aerial). Food refers to the main diet of a species. We derived dietary data from literature and from our ongoing studies in Panamá and Brazil (see Bonaccorso 1979; Handley et al. 1991; Kalko et al. 1996a).

Results

The combined captures of bats in the parallel ground and canopy mistnetting in the forests at Belém in 1968 and 1969 totaled 1871 individuals of 49 species, representing 33 genera and three families (Appendix 1). At ground level, 1102 individuals of 35 species were caught in 1110 mistnet hours (mnh), while in the canopy, 769 individuals of 41 species were taken in 845.5 mnh. Canopy net hours are less because of time lost while nets were down for removal of captures.

Eight species were captured exclusively in the ground nets and 14 only in the canopy (Table 1). The combination of ground and canopy nets yielded more species than either method alone. About half of the species ($n = 24$) were captured at numbers too low ($n < 6$) to allow statistical testing of our assignment to the categories canopy, ground, or none (no preference). Among the better-sampled species with six or more individuals captured ($n = 25$), capture frequency was significantly higher for six species in the canopy and for four species at ground level (Table 1). Overall, mean capture frequency was slightly lower in the canopy (0.9 bats mnh^{-1}) than at ground level (1 bat mnh^{-1}).

The rank-abundance patterns of ground and canopy samples are very similar. Each is characterized by a few dominant species, followed by a long tail of less commonly and rarely caught species (Figure 1). Frugivores dominate the samples in numbers of species and capture frequency (Table 1). In both ground and canopy nets, ranks one to seven are taken exclusively by frugivores. However, the sequence of species differs. Ground net samples are dominated by *Artibeus jamaicensis* (Leach), followed by *Carollia perspicillata* (Linnaeus), *Rhinophylla pumilio* (Peters), *Artibeus lituratus* (Olfers), *Carollia brevicauda* (Schinz), *Vampyrodes caraccioli* (Thomas), and *Artibeus obscurus* (Schinz) whereas in the canopy

Table 1. Distribution of bats in ground and canopy nets by guild. The guilds are adopted from Kalko et al. (1996a) and Schnitzler and Kalko (1998). Among guilds, the distribution in canopy nets is given as % of the total capture frequency (log (bats/100 mnh) +1) for each species. 1 mnh = one 12-m mistnet open for one hour. Species are arranged within guilds in order of increase in % of canopy captures. Numbers are rounded to the nearest whole digit. To detect significant differences between captures at the two forest levels, a two-tailed Student's t-test has been applied (d.f. = 96). Depending on the p value, total number of individuals sampled, and % canopy captures, species are assigned to the categories 'canopy' or 'ground' ($p < 0.05^*$; $n \geq 6$), 'none' ($p \geq 0.05$; $n \geq 6$) or '–' ($n < 6$).

Guild	Canopy (%)	*t*	*p*	*n* (total)	Assignment
'Background cluttered space, aerial insectivore' (10 species)					
Lasiurus egregius	0	1.00	0.32	1	–
Myotis riparius	36	1.35	0.18	8	none
Peropteryx macrotis	57	0.25	0.80	4	–
Cormura brevirostris	100	1.00	0.32	1	–
Eptesicus chiriquinus	100	1.15	0.25	2	–
Rhynchonycteris naso	100	1.00	0.32	1	–
Saccopteryx bilineata	100	1.16	0.25	2	–
Saccopteryx canescens	100	1.19	0.24	2	–
Saccopteryx gymnura	100	1.42	0.16	2	–
Saccopteryx leptura	100	2.47	0.01*	6	canopy
'Highly cluttered space, gleaning insectivore' (8 species)					
Tonatia silvicola	0	1.29	0.20	2	–
Trachops cirrhosus	0	2.47	0.02*	16	ground
Micronycteris minuta	15	0.89	0.38	3	–
Glyphonycteris daviesi	51	0.03	0.97	2	–
Tonatia saurophila	65	1.18	0.24	26	none
Mimon crenulatum	96	1.27	0.20	3	–
Tonatia brasiliensis	100	1.00	0.32	1	–
Tonatia carrikeri	100	1.21	0.23	2	–
'Highly cluttered space, gleaning carnivore' (2 species)					
Chrotopterus auritus	0	1.00	0.32	1	–
Vampyrum spectrum	100	1.41	0.16	2	–
'Highly cluttered space, gleaning sanguivore' (2 species)					
Desmodus rotundus	35	0.73	0.46	29	none
Diaemus youngi	100	2.31	0.02*	6	canopy
'Highly cluttered space, gleaning frugivore' (19 species)					
Carollia brevicauda	13	3.87	0.00*	49	ground
Carollia perspicillata	18	5.06	0.00*	352	ground
Artibeus obscurus	43	0.63	0.53	43	none
Rhinophylla pumilio	46	0.50	0.61	98	none
Artibeus jamaicensis	48	0.21	0.83	586	none
Platyrrhinus helleri	59	0.36	0.72	10	none
Vampyrodes caraccioli	60	0.68	0.50	74	none
Artibeus cinereus	62	0.77	0.44	21	none
Artibeus gnomus	63	0.92	0.36	21	none
Sturnira tildae	68	0.51	0.70	4	–
Uroderma magnirostrum	69	0.49	0.62	2	–

Table 1 continued.

Guild	Canopy (%)	*t*	*p*	*n* (total)	Assign-ment
Mesophylla macconnelli	73	0.92	0.35	5	–
Sturnira lilium	73	1.35	0.18	23	none
Vampyressa bidens	79	1.71	0.09	38	none
Artibeus lituratus	86	4.04	0.00*	301	canopy
Uroderma bilobatum	87	2.12	0.04*	28	canopy
Chiroderma trinitatum	92	1.99	0.04*	9	canopy
Chiroderma villosum	100	2.31	0.02*	6	canopy
Ametrida centurio	100	0.00	1.00	1	–
'Highly cluttered space gleaning nectarivore' (5 species)					
Choeroniscus minor	0	1.00	0.32	3	–
Lichonycteris degener	0	1.00	0.32	1	–
Lonchophylla thomasi	20	1.56	0.12	13	none
Glossophaga soricina	72	1.08	0.28	15	none
Lionycteris spurrelli	100	1.00	0.32	1	–
'Highly cluttered space gleaning omnivore' (3 species)					
Phyllostomus elongatus	0	2.52	0.01*	11	ground
Trinycteris nicefori	0	1.17	0.24	3	–
Phyllostomus hastatus	32	0.89	0.37	13	none

rigs *A. lituratus* is first, followed by *A. jamaicensis*, *V. caraccioli*, *C. perspicillata*, *R. pumilio*, *Vampyressa bidens* (Dobson), and *Uroderma bilobatum* (Peters).

Handley's (1967) brief study in APEG in 1965 revealed patterns of species composition similar to those we found in 1968–1969. Presumably due to lower capture effort, fewer species ($n = 25$) were detected in 1965 than in the present study ($n = 49$ species). With the exception of *Lampronycteris brachyotis* (Dobson) and *Rhinophylla fischerae* (Carter), all species mist-netted in APEG in 1965 were also found in 1968–1969. Additional ground mistnetting in open sites in all three years and collecting by P. S. Humphrey and others in and near our three forest sites increased the number of species known to inhabit the area to 65. Details about this part of the study will be given elsewhere.

Bat communities in Neotropical lowland forests are composed of bats representing ten guilds (Kalko et al. 1996a; Schnitzler & Kalko 1998). We caught bats from seven of the guilds in the Belém forests (Table 1). The missing guilds are those in which bats forage for insects in open spaces far away from vegetation and high above the ground ('uncluttered space, aerial insectivores'), hunt insects on the wing within vegetation close to obstacles ('highly cluttered space, aerial insectivores'), and forage mostly over water to catch mainly fish ('background cluttered space, gleaning piscivores').

We found ten species of the guild that catches insects on the wing in semi-open spaces close to vegetation ('background cluttered space, aerial insectivores') (Table 1). These bats belong to two families: sheath-tailed bats (Emballonuridae) represented by four genera, and evening bats (Vespertilionidae) represented by three genera (Appendix 1). Six of the seven emballonurids were caught only in canopy nets and one (*Peropteryx macrotis* Wagner) at both levels. Spatial distribution was clear enough and sample size was large enough for one species, *Saccopteryx leptura* (Schreber), to designate it as a canopy bat. Among the three vespertilionids, one was taken in a canopy rig (*Eptesicus chiriquinus* Thomas), one in a ground net (*Lasiurus egregius* Peters), and the third, *Myotis riparius* (Handley), for which there were eight captures, showed a non-significant tendency towards the lower levels of the forest.

We caught eight species of forest bats that mainly glean insects from surfaces ('highly cluttered space, gleaning insectivores') (Table 1). They represent five genera of the family Phyllostomidae (Appendix 1). Two species were caught exclusively in ground nets (*Trachops cirrhosus* Spix; *Tonatia silvicola* d'Orbigny) and two in the canopy rigs (*Tonatia brasiliensis* Peters; *Tonatia carrikeri* J. A. Allen). However, capture rates were very low ($n < 6$) for six of the eight species. In this guild, the species encountered most frequently were the fringe-lipped bat *T. cirrhosus* (exclusively in the understory) and the round-eared bat *Tonatia saurophila* (Koopman & Williams) (found in both strata but more often in the canopy).

We caught three individuals of 'highly cluttered space, gleaning carnivores', one woolly false vampire *Chrotopterus auritus* (Peters) in an understory net, and two great false vampires *Vampyrum spectrum* (Linnaeus) in canopy nets (Appendix 1, Table 1). Sample size was too small to assign either species with significance to ground or canopy category.

We found two species of blood-drinking vampires ('highly cluttered space, gleaning sanguivores') (Appendix 1, Table 1). The common vampire *Desmodus rotundus* (E. Geoffroy & St.-Hilaire) was encountered mostly in ground nets but also in the canopy, whereas the bird vampire *Diaemus youngi* (Jentink) was only caught in canopy nets.

The fruit-eating phyllostomids ('highly cluttered space, gleaning frugivores') proved to be the largest guild, with 19 species and about 90% ($n = 1671$) of all individuals captured (Appendix 1, Table 1). Most of the species ($n = 15$) were taken in large numbers ($n \geq 6$). Except for two species that were caught only in canopy nets (*Chiroderma villosum* Peters; *Ametrida centurio* Gray), all of the frugivorous bats were found at both forest levels. However, capture frequencies varied widely among species. With more than 80% of their captures in ground nets, the short-tailed fruit bats, *C. brevicauda* and *C. perspicillata*, showed a significant preference for the understory. Seven species with high capture rates ($n \geq 6$), including *R. pumilio*, *Platyrrhinus helleri* (Peters), *V. caraccioli* and four species of *Artibeus*, were caught in nearly equal proportions (40–65%) in ground and canopy nets and thus were assigned to the category 'none' (no preference). With more than 80% of their captures in the canopy rigs, the four frugivores *A. lituratus* (third highest in overall capture rate), *U. bilobatum*, *Chiroderma trinitatum* (Goodwin), and *C. villosum* were found significantly more often in the upper levels of the forest (Table 1). In contrast, *A. jamaicensis*, a close relative of *A. lituratus*, with the highest capture rate in our study (about one-third of all captures), was encountered almost equally in ground and canopy nets.

We caught five species of long-tongued, mostly nectar-drinking phyllostomids of four genera ('highly cluttered space gleaning nectarivores') (Appendix 1, Table 1). One species, *Lionycteris spurrelli* (Thomas), was caught only in a canopy net and two species (*Lichonycteris degener* Miller; *Choeroniscus minor* Peters) were exclusively in ground nets. Only two of the five species had capture rates $n \geq 6$. Of those, *Lonchophylla thomasi* (J.A. Allen) was mostly in understory nets and *Glossophaga soricina* (Pallas) was mostly in canopy nets, but differences in their capture rates between ground and canopy were not significant.

Finally, we captured three phyllostomids that feed on a variety of foods including fruit and insects and sometimes small vertebrates, pollen, and nectar ('highly cluttered space, gleaning omnivores') (Appendix 1, Table 1). We took two of them, *Phyllostomus elongatus* (E. Geoffroy & St.-Hilaire) and *Trinycteris nicefori* (Sanborn), only in ground nets. The greater spear-nosed bat, *Phyllostomus hastatus* (Pallas), one of the largest Neotropical bats, did not show a significant preference for canopy or ground nets.

Recapture data

From 1803 bats that we marked during the study in 1968 and 1969, 79 individuals (4.4% of all bats marked) of 12 species were later recaptured (Table 2). Sixty-six individuals were recaptured only once, 12 twice, and one three times (Table 2). Most recaptures occurred within 200 m of the original marking site, but eighteen individuals moved between forest blocks with maximum distances of 1000 m between capture and recapture sites. Because most of the bats were marked with ephemeral wing scars that did not carry over from one year to the next, we found only four bats in 1969 that had been marked in 1968.

For individual bats (Table 2) the recapture data are consistent with the use of forest strata proposed in our species-level classification (Table 1). Individuals of three species designated as predominantly ground-dwelling (two frugivores *C. perspicillata* and *C. brevicauda* and the gleaning insectivore, *T. cirrhosus*) were also recaptured mostly in the understory (Table 2). Two frugivores (*A. jamaicensis*, *R. pumilio*) that were regularly mistnetted at both canopy and ground levels (Table 1), showed similar tendencies on an individual basis (Table 2), indicating that they frequently move between canopy and ground levels. The mostly ground-dwelling sanguivore *D. rotundus* and the canopy frugivore *A. lituratus* were also recaptured throughout the forest profile, suggesting flexibility in habitat use among individuals. Five species, including four frugivores (*A. obscurus*, *P. helleri*, *V. bidens*, and *C. trinitatum*) and one nectarivore (*L. thomasi*), had only one recapture each, three of them at the mark level and two at a different level.

Discussion

The general patterns of community composition and rank abundance of the bats obtained at ground and canopy levels in our netting study at Belém, Brazil, resemble results of studies of bats at ground level elsewhere in the Neotropics (e.g., Ascorra et al. 1996; Brosset & Charles-Dominique 1990; Brosset et al. 1996; Findley 1993; Fleming 1986; Fleming et al. 1972; Kalko et al. 1996a; dos Reis & Muller 1995; Simmons & Voss 1998; Willig 1986; Wilson et al. 1996). Frugivorous phyllostomids dominate local communities, both in terms of species and number of individuals. Gleaning insectivorous and carnivorous phyllostomids are less common.

Table 2. Recapture data of marked bats. Indiv. = total number of recaptured individuals. Recap. = total number of recaptures; one time, two times, three times = number of time (s) individuals were recaptured. can-can = number of canopy captures subsequently recaptured in canopy nets; can-gro = number of canopy captures subsequently recaptured in ground nets; gro-can = number of ground captures subsequently recaptured in canopy nets; gro-gro = number of ground captures subsequently recaptured in ground nets.

Guild	Indiv. (total)	Recap. (total)	one time	two times	three times	can-can	can-gro	gro-can	gro-gro
'Highly cluttered space, gleaning insectivore'									
Trachops cirrhosus	2	2	2						2
'Highly cluttered space, gleaning sanguivore'									
Desmodus rotundus	5	6	4	1			1	1	4
'Highly cluttered space, gleaning frugivore'									
Carollia brevicauda	4	5	3	1					5
Carollia perspicillata	26	30	22	4		1		4	25
Artibeus obscurus	1	1	1				1		
Rhinophylla pumilio	13	17	9	4		2	3	8	4
Artibeus jamaicensis	19	23	16	2	1	6	6	8	3
Platyrrhinus helleri	1	1	1					1	
Vampyressa bidens	1	1	1			1			
Artibeus lituratus	5	5	5		2	1	2		
Chiroderma trinitatum	1	1	1		1				
'Highly cluttered space, gleaning nectarivore'									
Lonchophylla thomasi	1	1	1						1

In our study, we matched numbers of ground and canopy nets at three forest sites to determine species diversity and habitat use by the local bat community. The canopy nets caught more species but the combined catches included a greater number of species than did either high or low nets alone. Together with the few studies in the Neotropics that have used both ground and canopy nets, these findings demonstrate that the combination of netting levels increases the effectiveness of local inventories, making them more comprehensive (e.g., Ascorra et al. 1996; Bernard 1997; Bonaccorso 1979; Brosset & Charles-Dominique 1990; Handley 1967; Simmons & Voss 1998).

Unfortunately, small sample size is a common problem in studies of Neotropical forest bats and leaves conclusions about many species problematical. For instance, about half of all species in our study ($n = 24$) were captured in small numbers ($n < 6$). To minimize bias due to small sample size, we tested for statistical significance only in the more abundant species ($n \geq 6$). Lack of statistical power in our study, however, need not imply lack of stratification in species with small samples. For instance, although most emballonurids (aerial insectivores) were rarely caught ($n < 6$), most of them (6 of 7 species) were captured only in the canopy (Table 1). Thus, if data were grouped at the family level, those emballonurids would be classified as canopy bats. However, as we point out later, sampling bias also contributed to this pattern.

Vertical stratification

Our results demonstrate significant differences in capture rates between ground and canopy nets for ten of the better-sampled species, suggesting differences in habitat use and vertical stratification. Four species (two frugivores: *C. perspicillata*, *C. brevicauda*; one gleaning insectivore: *T. cirrhosus*; and one gleaning omnivore: *P. elongatus*) showed clear preferences for the understory ($p < 0.05$; Table 1). The other six species, including the blood-drinking bird vampire, *D. youngi*; four frugivores, *A. lituratus*, *U. bilobatum*, *C. trinitatum*, and *C. villosum*; and an aerial insectivore *S. leptura*, were caught at significantly higher frequencies in the canopy.

Many ($n = 15$) of the better sampled species in our study, however, were caught frequently at both ground and canopy levels, implying that they use all forest strata. These results question the importance of vertical stratification as a major component in structuring this community. Compared with birds, vertical distribution appears to be less pronounced in bats. Field studies in temperate and tropical regions show that forest bird communities are highly structured, with a strong vertical component (e.g., Brosset 1996; Cody 1996; Erard 1987, 1990). Most notably, many sibling species with similar morphology inhabit different strata, resulting in distinct spatial segregation of their niches.

Despite some differences in community composition due to geographic distributional limits of species and to differing netting strategies, general patterns of vertical distribution of bats at our Belém study site conform well with studies near Manaus, Brazil (Bernard 1997), in French Guiana (Brosset & Charles-Dominique 1990; Simmons & Voss 1998), and on Barro Colorado Island (BCI), the field station of the Smithsonian Tropical Research Institute in Panamá (Bonaccorso 1979).

At Belém and on BCI, *A. jamaicensis* was the most abundant bat, registering more than 60% of all captures, and high capture rates at both ground (52%) and canopy (48%) levels. *A. jamaicensis* was less common at the study sites in French Guiana and near Manaus, possibly correlated with a lower abundance of figs (Moraceae) at those sites, compared with BCI and Belém. Comparison of captures in nets at various levels in the forest confirm the preference of short-tailed fruit-bats (*Carollia* sp.) and gleaning insectivores/carnivores for the understory (Bernard 1997; Bonaccorso 1979; Brosset & Charles-Dominique 1990; Simmons & Voss 1998). A group of frugivores, particularly *A. lituratus*, was encountered mostly at the higher levels of the forest. Overall, our results at Belém support Bonaccorso's (1979) classification of fruit-eating bats as ground or canopy frugivores on the basis of predominant diet.

Comparison with other studies points to flexible use of forest strata within species as well as between closely-related species. For instance, Bonaccorso (1979) and Bernard (1997) found *P. hastatus*, a large omnivore, mostly in subcanopy nets and in canopy nets respectively. In contrast, we captured *P. hastatus* regularly in the understory during our long-term project on BCI (Kalko et al. 1996a) and together with its South American congener, *P. elongatus* in ground nets at Belém. The preference of *P. elongatus* for the understory is supported by Bernard (1997) near Manaus and Simmons and Voss (1998) in French Guiana.

For several reasons we believe classification of many taxa as ground or canopy inhabitants is still premature: a) capture rates of many species are very low, b) capture effort and mistnet configuration differ greatly in most studies that compare ground and canopy netting results, and c) placement of canopy mistnets (in openings and gaps *vs.* under closed canopy) varies among and within studies. Furthermore, classification of bats as ground and canopy species should not imply that they are limited to specific levels of the forest. In fact, as shown here with our mistnetting data on species as well as with the recapture data of individuals, most forest bats are rather free-ranging and use all levels of the forest. This is well illustrated by *A. jamaicensis* which occurs in great numbers at our study sites near Belém and on BCI. It feeds mostly on canopy fruit and thus is classified as a canopy frugivore (Bonaccorso 1979), but it often processes fruit and roosts in lower strata. Consequently it is commonly caught in both canopy and ground nets. It frequently flies so low that it is captured in the lowest shelf of ground nets (Kalko & Handley, pers. obs.). Similar observations have been reported by Ascorro et al. (1996), Bernard (1997), Brosset & Charles-Dominique (1990) and Simmons & Voss (1998). They showed rather high variability in the use of forest strata, with captures at both ground and canopy levels, for a variety of commonly mistnetted frugivorous species such as *A. lituratus*, *A. jamaicensis*, *A. obscurus*, and *R. pumilio*. For such species, assignments to forest levels on the basis of mistnetting results is clearly impossible.

Diet, foraging behavior, and roost sites

To interpret the observed patterns of habitat use by forest bats we must understand the underlying factors that cause variation. Most differences can be best explained by diet, foraging behavior, and roost site selection (see Law 1996; Wunder & Carey 1996). To illustrate this point, we selected several well-sampled species of frugivores, insectivores, carnivores, and sanguivores to serve as examples.

The short-tailed fruit-eating bats, *C. brevicauda* and *C. perspicillata*, which we caught mostly in ground nets (87% and 82% respectively), feed primarily on understory plants such as *Piper, Solanum*, and *Vismia* (e. g., Bonaccorso 1979; Charles-Dominique 1986; Fleming 1988). They carry fruit to temporary feeding roosts in the understory and commonly have day roosts near the ground in hollow trees. Most of the life cycle of these bats is completed in the lower stratum of the forest, but we must note that they also feed occasionally on canopy fruits (Bonaccorso 1979; Kalko et al. 1996a). This might explain the occasional captures of *Carollia* in our canopy rigs at Belém, but more likely those were bats that slipped in unnoticed while nets were down for removal of real canopy captures.

In contrast to *Carollia*, the frugivorous *A. lituratus*, *V. caraccioli*, *P. helleri*, *V. bidens*, and *U. bilobatum*, which we caught mostly in the canopy rigs, feed primarily on canopy fruits such as figs (*Ficus*: Moraceae) (e. g., Bonaccorso 1979; Gardner 1977; Handley et al. 1991; Kalko et al. 1996a,b) but their dining roosts are in the subcanopy (Handley et al. 1991). Day roost of *A. lituratus* are occasionally in foliage in the subcanopy but usually in dark recesses in the canopy (Morrison 1980). *V. caraccioli* also roosts in foliage in the subcanopy (Morrison 1980). Day roosts of *U. bilobatum* are situated mostly in the lower forest stratum, under palm fronds, cut leaves, and eaves of houses (e.g., Kalko & Handley, unpubl. data; Kunz et al. 1994). These species probably commute between day roosts and feeding areas mostly in the upper level of the forest. As with the groundstorey frugivores, separation between ground and canopy levels is not complete. This variability is well reflected in our capture data, which show that all of the better sampled canopy frugivores were also encountered in the ground nets, usually in lower numbers.

Distribution of food and roost sites may also affect habitat use of fruit-eating bats such as *R. pumilio*, *Artibeus cinereus* (Gervais), *Artibeus gnomus* (Handley), and *A. obscurus*, which we caught in moderate numbers in both forest levels. They are seen as fruit generalists feeding on a variety of understory and canopy fruit. Simmons & Voss (1998) found *A. obscurus* roosting under large pieces of exfoliating bark on tree trunks and under an unmodified leaf. The other two species of *Artibeus* and *R. pumilio* are reported to roost beneath cut leaves in the understory (e.g., Charles-Dominique 1993, Simmons and Voss 1998).

Because of its ubiquity and abundance, *A. jamaicensis* is the most notable frugivore that we caught in high numbers at both ground and canopy levels. It feeds mostly on canopy fruits but processes them in temporary dining roosts in the forest understory. Its maternity roosts are in tree holes, caves, and houses. Bachelors and subadults roost mostly in foliage in the understory. From day roosts and dining roosts, *A. jamaicensis* commutes to fruit trees through lower levels of the forest (Handley et al. 1991). Flying low in the forest may be advantageous for evading potential predators such as owls (Bonaccorso 1979; Handley et al. 1991), and it may facilitate detection of trees with ripe fruit. Perhaps scent cues from ripe fruit are easier to detect in the forest interior near the ground than above the canopy because of high humidity, relatively turbulence-free air, and fallen ripe fruit on the forest floor. Scent is an important cue for many fruit-eating bats for detection and localization of ripe fruit (Kalko *et al.* 1996b; Schnitzler & Kalko 1998).

Availability of food and roost sites may also impact the distribution and abundance of the well sampled gleaning insectivores, *T. cirrhosus* and *T. saurophila*. We caught *T. cirrhosus* exclusively in ground nets. It roosts in hollow tree trunks and roots, and hunts mostly for large insects and frogs (e.g., Bonaccorso 1979; Kalko et al. 1996a). Our radiotracking studies on BCI show that it forages along small streams and in swampy areas (Kalko et al. 1999). In contrast, *T. saurophila* may hunt throughout the forest profile. In Belém, we caught it in both canopy and ground levels, on BCI it is commonly caught in ground nets (Kalko et al. 1996a), and Bonaccorso (1979) found it in both understory and subcanopy. It roosts in hollow trees, mostly at ground level but occasionally higher up (Kalko et al. 1996a; Spehn pers. com.). It feeds mostly on large insects. Its close relative *T. bidens* preys on small birds as well as insects (Martuscelli 1995).

Distribution of prey (i.e., roosting birds) also may explain the captures of six blood-drinking bird vampires (*D. youngi)* in canopy rigs at Belém. Without canopy mistnetting, this species would not have been detected. On the other hand, most of the captures of the common vampire *D. rotundus* were at ground level. Its food consists largely of blood of mammals, which it approaches on or near the ground.

Factors affecting mistnet captures

As we have shown, number of species, capture frequencies, and habitat use are associated with ecological factors including diet, foraging behavior, and roost selection. Capture results are also influenced by methodology. Placement of mistnets, capture effort, time of year, weather, and moon cycle influence capture rates of bats and must be taken into account when interpreting results of mistnetting studies. For instance, the fruit-eating *C. villosum* and *U. bilobatum* were caught in small numbers, mostly in canopy rigs at Belém and near Manaus (Bernard 1997) and in subcanopy nets during Bonaccorso's (1979) study on BCI. With this evidence, *C. villosum* and *U. bilobatum* could be considered to be rare canopy species. However, in our long-term study on BCI, both species were common to abundant in our ground netting samples. In fact, *U. bilobatum* was the third most abundant species, with 1957 captures, and *C. villosum* was fifth most abundant, with 722 captures (Handley et al. 1991; Kalko et al. 1996a). This discrepancy is in part the result of different sampling regimes. Bonaccorso (1979) sampled monthly on a grid and Bernard (1997) sampled on a monthly regime at fixed sites. During our study on BCI, most sampling was targeted on fruiting trees, particularly figs, which form a large portion of the diets of *C. villosum* and *U. bilobatum*. Such a netting strategy is more likely to coincide with flocks of bats. Thus, on several nights on BCI, surges of *C. villosum* (up to 90 individuals per night) and *U. bilobatum* (up to 138 individuals per night) were caught in ground nets around trees with ripe fruit (Handley et al. 1991). Year-round data from BCI and nearby mainland sites indicate further that numbers of both species may fluctuate seasonally (Bonaccorso 1979; Fleming et al. 1972; Handley et al. 1991; Kalko et al. 1996a). There may be similar seasonal fluctuations at Belém.

To obtain a comprehensive sample of phyllostomids with mistnetting requires sufficient time, since number of species detected increases with sampling effort (Kalko et al. 1996a; Simmons & Voss 1998; Voss & Emmons 1996; Wilson et al. 1996). For instance, on BCI it took an average of about 30 mistnetting nights with ten 12-m mistnets open for at least half of the night to record all of the common and some of the less common core species (Kalko et al. 1996a). With substantial capture effort, mistnetting is likely to fairly accurately portray distribution and relative abundance of the phyllostomid fraction of a Neotropical bat community.

Another factor that affects foraging behavior and ultimately capture rates of bats is moon phase. Some species, particularly canopy frugivores, exhibit lunarphobia (e.g., Morrison 1978, 1980). These bats drastically reduce activity around full moon, presumably to avoid predators such as owls, but other species such as the short-tailed fruit-bats *(Carollia)* that fly mainly in the understory, are less affected by moonlight (e.g., Fleming & Heithaus 1986; Thies 1998).

Capture frequency, sampling bias, and inventory methods

Characteristically, rank abundance curves of Neotropical bats based on mistnetting are composed of a few frequently netted species and a large number of rarely caught species (e.g., Fleming 1986; Kalko et al. 1996a; Simmons & Voss 1998). To better understand how accurately capture frequencies reflect relative abundance, we discuss possible causes for the rareness of species in the samples.

Some species such as top predators (*C. auritus*, *V. spectrum*) are intrinsically rare. Presumably because of their rather sedentary feeding mode, small home ranges, and specific habitat, food, and roost requirements, gleaning insectivores are less commonly caught, and presumably are less abundant, than the highly mobile and species-rich guild of Neotropical frugivores (e.g., Bonaccorso 1979; Fleming et al. 1972; Kalko 1997; Kalko et al. 1999; Simmons & Voss 1998).

Diversity and abundance patterns of bats in local communities also reflect differences in ecological conditions. Lack of adequate roost sites and foraging habitats may be responsible in part for the complete absence, in our mistnetted sample, of five families of bats which based on their distribution patterns, should be present in our study area at Belém. These include mustached bats (Mormoopidae), bulldog bats (Noctilionidae), funnel-eared bats (Natalidae), smoky bats (Furipteridae), and disk-winged bats (Thyropteridae). Natalids and most mormoopids roost preferentially in caves, but there are no caves in the immediate vicinity of our study area. We do not know why the mustached bat, *Pteronotus parnellii* (Gray), is missing. Although this species is well known for its cave roosting habits, it is also common in forests without caves (Kalko et al. 1996a; Simmons & Voss 1998). Disk-winged bats are seldom captured in mistnets, but we found one roosting in a furled *Heliconia* leaf in our study area. The absence of the insect- and fish-eating bulldog bats in our samples is not surprising since we set no mistnets over larger streams.

Sampling method is clearly the primary reason for the total absence in our study of the guild of open space aerial insectivores, and the low number of captures in the guild of 'background cluttered space aerial insectivores'. Firstly, our high nets were set under the closed canopy, so they did not capture aerial insectivores that were foraging above the canopy and in forest gaps. In contrast, nets set high in forest gaps and in road corridors in the rain forest in French Guiana and near Manaus, Brazil, sampled aerial insectivores, including free-tailed bats (Molossidae) and a ghost-bat *Diclidurus scutatus* (Peters) (Emballonuridae) foraging in open space (Bernard 1997; Brosset & Charles-Dominique 1990; Simmons & Voss 1998).

Secondly, even under the most favorable circumstances, mistnetting adequately samples only part of local bat faunas. In the Neotropics, that part is the phyllostomids. Using their sophisticated echolocation system, aerial insectivores readily detect and avoid mistnets. Moreover, many open space aerial insectivores fly so high above the ground or canopy that they are out of reach of mistnets, even when they are set high as in the study of Simmons & Voss (1998). At our study site, based on patterns of geographic distribution, we may have missed as many as 21 species of free-tailed bats (Molossidae) and 8 species of sheath-tailed bats (Emballonuridae) of the guild of 'open space, aerial insectivores'. In the guild of 'background cluttered space, aerial insectivores', we caught only three of the 15 species that are likely to be present.

To overcome this sampling limitation, ultrasound detectors are being used increasingly for inventories of bats and in studies of habitat use (e.g., Fenton & Bell 1981; Fenton et al. 1983; Mills et al. 1996; O'Farrell & Miller 1999). Based on characteristic signal patterns and call structure, many Neotropical aerial insectivores can be identified to genus and often to species (e.g., Kalko 1995; O'Farrell & Miller 1997, 1999). As a prerequisite, a reliable identification key to signals must be compiled. To adequately interpret the acoustic data, a thorough understanding of call variability within and among species in response to environmental conditions (e.g., distance to obstacles and conspecifics, height over ground, foraging mode) is a necessity (e.g., Barclay 1999; Fenton 1990; Neuweiler 1983; Schnitzler & Kalko 1998).

Although acoustic monitoring is a powerful tool, it must be used with care for inventorying bats (see Barclay 1999; O'Farrell et al. 1999). Effective acoustic monitoring requires appropriate equipment, knowledge of bioacoustics and echolocation behavior, as well as detailed field observations. Signal design may vary with habitat structure (e.g., Fenton 1995; Neuweiler 1983; Schnitzler and Kalko 1998), and in some instances, signal shape may be almost identical in morphologically similar bats that live under similar ecological conditions. This makes species identification based solely on differences in signal structure difficult or impossible. Furthermore, acoustic monitoring is not feasible for species such as 'whispering' bats, phyllostomids which emit very faint and uniform signals (e.g., Belwood 1988; Schnitzler & Kalko 1998).

With ultrasonic-sampling methods in place, several studies have described habitat use and resource partitioning of aerial insectivores. First results recognize horizontal partitioning as well as vertical stratification as important factors structuring bat communities (e.g., Aldridge & Rautenbach 1987; Carmel & Safriel 1998; Crome & Richards 1988; Fullard et al. 1991; McKenzie & Rolfe 1986; McKenzie et al. 1995). These studies show that habitat complexity, such as degree of openness of the forest understory and canopy, and size and number of forest gaps, profoundly influence community structure and diversity of aerial insectivores.

Few studies have given descriptions of echolocation behavior and habitat use of Neotropical aerial insectivores (e.g., Barclay 1983; Bradbury & Vehrencamp 1976; Kalko 1995; O'Farrell & Gannon 1999; O'Farrell & Miller 1997, 1999). Those studies and first results from our own ongoing studies in Panamá, Venezuela, and Brazil show that diversity and relative abundance of aerial insectivores is much greater in local communities than mistnetting has indicated. For instance, although mistnets do not catch many emballonurids such as *Saccopteryx* sp., with acoustic monitoring, we have found them to be common and foraging in the understory as well as beneath the canopy (see Bradbury & Vehrencamp 1976; Kalko 1995).

Implications of bat studies for conservation

The indispensable roles tropical bats play in maintaining forest diversity and promoting regeneration processes through seed dispersal, pollination, and reduction of insect populations is being increasingly recognized (e.g., Findley 1993; Fleming 1993; Foster et al. 1986; Kalko 1997; Parotta 1995). However, little is known of effects of habitat alterations on structure and function of bat communities in tropical forests. The intense use of all forest levels from groundstorey to canopy level by bats as seen in our study at Belém, and in the comprehensive studies of Bonaccorso (1979), Brosset & Charles-Dominique (1990), Bernard (1997), Cosson et al. (1999) and Simmons & Voss (1998) suggest that the integrity and complexity of a forest are important factors influencing composition, abundance, and probably also function of bat communities.

Changes in availability and abundance of food resources, exemplified by the explosion of secondary growth shrubs at forest edges, the reluctance of gleaning insectivores to cross open spaces, and the specialized roosting habits of some species, are likely to contribute to patterns stemming from habitat disturbance (e.g., Brosset et al. 1996; Fenton et al. 1992; Charles-Dominique 1986; Kalko et al. 1999). Similar patterns have been found in a study in the temperate zone, where various species of insectivorous bats reacted differently to habitat patchiness and some species avoided open habitats (e.g., Johannson & de Jong 1996). However, it is still unclear what role forest canopies play in these scenarios.

Published accounts of studies with mistnetting at ground level comparing bats of habitats with varying degrees of disturbance in the Neotropics point towards distinct differences in species composition and relative abundance of phyllostomids. (e.g., Ascorra et al. 1996; Brosset et al. 1996; Estrada et al. 1993; Fenton et al. 1992; Simmons & Voss 1998; Vaughan & Hill 1996; Wilson et al. 1996). Gleaning insectivores are mostly associated with less-disturbed and pristine forest habitats. A high abundance of ground story frugivores such as the species of *Carollia* is indicative of disturbed areas.

Contrary to the observation that increased structural diversity leads to a much higher diversity in species of birds, studies comparing species richness and activity of aerial insectivorous bats in habitats with differing degrees of complexity and disturbance in savannah/woodlands in Africa and in Queensland, Australia found less pronounced differences in species richness and activity than was expected (Cumming et al. 1997; Fenton et al. 1998; Kutt 1995; Rautenbach et al. 1996). Fenton et al. (1998) concluded that aerial insectivores probably are not good indicators of habitat disturbance, but a study in Australia pointed toward possible correlations between diversity and distribution of aerial insectivores and environmental disturbances (McKenzie & Start 1989).

Overall, studies of effects of habitat alterations on bats are few, and our notions of factors that shape bat communities in space and time are just beginning to take form. Notably lacking are studies of the role of forest canopies in structuring bat communities. Promising approaches for future studies are falling into place: (1) Advances in sampling techniques to achieve comprehensive inventories, using a combination of methods such as ground and canopy mistnetting, acoustic monitoring, and searches for roost sites (see Fenton 1997; O'Farrell & Gannon 1999; Simmons & Voss 1998; Voss & Emmons 1996). (2) Easier access to places that have been difficult to sample, including access to the canopy by means of walkways and crane systems (see also papers on other taxa in this volume). (3) Standardization of capture protocols to reduce the effects of some of the variables influencing capture success (e.g., differences in capture effort and mistnet configuration) to facilitate comparison between sites. (4) Behavioral studies of selected species leading to a better understanding of their ecological requirements. (5) An increasing understanding of the association between environmental factors such as distribution and availability of resources (i.e., food, roost sites), habitat complexity, and disturbance level on the structure and dynamics of communities. (6) Application of modern modeling tools such as GIS that allow predictions of distribution, abundance, and community patterns (see Yom-Tov & Kadmon 1998).

These approaches are crucial given the ever increasing loss of undisturbed habitats, the lack of adequate information about the role of tropical forest canopies for the maintenance of diversity, and the need to develop and apply conservation-oriented plans for land use and management of protected areas.

Acknowledgements

For generous support in the field we are indebted to John P. Woodall (former director of Belem Virus Laboratory, Rockefeller Foundation), Thomas H. C. Aitken (formerly Rockefeller Foundation), Yoshika Oniki (formerly Universidade de Sao Paulo), Philip S. Humphrey (formerly Smithsonian Institution), Thomas E. Lovejoy III (formerly Yale University), and Lester A. Hart (formerly Virginia Tech). We thank Sonja Tejada for her untiring work in computerizing the data, Darelyn Handley for constantly monitoring and advising on the manuscript, Bob Stallard (USGS, Boulder, CO) for valuable discussion and suggestions on the manuscript, Enrico Bernard (presently University of York, Canada) for stimulating discussions about the results of his masters' thesis, Björn Siemers for helpful comments, and H.-U. Schnitzler (University of Tübingen, Germany) for his continuous support of work and thoughts. Financial support came from an habilitation stipend from the Deutsche Forschungsgemeinschaft (DFG) to Kalko, a DFG grant Schn 138/22-1 to Schnitzler and Kalko, Smithsonian Institution Research Awards to Handley, and from Soundprints, Inc., through the Smithsonian Office of Product Development and Licensing.

References

Aldridge, H. D. J. N. & Rautenbach, I. R. 1987. Morphology, echolocation and resource partitioning in insectivorous bats. J. Exp. Biol. 128: 419–425.

Ascorra, C., Solari, S. & Wilson, D. E. 1996. Diversidad y ecología de los quirípteros en Pakitza. Pp. 593–612. In: Wilson, D. E. & Sandoval, A. (eds), Manu. The biodiversity of Southeastern Peru. Editorial Horizone, Lima.

Barclay, R. M. R. 1983. Echolocation calls of emballonurid bats in Panama. J. Comp. Physiol. 151: 51–520.

Barclay, R. M. R. 1999. Bats are not birds – a cautionary note on using echolocation calls to identify bats: a comment. J. Mamm. 80: 290–296.

Bell, H. L. 1982. A bird community of lowland rain forest in New Guinea. III. Vertical stratification of the avifauna. Emu 82: 143–162.

Belwood, J. J. 1988. Foraging behavior, prey selection, and echolocation in phyllostomine bats (Phyllostomidae). Pp. 610–605. Nachtigall P. E. & P. W. B. Moore (eds), Animal sonar: processes and performance. Plenum Press, New York.

Bernard, E. 1997. Vertical stratification of bat communities in tree fall gaps in primary forest of Central Amazonia. Pp. 1–80. Unpublished MSc thesis, Brazilian National Institute for Research in the Amazon – INPA, Manaus.

Bonaccorso, F. J. 1979. Foraging and reproductive ecology in a Panamanian bat community. Bulletin of the Florida State Museum, Biol. Sci. 4: 359–408.

Bradbury, J. W. & Vehrencamp, S. L. 1976. Social organization and foraging in emballonurid bats. I. Field studies. Behav. Ecol. Sociobiol. 1: 337–381.

Brosset, A. 1996. Role of sibling species in the dynamics of the forest-bird community in M'Passa (Northeastern Gabon). Pp. 251–290. In: Cody, M. L. & Smallwood, J. A. (eds), Long-term studies of vertebrate communities. Academic Press, Los Angeles.

Brosset, A. & Charles-Dominique, P. 1990. The bats from French Guiana: a taxonomic, faunistic and ecological approach. Mammalia 54: 509–560.

Brosset, A., Charles-Dominique, P., Cockle, A., Cosson, J. F. & Masson, D. 1996. Bat communities and deforestation in French Guiana. Can. J. Zool. 74: 1974–1982.

Carmel, Y. & Safriel, U. 1998. Habitat use by bats in a Mediterranean ecosystem in Israel – conservation implications. Biol. Cons. 84: 245–250.

Charles-Dominique, P. 1986. Inter-relations between frugivorous vertebrates and pioneer plants: *Cecropia*, birds and bats in French Guiana. Pp. 119–136. In: Estrada, A. & Fleming, T. H. (eds), Frugivores and seed dispersal. Junk, Dordrecht, Boston.

Charles-Dominique, P. 1993. Tent use by the bat *Rhinophylla pumilio* (Phyllostomidae: Carolliinae) in French Guiana. Biotropics 25: 111–116.

Cody, M. L. 1996. Bird communities in the Rocky Mountains. Pp. 291–342. In: Cody, M. L., & Smallwood, J. A. (eds), Long-term studies of vertebrate communities. Academic Press, Los Angeles.

Cosson, J. F. 1995. Captures of *Myonycteris torquata* (Chiroptera: Pteropodidae) in forest canopy in South Cameroon. Biotropics 27: 395–396.

Cosson, J. F., Pons, J. M., & Masson, D. 1999. Effects of forest fragmentation on frugivorous and nectarivorous bats in French Guiana. J. Trop. Ecol. 15: 515–534.

Crome, F. H. & Richards, G. C. 1988. Bats and gaps: microchiropteran community structure in a Queensland rain forest. Ecol. 69: 1960–1969.

Cumming, D. H. M., Fenton, M. B., Rautenbach, I. L., Taylor, R. D., Cumming, G. S., Cumming, M. S., Dunlop, J. M., Ford, A. G., Hovorka, M. D., Jonston , D. S., Kalcounis, M., Mahlangu, Z. & Portfors, C. V. R. 1997. Elephants, woodlands and biodiversity in southern Africa. South African J. Sci. 93: 231–236.

Erard, C. 1987. Ecologie et comportment des Gobemouches (Aves: Muscicapinae, Platysteirinae, Monarchinae) du Nord-Est du Gabon. Vol. I. Morphologie des espèces et organisation du peuplement. Mém. Mus. Natl. Hist. Nat. Ser. A (Paris) 138: 1–256.

Erard, C. 1990. Ecologie et comportment des Gobemouches (Aves: Muscicapinae, Platysteirinae, Monarchinae) du Nord-Est du Gabon. Vol. II. Organisation sociale et reproduction. Mém. Mus. Natl. Hist. Nat. Ser. A (Paris) 146: 1–233.

Estrada, A., R. Coates-Estrada & Meritt, D. 1993. Bat species richness and abundance in tropical rain forest fragments and in agricultural habitats at Los Tuxtlas, Mexico. Ecography 16: 309–318.

Fenton, M. B. 1990. The foraging behavior and ecology of animaleating bats. Can. J. Zool. 68: 411–422.

Fenton, M. B. 1995. Natural history and biosonar signals. Pp. 17–86. In: Fay, R. R. & A. N. Popper (eds), Hearing by bats. Springer Handbook of auditory research. Springer-Verlag, Heidelberg.

Fenton, M. B. 1997. Science and the conservation of bats. J. Mamm. 78: 1–14.

Fenton, M. B., Acharya, L., Audet, D., Hickey, M. B. C., Merriman, C., Obrist, M. K., Syme, D. M. & Adkins, B. 1992. Phyllostomid bats (Chiroptera: Phyllostomidae) as indicators of habitat disruption in the Neotropics. Biotropics 24: 440–446.

Fenton, M. B. & Bell, P. G. 1981. Recognition of species of insectivorous bats by their echolocation calls. J. Mamm. 62: 233–243.

Fenton, M. B., Cumming, D. H. M., Rautenbach, I. L., Cumming, G. S., Cumming, M. S., Ford, G., Taylor, R. D., Dunlop, J., Hovorka, M. D., Johnston, D. S., Portfors, C. V., Kalcounis, M. C. & Mahlanga, Z. 1998. Bats and the loss of tree canopy in African woodlands. Cons. Biol. 12: 399–407.

Fenton, M. B., Merriam, H. G. & Holroyd, G. L. 1983. Bats of Kootenay, Glacier, and Mount Revelstoke national parks in Canada: identification by echolocation calls, distribution, and biology. Can. J. Zool. 61: 2506–2508.

Findley, J. S. 1993. Bats: a community perspective. Cambridge, University Press, Cambridge, 167 pp.

Fleming, T. H. 1986. The structure of Neotropical bat communities: a preliminary analysis. Rev. Chil. Hist. Nat. 59: 135–150.

Fleming, T. H. 1988. The short-tailed fruit bat: a study in plant-animal interactions. University of Chicago Press, Chicago, 365 pp.

Fleming, T. H. 1993. Plant-visiting bats. Am. Sci. 81: 460–467.

Fleming, T. H. & Heithaus, E. R. 1986. Seasonal foraging behavior of the frugivorous bat *Carollia perspicillata*. J. Mamm. 67: 660–671.

Fleming, T. H., Hooper, E. T. & Wilson, D. E. 1972. Three Central American bat communities: structure, reproductive cycles and movement patterns. Ecol. 53: 555–569.

Foster, R. B., Javier, A. C. & Tatzyana, S. W. 1986. Dispersal and sequential plant communities in Amazonian Peru floodplain. Pp. 357–370. In: Estrada, A. & Fleming, T. H. (eds), Frugivores and seed dispersal. Junk, Dordrecht.

Francis, C. M. 1989. A comparison of mistnets and two designs of harp traps for capturing bats. J. Mamm. 70: 865–870.

Francis, C. M. 1990. Trophic structures of bat communities in the understory of lowland dipterocarp rain forest in Malaysia. J. Trop. Ecol. 6: 421–431.

Francis, C. M. 1994. Vertical stratification of fruit bats in lowland rain forests of Malaysia. J. Trop. Ecol. 10: 523–530.

Fullard, J. H., Koehler, C., Surlykke, A. & McKenzie, A. L. 1991. Echolocation ecology and flight morphology of insectivorous bats (Chiroptera) in South-western Australia. Australian J. Zool. 39: 427–438.

Gardner, A. L. 1977. Feeding habits. Pp. 1–364. In: Baker, R. J., Jones, J. K., Jr. & Carter, D. C. (eds), Biology of bats of the New World family Phyllostomidae, Part II. Spec. Publ. Mus. Texas Tech. Univ., p. 13.

Gaskell, B. H. 1984. Flying fruit bat faunas of the upper canopy in two palaeotropical rain forests. Pp. 303. In: Chadwick, A. C. & Sutton, S. L. (eds), Tropical forests: The Leeds Symposium, Leeds Phil. Lit. Soc. Leeds.

Handley, C. O., Jr. 1967. Bats of the canopy of an Amazonian forest. Atas do Simpósio sôbre a Biota Amazônica 5: 211–215.

Handley, C. O. Jr., Wilson, D. E. & Gardner A. L. 1991. Demography and natural history of the Common Fruit Bat, *Artibeus jamaicensis*, on Barro Colorado Island, Panamá. Smithsonian Contributions to Zoology 511, pp. 173.

Humphrey, P. S., Bridge, D. & Lovejoy, T. E. III. 1968. A technique for mist-netting in the forest canopy. Bird-banding 39: 43–50.

Ingle, N. R. 1993. Vertical stratification of bats in a Fallopian rain forest. Asia Life Sci. 2: 215–222.

Johannson, M. & Jong, J.de. 1996. Bat species diversity in a lake archipelago in central Sweden. Biodiv. Cons. 5: 1221–1229.

Kalko, E. K. V. 1995. Echolocation signal design, foraging habitats, and guild structure in six Neotropical sheath-tailed bats (Emballonuridae). Symp. Zool. Soc. London 67: 259–273

Kalko, E. K. V. 1997. Diversity in tropical bats. Pp. 13–43. In: Ulrich, H. (ed.), Tropical biodiversity and Systematics. Zool. Forschungsinstitut und Museum Alexander Koenig, Bonn.

Kalko, E. K. V., Handley, C. O. & Handley, D. 1996a. Organization, diversity, and long-term dynamics of a Neotropical bat community. Pp. 503–553. In: Cody, M. & Smallwood, J. (eds), Long-term studies in vertebrate communities. Academic Press, Los Angeles.

Kalko, E. K. V., Herre, E. A. & Handley, C. O. 1996b. The relation of fig fruit syndromes to fruit-eating bats in the New and Old World tropics. J. Biogeogr. 23: 565–576.

Kalko, E. K. V., Krull, D., Handley, C. O. & Schnitzler, H. U. 1999. Foraging areas, activity patterns, and roost sites of two Neotropical gleaning bats, *Tonatia silvicola* and *Trachops cirrhosus* (Phyllostomidae). Biotropics 31: 344–353.

Koopman, K. F. 1993. Order Chiroptera. Pp. 137–241. In: Wilson, D. E. & Reeder, D. M. (eds), Mammal species of the World, a taxonomic and geographic reference, 2nd edition, Smithsonian Institution Press, Washington, D.C.

Kunz, T. H., Fujita, M. S., Brooke, A. P. & McCracken, G. F. 1994. Convergence in tent architecture and tent-making behavior among Neotropical and palaeotropical bats. J. Mamm. Evol. 2: 57–78.

Kutt, A. S. 1995. Activity and stratification of microchiropteran bat communities in thinned, unthinned, and old lowland regrowth forest, East Gippsland. Victorian Nat. 112: 86–92.

Law, B. S. 1996. The ecology of bats in south-east Australian forests and potential impacts of forestry practices: a review. Pacific Conservation Biology 2: 363–374.

Lovejoy, T. E. III. 1967. Birds and viruses in lower Amazonia. EBBA News 30 (3), 8.

Martuscelli, P. 1995. Avian predation by the round-eared bat (*Tonatia bidens*, Phyllostomidae) in the Brazilian Atlantic forest. J. Trop. Ecol. 11: 461–464.

McKenzie, N. L., Gunnell, A. C., Yani, M. & Williams, M. R. 1995. Correspondence between flight morphology and foraging ecology in some palaeotropical bats. Australian J. Zool. 43: 241–257.

McKenzie, N. L. & Rolfe, J. K. 1986. Structure of bat guilds in the Kimberley mangroves, Australia. J. Anim. Ecol. 55: 401–420.

McKenzie, N. L. & Start, A. N. 1989. Structure of bat guilds in mangroves: environmental disturbances and determinism. Pp. 167–178. In: Morris, D. W., Abramsky, Z., Fox, B. J. & Willig, M. R. (eds), Patterns in the structure of mammalian communities. Texas Tech University Press, Lubbock.

Mills, D. J., Norton, T. W., Parnaby, H. E., Cunningham, R. B. & Nix, H. A. 1996. Designing surveys for microchiropteran bats in complex forest landscapes – a pilot study from south-east Australia. Forest Ecol. Manag. 85: 149–161.

Morrison, D. W. 1978. Lunar phobia in a Neotropical fruit bat, *Artibeus jamaicensis* (Chiroptera: Phyllostomidae). Anim. Behav. 26: 852–855.

Morrison, D. W. 1980. Foraging and day-roosting dynamics of canopy fruit bats in Panamá. J. Mamm. 61: 20–29.

Neuweiler, G. 1983. Echolocation and adaptivity to ecological constraints. Pp. 280–302. In: Huber, F. & Markl, H. (eds), Neuroethology and behavioral physiology. Springer-Verlag, Berlin.

Norberg, U. M. & Rayner, J. M. V. 1987. Ecological morphology and flight in bats (Mammalia: Chiroptera): wing adaptations, flight performance, foraging strategy and echolocation. Phil. Trans. Roy. Soc. Lond. 316: 335–427.

O'Farrell, M. J., Corben, C., Gannon, W. L. & Miller, B. W. 1999. Confronting the dogma: a reply. J. Mamm. 80: 297–302.

O'Farrell, M. J. & Gannon, W. L. (1999) Qualitative identification of free-flying bats using the Anabat detector. J. Mamm. 80: 11–23.

O'Farrell, M. J. & Miller, B. W. 1997. A new examination of echolocation calls of some Neotropical bats (Emballonuridae and Mormoopidae). J. Mamm. 78: 954–963.

O'Farrell, M. J. & Miller, B. W. 1999. Use of vocal signatures for the inventory of free-flying Neotropical bats. Biotropics 31: 507–516.

Parrotta, J. 1995. Influence of overstory composition on understory colonization by native species in plantations on a degraded tropical site. J. Veg. Sci. 6: 627–636.

Pearson, D. L. 1971. Vertical stratification of birds in a tropical dry forest. Condor 73: 46–55.

Pearson, D. L. 1975. The relation of foliage complexity to ecological diversity of three Amazonian bird communities. Condor 77: 453–466.

Rautenbach, I. L., Fenton, M. B. & Whiting, M. J. 1996. Bats in riverine forests and woodlands: a latitudinal transect in southern Africa. Can. J. Zool. 74: 312–322.

Reis, N. R. dos & Muller, M. F. 1995. Bat diversity of forests and open areas in a subtropical region of South Brazil. Ecol. Austr. 5: 31–36.

Root, R. B. 1967. The niche exploitation pattern of the blue-gray gnatcatcher. Ecol. Monogr. 37: 317–350.

Roth, R. 1976. Spatial heterogeneity and bird species diversity. Ecol. 57: 773–782.

Schnitzler, H. -U. & Kalko, E. K. V. 1998. How echolocating bats search for food. Pp. 183–196. In: Kunz, T. H. & Racey, P. A. (eds), Bats: phylogeny, morphology, echolocation, and conservation biology. Smithsonian Institution Press, Washington, D. C.

Simmons, N. B. & Voss, R. S. 1998. The mammals of Paracou, French Guiana: a Neotropical lowland rain forest fauna. Part 1. Bats. Bull. Am. Mus. Nat. Hist. 237: 1–219.

Thies, W. 1998. Resource and habitat use in two frugivorous bat species (Phyllostomidae: *Carollia perspicillata* and *C. castanea*) in Panamá: Mechanisms of coexistence. Pp. 1–181. PhD thesis, University of Tübingen.

Vaughan, N. & Hill, J. E. 1996. Bat (Chiroptera) diversity and abundance in banana plantations and rain forest, and three new records for St. Vincent, Lesser Antilles. Mammalia 60: 441–447.

Voss, R. S. & Emmons, L. E. 1996. Mammalian diversity in Neotropical lowland rain forests: a preliminary assessment. Bull. American Mus. Nat. Hist. 230: 1–115.

Willig, M. R. 1986. Bat community structure in South America: a tenacious chimera. Rev. Chil. Hist. Nat. 59: 151–168.

Wilson, D. E., Ascorra, C. F. & Solari, T. S. 1996. Bats as indicators of habitat disturbance. Pp. 613–625. In: Wilson, D. E. & Sandoval, A. (eds), Manu. The biodiversity of Southeastern Peru. Editorial Horizonte, Lima.

Wunder, L. & Carey, A. B. 1996. Use of the forest canopy by bats. Northwest Sci. 70: 79–85.

Yom-Tov, Y. & Kadmon, R. 1998. Analysis of the distribution of insectivorous bats in Israel. Diversity Distrib. 4: 63–70.

Appendix 1. Checklist of bats netted in the APEG forest reserves, Belém, Brazil, 1965–1969. Order of families and subfamilies follows Koopman (1993). Genera and species are listed alphabetically.

Emballonuridae
Cormura brevirostris
Peropteryx macrotis
Rhynchonycteris naso
Saccopteryx bilineata
Saccopteryx canescens
Saccopteryx gymnura
Saccopteryx leptura
Phyllostomidae
Phyllostominae
Chrotopterus auritus
Glyphonycteris daviesi
Lampronycteris brachyotis
Micronycteris minuta
Mimon crenulatum
Phyllostomus elongatus
Phyllostomus hastatus
Tonatia brasiliensis
Tonatia carrikeri
Tonatia saurophila
Tonatia silvicola
Trachops cirrhosus
Trinycteris nicefori
Vampyrum spectrum
Lonchophyllinae
Lionycteris spurrelli
Lonchophylla thomasi
Glossophaginae
Choeroniscus minor
Glossophaga soricina
Lichonycteris degener
Carolliinae
Carollia brevicauda
Carollia perspicillata
Rhinophylla fischerae
Rhinophylla pumilio
Stenodermatinae
Ametrida centurio
Artibeus cinereus
Artibeus gnomus
Artibeus jamaicensis
Artibeus lituratus
Artibeus obscurus
Chiroderma trinitatum
Chiroderma villosum
Mesophylla macconnelli
Platyrrhinus helleri
Sturnira lilium
Sturnira tildae
Uroderma bilobatum
Uroderma magnirostrum
Vampyressa bidens
Vampyrodes caraccioli
Desmodontinae
Desmodus rotundus
Diaemus youngi
Vespertilionidae
Eptesicus chiriquinus
Lasiurus egregius
Myotis riparius

Plant Ecology **153:** 335–345, 2001.

Reproductive ecology of tropical forest trees in logged and fragmented habitats in Thailand and Costa Rica

Jaboury Ghazoul & Moray McLeish*
*TH Huxley School of Environment, Earth Science and Engineering, Imperial College of Science, Technology and Medicine, Silwood Park, Ascot, Berks SL5 7PY, UK (e-mail: j.ghazoul@ic.ac.uk); (*Current address: Institute of Ecology and Resource Management, The University of Edinburgh, West Mains Rd., Edinburgh EH9 3JG)*

Key words: Anacardium excelsum, Canopy biology, Costa Rica, Dry deciduous Forest, Dipterocarpaceae, Forest fragmentation, Functional redundancy, Logging, Pollination, Tree regeneration, *Trigona*, *Shorea siamensis*, Thailand

Abstract

Invertebrates mediate several important ecological processes, including pollination and seed predation, and events that affect invertebrate diversity or behaviour can potentially disrupt forest regeneration processes. This study investigates the impact of logging in Thailand and forest fragmentation in Costa Rica on the pollination and seed production of two self-incompatible forest trees. Logging in a dry deciduous dipterocarp forest in Thailand resulted in reduced densities of the common dipterocarp tree *Shorea siamensis* and variably isolated individual trees. The number of flower visits to *S. siamensis* by pollinating *Trigona* bees was not affected by logging disturbance. However, pollinators did spend longer periods of time foraging in the canopies of isolated trees which were more prevalent in logged areas where tree density had been reduced. Consequently, at the logged site few cross-pollinations were effected and fruit set of *S. siamensis* was considerably lower than at nearby unlogged sites where distances between flowering conspecifics were smaller. Reduced fruit set has long-term implications for the recovery of *S. siamensis* populations in disturbed areas, and local population genetic structure is likely to be affected as reduced outcrossing rates among trees in disturbed regions results in relatively inbred seed. In Costa Rica forest fragmentation has restricted the once widespread tree *Anacardium excelsum* to forest patches located in an agriculturally-dominated landscape. As with *S. siamensis*, the abundance of pollinators, also *Trigona* bees, in the canopies of *A. excelsum* was largely unaffected by fragment size. Nevertheless, pollination success and seed production was positively correlated with fragment size. We propose that small bees rarely move between forest fragments and gene exchange through pollination occurs predominantly among trees within fragments and, together with likely low genetic variability in small fragments, that this contributes to the observed reduced fertilisation and seed set of *A. excelsum*. Thus increased tree isolation tree through selective logging or habitat fragmentation by forest clearance can result in reduced seed set due to changes in the foraging patterns of poorly mobile pollinators. Even if population sizes of the pollinators are maintained following environmental perturbation, this study shows that disturbance may disrupt pollination processes through changes in pollinator foraging behaviour. More attention needs to be focussed on changes in the behaviour of species involved in key ecological interactions following disturbance events in tropical forests.

Introduction

Anthropogenic disturbance through tree extraction and forest fragmentation is affecting forests globally, but its biological impacts is most severe in tropical forests where community relationships and species connectivities are complex and poorly understood. Recent research on the impact of anthropogenic disturbance on tropical forest dynamics has focussed on ecological interactions between species that contribute to ecosystem function and structure including, for example, plant regeneration (Aizen & Feinsinger 1994; Ghazoul et al. 1998; Ghazoul, in press). However, the impacts of disturbance on the reproductive output of tropical forest trees remain largely unknown due to the complexity of plant-pollinator interactions and the practical difficulties of quantifying these interactions in the forest canopy.

Tropical trees are mostly self-incompatible and are generally dependent on animal-mediated pollination for seed production (Bawa 1974, 1990). Disturbance that affects vectors of pollen transfer can therefore have an important impact on the reproductive output of tree species. Pollination processes can be disrupted by a decline in pollinator abundance caused, for example, by environmental change (Vinson et al. 1993), resource availability (Schaal 1980; Jennersten 1988), decline in pollinator visitation frequency due to changes in the distribution of floral resources (Real et al. 1983; Rathcke 1983), or competitive exclusion from floral resources by inefficient pollinating species (Huryn 1997). Different intensities of extraction of *Shorea siamensis* Miq. trees between sites in a dry deciduous dipterocarp forest in Thailand in the 1980s provides an opportunity to monitor the impacts of disturbance on the reproductive output of this species. Likewise, fragmentation of seasonally dry forest in Guanacaste, Costa Rica, resulting in differential isolation of *Anacardium excelsum* Skeels trees allows an investigation of fragmentation impacts on tree reproductive output and comparisons with the ecologically similar *Shorea siamensis*.

Factors likely to determine the scale of indirect community responses to disturbance are:

- the number of interactions between the target species and other species in the community,
- the strength of those interactions, and
- the relative abundance of the target species in the community.

Thus communities that are most susceptible to environmental perturbations are those where there is little functional redundancy and where, consequently, the decline of or loss of one species has consequences for many others. Where there is much functional redundancy the effects of perturbation may still be severe if the species directly affected by the disturbance are very dominant in the community, either in size or abundance. Trees are generally not dependent on only a few organisms through strong mutualistic interactions (above ground at least) and, for example, rely on a broad suite of pollinators for their reproduction (even though one or a few are usually numerically dominant). Nevertheless, as dominant members of the forest community trees support and are part of numerous weak interactions. Thus, while the direct effects of anthropogenic disturbance may be relatively simple to predict there may be much wider indirect effects that are more difficult to track.

In this paper we investigate how the reproductive ecology of two canopy trees, *S. siamensis* (Dipterocarpaceae) in Thailand and *A. excelsum* (Anacardiaceae) in Costa Rica, is disrupted as a result of selective logging or fragmentation respectively. The aims of this study are to determine the impact of extraction of *S. siamensis* and isolation by fragmentation of *A. excelsum* on the behaviour of their pollinators and, consequently, their potential reproductive output.

Methods

Species and study site

Shorea siamensis

The dry deciduous forest (DDF) of Thailand is dominated by four tree species belonging to two genera, *Shorea* and *Dipterocarpus* both within the family Dipterocarpaceae. Prior to a national logging ban in 1989 *S. siamensis* and *S. obtusa* Wall. ex Bl. were extracted commercially from many forests in Thailand. Since then illegal small-scale extraction of *Shorea* trees for domestic use continues throughout Thailand, particularly where villages occur near protected forest areas. Resultant degraded forests are often characterized by greatly reduced densities of *Shorea* spp. but little change to densities of other tree species.

Shorea siamensis was studied at three sites of differing disturbance intensity in DDF at Huay Kha Khaeng Wildlife Sanctuary (HKK), Uthai Thani Province, in western Thailand (15° 36′ N 99° 20′ E). Anthropogenic disturbance over two decades has created a gradient of increasing disturbance intensity from the sanctuary interior to its boundary. Three sites were selected to represent this gradient. An undisturbed site (U) was located 9 km within the sanctuary where human disturbance of any kind is strictly prohibited and adequately enforced. The density of *S. siamensis* at this site was 205 trees ha^{-1} (Ghazoul et al. 1998). *S. siamensis* trees had been extracted in small numbers from the moderately-disturbed site (M) prior to the relocation of villages away from this area in 1990. Since then this site, which is located 3 km within the park, has been subject to the same strict controls as the undisturbed site and *S. siamensis* density was 86 trees ha^{-1}. The disturbed site (D) lies outside the sanctuary but within its buffer zone. Prior to the creation of the buffer zone in 1990, this area was subject to widespread extraction of *S. siamensis* trees resulting in current densities of 22 trees ha^{-1}.

All sites are located within the same forest basin with similar elevations and topography. The soil is homogeneous throughout this basin and no significant differences between sites were apparent in depth, pH, organic content, or water, potassium and phosphorous availability, although nitrogen did occur at a significantly higher concentration at site D (J. Ghazoul, unpubl. data). Grazing and small-wood collection are prohibited in the sanctuary but limited grazing is permitted in the buffer zone and may contribute to the relatively high soil nitrogen content in this area. Nevertheless, soil resources were assumed to be equally available among sites.

In HKK *S. siamensis* produces small yellow flowers from mid February to mid March. These self-incompatible flowers are pollinated by small bees over 95% of which are *Trigona* spp. (Ghazoul et al. 1998). The flowers are partially self-incompatible and therefore largely dependent on the transfer of pollen from neighbouring trees by pollinators for successful seed production (Ghazoul et al. 1998). In common with most other trees in this forest type *S. siamensis* is deciduous and sheds its leaves in November for a period of two to three months. The study was carried out during the flowering seasons of these trees in 1996 and 1997.

Table 1. Site description of the fragments where studies on *Anacardium excelsum* were conducted. Sites listed in increasing order of disturbance from the top to bottom. The isolation codes L, M and H refer to Low (< 500 m between forest fragments), Moderate (0.5–2 km between forest fragments) and High isolation (> 2 km between forest fragments). The intervening landscape matrix was predominantly sugar cane fields or grass pasture. The disturbance rating is a subjective measure that ranks sites from 1 (least disturbed) to 12 (most disturbed) based on current disturbances such as grazing and fires as well as fragment size and isolation.

Sites	Size (ha)	Isolation	Disturbance rating
Lomas	>500	L	1
Marcela	48	L	2
Corobici	44	M	3
Gotera	20	M	4
Jonny	16	H	5
Bosque Duquesa	13.5	H	6
Avispa	10	H	7
Rio Reventado	8.4	H	8
Toronja	5	H	9
Rodeo	1.25	M	10
Tres Espavels	0.3	H	11

Anacardium excelsum

This tree attains heights of between 20–40 m and, until recent years, was widely distributed in forests throughout Guanacaste province in NW Costa Rica. Extensive conversion of forest to cattle pasture and sugar cane plantations has resulted post-clearance distribution of *A. excelsum* that is very much more fragmented and divided than its pre-clearance distribution. Remaining forest is restricted to relatively small fragments (from a few trees up to about 75 ha of forest) that occur either in patches within a pasture or sugar cane matrix, or in linear strips along river courses. Fragments were selected to include a broad range of fragment sizes (Table 1). All these fragments are found in an area of about 1500 km^2 roughly centred on the town of Cañas (10°27′ N and 85°06′ W). Past disturbances to the forest fragments include grazing of cattle in the fragments and collection of wood for timber. These disturbances appear to have mostly ceased except in some fragments where grazing still occurs (Table 1).

Anacardium excelsum has a flowering period which, in 1998 and 1999, was from late January to mid March. Fruit development proceeding from early March through to fruit maturation and dispersal in May. A mass of tiny (corolla diameter about 3 mm) white flowers are produced in inflorescences that can support tens to hundreds of flowers at any one time. The tree bears two flower types, hermaphrodite and male only (i.e., pollen-bearing but lacking ovules). Both these flower types occur on all trees. Pollination is mostly by small *Trigona* bees, but the flowers are also visited by larger bees (Anthophoridae, *Xylocopa* sp., and *Apis mellifera*), a variety of small wasps, flies and, occasionally, butterflies and beetles. The flowers are partially self-incompatible, with some 5% of selfed flowers setting fruit compared with 25% fruit set from cross-pollinated flowers (Ghazoul, unpubl. data).

Pollinator activity

Methods used to record the activity of bee pollinators at *S. siamensis* trees are detailed in Ghazoul, et al. (1998). In particular, the movement of bees from one *S. siamensis* canopy to either other *S. siamensis* canopies or to the nest was noted. Using binoculars, individual bees leaving a particular tree canopy (the focal observation tree) could be easily followed to distances of 60 m or more. Bees leaving tree canopies for the nest were observed to rise from the canopy in a spiralling flight of increasing diameter before flying directly towards the nest without further flower visits. Such a characteristic spiralling flight, which was not observed on flights between tree canopies, was

Table 2. Visits by insects to inflorescences at *A. excelsum* trees during two-minute observation periods.

Site	Flower visitors observed in 2 min (± SE)	Proportion of Trigona bees among all flower visitors (%)	Disturbance rank
Marcela	7.8 ± 1.8	56	2
Toronja	7.5 ± 1.8	55	9
Rio Reventado	7.0 ± 1.5	41	8
Lomas Barbudal	6.4 ± 1.7	54	1
Corobici	6.3 ± 1.4	84	3
Bosque Duquesa	5.7 ± 1.3	53	6
Gotera	5.2 ± 0.7	78	4
Rodeo	4.9 ± 1.1	65	10
Jonny	4.2 ± 0.7	63	5
Avispa	4.7 ± 1.3	45	7
Tres Espavels	3.0 ± 0.8	17	11

Table 3. Pollination of *Shorea siamensis* flowers from trees at three sites of differing disturbance intensities. Percentage data are back-transformed from an arcsin-transformation with confidence limits in brackets, except for values for the occurrence of pollen tubes which are presented untransformed (± standard error) as these data are used later to calculate expected values of fruit set. Single classification ANOVA conducted on arcsin transformed data. Table adapted from Ghazoul, et al. (1998), reprinted with permission from the British Ecological Society.

	D	M	U	*p*-value	*F*	d.f.
No. of trees	23	23	20			
Pollinated flowers (%)	90.5 ± 1.2	91.3 ± 1.1	96.0 ± 0.9	< 0.001	11.04	2, 63
Flowers pollinated with >5 pollen grains (%)	62.9 ± 2.8	58.8 ± 3.3	79.1 ± 3.0	< 0.001	13.29	2, 63
Styles with pollen tubes as % of flowers with >5 pollen	15.3 ± 3.1	19.5 ± 3.8	29.2 ± 4.1	0.013	4.69	2, 63

Table 4. Fruiting success of *Shorea siamensis* trees at three sites of differing disturbance intensities. Data recorded as percentage (± SE) of flowers that develop into fruit. ANOVA on arcsin-transformed percentage data. Results from Tukey's Honest Significant Differences test for unequal sample sizes at $p < 0.05$ are shown in brackets. Table from Ghazoul et al. (1998), reprinted with permission from the British Ecological Society.

		D	M	U	*p*-value	*F*	d.f.
No. of trees	1996	10	14	15			
	1997	6	16	15			
Initial fruit set	1996	5.9 ± 1.5	6.9 ± 1.3	8.1 ± 1.3	n.s.	2.06	38
(%)	1997	–	31.5 ± 2.5	22.5 ± 3.0	0.031	2.27	29
Mature fruit set	1996	0.7 ± 0.1 (a)	2.2 ± 0.3 (b)	2.5 ± 0.4 (b)	0.001	13.82	35
(%)	1997	1.5 ± 0.7 (a)	5.5 ± 0.9 (b)	5.5 ± 0.9 (b)	0.009	5.46	35
Survival to maturity	1996	11.6 ± 5.3 (a)	32.2 ± 5.5 (b)	31.5 ± 3.3 (b)	0.018	4.54	35
(%)	1997	–	17.8 ± 2.6	25.0 ± 3.5	n.s.	1.70	29

thereafter assumed to indicate flights to the nest even though the nest may not have been within site of the observers. Outcrossing potential was estimated by calculating the flights to conspecific trees as a proportion of total flights recorded from the tree. This was correlated with a proximity index for each focal tree as follows:

$$\Sigma(1/x_{i-j}), \tag{1}$$

where x_{i-j} is the distance in metres to each conspecific visited by bees from the focal tree. By summing the reciprocal of the distances this index considers both the number of surrounding flowering trees and their distance, weighted in favour of the nearest neighbours. Thus a high index value indicates low isolation and many close neighbours.

Pollinator activity on *A. excelsum* inflorescences was recorded by randomly selecting at least ten inflorescences on each *A. excelsum* tree and, with the aid of telescope and binoculars, counting the number of flower visitors in two-minute observation periods.

Stigma pollen counts and pollen tube growth

Pollination effectiveness of *S. siamensis* and *A. excelsum* was determined by counting the number of pollen grains on stigmas and pollen tubes growing in the styles of 2–3 day old flowers collected from trees at all sites. Both pollinated and unpollinated flowers of persisted on the panicles for some days after anthesis (4–5 days for *S. siamensis* and 3–4 days for *A. excelsum*). Flower age could be determined from the appearance of the styles and stigmas, which were pale and firm at on the first days but withered gradually thereafter. Recording pollen transfer on 2–3 day-old flowers minimised errors arising from either insufficient time for pollination or abortion of unpollinated flowers. At least 70 flowers were collected from each of 23 *S. siamensis* trees at D and at M, and 20 trees at U. Pollen on stigmas was counted at 80x magnification after staining with a few drops of 1% Nile-blue solution. *S. siamensis* pollen was distinguished from other pollen types by size, and by comparison to a pollen reference collection. Similar numbers of flowers were collected from *A. excelsum* trees at each of the fragments in Costa Rica to determine the proportion of flowers containing pollen tubes. Pollen tubes in the styles was used to assess pollination success and compatibility of the pollen received. Styles of *S. siamensis* and *A. excelsum* fixed in Formalin-Acetic-Alcohol (FAA) were washed and softened in 8 molar sodium hydroxide for five hours. After rinsing in water, specimens were stained for four hours in 0.1% aniline blue solution in 0.1 molar potassium phosphate, before mounting, dissection and examination with fluorescence microscopy.

Pollination success and fruit production

Fruit production of both *S. siamensis* and *A. excelsum* can be directly equated with seed production as fruits contain only a single seed. Thus fruit set of *S. siamensis* was estimated by counting corollas and fruit collected in a single 1 m^2 circular tray positioned 1m above the ground under each of 10 trees at D, 14 trees at M, and 15 trees at U in 1996 and 6, 16 and 16 trees in 1997. Every two or three days the corollas and fruit in the trays were removed and counted. Fruit set values were determined by dividing the number of mature fruit by the number of shed corollas.

A. excelsum seed set was determined at two stages, initial and mature seed set. Initial seed set was quantified by recording the number of flowers that developed into fruit on tagged inflorescences two weeks after anthesis. These results give a relative measure of seed production among fragments prior to the impact of pre-dispersal mortality due to seed or fruit abortion or seed predation. Secondly, mature fruit fallen in 4 m^2 areas on the forest floor beneath the canopies of randomly selected *A. excelsum* trees were counted shortly after fruit fall in June and July to estimate relative reproductive output of trees among fragments.

Results

Pollinator activity

Bees visited many flowers within a single *S. siamensis* tree before departing so most flowers received selfed pollen. Bees foraging at isolated *S. siamensis* trees were relatively more likely upon leaving the tree to return to the nest rather than move to another tree (Figure 1; multiple regression: partial correlation = 0.84, $t_{(28)} = 8.267, p < 0.001$). Conversely, there was more inter-tree movement among trees which have a high proximity index. At very isolated trees the high percentage of flights to the nest indicates that bees left the tree only when they had achieved a maximal pollen load.

Small *Trigona* bees were the principal flower visitors (and presumed pollinators) at inflorescences of *A. excelsum* accounting for 57% of all visits. There

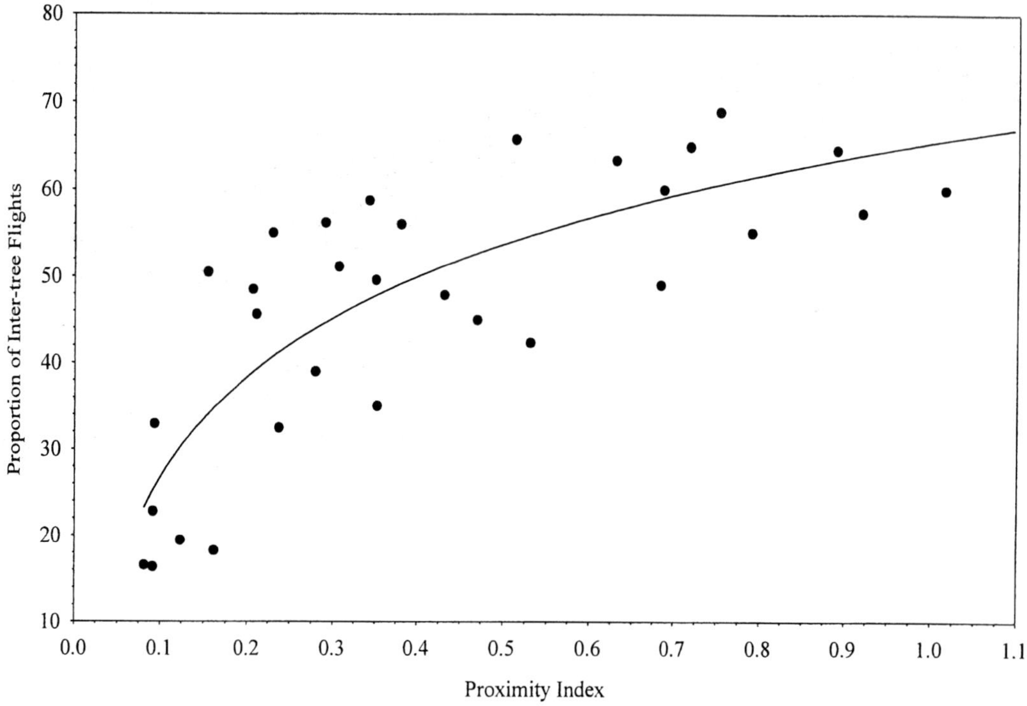

Figure 1. Bee inter-tree flights as a proportion of all flights from *S. siamensis* trees along an isolation gradient. $Y = 65.39 + 38.66 \log X$, $r^2 = 0.521$ (t-test, $t = 5.63$, $p < 0.001$, $N = 31$). The proximity index, along the x-axis, is calculated as $\Sigma(1/x_{i-j})$ where x_{i-j} is the distance in metres to each conspecific tree visited by bees from a focal tree. By summing the reciprocal of the distances this index is a measure of both the number of surrounding flowering trees and their distance, weighted in favour of the nearest neighbours. Thus a high index value indicates low isolation and many close neighbours. Figure from Ghazoul et al. (1998) reprinted with permission from the British Ecological Society.

was no significant difference in the abundance of flower visitors at inflorescence across sites (Table 2; ANOVA: $F_{(10,120)} = 1.25$, p = n.s.) although it is interesting to note that the smallest and most disturbed fragment, Tres Espavels, had the least number of flower visitors, largely due to the decline of visits by *Trigona* bees which accounted for only 17% of visits compared to over 40% at all other sites.

Pollination success and fruit production

After controlling for year and site, fruit set of *S. siamensis* was negatively correlated with the degree of tree isolation (ANCOVA within cells regression, $F = 7.837$, d.f. = (1, 67), $p = 0.007$; Figure 2). Thus increasing isolation has a negative impact on fruit production, and differences among sites in pollen compatibility as measured by pollen tube growth (Table 3) and fruit maturation after fruit initiation (Table 4) reflect this density effect. Given similar resource availability across sites and observed patterns of pollinator behaviour, the emerging explanation for differential fruiting success is that there is a reduction in the transfer of compatible cross pollen, which determines fruit set, at low tree densities.

Table 5. Pollination success of *Anacardium excelsum* trees in different-sized fragments in Guanacaste. Pollination success is determined by the presence of pollen tubes in the styles of flowers.

Site	N trees	Pollinated flowers (% ± SE)	Disturbance rank
Lomas Barbudal	8	21.2 ± 4.4	1
Bosques Duquesa	3	15.3 ± 4.5	6
Rio Reventado	7	14.6 ± 4.9	8
Jonny	4	13.4 ± 6.0	5
Gotera	6	11.8 ± 2.2	4
Avispa	5	11.2 ± 3.8	7
Marcella	3	8.8 ± 1.9	2
Corobici	6	4.2 ± 2.2	3
Rodeo	3	3.1 ± 1.9	10
Toronja	5	2.5 ± 2.4	9
Tres Espavels	3	1.5 ± 1.2	11

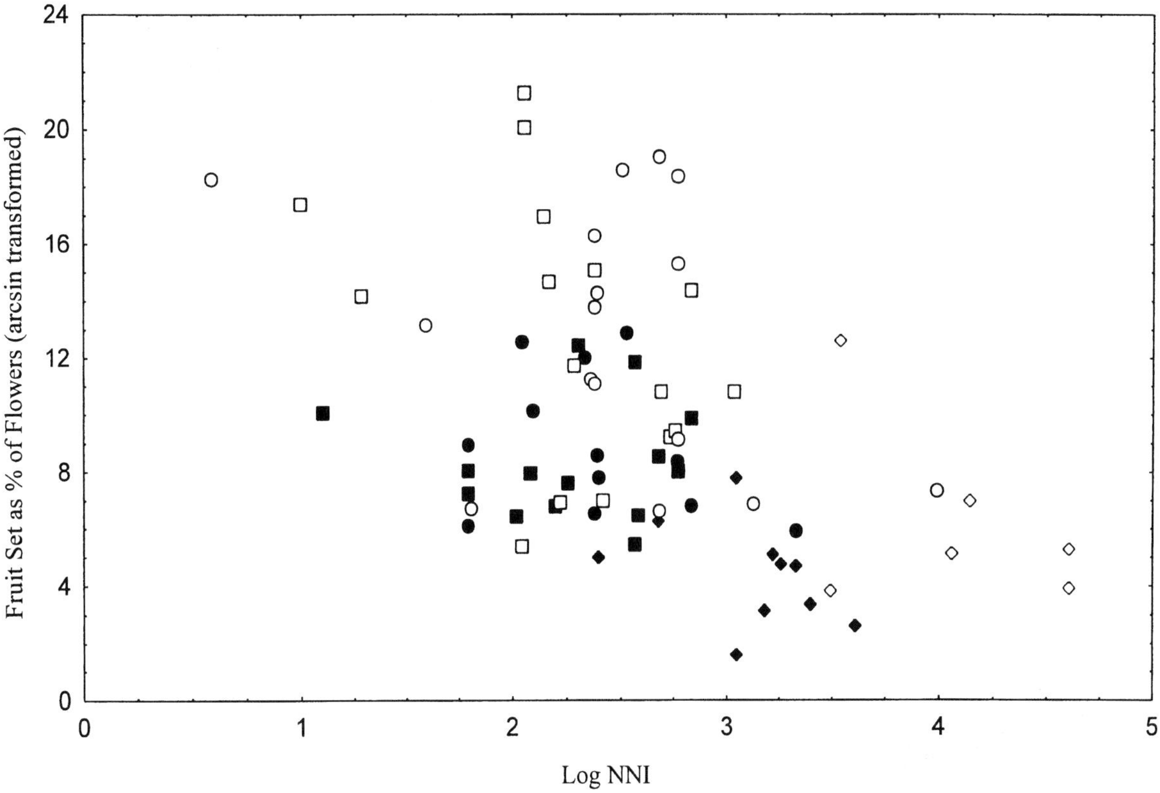

Figure 2. Fruit set as a function of nearest neighbour index. The nearest neighbour index (NNI) is calculated as the square root of the product of the first and second nearest flowering neighbour distances from a focal tree. 1996 values represented by solid symbols, 1997 values by open; circles represent undisturbed site; squares, moderately-disturbed; and diamonds, disturbed site. ANCOVA within cells regression: $F = 7.837$, d.f. $= (1, 67)$, $p = 0.0067$. $y = 17.19 - 2.91x$. The relationship between mature fruit set and NNI is determined with a two-way analysis of covariance with arcsin-transformed mature fruit set values as the dependent variable, year and site as the independent factors and log transformed NNI values as the covariate. To satisfy the assumptions of ANCOVA data were tested for normality and, where required, parallelism of regression lines among groups. Figure from Ghazoul et al. (1998) reprinted with permission from the British Ecological Society.

Table 6. Initial and mature seed production as a percentage of the flowers on tagged inflorescences of *Anacardium excelsum* in Guanacaste fragments. Means reported with standard errors. Note no data from Rodeo.

Site	Initial seed production as % flowers (N inflorescences)	Mature seed set Seeds/4 m^2 (N trees)	Disturbance rank*
Lomas	16.9 ± 0.9 (141)	109.7 ± 7.2 (30)	1
Jonny	15.5 ± 1.0 (44)	165.6 ± 21.0 (21)	5
Bosque Duquesa	14.2 ± 1.3 (42)	91.0 ± 16.9 (21)	6
Gotera	11.8 ± 1.6 (36)	170.3 ± 14.9 (30)	4
Corobici	9.7 ± 1.4 (22)	88.4 ± 7.2 (30)	2
Marcela	8.8 ± 1.0 (22)	83.6 ± 10.9 (30)	3
Rio Reventado	8.6 ± 0.8 (90)	73.1 ± 9.7 (30)	8
Avispa	5.4 ± 1.0 (47)	29.6 ± 4.8 (30)	7
Toronja	3.7 ± 0.5 (58)	99.0 ± 31.0 (6)	9
Tres Espavels	3.6 ± 0.8 (47)	39.2 ± 16.4 (3)	11

*Where 1 = least disturbed fragment (see Table 1).

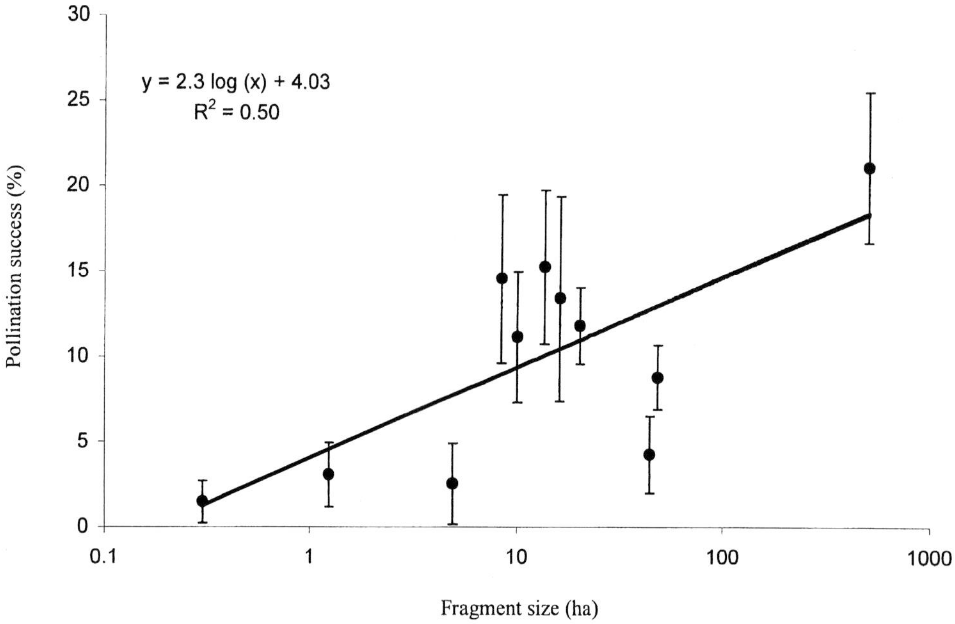

Figure 3. Pollination success of *Aracardium excelsum* trees in forest fragments of varying sizes. Pollination success quantified as the proportion of flowers in which pollen tubes were detected.

Pollination success, recorded as proportion of flowers containing pollen tubes, of *A. excelsum* in Costa Rica was significantly different among the fragments (Table 5; Kruskal-Wallis test: $H_{(10,53)} = 21.46$, $p = 0.011$) and positively correlated with fragment size (Figure 3; $r^2 = 0.513$, $F_{(1,9)} = 9.47$, $p < 0.013$). Thus the flowers of trees occurring in small fragments appear, on average, to receive fewer compatible pollen transfers than those of larger fragments.

There were significant differences between sites in reproductive success of *A. excelsum* measured at early stages of fruit development (Table 6; ANOVA: $F_{(9,539)} = 24.66$, $p < 0.001$) and initial seed set is positively correlated with fragment size (Figure 4; Spearman's correlation coefficient = 0.758, $t_{(9)} = 3.283$, $p = 0.011$). Mature seed production of *A. excelsum* trees also varied among sites (Table 6; ANOVA: $F_{(9,263)} = 18.88$, $p < 0.001$) but was not significantly correlated with fragment size (Spearman's Rank Order Correlation: $R = 0.442$, $t_{(8)} = 1.395$, $p = 0.20$).

Discussion

This study describes the impacts on the reproductive ecology of two trees in seasonally dry tropical forests. Both *Shorea siamensis* in Thailand and *Anacardium excelsum* in Costa Rica show reduced reproductive output as a result of anthropogenic disturbances that affect the behaviour of their pollinators and thus the fertilisation of their ovules. Despite the similar effects observed for both these trees, the disturbances to the forests in which they occur are qualitatively different.

Extraction of *S. siamensis* trees through selective logging reduces the densities of the local *S. siamensis* population but does not affect the area of land under forest cover. The decline in the population density of *S. siamensis* increases the average distance between nearest flowering neighbours. Seed production, as a function of flower number, declines where tree densities are low due to a change in the behaviour of pollinators such that gene flow between trees is reduced. Reproductive success of *S. siamensis* in low density situations is therefore limited by the transfer of compatible pollen, as mediated by density-dependent foraging behaviour of its insect pollinators. This demonstrates the Allee effect (Allee 1949; Burgman, Ferson & Akçakaya 1992) where, in this case, extraction of trees reduces population viability disproportionately to population size by reducing the individual reproductive output of remaining trees.

The principal disturbance regime in Costa Rica differs from HKK in Thailand in that forest clearance and conversion to other land uses fragments the forest into a series of more or less isolated patches. Thus although

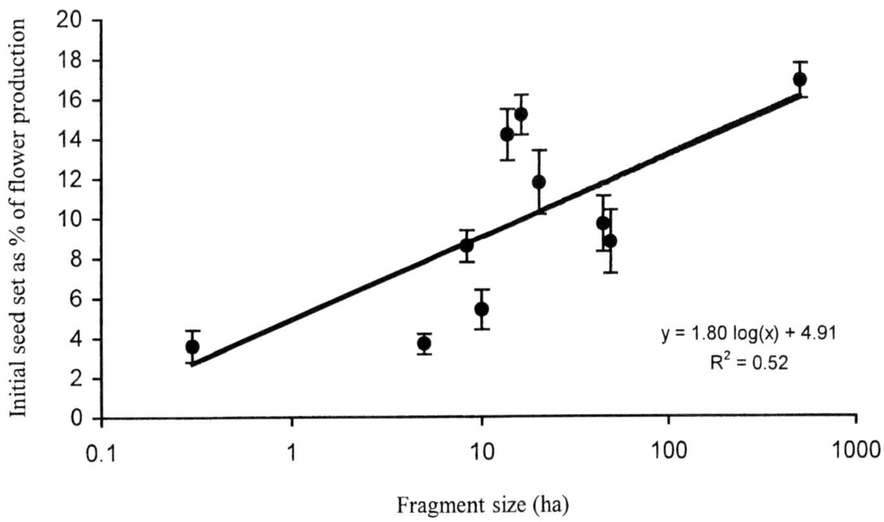

Figure 4. Initial seed set (two weeks after flowering) of *A. excelsum* trees occurring in variably-sized fragments.

the trees now occur in a patchy landscape, the density of *A. excelsum* trees remains unchanged. Low initial seed set at the smaller fragments may be due either to a decline in the abundance of pollinators and pollinator visits to flowers (e.g., Somanathan & Borges 2000), to changes in the foraging behaviour of pollinators unrelated to their abundance (as shown here for *S. siamensis*; see also Ghazoul, in press), or to some other factors unrelated to the pollination process (e.g., Molano Flores et al. 1999). The significant correlation between pollination success and fragment size and the correspondence of pollination success with initial seed production (Tables 4 and 5) indicates that, as for *S. siamensis*, seed set of *A. excelsum* is determined to some extent by pollination. However, forest fragmentation does not seem to have impacted the abundance of the pollinators themselves (Table 2), except in the case of the smallest fragment Tres Espavels where the abundance of *Trigona* bees is markedly reduced. Why should fragment size affect pollination success if tree density remains unchanged and pollinators themselves are numerically unaffected? It has long been known that genetic variability is spatially structured (isolation by distance) in many plant populations with near neighbours being, on average, more closely related than widely separated individuals (Epperson 1993). Where fragments are so small that they contain only a few trees the genetic variability among these trees is consequentially low and their relatedness is likely to be high. If pollinators rarely move between fragments then outcrossing among these trees draws from a gene pool that is limited both by the number of trees contributing to it and also by their genetic similarity. Thus even when pollen is being transferred among trees it is likely that progeny will be more inbred in small than in large fragments leading to fertilisation failure or seed abortion during development. Exactly this effect has been recently shown for *Dombeya acutangula*, an endemic tree occurring in remnant forests in the Mascarene archipeligo (Gigord et al. 1998). At present we are unable to determine whether this is the real explanation for our observations, but work on the genetic variability of the progeny that will confirm or reject this hypothesis is underway (Jonathan Cornelius, in prep.). Mature seed set is not significantly correlated with fragment size, indicating that other factors such as competition for resources or seed predation also contribute to seed mortality during development.

The spatial pattern of *S. siamensis* trees affected pollinator foraging patterns such that there was more pollinator movement between trees at high tree densities and, conversely, potentially less outcrossing when trees were relatively isolated. Handel (1982) has demonstrated that insects tend to move to the nearest neighbouring plant, while Rasmussen & Brødsgaard (1992) showed that pollinators which have no difficulty crossing a gap between patches will nevertheless remain within local patches and preferentially visit nearby plants within the patch. Likewise, Fenster (1991) and Smithson & Macnair (1997) showed that bumblebee and honeybee flight distances are inversely related to flowering plant density. The same principle may apply to forest fragments in Costa Rica where small bees such as *Trigona* spp. are less likely to move between fragments and are therefore vectors of gene flow only within the fragment in which they occur. For large fragments such as Lomas Barbudal and Marcela this is unlikely to present problems of inbreeding owing to the greater abundance of trees in these fragments and thus a broader genetic base. In small fragments, however, gene flow may be limited only to the trees present within the fragment. That forest fragmentation restricts pollinator movement has been previously demonstrated for male euglossine bees (Powell & Powell 1987), and conversion of forest to a series of forest patches within an agricultural matrix has been shown to depress the reproductive success of other trees (*Dombeya acutangula*, Gigord et al. 1999), understorey herbs (Jules 1998), and orchids (Parra Tabla et al. 2000).

The quality, in terms of the extent of gene flow, of the pollination vector is an important consideration when evaluating the impact of habitat fragmentation on plant reproductive output. For example, *Syzigium cormiflorum*, a tree occurring in the highly fragmented Australian tropical rainforests, receives pollen of high quality and across fragments owing to the mobility of the blossom bat pollinator in the fragmented landscape (Law & Lean 1999). Despite pollination by small insects substantial gene flow among a fragmented population of the dry grassland species *Filipendula vulgaris* is achieved because a large proportion of seeds are derived through wind pollination (Weidema et al. 2000). Changes in butterfly flight patterns reduced pollination of *Dipterocarpus obtusifolius*, a sympatric tree with *S. siamensis* at HKK, but did not result in reduced fruit set owing to compensatory pollination by nocturnal moths (Ghazoul, in review). In these cases an alternative pollination vector effectively insures the plants against environmental change. Both *S. siamensis* and *A. excelsum* are pollinated exclusively by small insects and therefore lack the insurance of diversity of pollinator types.

This study demonstrates that logging without loss of area under forest cover as well as forest clearance without changes to tree density within the resulting fragments, can both disrupt plant reproductive processes by causing changes in the foraging behaviour of pollinators and, consequently, the quality of gene flow. Both *S. siamensis* and *A. excelsum* have reduced seed set as a result of reduced pollination efficiency in the anthropogenically disturbed environment. Yet simply monitoring changes in the abundance of pollinators would have provided no indication of these results implying that predicting of future community composition and the functioning of ecological processes using measures of current biological diversity can be misleading. Thus species counts fail to account for changes in animal or plant behaviour which may have greater relevance to ecological dynamics and, ultimately, management for conservation (Lugo 1995). Process-orientated studies will improve management policies by focussing attention on species most relevant to the maintenance of conditions necessary for long-term viability.

Acknowledgements

Mr Chatchawan Pisdamkham, Mr Preecha and the staff of Huay Kha Khaeng Wildlife Sanctuary gave permission, support and encouragement throughout our stay at the Huay Kha Khang Wildlife Sanctuary. Field assistance in Thailand was provided by Katharine Liston and Andy Macgregor, and in Costa Rica by Manuel Sojo. Jim Coles, Kowit Chaisurisri of the ASEAN Forest Tree Seed Centre, Muak Lek, Thailand, and Jonathan Cornelius at CATIE, Turrialba, Costa Rica provided ongoing support for this project. This paper describes research from a CIFOR project 'In situ conservation of plant genetic resources' in Thailand in collaboration with Dr Chaweewan Hutacharern of the Royal Forest Department in Thailand and funded by the Department for International Development UK (DFID), and a second CIFOR project in Costa Rica 'Impacts of forest fragmentation on the regeneration of forest trees' also funded by DFID. Permission to conduct research in Thailand was granted by the National Research Council of Thailand.

References

Aizen, M. A. & Feinsinger, P. 1994. Forest fragmentation, pollination and plant reproduction in a chaco dry forest, Argentina. Ecology 75: 330–351.

Allee, W. C. 1949. Group survival value for *Philodina roseola*, a rotifer. Ecology 30: 395–397.

Bawa, K. S. 1974. Breeding systems of tree species of a lowland tropical community. Evol. 28: 85–92.

Bawa, K. S. 1990. Plant-pollinator interactions in tropical rain forests. Ann. Rev. Ecol. Syst. 21: 399–422.

Burgman, M. A., Ferson, S. & Akçakaya, H. R. 1992. Risk assessment in conservation biology. Chapman and Hall, London.

Epperson, B. K. 1993. Recent advances in correlation studies of spatial patterns of genetic variation. Evol. Biol. 27: 95–155.

Fenster, C. B. 1991. Gene flow in *Chamaecrista fasciculata* (Leguminosae) I. Gene dispersal. Evolution 45: 398–409.

Ghazoul, J. In review. Indirect effects of logging on the reproductive ecology of a butterfly-pollinated tree. J Ecol.

Ghazoul, J., Liston, K. A. & Boyle, T. J. B. 1998. Disturbance-induced density-dependent seed set in *Shorea siamensis* (Dipterocarpaceae), a tropical forest tree. J. Ecol. 86: 462–474.

Gigord, L., Lavigne, C. & Shykoff, J. A. 1998. Partial self-incompatibility and inbreeding depression in a native tree species of La Reunion (Indian Ocean). Oecologia 117: 342–352.

Gigord, L., Picot, F. & Shykoff, J. A. 1999. Effects of habitat fragmentation on *Dombeya acutangula* (Sterculiaceae), a native tree on La Reunion (Indian Ocean). Biol. Cons. 88: 43–51.

Handel, S. N. 1982. Dynamics of gene flow in an experimental population of *Cucumis melo* (Cucurbitaceae). Am. J. Bot. 69: 1538–1546.

Huryn, V. M. B. 1997. Ecological impacts of introduced honey bees. Q. Rev. Biol. 72: 275–297.

Jennersten, O. 1988. Pollination in *Dianthus deltoides* (Caryophyllaceae): effects of habitat fragmentation on visitation and seed set. Cons. Biol. 2: 359–366.

Jules, E. S. 1998. Habitat fragmentation and demographic change for a common plant: *Trillium* in old-growth forest. Ecology 79: 1645–1656.

Law, B. S. & Lean, M. 1999. Common blossom bats (*Syconycteris australis*) as pollinators in fragmented Australian tropical rainforest. Biol. Cons. 91: 201–212.

Lugo, A. E. 1995. Management of tropical biodiversity. Ecol. Appl. 5: 956–961.

Molano Flores, B., Hendrix, S. D. & Heard, S. B. 1999. The effect of population size on stigma pollen load, fruit set, and seed set in *Allium stellatum* Ker. (Liliaceae). Int. J. Plant Sci. 160: 753–757.

Parra Tabla, V., Vargas, C. F., Magana Rueda, S. & Navarro, J. 2000. Female and male pollination success of *Oncidium ascendens* Lindey (Orchidaceae) in two contrasting habitat patches: forest vs. agricultural field. Biol. Cons. 94: 335–340.

Powell, A. H. & Powell, G. V. N. 1987. Population dynamics of male euglossine bees in Amazonian forest fragments. Biotropica 19: 176–179.

Rasmussen, I. R. & Brødsgaard, B. 1992. Gene flow inferred from seed dispersal and pollinator behaviour compared to DNA analysis of restriction site variation in a patchy population of *Lotus corniculatus* L. Oecologia 89: 277–283.

Rathcke, B. 1983. Competition and facilitation among plants for pollination. Pp. 305–329. In: Real, L. A. (ed), Pollination biology. Academic Press, London.

Real, L. A., Otte, J. & Silverfine, E. 1983. On the trade-off between the mean and variance in foraging: an experimental analysis with bumblebees. Ecology 63: 1617–1623.

Schaal, B. A. 1980. Measurement of gene flow in *Lupinus texensis*. Nature 284: 450–451.

Smithson, A. & Macnair, M. R. 1997. Density-dependent and frequency-dependent selection by bumblebees *Bombus terrestris* (L.) (Hymneoptera: Apidae). Biol. J. Linn. Soc. 60: 401–417.

Somanathan, H & Borges, R. M. 2000. Influence of exploitation on population structure, spatial distribution and reproductive success of dioecious species in a fragmented cloud forest in India. Biol. Cons. 94: 243–256.

Vinson, S. B., Frankie, G. W. & Barthell, J. 1993. Threats to the diversity of solitary bees in a neotropical dry forest in Central America. Pp. 53–81. In: LaSalle, J. & Gauld, I. D. (eds), Hymenoptera and biodiversity. CAB International, Wallingford.

Weidema, I. R., Magnussen, L. S. & Philipp, M. 2000. Gene flow and mode of pollination in a dry-grassland species, *Filipendula vulgaris* (Rosaceae). Heredity 84: 311–320.

Plant Ecology **153:** 347–359, 2001.

Edge effects and tropical forest canopy invertebrates

Andrew Foggo[1], Claire M.P. Ozanne[2], Martin R. Speight[3] & Clive Hambler[3]
[1] *Plymouth Environmental Research Centre & Department of Biological Sciences, University of Plymouth, Drake Circus, Plymouth PL4 8AA, UK;* [2] *School of Life Sciences, Roehampton Institute London, West Hill, London SW15 3SN, UK;* [3] *Department of Zoology, South Parks Road, Oxford, OX1 3PS, UK*

Key words: Biodiversity, Deforestation, Ecological process, Edge function, Fragmentation, Review

Abstract

The term edge effect can be used to encompass a wide range of both biotic and abiotic trends associated with boundaries between adjacent habitat types, whether these be natural or anthropogenic. Edge effects have been shown to represent significant forces affecting both faunal and floral assemblages in fragmented ecosystems. Specific studies of faunal assemblages associated with habitat edges have revealed trends at all levels of biological organisation from individuals to communities.

Studies of edge effects on invertebrates in tropical forests have been relatively scarce. In this paper we review the nature and organisation of edge effects, focusing upon the processes which may lead to detrimental consequences for both forest canopy invertebrates and the forests themselves. We present as a case study data illustrating the very large amount of variance (over 50%) in community structure that is predicted simply by abiotic (microclimatic) variables in both a tropical and a temperate forest edge. We summarise major features of edge effects amongst forest invertebrates, stress the inter-relatedness of edge and canopy biology, and present an agenda for study of the canopy as an edge.

Introduction

'Edge effects' in landscape ecology are essentially the biotic and abiotic contrasts between adjacent habitat types. These edge effects are causal mechanisms influencing behaviour, distribution, and abundance of species and other higher order assemblages. They may therefore exert an influence that extends beyond the limits of the physical edge itself. In most terrestrial ecosystems, edges can be defined as the physical boundaries between plant community types (Samways 1994), essentially ecotones characterised by changes in factors such as floral structure, composition, and microclimate.

The term 'edge effect' has strong connotations of anthropogenesis, for example through tree planting or felling, or other habitat modification such as clearance for urban or industrial development that makes edges relatively abrupt. Ecologists historically used the term 'edge effects' to refer to regions of high species richness at habitat boundaries resulting from the mixing of communities (Leopold 1933). However usage of the term has expanded considerably, and it is now employed in too wide a sense, referring to both patterns and processes associated with landscape boundary features stretching from microclimate to distribution and abundance of biota and the nature of interactions between them.

Edge effects are likely to be highly relevant to canopy biology, since the canopy itself is a form of community edge which may suffer from 'dorsal' edge effects such as acid rain (Soulé 1986). Moreover, fragmentation and gap-formation may create microclimatic and biotic conditions near ground level that are more usually associated with the canopy, and species which normally reside in the canopy may move nearer to the ground (Davis & Sutton 1998).

In this paper we consider what is known of edge effects in tropical forests, and make tentative links to what is known of edge effects in canopies, which have been studied more intensively in temperate forests. We draw on case studies of edge effects in tropical and coniferous woodland canopy to illustrate the types of effect researchers might find generally present in forests. We identify research likely to yield productive comparisons between edges and the canopy.

The forest canopy as an edge

The canopy layer of a forest interfaces with three distinct habitats: the understorey, the external matrix which can include any other vegetation type abutting the forest, even another forest type or a plantation, and finally the atmosphere. The boundary of greatest contrast is that between the biosphere and the atmosphere where a dorsal edge results in vertical edge effects. These vertical effects can be clearly seen in patterns of microclimate variation, and in invertebrate and epiphyte distribution (Camargo & Kapos 1995; Evans & Hambler 1995; Davis & Sutton 1998; Schowalter & Ganio 1998).

The dorsal or canopy/atmosphere edge will be subject to natural and anthropogenic influences, for example changes in vapour pressure deficit due to wind speed (Camargo & Kapos 1995) and in chemical loading from pollutants such as acid rain (Shibata et al. 1995). Physical conditions at this edge can be extreme with higher wind speeds than within the canopy, greater insolation and considerable chemical fluxes across the boundary, e.g., CO_2 and isoprene (Jiang et al. 1997; Lerdau & Keller 1997). Invertebrates able to exploit this edge would be expected to either be tolerant of the physical conditions or to be mobile and able to escape extremes.

The physical structure (contiguity) of the forest dorsal edge may be more regular and less frequently interrupted than the external edge of a forest. At external edges tree fall is amplified creating gaps (Lovejoy et al. 1986), and herbs and grasses have seasonal cycles. However dorsal edges will also will be punctuated by canopy gaps caused by tree fall. At such canopy gaps the canopy/atmosphere edge will give way to a type of edge where the forest abuts vegetation of the gap: a forest/matrix edge. Canopy gaps have been little studied, however it is highly likely that they will have a characteristic set of conditions delimited by the forest type and gap size (Duncan et al. 1998; Birnbaum 2001). This could lead to a specialist fauna, perhaps a subset of those generally able to exploit edge habitats. Gap dynamics become particularly significant in the canopies of managed tropical forests where logging is typically selective and gaps of varying sizes are created by extraction.

External edges (canopy/matrix) are also important since they occur naturally for example along rivers, and are generated when logging roads are laid down, frequently leading to an increase in linear patterns of disturbance caused by timber extraction. The degree of contrast and therefore edge effect at external edges and at the canopy/understorey boundary is dependent upon the nature of the understorey vegetation or the external habitat. In unmanaged forests greater mixing of organisms such as Collembola between the canopy and the external matrix and with the understorey is likely (Bowden et al. 1976) as the boundary is soft. In managed forests and plantations the boundary with the matrix can be very abrupt and therefore more similar to that of the canopy-atmosphere edge. The degree of contrast between habitats and the strong three dimensional structure of the canopy are both significant factors that influence the pattern of edge effects, both in terms of the distribution of organisms and the processes by which they interact.

Patterns and processes at edges

Some authors have attempted to promote a differentiation of terms between the factors underlying patterns in the distribution and abundance of biota, and the distributional patterns themselves, ascribing the term *edge effect* to the former and *edge function* to the latter (Laurance 1991; Laurance & Yensen 1991). However this dichotomy has not been broadly adopted; the edge *function* of one species may in itself constitute an edge *effect* for another species and so on, making the distinction between pattern and process somewhat blurred and at least arbitrary. Broadly, ‘low-order’ effects are relatively direct influences of edges on microclimate or biota, whilst ‘high-order’ effects are consequent to these through community interactions.

Murcia (1995) proposes three categories of edge effects influencing fragmented forests that reiterate the hierarchy originally proposed by Lovejoy et al. (1986). Firstly there are abiotic effects including patterns in factors such as insolation, relative humidity, soil moisture, and temperature. Secondly come direct biological effects which are usually simple responses of organisms to abiotic patterns, such as increased growth and reproduction of both sapling canopy and understorey plants in response to elevated insolation. Finally there are indirect biological effects which are those changes that cascade through communities in response to interactions between organisms, and which are driven by initial changes elicited in response to abiotic patterns.

Organisms such as Psocoptera and Collembola, with specific microclimatic requirements, may show strong responses to abiotic factors (e.g., Ozanne et al. 1997) and may thus act as foci around which higher order edge effects proliferate. This is however dependent upon the functional importance of these organisms in the communities in question. The relationships between biotic and abiotic edge effects are thus complex, with higher-order biotic edge effects being consequences of both abiotic trends and the edge responses of the biota themselves. Most organisms are sensitive to suites of edge effects, both abiotic and biotic (see Tables 2 and 3), and their action may be synergistic. The influence of individual edge effects upon the biota may be masked by other, confounding influences, and therefore difficult to determine. This dilemma further strengthens the need for investigations into the operation and importance of mechanisms underlying edge effects.

Any species may exhibit variable sensitivity to an edge depending upon its size, sex, or life-history stage (Ozanne et al. 1997). Species of both plants and animals appear to exhibit the full range of responses to habitat edges from totally insensitive to highly sensitive. In addition, the nature of an edge may change with time. 'Creeping' or 'advancing' edges, in which the edge-like conditions move progressively into the core habitat through auto-catalytic effects (Soulé 1986; Murcia 1995) are highly significant features in landscape ecology and may often represent the future condition of present sharp, well-defined edges.

Edge effects have two definitive properties: penetration distance of the effect (perpendicular distance from the edge at which consistently measurable differences from interior conditions persist), and intensity (magnitude of perturbation compared to the forest interior at any given distance from the edge) (Laurance & Yensen 1991; Chen et al. 1995). Some authors however confuse the two (see Murcia 1995). Intensity of edge effect may modulate across the distance of measurable penetration. Indeed few abiotic patterns have been demonstrated to show constant or simple linear relationships with distance from the edge (Kapos 1989; Ozanne et al. 1997), and thus edge effects at the community level are usually not unimodal (Didham 1998). This latter trend is a likely product of the integration of various penetration distances and intensities for edge effects upon different species and their subsequent ramifications. Penetration distance *per se* is therefore not a good predictor of the influence of edge effects at any given distance, and measured deviation from core conditions can only be regarded as a surrogate in interpretation of the impact of an edge effect upon the biota. The general properties of edge effect intensity might thus be best measured at the community level.

Edges, fragmentation, and the importance of scale

Studies of edges and their effects are intrinsically linked to those of habitat fragmentation (see Soulé 1986; Didham et al. 1996). Edges are a consequence of fragmentation, and also significantly contribute to its impacts. In some cases edges may dramatically alter the character of the remaining habitat (Murcia 1995). Habitat fragmentation usually involves one habitat that is more natural whilst the other is more anthropogenic, caused by activities such as burning or clear felling; this combination of habitats produces acutely defined 'edges'. At smaller disturbance scales edges may be less well defined, and their effects overlap with those intrinsic to gap phenomena (but see Vitt et al. 1998).

Several recent reviews have dealt with the implications of habitat fragmentation (Soulé 1986; Saunders et al. 1991; Andrén 1994; Didham et al. 1996), but only one has concerned itself specifically with edge effects (Murcia 1995). Although there are apparent conflicts amongst the findings of some studies (see Baldi 1996) there exists some consensus on the nature of edge effects and on the types of mechanisms responsible for their impact upon the biota. Some simple considerations, such as patterns in microclimate, can be shown to be highly probable determinants of abundance and distribution in certain faunal groups (Dyer & Landis 1997; Ozanne et al. 1997). Evidence for this, however, is largely correlative, and rigorous testing of hypotheses of causality arising from these correlations is lacking.

Several key attributes of habitats and their associated assemblages of organisms appear to render them edge sensitive. High host or habitat specificity, high proportions of indigenous or endemic species, and low historical anthropogenic influence upon the landscape all contrive to render organisms edge effect-prone (Parsons 1991; Stouffer & Bierregaard 1995; Basset 1996; Didham et al. 1996; Turner 1996). Of these, the two former attributes are certainly true for many

Table 1. Abiotic edge effects summarised. Columns indicate whether the character described shows an increase or decrease at the habitat edge compared to the core.

Character	Increased	Decreased
Humidity		×
Vapour pressure deficit	×	
Air temperature	× (day)	× (night)
Soil temperature	× (day)	× (night)
Total light levels	×	
Short wave radiation	×	
Soil moisture		×
Net precipitation (after canopy interception)		×
Agrochemical intrusion	×	
Deposition of pollutants	×	
Soil nutrient availability	×	
Gap formation	×	

tropical forest canopy invertebrates, perhaps more so than for any other group of terrestrial organisms. Despite recent suggestions that large areas of the planet's tropical rain forests have long been subject to anthropogenic disturbance (Lopez Zent 1998), they are likely to have experienced considerably less disturbance compared to highly impacted temperate forests, at any chosen spatial scale.

A review of edge effects of significance for tropical forest canopy invertebrates

The extent of knowledge

A review of literature covered by the Science Citation Index database (1981–1999) reveals an extensive suite of edge effects found in general forest ecosystems; these are summarised in Tables 1–3. Many of the studies that contribute to the table were conducted not in the tropics but in old-growth temperate forests, and very few concern themselves with the invertebrate fauna. The review found few published data explicitly linking edge effects and tropical forest canopy invertebrates. In the absence of empirical studies of the tropical canopy invertebrate fauna, extrapolation from the data available for other groups of invertebrates and other habitats provides the only means of predicting the responses of these organisms to edge effects.

Examples of edge-effects in the canopy

Edge effects impact organisms at all levels of biological organisation from the individual, through populations to the community. Several of the studies of edges are of species inhabiting the foliage, trunks and branches of trees, and the spaces between them, and are included here as influences in the 'canopy' community although their vertical (as opposed to horizontal) extent has rarely been studied.

Size, sex, and life history stage-related differences have all been noted near edges for invertebrates in the canopies of temperate forest (Ozanne et al. 1997). Behavioural as well as physiological and distributional responses have been noted, with edge 'permeability' to mobile species being a significant component of edge influence upon organisms dispersing between habitats.

Significant effects of edge *per se* upon invertebrates have been recorded in a variety of different ecosystems. These include tropical forest canopies (Malcolm 1997) and floors (Didham et al. 1998a), temperate forest canopies (Toda 1992; Ozanne et al. 1997) and floors (Bedford & Usher 1994; Davies & Margules 1998), temperate scrub ecosystems (Suarez et al. 1998), orchards (Brown & Lightner 1997) and birch dominated islands in freshwaters (Cappuccino & Martin 1997).

Several authors have commented upon the significance of edges for fragmentation-related changes in invertebrate assemblages in forests (e.g., Didham et al. 1996; Ingham & Samways 1996, Davis & Sutton 1998). These include influences of predators and pathogens upon the populations (Roland 1993; Rothman & Roland 1998). Differences in invertebrate assemblages found in forest gaps compared to interiors have been recorded in other studies of these environments; edges and indeed forest canopies share many key environmental characteristics, such as elevated light levels and temperatures, with gaps (Schowalter 1994; Wood & Gillman 1998).

Surrogates which may indicate invertebrate assemblage changes at edges or in gaps, have also been recorded. These include changes in pollination processes (Aizen & Feinsinger 1994), increased herbivory levels (Khan & Tripathi 1991; Nicholsorians 1991) and altered seed predation or dispersal patterns by vertebrates (Greig 1993).

Table 2. Biotic edge effects upon plants, lichens and fungi summarised. Columns indicate whether the character described shows an increase or decrease at the habitat edge compared to the core, or whether there is a context-specific qualitative change.

Character	Increased	Changed	Decreased
Tree densities	×		×
Mean tree height			×
Total basal area of stems	×		
Tree species richness	×		
Understorey cover	×		
Subcanopy cover			×
Canopy cover			×
Seed predation	×		
Tree growth rates	×		×
Seed dispersal	×		
Seed germination			×
Post-dispersal seed predation			×
Pollination success			×
Seedling density			×
Seedling establishment			×
Abundance of disturbance adapted plants	×		
Variation in plant isotopic composition	×		
Herbivory levels	×		
Windthrow	×		
Overall plant mortality rates	×		
Species turnover	×		
Overall species richness	×		
Species composition		×	
Leaf litter decomposition rates		×	
Lichen growth rates	×		
Fungal species richness			×

Table 3. Biotic edge effects upon animals summarised. Three columns indicate whether the character described shows an increase or decrease at the habitat edge compared to the core, or whether there is a context-specific qualitative change.

Character	Increased	Changed	Decreased
Bird species turnover	×		
Animal species richness	×		
Animal growth rates	×		
Animal densities		×	
Predator/prey ratios	×		×
Nest/brood parasitism	×		
Nest predation	×		×
Predation upon parasitoids	×		
Mortality due to pathogens	×		
Animal body size spectra		×	
Animal community composition		×	
Human impacts (e.g., firewood gathering)	×		
Impact from grazing stock	×		

Data deficiencies

Although studies of edges and their biotic and abiotic consequences have proliferated recently, major gaps remain in the literature. Certain taxonomic groups (particularly birds, mammals, vascular plants), certain forest types (e.g., temperate conifer plantations, Amazonian high forest), and certain vertical strata (ground-layer, lower canopy) of the forest are relatively well known, yet there is deficiency in important areas for ecology and for biodiversity conservation. This makes generalisation about the extent and importance of edge effects difficult and risky.

The lack of data concerning fragmentation effects upon rain forest canopy invertebrates is notable considering the postulated importance of these organisms as a component of the planet's biodiversity (Janzen 1987). Problems of sampling, access, logistics and taxonomy within this speciose and abundant group of organisms perhaps explain this deficiency. However, primary tropical forests continue to become fragmented, and edge effects are becoming an increasingly dominant feature of all types of forest ecosystem in the tropics (Ranta et al. 1998). There is also evidence that selective logging has impacts upon the physical environment that make anthropogenic forest gaps more influential upon the surrounding forest than natural ones (Malcolm 1998), with potentially greater edge effects (see Vitt et al. 1998). The need for studies of both simple patterns and mechanisms of edge effects in tropical forest invertebrates is therefore pressing.

Mechanisms

As stated earlier, the specific mechanisms by which edges influence the biota are not well documented. First-order mechanisms derived from organisms' responses to abiotic factors are best understood. Evidence exists for both positive and negative influences of first order abiotic factors (Lovejoy et al. 1986) such as light levels (Ozanne et al. 1997; Vitt et al. 1998), temperature (Nelson & Campbell 1993; Didham et al. 1998a) and humidity (Parsons 1991; Jose et al. 1996). Some supporting evidence exists for higher order biotic edge effects, for example in the cases of host plant effects upon herbivores (MacGarvin et al. 1986; Warren 1987), and prey availability effects upon predators and parasitoids (Lovejoy et al. 1986; Roland 1993; Burke & Nol 1998; De Casenave et al. 1998; Peltonen & Heliovaara 1998). However, the relative importance of the different orders of effect is as yet unknown.

Tropical forest trees, particularly those in the lowlands show low levels of self-pollination and high levels of dioecy (Bawa et al. 1985). They are often long-lived and spatially dispersed, with strong dependence upon specialist animal pollinators and dispersers (Regal 1982; Howe & Westerley 1988; Bawa 1990). Pollinator specificity is particularly high in keystone forest species such as figs (Bawa 1990). However, some studies suggest that despite this high specificity and the small size of the fig-wasp vector, within-patch pollen dispersal is efficient and may be more robust to effects of fragmentation than predicted (Nason & Hamrick 1997; but see Anstett et al. (1997) for possible deleterious effects of declining fig population sizes upon the mutualism). These mutualist traits place both canopy and understorey species, including fugitives such as *Cercropia* (Garwood 1989), in a perilous situation should their specialist pollinators or dispersers be negatively influenced either through abiotic or knock-on edge effects (Bond 1995). This further stresses the importance of building an understanding of the mechanisms of higher-order edge effects. With host plants at risk, the knock-on effects of pollinator or disperser failure for other organisms could be drastic, invoking the kind of extinction crisis envisaged by Janzen (1987).

The relative importance of low-order (abiotic) edge effects: case studies

Method

As a test of the relative importance of low- and high-order edge effects in tropical forests, we conducted a comparative study of a tropical forest edge with a temperate one. For the tropical system we analysed an as yet unpublished dataset describing dry-season invertebrate community patterns at the morphospecies level, across a rain forest-wet sclerophyll forest transition in Northern Queensland, Australia, 17°30′ N, 145°45′ E (Watson et al., unpubl. data). We then compared these results with those from analysis of a dataset describing edge responses in temperate canopy invertebrate communities of Norway Spruce plantations in the UK (Ozanne et al. 1997).

The tropical data describe samples of both ground (pitfall) and understorey (flight intercept) fauna taken at nine stations staggered along five replicate 160 m transects across the abrupt fire-defined boundary between the two forest types. This boundary is char-

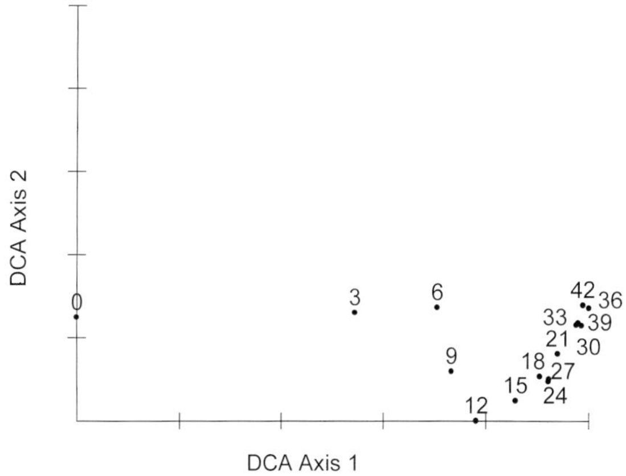

Figure 1. Detrended Correspondance Analysis of mean faunal abundances along the edge-core transect in the Norway Spruce plantation. Sampling stations are labelled with the distance from the edge of the plantation in metres.

acterised by a rapid change from light, open *Eucalyptus/Casuarina* dominated sclerophyll forest to dense secondary rain forest over the space of 2–3 m (Laurance & Harrington 1997). The temperate data describe insecticide knock-down samples of canopy invertebrates from three transects running from the edge to the core of the Norway Spruce plantation.

To compare the edge effects in these examples from tropical and temperate regions, we addressed several specific questions:
(1) Are there comparable multivariate patterns in the sampled fauna across the edges?
(2) Are there comparable multivariate patterns in microclimate across the edges?
(3) Do microenvironmental data (and hence low-order edge effects) correlate well with the faunal assemblages of the two communities?

Data were at species level for all temperate samples, and morphospecies for tropical samples. Basic microclimate data describing light levels, ground level relative humidity and temperature were available for both data sets.

We firstly conducted detrended correspondence analysis (DCA) (Hill 1979) of the mean abundances of all taxa at each transect point in each habitat. Mean values of log transformed microclimate variables were also subjected to DCA.

In order to investigate correlation between patterns in the highly autocorrelated microclimate variables and those in the fauna, we used the values for the first microclimate DCA axis at each transect point. Differences between the values for the microclimate DCA axis at each transect point were then correlated with the differences in community composition calculated using the Bray-Curtis similarity coefficient on 4th-root transformed abundances (Clarke & Warwick 1994). This process avoids problems with highly autocorrelated environmental variables in techniques such as canonical correspondence analysis; in addition the Bray-Curtis coefficient is highly appropriate for data describing organismal abundances where data are highly non-normally distributed (Clarke & Warwick 1994). The analysis was conducted using the subroutine BIOENV (Clarke & Ainsworth 1993) within the multivariate analysis software PRIMER (Plymouth Marine Laboratory, Natural Environment Research Council). The routine produces a Harmonic Spearman Rank correlation between the distance matrix for environmental variables at all combinations of sites and that of the similarly structured Bray-Curtis similarities.

Results

The invertebrate communities of these two study areas differ considerably in their fauna. The tropical sites were considerably more species rich and equitable (see results) (Simpson's $D = 0.15$) than the temperate ones (Simpson's $D = 0.63$) which were numerically dominated by a few species of Collembola and Psocoptera known to exhibit strong responses to abiotic parameters such as humidity (Bowden et al. 1976).

Results of DCA of faunal data confirmed that the first two axes extracted explained a satisfactory proportion of the overall variation for both habitats (70.26% for the tropical dataset and 75.4% for the temperate). Results of DCA of microclimate data were also satisfactory, the first axis representing 99.6% of variation in the tropical dataset and 99.5% in the temperate.

Clear and similar trends across both the tropical forest boundary and the temperate plantation edge to core transects are apparent from Figures 1 and 2. Patterns in microclimate also show great similarity (Figures 3 and 4). The tropical community showed a greater correlation with microclimate (Harmonic $\rho = 0.74$) than the temperate (Harmonic $\rho = 0.52$).

Discussion

In both temperate and tropical cases, the more light and open the forest canopy (sclerophyll samples from

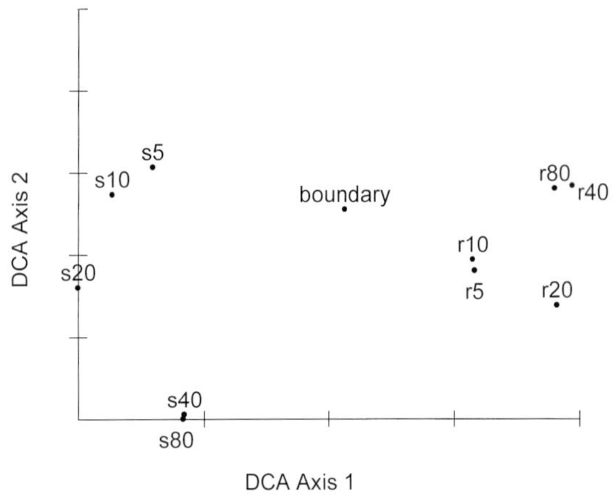

Figure 2. Detrended Correspondance Analysis of mean faunal abundances along the sclerophyll-rain forest transect in Queensland. Sampling stations are labelled with a letter (s = sclerophyll, r = rain forest) and a distance in metres from the boundary between the two forest types.

Queensland and Norway Spruce samples close to the plantation edge), the more variability there was in the faunal composition. Correlations between community and abiotic patterns were strong in both habitats. The temperate community chosen might be expected to be dominated by low-order edge effects due to the community dominance of mycetophages and xylomycetophages with documented responses to microclimate, and indeed for microclimate to explain 50% of the variation in the pattern does indicate that low-order edge effects might dominate. However microclimate patterns are more highly correlated with the fauna in the tropical samples, suggesting that in the case of the rain forest-sclerophyll boundary, there may also be a predominance of low-order effects.

Such studies applied to forest canopies would provide a valuable insight into the sensitivity of canopy invertebrates to microclimatic patterns, and hence could represent a useful test of predictions about the edge-like nature of forest canopies (*sensu* Davis & Sutton 1998).

Whilst it is not possible to draw any unequivocal conclusions about mechanisms underlying edge effects from such correlative studies, it is clear that there is good evidence for significant low-order edge effects in the two communities studied. Should such low-order edge effects prove to be predominant in invertebrate communities, then there may be considerable implications as far as mitigation of edge effects is concerned – for example the plantation of 'buffer' forests around reserves.

The significance of edge effects for habitat exploitation and management

Three forms of exploitation of tropical rain forests have particularly wide implications for future edge effects. These are: (1) clearance or clearfelling leading to fragmentation or increased edge lengths of remaining forests; (2) plantation forestry; (3) selective felling and extraction of individual trees within areas of forest.

Forest clearance

We believe that edge effects must be considered at the planning stage of clearance whenever possible, and also factored into management plans where appropriate. Choice of orientation of fragments relative to environmental gradients such as prevailing winds may help to control the significance of lower-order effects from which other changes may stem. Some edge effects may be mitigated through management, lower order effects more so, for example through shelter-belt planting or encouragement of edge advancement (*sensu* Murcia 1995). However it must be emphasised that such general statements must be scrutinised in the context of the target taxa, which may have unique responses to different management provisions, as illustrated by the case studies described earlier. Higher order effects are likely to be more insidious and very difficult to mitigate, especially as they are as yet poorly understood. It is simply not wise as yet to speculate upon the role that management may take in controlling for these effects.

Other effects of fragmentation and edge effects in the ecology of tropical forests are also relevant to their long-term future in conservation and as resources for exploitation. For example the demonstrated influences of edges and fragmentation increasing the frequency and intensity of outbreaks of defoliating insects or pathogenic infections are now recognised. These aspects of forest and canopy ecology are discussed in the following section from the point of view of anthropogenic forests-plantations.

Plantation forestry

The highly ordered nature of forest plantation and the need to maintain access for sanitation, management and extraction of timber means that edges are a prominent feature of these habitats. Whilst it is likely that plantations will never be able to support the full range

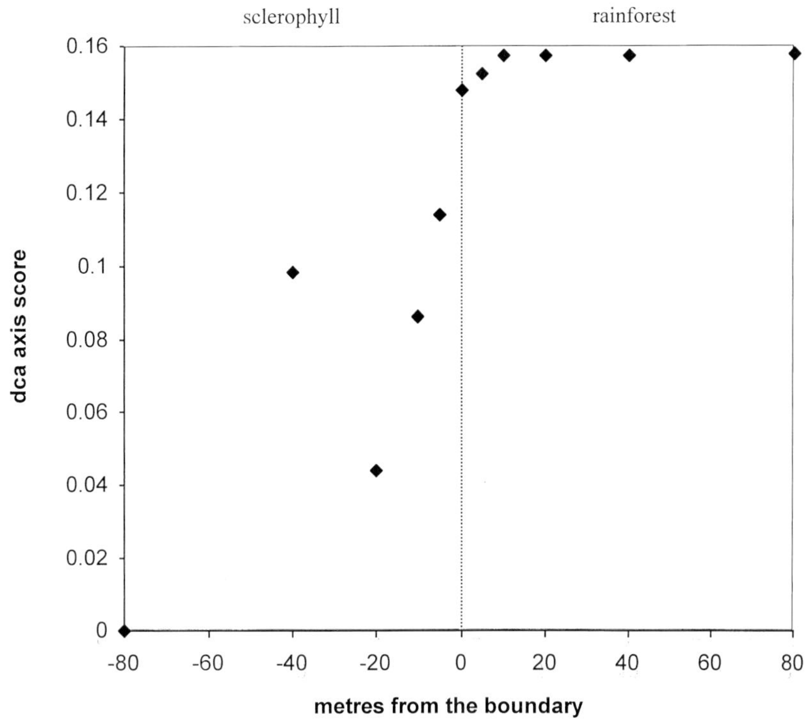

Figure 3. First axis scores from Detrended Correspondance Analysis of the microclimate data from the sclerophyll-rain forest transect in Queensland, plotted against distance in metres from the boundary of the two forest types (0 = habitat boundary, negative values occur in sclerophyll forest, positive in rain forest).

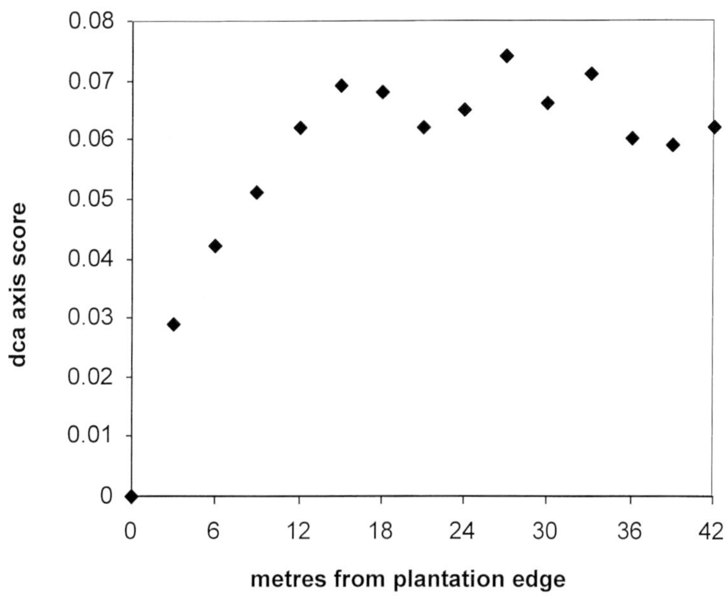

Figure 4. First axis scores from Detrended Correspondance Analysis of the microclimate data from the edge-core transect in the Norway Spruce plantation, plotted against distance in metres from the plantation edge.

of invertebrate species that are encountered in primary or even secondary forest, high species richness of some taxa have been encountered in the canopies of tropical plantations of frequently planted exotics (Chey et al. 1997). The potential therefore exists for plantations to act as reservoirs of species that might recolonise regenerating forests after clearfelling or selective logging. However, there are also two potentially negative impacts of plantation forestry for edge effects in both other plantations and remaining natural forests. Firstly, several studies have demonstrated differences between the responses of invertebrate forest pests and their predators and pathogens to fragmentation and edges (Cudjoe et al. 1993; Roland 1993; Roland & Taylor 1997; Rothman & Roland 1998), increasing the probability of pest outbreaks escaping natural control mechanisms in fragmented landscapes (Kruess & Tscharntke 1994). Some pathogens display positive, rather than negative, edge effects (Nelson & Campbell 1993), suggesting that patterns of outbreak of tree diseases may be changed by fragmentation and edge effects also. Secondly, plantations may result in the provision of high areas of edge for proliferation of eurytopic species, increasing the rain of migrants impinging upon the edges (including the dorsal surface of the canopy) of natural forest. If higher-order edge effects are proven to be of significance in tropical invertebrate communities, then the effects of this increased abundance of dispersing non-forest interior species may further enhance edge effects in the remaining natural systems.

Selective extraction

Gap formation is a natural process in tropical forests, and indeed is one mechanism thought to underpin the biodiversity of the ecosystem (Connell 1978). However, even naturally occurring gaps have been found to have a distinct fauna that resembles that of edges in having higher proportions of eurytopic, weedy or thermophilous species (Schowalter 1994; Wood & Gilman 1998). Canopy species may move nearer to the ground where the canopy is reduced or modified, even by small gaps (Davis & Sutton 1998). Harvesting gaps do not mimic natural tree fall gaps closely, and distinct faunistic patterns have been demonstrated for taxa associated with the altered thermal regimes associated with them (Vitt et al. 1998). As long as these 'gap edge effects' have low penetration distance, and the density of extraction sites is controlled, the deleterious effects of selective logging for the forest fauna and flora will be reduced. However, recent simulations based on empirical studies indicate that abiotic perturbations originating from selective logging gaps can penetrate far into the forest canopy due to the additive nature of edge effects (Malcolm 1998). The impact that such effects will have upon the forest canopy biota can only be speculated upon at this time. However, if higher-order effects also ramify from these abiotic patterns, and the density of logging sites is not controlled accordingly, then selective extraction could exert a strong influence on the forest ecosystem.

An agenda for canopy research into edge effects

We have illustrated the large deficiencies in data on edges, whether studying the canopy or other habitats. We therefore suggest some priorities for future research:

(1) Studies of the microclimate of canopies in different vertical strata, and comparison with the microclimate at different horizontal distances from the forest edge. The interactions between the dorsal and horizontal edge effects may be interesting. We predict that whether the canopy or the ground layer is more adversely affected by edge effects will largely depend on the size of gaps and the contrast between the microclimate within gaps and that above the canopy.

(2) Studies of the vertical distribution of organisms in the canopy in relation to microclimate and biotic factors. The nearer one is to the edge of the forest patch, the nearer to the ground some of the upper canopy species may move – although at the extreme edge of a forest the microclimatic characteristics may be more inimical to invertebrates and other sensitive taxa than in tree tops.

(3) Studies of groups likely to be highly responsive to microclimate through indirect influences, e.g., on food availability. Species of invertebrates feeding on epiphytic, moisture requiring plants such as many orchids, ferns, lichens, and mosses would be predicted to be influenced by edge formation, and communities resembling those of the upper canopy may start to occur nearer the ground.

(4) Studies of dorsal edge effects, such as acid rain, on taxa predicted to be sensitive – e.g, lichens and lichen-feeding invertebrates.

(5) Studies of edge effects due to gaps created by selective logging, and of how gap geometry, distribution and temporal synchrony influence them.

(6) Studies of the biotic and abiotic similarity of canopies to open and gap habitats such as savannah, and their importance for dispersal amongst species of open habitat (Walter 1984).

Conclusion

Studies of the role of edge effects in the ecology of rain forest invertebrates are few and the canopy fauna has been largely neglected, principally for logistical and taxonomic reasons. General studies of edge effects indicate that responses of rain forest canopy invertebrates may occur at every level, from community composition to the demographic structure of populations, and may adversely affect the whole community through processes such as the breakdown of mutualisms or altered predator-prey interactions. Edge effects themselves are hierarchical and probably cascade in response to certain primary changes in community composition, with functional roles of species affected (either positively or negatively) determining the consequences of changes in their abundance and distribution.

The process of habitat fragmentation inherently increases the significance of edge effects, and the prevalence of the mobile, opportunistic species that characterise edge faunas. Fragmentation and the increased prevalence of edge effects may both be linked to increased extinction risks in sensitive taxa (*sensu* Didham et al. 1998b). The exploitation of tropical forests at any scale is likely to lead to increased significance of edge effects, possibly accompanied by increased risks of pest outbreak. Communities typical of the high canopy (dorsal edge) of the forest may move, in whole or in part, to lower regions of the canopy near the edge of habitat patches (Davis & Sutton 1998).

As disturbance of the forest environment escalates, habitat matrices will increasingly fill with species adapted to exploiting change (Myers 1996); mitigation of the effects of change can only be achieved through a more complete knowledge of the mechanisms by which they occur.

Acknowledgements

The authors wish to thank the students and technical staff who have assisted in various stages of this work over the last seven years. Thanks in particular to Paul Embden and Steve Roberts in Oxford, Pete Smithers in Plymouth, Paul G. Watson and the other members of the Oxford University Expedition to Queensland 1996, and to Drs Nigel Stork, Graham Harrington and Roger Kitching for their hospitality in Queensland. Final thanks go to the editors of this volume for their diligent attention to detail.

References

Aizen, M. A. & Feinsinger, P. 1994. Forest fragmentation, pollination, and plant reproduction in a chaco dry forest, Argentina. Ecology 75 (2): 330–351.

Andrén, H. 1994. Effects of habitat fragmentation on birds and mammals in landscapes with different proportions of suitable habitat: a review. Oikos 71: 355–366.

Anstett, M. C., Hossaert-McKey, M. & McKey, D. 1997. Modelling the persistence of small populations of strongly interdependent species: figs and fig wasps. Cons. Biol. 11 (1): 204–213.

Baldi, A. 1996. Edge effects in tropical versus temperate forest bird communities: Three alternative hypotheses for the explanation of differences. Acta Zool. Acad. Sci. Hun. 42 (3): 163–172.

Basset, Y. 1996. Local communities of arboreal herbivores in Papua New Guinea: Predictors of insect variables. Ecology 7 (6): 1906–1919.

Bawa, K. S. 1990. Plant-pollinator interactions in tropical rain forests. Ann. Rev. Ecol. Syst. 21: 399–422.

Bawa, K. S., Perry, D. R. & Beach, J. H. 1985. Reproductive biology of tropical lowland rain forest trees. I. Sexual systems and incompatibility mechanisms. Am. J. Bot. 72: 331–345.

Bedford, S. E. & Usher, M. B. 1994. Distribution of arthropod species across the margins of farm woodlands. Agric. Ecos. Env. 48 (3): 295–305.

Birnbaum, P. 2001. Canopy surface topography in a French Guiana forest and the folded forest theory. Plant Ecol. 153: 293–300 (this volume).

Bond, W. J. 1995. Assessing the risk of plant extinction due to pollinator and disperser failure. Pp. 131–146. In: Lawton, J. H. & May, R. M. (eds), Extinction rates. Oxford University Press, Oxford.

Bowden, J., Haines, I. H. & Mercer, D. 1976. Climbing Collembola. Pedobiologia 16: 298–312.

Brown, M. W. & Lightner, G. W. 1997. Recommendations on minimum experimental plot size and succession of aphidophaga in West Virginia, USA, apple orchards. Entomophaga 42 (1–2): 257–267.

Burke, D. M. & Nol, E. 1998. Edge and fragment size effects on the vegetation of deciduous forests on Ontario, Canada. Nat. Areas J. 18 (1): 45–53.

Camargo, J. L. C. & Kapos, V. 1995. Complex edge effects on soil moisture and microclimate in central Amazonian forest. J. Trop. Ecol. 11: 205–111.

Cappuccino, N. & Root, R. B. 1992. the significance of host patch edges to the colonisation and development of *Corythucha marmorata* (Hemiptera, Tingidae). Ecol. Ent. 17 (2): 109–113.

Chen, J. Q., Franklin, J. F. & Spies, T. A 1995. Growing-season microclimatic gradients from clear-cut edges into old-growth douglas-fir forests. Ecol. Appl. 5 (1): 74–86.

Chey, V. K., Holloway, J. D. & Speight, M. R. 1997. Diversity of moths in forest plantations and natural forests in Sabah. Bull. Ent. Res. 87 (4): 371–385.

Clarke, K. R. & Ainsworth, M. 1993. A method of linking multivariate community structure to environmental variables. Mar. Ecol. Prog. Ser. 92: 205–219.

Clarke K. R. & Warwick, R. M. 1994. Change in marine communities: an approach to statistical analysis and interpretation. Natural Environment Research Council, Plymouth Marine Laboratory, Plymouth.

Connell, J. H. 1979. Tropical rain forests and coral reefs as open nonequilibrium systems. Pp. 141–163. In: R. M. Anderson, B. D. Turner & L. R. Taylor (eds.), Population Dynamics. Blackwell Scientific, Oxford.

Cudjoe, A. R, Neuenschwander, P. & Copland, M. J. W. 1993. Interference by ants in biological-control of the cassava mealybug *Phenacoccus manihoti* (Hemiptera, Pseudococcidae) in Ghana. Bull. Ent. Res. 83 (1): 15–22.

Davis A. J. & Sutton, S. E. 1998. The effects of rain forest canopy loss on arboreal dung beetles in Borneo: implications for the measurement of biodiversity in derived tropical ecosystems. Diversity Distrib. 4: 167–173.

Davies, K. F. & Margules, C. R. 1998. Effects of habitat fragmentation on carabid beetles: experimental evidence. J. Anim. Ecol. 67 (3): 460–471.

DeCasenave, J. L., Pelotto, J. P., Caziani, S. M., Mermoz, M. & Protomastro, J. 1998. Responses of avian assemblages to a natural edge in a Chaco semiarid forest in Argentina. Auk 115 (2): 425–435.

Didham, R. K. 1998. Altered leaf-litter decomposition rates in tropical forest fragments. Oecologia 116 (3): 397–406.

Didham, R. K. Ghazoul, J., Stork, N. E. & Davis, A. J. 1996. Insects in fragmented forests: a functional approach. Trends Ecol. Evol. 11 (6): 255–260.

Didham, R. K., Hammond, P. M., Lawton, J. H., Eggleton, P. & Stork, N. E. 1998a. Beetle species responses to tropical forest fragmentation. Ecol. Monog. 68 (3): 295–323.

Didham, R. K., Lawton, J. H., Hammond, P. M. & Eggleton, P. 1998b. Trophic structure stability and extinction dynamics of beetles (Coleoptera) in tropical forest fragments. Phil. Trans. Roy. Soc. London B 353 (1367): 437–451.

Duncan, R. P., Buckley, H. L., Ulrich, S. C., Stewart, G. H. & Geritzlehner, J. 1998. Small-scale species richness in forest canopy gaps: The role of niche limitation versus the size of the species pool. J. Veg. Sci. 9: 455–460.

Dyer, L. E. & Landis, D. A. 1997. Influence of noncrop habitats on the distribution of Eriborus terebrans, (Hymenoptera: Ichneumonidae) in cornfields. Ecol. Ent. 26 (4): 924–932

Evans, K. L. & Hambler, C. 1995. The microhabitat of *Tuberta maerens* (Araneae, Aglenidae). Bull. Brit. Arach. Soc. 10 (3): 101–103.

Garwood, N. C. 1989. Tropical seed banks: a review. Pp. 149–210. In Leck, M. A., Parker, V. T. & Simpson, R. L. (eds), Ecology of seed banks. Academic Press, New York.

Greig, N. 1993. Predispersal seed predation on 5 piper species in tropical rain forest. Oecologia 93 (3): 412–420.

Hill, M. O. 1979. DECORANA: a FORTRAN program for detrended correspondence analysis and reciprocal averaging. Section of Ecology and Systematics, Cornell University, Ithaca, New York.

Howe, H. F. & Westerley, L. C. 1988. Ecological relationships of plants and animals. Oxford University Press, Oxford.

Ingham, D. S. & Samways, M. J. 1996. Application of fragmentation and variegation models to epigaeic invertebrates in South Africa. Cons. Biol. 10 (5): 1353–1358.

Janzen, D. H. 1987. Insect diversity of a Costa Rican dry forest: why keep it and how? Biol. J. Linn. Soc. 30: 343–356.

Jiang, G., Han, X. Zhou, G. 1997. Changes of atmospheric CO_2, photosynthesis of the grass layer and soil CO_2 evolution in a typical temperate deciduous forest stand in the mountainous areas of Beijing. Acta Bot. Sinica 39: 653–660.

Jose, S., Gillespie, A. R., George, S. J. & Kumar, B. M. 1996. Vegetation responses along edge-to-interior gradients in a high altitude tropical forest in peninsular India. For. Ecol. Man. 87 (1–3): 51–62.

Kapos, V. 1989. Effects of isolation on the water status of forest patches in the Brazilian Amazon. J. Trop. Ecol. 5 (2): 173–185.

Khan, M. L. & Tripathi, R. S. 1991. Seedling survival and growth of early and late successional tree species as affected by insect herbivory and pathogen attack in subtropical humid forest stands of north-east India. Acta Oecol.-Int. J. Ecol. 12 (5): 569–579.

Kruess, A. & Tscharntke, T. 1994. Habitat fragmentation, species loss, and biological-control. Science 264 (5165): 1581–1584.

Laurance, W. F. 1991. Edge effects in tropical forest fragments – application of a model for the design of nature-reserves. Biol. Cons. 57 (2): 205–219.

Laurance, W. F. & Yensen, E. 1991. Predicting the impacts of edge effects in fragmented habitats. Biol. Cons. 55 (1): 77–92.

Laurance, W. F. & Harrington, G. N. 1997. Ecological associations of feeding sites of feral pigs in the Queensland wet tropics. Wild. Res. 24 (5): 579–590.

Leopold, A. 1933. Game management. Charles Scribner & Sons, New York.

Lerdau, M. & Keller, M. 1997. Controls on isoprene emission from trees in a subtropical dry forest. Plant Cell Env. 20: 569–578.

Lopez Zent, E. 1998. Creative perspective of environmental impacts by native Amazonian human populations. Interciencia 23 (4): 232–242.

Lovejoy, T. E., Bierregaard, R. O., Rylands, A. B., Malcolm, J. R., Quintela, C. E., Harper, L. H., Brown, K. S.Jr., Powell, A. H., Powe, G. V. N., Shubart, H. O. R. & Hays, M. B. 1986. Edge and other effects of isolation on Amazon Forest fragments. Pp. 275–285. In: M. E. Soulé (ed.), Conservation biology: the science of scarcity and diversity. Sinauer, Sunderland.

MacGarvin, M., Lawton, J. H. & Heads, P. A. S. 1986. The herbivorous insect communities of open and woodland bracken – observations, experiments and habitat manipulations. Oikos 47 (2): 135–148.

Malcolm, J. R. 1997. Insect biomass in Amazonian forest fragments. Pp. 510–533. In: N. E. Stork, N. E., Adis, J. & Didham, R. K. (eds), Canopy arthropods. Chapman & Hall, London

Malcolm, J. R. 1998. A model of conductive heat flow in forest edges and fragmented landscapes. Clim. Chan. 39 (2–3): 487–502.

Murcia, C. 1995. Edge effects in fragmented forests – implications for conservation. Trends Ecol. Evol. 10 (2): 58–62.

Myers, N. 1996. Two key challenges for biodiversity: Discontinuities and synergisms. Biodiv. Cons. 5 (9): 1025–1034.

Nason, J. D. & Hamrick, J. L. 1997. Reproductive and genetic consequences of forest fragmentation: two case studies of neotropical canopy trees. J. Hered. 88 (4): 264–276.

Nelson, S. C. & Campbell, C. L. 1993. Comparative spatial-analysis of foliar epidemics on white clover caused by viruses, fungi, and a bacterium. Phytopathology 83 (3) 288–301.

Nicholsorians, C. M. 1991. The effects of light on foliar chemistry, growth and susceptibility of seedlings of a canopy tree to an attinine ant. Oecologia 86 (4): 552–560.

Ozanne, C. M. P., Hambler, C. Foggo, A. & Speight, M. R. 1997. The significance of edge effects in the management of forests for invertebrate biodiversity. Pp. 535–550. In: Stork, N. E., Adis, J.

& Didham, R. K. (eds), Canopy arthropods. Chapman & Hall, London .

Parsons, P. A. 1991. Biodiversity conservation under global climatic-change – the insect *Drosophila* as a biological indicator. Glob. Ecol. Biogeog. Lett. 1 (3): 77–83.

Peltonen, M. & Heliovaara, K. 1998. Incidence of *Xylechinus pilosus and Cryphalus saltuarius* (Scolytidae) in forest-clearcut edges. For. Ecol. Manag. 103 (2–3): 141–147.

Ranta, P., Blom, T., Niemela, J., Joensuu, E. & Siitonen, M. 1998. The fragmented Atlantic rain forest of Brazil: size, shape and distribution of forest fragments. Biodiv. Cons. 7 (3): 385–403.

Regal, P. J. 1982. Pollination by wind and animals: ecology of geographic patterns. Ann. Rev. Ecol. Sys. 13: 497–524.

Roland, J. 1993 Large-scale forest fragmentation increases the duration of tent caterpillar outbreak. Oecologia 93 (1): 25–30.

Roland, J. & Taylor, P. D. 1997. Insect parasitoid species respond to forest structure at different spatial scales. Nature 386 (6626): 710–713.

Rothman, L. D. & Roland, J. 1998. Forest fragmentation and colony performance of forest tent caterpillar. Ecography 21 (4): 383–391.

Samways, M. J. 1994. Insect conservation. Biology. Chapman & Hall, London.

Saunders, D. A., Hobbs, R. J. & Margules, C. R. 1991. Biological consequences of ecosystem fragmentation: a review. Cons. Biol. 5: 18–32.

Schowalter, T. D. 1994. Invertebrate community structure and herbivory in a tropical rain forest canopy in Puerto-Rico following hurricane Hugo. Biotropica 26 (3): 312–319.

Schowalter, T. D.& Ganio, L. M. 1998. Vertical and seasonal variation in canopy arthropod communities in an old-growth conifer forest in southwestern Washington, USA. Bull. Ent. Res. 88 (6): 633–640.

Shibata, H., Satoh, F., Tanaka, Y. & Sakuma, T. 1996. The role of organic horizons and canopy to modify the chemistry of acidic deposition in some forest ecosystems. Water Air Soil Pollut. 85 (3): 1119–1124.

Soulé, M. J. 1986 (ed) 1986. Conservation, the science of scarcity and diversity. Sinauer, Sunderland.

Stouffer, P. C. & Bierregaard, R. O. 1995. Use of Amazonian forest fragments by understorey insectivorous birds. Ecology 76 (8): 2429–2445.

Suarez, A. V., Bolger, D. T. & Case, T. J. 1998. Effects of fragmentation and invasion on native ant communities in coastal southern California. Ecology 79 (6): 2041–2056.

Toda, M. J. 1992. 3-dimensional dispersion of drosophilid flies in a cool temperate forest of northern Japan. Ecol. Res. 7 (3): 283–295.

Turner, I. M. 1996. Species loss in fragments of tropical rain forest: a review of the evidence. J. Appl. Ecol. 33 (2): 200–209.

Walter, P. 1984. Contribution à la connaissance des Scarabéides coprophages du Gabon (Col.) 2. Présence de populations dans la conopée de la forêt gabonaise. Bull. Soc. Ent. France 88: 514–521.

Warren, M. S. 1987. The ecology and conservation of the heath fritillary butterfly, *Mellicta athalia* II & III. J. Appl. Ecol. 24: 483–514.

Wood, B. & Gillman, M. P. 1998. The effects of disturbance on forest butterflies using two methods of sampling in Trinidad. Biodiv. Cons. 7 (5): 597–616.

Vitt, L. J., Avila Pires, T. C. S., Caldwell, J. P. & Oliveira, V. R. L. 1998. The impact of individual tree harvesting on thermal environments of lizards in Amazonian rain forest. Cons. Biol. 12 (3): 654–664.

Plant Ecology **153:** 361–367, 2001.

Enhancement of forest canopy research, education, and conservation in the new millennium

Nalini M. Nadkarni
The Evergreen State College, Olympia, WA 98505, USA (e-mail: Nadkarnn@evergreen.edu)

Key words: Conservation, Forest Canopy, Graduate Study, Rain forests

Abstract

Study of the forest canopy has reached a critical stage in its development from a young 'frontier' area of study to a vibrant and coalescing field of investigation and communication. Many current environmental and social issues at global scales (e.g., environmental change, acid deposition, loss of biodiversity) are related directly to our knowledge of forest canopies. I present six activities that are needed to help the vibrant and growing field of forest canopy studies progress efficiently. Enabling canopy researchers to communicate with each other and with those outside our field is an important element to address these issues. The establishment of a graduate-level training program is also a high priority to generate and maintain a healthy discipline. Formal procedures to identify particular forest sites of critical concern should be initiated, and these should include communication of prioritized sites to conservation groups and policy makers. Instilling a sense of wonder and appreciation for organisms and interactions in non-scientists is another important avenue for forest canopy conservation.

Introduction

The forest canopy has been termed 'the last biotic frontier', and is one of the richest but most poorly studied habitats in the biosphere (Lowman & Nadkarni 1995). Canopy study is becoming a data-rich discipline that bears on many fields of science and environmental issues. Canopy-dwelling plants (epiphytes) constitute up to half the total plant diversity of some wet tropical forests and provide crucial resources for a host of arboreal birds and mammals (Nadkarni & Matelson 1988). Recent research from numerous tropical and temperate canopies document that canopy invertebrates exhibit a tremendous amount of diversity and 'endemism' in the canopy (Winchester 1993; Basset 1993; Stork et al. 1997; Fagan & Winchester 1997). Canopy structural elements such as foliage and twigs account for a tremendous 'sieving' effect of fog in some forests, causing wind-borne precipitation and its accompanying nutrients and pollutants to be deposited locally (Coxson & Nadkarni 1995). Because canopy organisms dwell at the atmosphere-forest interface, they can serve as indicator organisms to monitor changes in global climate and atmospheric conditions (Lugo & Scatena 1992; Benzing 1998). The bulk of the photosynthetic machinery of the biosphere is located within forest canopies, so understanding their physiology and interactions has tremendous importance on issues relating to global carbon budgets and global environmental change.

Studies from multiple scientific disciplines constitute the field of canopy research: e.g., forest ecology, meteorology, zoology, geography, and conservation biology. Recent technological applications for access to the canopy such as the 'canopy raft' and the canopy crane, have allowed researchers to record and interpret larger amounts of meaningful canopy data. In the last decade, a burgeoning of scientific and popular interest in the canopy has occurred (Parker et al. 1992; Nadkarni & Lowman 1995). Interdisciplinary research groups are just now coalescing to approach canopy questions from new and different spatial scales. Heightened public interest in biodiversity, global climate change, and tropical deforestation has generated books, symposia, popular articles, and films about the canopy.

Both the types and amounts of canopy structure data are changing rapidly. In the past, scientists working alone with simple rope-climbing techniques generated studies that produced fairly small data sets. However, recent access innovations permit multiple teams of scientists to work within the same volume of the canopy. Canopy scientists have to deal with more data, new kinds of data, and the need to share data. Data collected by canopy research teams will be useful to other scientists (e.g., geographers, land use managers), just as data emanating from allied fields could aid forest canopy researchers.

Historically, canopy scientists have been notorious for independent ways of taking, storing, analyzing data and communicating their results in ways that are not conducive to comparative studies. As a first step to increase the capacity of canopy researchers to improve this situation, my colleagues and I carried out a survey of canopy researchers (Nadkarni & Parker 1994; Stork & Best 1994) to identify the major obstacles impeding forest canopy research. The surveys indicated that the field is developing rapidly but is fragmented, due to the multidisciplinary nature of the field and the historically isolated nature of the researchers. Three major obstacles were identified: dearth of communication networks among canopy researchers, lack of formal training for future canopy researchers, and problems with data management. More specifically, the latter obstacle involved the lack of uniformity in collecting, processing, and analyzing canopy data, the dearth of data archives, and the inability to link data for comparative research.

In this paper, I outline the dynamics of the field as it has developed and discuss six ways in which canopy researchers might make the pathway of development of this emerging field more efficient and useful to themselves, science, and society. I place the development of training programs, tools for analysis, and pathways of communication for canopy scientists in the context of other emerging and interdisciplinary scientific fields.

Progression of development of scientific fields and canopy studies

Scientific fields go through stages of maturity, as do societies, people, and plants. A scientific field in its infancy is typified by descriptive studies by individual scientists who identify phenomena and document patterns. As a field matures, investigations involve multiple researchers who address process-oriented questions to explain the observed patterns. A sign of maturity for a field is when its scientists can validate predictive models and relate findings to those of other fields. At that stage, scientists can respond to the 'what if' questions about pressing societal concerns posed by policy-makers (Figure 1).

At the turn of the 20th century, canopy science is at an early stage of development. It has been characterized by isolation of researchers, lack of harmonized methods, and the perception that questions addressed are 'obscure'. In the last decade, however, canopy studies have developed rapidly due to six activities:

1. establishment of formal and informal networks within and outside canopy research (e.g., the International Canopy Network (Nadkarni et al. 1997), the European Science Foundation's Tropical Canopy Research Programme);
2. growing realization that canopy studies yield important insights into issues of global environmental issues such as climate change, maintenance of biodiversity, and the creation of sustainable forest management practices;
3. a growing information base from which to draw theory and synthetic works,
4. increased use of experimental approaches to research;
5. active collaborations among canopy scientists and database and informatics scientists to better visualize, process, and archive canopy datasets;
6. heightened awareness of the general public to the excitement and importance of canopy studies due to efforts of educators and media attention.

Recommendations for enhancement of forest canopy activities

Canopy researchers should take advantage of these growth-promoting elements to further their own field, enhance other fields of science, and contribute to environmental issues facing our society. I propose and describe six actions that will address these obstacles and enhance the role of canopy studies in global and political realms.

1. *Strengthen formal and informal communication networks within canopy studies and to allied fields*

In response to the 1994 canopy survey, which identified lack of communication networks as a major

Relevance to Compelling Social Questions ↓	Stage	Types of Studies	Data Characteristics	Logistics
	YOUNG ↓	Descriptive	Few datasets, individually designed	Personal, individual communication
				Technological tools
				Informal networks
	MATURING ↓	Process-oriented	Use of database models & technology	Formal communications
				Courses
				Graduate programs
	MATURE ↓	Predictive	Linked, harmonized, knowledge base with connectivity to many datasets	Jointly funded programs

Figure 1. Costs and benefits of database development and use by different scales and approaches of scientific projects. See text for explanation of categories.

obstacle to canopy studies, The International Canopy Network (ICAN) organization was created in 1994 to facilitate communication among individuals and institutions concerned with research, education, and conservation of organisms in tree crowns and forest canopies. Core activities of ICAN include maintenance of an electronic mail bulletin board, circulation of a quarterly newsletter and member directory, organization of canopy symposia, circulation of a membership directory, maintenance of a citations bibliographic database, and creation of instructional materials about forest canopies for children. The group was originally supported by research funds from the National Science Foundation, but was subsequently incorporated as a non-profit organization, an arrangement that enables it to continue beyond the life of the supporting short-lived grant. These actions have enhanced communication and promoted the efficiency of canopy research and thus largely solved a substantial impediment to the development of our field.

2. *Develop an international graduate program in canopy studies to ease the way for future canopy researchers*

The survey of canopy researchers highlighted the condition that forest canopy study lacks any formal graduate training program. Students need a program that trains them in a variety of disciplines, effectively crosses disciplines, and provides contact with the worldwide community of canopy scientists. Because of the relative youth of canopy studies, canopy researchers are often isolated in their institutions. This contrasts to colleagues in more established disciplines (e.g., marine biology, soil science), who have colleagues 'down the hall' with whom to confer and collaborate. Thus, a student wishing to gather the tools to become a successful researcher in forest canopy science must either rely upon a single mentor who works on the canopy or jump from one institution to another to gain his/her background. The result is a background that is either too narrow for the breadth of background that canopy studies demands, or a program that is fragmented and inefficient.

Goals for a formal graduate training program are to train scientists who can work across traditional disciplinary fields to understand the complex factors that explain forest canopies and their associated abiotic and biotic attributes. My colleagues and I recommend certain characteristics and components for such a program (Table 1). The program we envision consists of a multi-tiered institutional structure, trainees will matriculate at one of the several resident institutions. They would have periodic, structured access to a group (ca. 20–30) of 'Canopy Research Advisors' at other

MODE OF RESEARCH	USE OF DATABASE TOOLS	COST OF USE TO INDIVIDUAL	COST OF NON-USE	BENEFIT TO INDIVIDUAL	BENEFIT TO SCIENCE
INDIVIDUAL RESEARCHER	OPTIONAL	HIGH	LOW	LOW	LOW
CO-LOCATED STUDIES	OPTIONAL	HIGH	LOW	LOW	HIGH
COMPARATIVE STUDIES	INTEGRAL	HIGH	FATAL	HIGH	HIGH
COLLABORATIVE STUDIES	INTEGRAL	LOW	FATAL	HIGH	HIGH

Figure 2. Progression of development for emerging scientific fields.

Table 1. Characteristics and components of recommended graduate training program in canopy studies.

1.	Commitment of one or more 'home institutions' for coursework, administration, and guidance
2.	Flexible academic curriculum to provide 'tailored' interdisciplinary course input
3.	Summer field program at canopy research sites
4.	'Canopy walkabout' for exchange of ideas, approaches, and protocols
5.	Virtual and real seminars and workshops
6.	Funding from private, agency, and university sources

institutions around the world who have agreed to serve as mentors in this 'university without walls'. The program has three parts: coursework, research training at field sites associated with the resident institutions, and the 'canopy walkabout'. The latter is a way to expose students to a variety of field situations and research approaches; during the summer walkabout, trainees will work directly with a subset of Canopy Research Advisors to broaden their field research experiences. This will have the positive consequence of knitting together the scattered canopy community by direct interchanges of research experiences.

3. *Actively develop ties with database and computer scientists in order to develop harmonized tools to share data, maintain a data and dataset archives, database of scientific references, and recognized research protocols*

The relative youth of the field – with its lack of entrenched methods, legacy datasets, and conflicting camps of competing groups – provides a unique opportunity for integrating data management and analysis tools into the research process. The sociology of the discipline is conducive to sharing data; forest canopy researchers appear openly communicative and supportive of each others' work. Thus, the forest canopy studies serve as an excellent arena to generate database tools that could also serve other fields of ecology and science (Linsenmair & Stork 1994).

Since 1994, my colleagues and I have developed a history of querying the canopy community about the need for and value of database tools. In September 1993, Geoffrey G. Parker and I convened a workshop on canopy studies, as part of the Long-Term Ecological Research All-Scientists' Meeting. Participants developed the framework presented below to study forest canopy structure-function studies. At a follow-up meeting in 1994, a group of database scientists from the H.J. Andrews Experimental Forest and elsewhere expressed interest in collaborating with canopy scientists. There was also excitement at 'reviving' some of the canopy datasets previously collected.

After these meetings, canopy scientists undertook several exploratory data management projects and evaluation, including our building some prototype database tools. At workshops we organized at international canopy meetings in 1998, the canopy research community came to consensus that a variety of obstacles to database use still exist. These include non-overlapping disciplinary focus, sizes of project, and conceptual differences concerning canopy structure and function. Because informatics activity ultimately reflects the science, we concluded that a database cannot become an effective integrative tool until the science itself is integrated. Paradoxically, the science cannot easily become integrated without the use of database tools. Our reviews of tools applicable to canopy science discovered a wealth of software tools used in other disciplines for displaying information about complex structures, processes, and datasets, but the best of these were not easily portable to other disciplines. To explore why has database use in canopy science has not been generally successful, we revisited how canopy research is done and found that different 'research modes' are differentially amenable to the successful use of databases (Figure 2). 'Individual projects' involve a single research question and a small, usually single-investigator project. 'Co-located research' is a collection of projects that share the same study site or organism, but different researchers focus on different questions. 'Comparative projects' take place at different locations but are driven by an emergent question that requires comparative data, though not necessarily collected in standardized ways. 'Collaborative projects' (often typified as 'big science'), involving multiple researchers on several sub-questions that relate to a single overarching theme. In Figure 2, each row shows a different mode of conducting canopy research (distinguished by the scale of the activity) and each column the general importance, and cost of implementing or neglecting database activities.

Generally, the importance of database structures and cost of non-implementation rises with the scope of the project, while costs to the individual researcher falls. For individual projects, a large investment in database tools is cost-ineffective. On the other extreme, in collaborative projects, database activities are integral to the activities involved, so database tools are included from the inception, with the cost borne by the project as a whole, thus reducing the load on individual participants. Between these extremes are cases where the introduction of suitable database technology could have relatively high benefit and low cost. This analysis suggests that the case of co-located and especially comparative canopy studies could benefit from an investment in database activities. Canopy researchers have agreed that additional effort is warranted for efforts to produce a database for such intermediate cases.

4. *Establish specific criteria for standards that will help conservationists in decision-making on priorities for protected areas*

Although research projects, education/outreach programs, and communication networks serve important functions in forest conservation, activities that directly contribute to conservation of specific pieces of land are also needed. The ICAN has outlined a procedure which may serve this function (Table 2). The 'Adopt-a-Canopy' concept enlists the input of forest canopy researchers to identify critical sites (termed 'canopies of international significance', CIS), who base the listing of sites on ecologically sound criteria (e.g., the presence of endangered or endemic species, high beta diversity, indiginous human cultures). A committee then ranks these by conservation priority and makes recommendations for the amount and types of support needed to preserve or maintain the CIS. This may involve the establishment or maintenance of a research field station, the salary of field support staff, or installation of a canopy crane or computer system. An important step in this process is to communicate with established conservation groups (e.g., World Wildlife Fund, The Nature Conservancy) about these priorities. This committee should also work to raise funds through foundations, individuals, and agencies. Policy makers must then be contacted and convinced to implement the purchase and protection of these sites.

5. *Nurture contacts with policy makers to provide scientifically sound input to land use and environmental issues*

Links between scientists and policy makers are notoriously weak and sporadic. Because canopy research has direct bearing on a number of extremely important issues of environmental and social concern, many policy makers are interested in input from forest canopy researchers; a few communication pathways between canopy researchers and policy makers now exist. For example, an organization called the Union for Concerned Scientists (UCS, Cambridge, Massachusetts), has as its goal the bringing together of policy-makers and scientists to better resolve issues requiring scientifically sound input. One of the projects of the UCS is the Sound Science Initiative. In this program, a consortium of scientists from diverse disciplines and geographical areas and policy makers approach complex environmental problems. Their current topics include biodiversity and climate change, both of which relate directly to forest canopy studies.

Table 2. Recommended steps to initiate the 'Adopt-a-Canopy' concept, a practical program to effect forest canopy conservation.

1.	Assemble a committee of forest canopy researchers to develop criteria for canopies of international significance (CIS) (e.g., high beta diversity, endangered species, 'keynote' functions, protected status, indiginous human cultures)
2.	Solicit CIS nominations from the canopy researcher community; gather documentation for sites that meet stated criteria
3.	Select and prioritize sites, and make recommendations for appropriate support needed at each site
4.	Seek corporate, agency, and individual sponsors
5.	Communicate and collaborate with conservationists and policy makers

6. *Foster the imagination and support of the general public by providing educators and the media with sound information in accessible forms*

The world of the forest canopy holds such great biodiversity and such intriguing and little-known interactions that tremendous interest exists about canopy studies among scientists and non-scientists. In the last five years, a large number of articles, television programs, and films have been produced about forest canopy researchers and their research. Interest comes from a wide range of age groups, particularly children, who have a natural love of climbing trees and exploring little-known places.

Canopy researchers must take the time from their project work to provide information in accessible forms so that the general public can understand more about forest canopies and the importance of the interactions and organisms that live there. This could take the form of writing popular articles for childrens' magazines or a publication for lay adults, giving an informal talk at a local naturalist or bird-watching club or a class of students in a local elementary school, or providing expertise on local environmental issues that relate to forest canopies to politicians or land managers. Such 'indirect' activities, though rarely rewarded in the traditional academic system, are fundamental to improving societal attitudes about research and conservation, and will ultimately produce more support for forest canopy work.

Acknowledgments

I thank the European Science Foundation's Tropical Canopy Research Programme for support in participating in the workshop that inspired this paper. Geoffrey G. Parker, of the Smithsonian Environmental Research Center, participated in developing many of the ideas presented here, particularly in the analysis of the costs and benefits of database use for canopy studies. Judy B. Cushing, David Maier, and Lois Delcambre also contributed to the concepts in this paper. I thank the Evergreen State College for support of my forest canopy research projects. Many of the ideas presented in this paper were derived from research sponsored by the National Science Foundation with funding from the Database Activities Program (BIR 93-07771, BIR 96-30316, and a POWRE Grant Supplement), and the Long-term Ecological Research in Environmental Biology program (BIR 96-15341).

References

Benzing, D. 1998. Vulnerabilities of tropical forests to climate change: the significance of resident epiphytes. Climatic Change 39: 519–540.

Coxson, D. S. & Nadkarni, N. M. 1995. Ecological roles of epiphytes in nutrient cycles of forest ecosystems. Pp. 495–546. In: Lowman, M & Nadkarni, N. (eds), Forest canopies. Academic Press, San Diego.

Linsenmair, E. and Stork, N. 1994. Tropical canopy research. ESF Commun. 31: 4–7.

Lowman, M. D. & Nadkarni, N. M. 1995. Forest canopies. Academic Press. San Diego.

Lugo, A. E. & Scatena, F. 1992. Epiphytes and climate change research in the Caribbean: a proposal. Selbyana 13: 123–130.

Nadkarni, N. M.& Lowman, M. D. 1995. Canopy science: a summary of its role in research and education. Pp. 609–610. In: Lowman, M & Nadkarni, N. (eds), Forest canopies. Academic Press, San Diego.

Nadkarni, N. M. & Matelson, T. M. 1989. Bird use of epiphyte resources in neotropical trees. Condor 69: 891–907.

Nadkarni, N. M. & Parker, G. G. 1994. A profile of forest canopy science and scientists – who we are, what we want to know, and

obstacles we face: results of an international survey. Selbyana 15: 38–50.
Nadkarni, N. M., Parker, G. G., Ford, E. D., Cushing, J. B. & Stallman, C. 1997. The International Canopy Network: A pathway for interdisciplinary exchange of scientific information on forest canopies. Northwest Sci. Special Issue: 70: 104–108.
Parker, G. G., Smith, A. P. & Hogan, K. P. 1992. Access to the upper forest canopy with a large tower crane. BioScience 42: 664–670.

Subject Index